# MOLECULAR BIOLOGY OF DNA AND RNA

AN ANALYSIS OF RESEARCH PAPERS

# MOLECULAR BIOLOGY OF DNA AND RNA
## AN ANALYSIS OF RESEARCH PAPERS

**ILSE DOROTHEA RAACKE, Ph.D.**

Professor of Biology, Boston University,
Boston, Massachusetts

With 154 illustrations

**THE C. V. MOSBY COMPANY**
SAINT LOUIS   1971

Copyright © 1971 by The C. V. Mosby Company

All rights reserved. No part of this book may be reproduced in any manner without written permission of the publisher.

Printed in the United States of America

Standard Book Number 8016-4068-7

Distributed in Great Britain by Henry Kimpton, London

To Don, Kent, Evan, and Omi

# PREFACE

The general public envisions a scientist as being isolated in his laboratory, pursuing the "truth." However, this picture of science as a private pursuit of the individual is quite wrong. Science is not private, but public. Scientific discoveries become discoveries not when an individual thinks he has made them, but rather when they have been recognized as such by his peers and when they enter the fabric of generally accepted knowledge.

Scientific discoveries are communicated mainly through publication in a scientific journal. Therefore, learning to *read* scientific journals is essential for the developing scientist. One of the objectives of this book is to assist beginning students of molecular biology and biochemistry in acquiring this skill. To help them to focus on the substance and methodology of the research papers reprinted in this book, each paper has been provided with a set of questions and problems. Since complete coverage is not possible in this limited space, the material has been restricted to the structures of DNA and RNA and to the replication and transcription of the genetic material.

The papers have been selected especially to represent a cross section of those realistically encountered in the literature by the student; therefore they may not all be uniformly excellent, concise, well written, or even totally correct. Many of them are classics, but the information may be in need of reinterpretation in the light of later developments. Thus, one must continually evaluate the material one reads against other sources of information.

One should keep in mind that some of the material in an article may be skimmed over or skipped entirely, depending on the purpose for which a paper is read. On the average, one should spend about half an hour on a paper, and none should take more than an hour. If a student finds himself spending much more time than that on any one paper, he can be fairly certain he is delving into things he cannot profitably digest at his level of scientific prowess. It is important for him to develop good judgment as to what can be profitably appropriated at one time and what is best left until later. The student will be amazed to see how his understanding of the very same paper grows from one year to the next.

The questions on a paper facilitate its reading by making it more purposeful. It would therefore be helpful if the student were to read the questions before reading the article, but he should definitely not read the answers. Before he does that, he should make a serious effort to answer the questions himself. Most of the answers to the problems can be found in the articles themselves. The questions do, of course, require a normal amount of background information not only in molecular biology but also in allied fields such as genetics or biochemistry. Occasionally information from some article in the bibliography is required so that the student may become aware of the cumulative nature of the scientific literature. General information can be found in some of the standard laboratory handbooks such as *The Merck Index* or the *Handbook of Chemistry and Physics*.

The student should not expect to be able to answer all the questions. Some are designed to point out some wrong or obsolete feature in the paper, which the student could not be expected to recognize. Others are asked for the purpose of imparting some specific information through the answers. This is the other purpose of the book. Consequently the supplied answers are much more extensive than what one would expect from a student. They have been designed to provide not only general background information, but also many bits of practical information hard to find in print because they are part of laboratory "lore." In molecular biology there is an unusually close relationship between concepts and the methodology used to arrive at them. Therefore in this science, and more than in most, an appreciation of the practical aspects of the experiments is critical to a proper understanding of the subject.

An additional value of the book to beginning graduate students lies in that the questions asked and the calculations required are often those with which one becomes involved when interpreting and planning one's own experiments or when converting raw data into a more meaningful form.

The papers have been grouped by subject matter so that the book can be used in conjunction with a course in basic molecular biology. For those not taking a concomitant course, each chapter has been preceded by a short summary of basic information.

**Ilse Dorothea Raacke**

# READING A SCIENTIFIC PAPER: Advice to the student

Reading a scientific paper from beginning to end is rarely the best way to extract the maximum amount of information from the article. You will rarely have the time or interest to read straight through every paper you encounter. Fortunately, since the articles are usually written according to a specified formula, you often need not read the entire paper. The basic components of a paper are listed below, although specific names of sections may vary with the journals involved.

*Introduction.* Enough background information is included to enable someone from outside the immediate field to orient himself. Since the author has only a limited amount of space, this section presents reviews that are often veritable gems in conciseness.

*Experimental* or *materials and methods.* This contains the technical information on the methods used to obtain the results.

*Results.* This may be a very technical presentation of the results of the particular series of experiments, often expressed in graphs and tables.

*Discussion.* Here the author is expected to analyze and interpret his own experimental data and bring his results into context with the work of others. Often this part contains a verbal restating of the diagrammatic and tabular data presented under *results.*

*Summary* or *abstract.* This is a short description of experiments, results, and conclusions reported. It usually appears either underneath the title on the first page or at the end of the paper.

Every experienced reader has naturally developed his own individual style for dealing with the literature. At the beginning, however, it is easy to be overwhelmed. The following procedure has been found useful by my students in the past.

For general information you should read the *summary* first in order to know if the results obtained in the paper warrant a further investment of time. Next read the *introduction* and parts, or all, of the *discussion,* since they constitute the remainder of the information necessary for a general view of the topic.

In many cases it is possible to stop at this point, unless the paper is of special interest. If, however, you are curious as to the details of the methods used, the *experimental* or *methods* section would be next. By this point, you should have enough information about the outcome to judge whether the methods were adequate. If the paper is being read for technical reasons, this section should be read carefully, since the minutest detail is often most important for reproducing or revising the experiments. Finally the *results* can be read and evaluated intelligently as you examine every individual result against the methods used and the interpretation given it by the author.

I have designed this sequence after many years of reading scientific papers and learning that not all the information is gospel truth or of personal interest. I hope this procedure in *Molecular Biology of DNA and RNA* will help you to determine what is important, both in a research paper and in molecular biological research. It should prepare you for intelligent use of the literature.

# CONTENTS

## Chapter 1   THE STRUCTURE OF DEOXYRIBONUCLEIC ACID, 1

1. Molecular structure of nucleic acids; a structure for deoxyribose nucleic acid (J. D. Watson and F. H. C. Crick), **3**

2. Separation of microbial deoxyribonucleic acids into complementary strands (Rivka Rudner, John D. Karkas, and Erwin Chargaff), **5**

3. Strand separation and specific recombination in deoxyribonucleic acids: physical chemical studies (P. Doty, J. Marmur, J. Eigner, and C. Schildkraut), **12**

4. Studies on hybrid molecules of nucleic acids; I. DNA-DNA hybrids on nitrocellulose filters (J. Legault-Démare, B. Desseaux, T. Heyman, S. Séror, and G. P. Ress), **25**

Questions, 30

Answers, 32

## Chapter 2   REPLICATION OF DNA IN VIVO, 37

5. The replication of DNA in Escherichia coli (Matthew Meselson and Franklin W. Stahl), **39**

6. The chromosome of Escherichia coli (John Cairns), **49**

7. Isolation of the growing point in the bacterial chromosome (Philip C. Hanawalt and Dan S. Ray), **53**

8. Mechanism of DNA chain growth, I. Possible discontinuity and unusual secondary structure of newly synthesized chains (Reiji Okazaki, Tuneko Okazaki, Kiwako Sakabe, Kazunori Sugimoto, and Akio Sugino), **59**

9. Role of polynucleotide ligase in T4 DNA replication (Jean Newman and Philip C. Hanawalt), **66**

Questions, 69

Answers, 71

## Chapter 3  SYNTHESIS OF DNA IN VITRO, 77

10  Enzymatic synthesis of deoxyribonucleic acid; I. Preparation of substrates and partial purification of an enzyme from Escherichia coli (I. R. Lehman, Maurice J. Bessman, Ernest S. Simms, and Arthur Kornberg), 78

11  Enzymatic synthesis of deoxyribonucleic acid; XXVIII. The pyrophosphate exchange and pyrophosphorolysis reactions of deoxyribonucleic acid polymerase (Murray P. Deutscher and Arthur Kornberg), 91

12  Enzymatic joining of polynucleotides, V. A DNA-adenylate intermediate in the polynucleotide-joining reaction (Baldomero M. Olivera, Zach W. Hall, and I. R. Lehman), 105

13  Enzymatic synthesis of DNA, XXIV. Synthesis of infectious phage $\phi$X174 DNA (Mehran Goulian, Arthur Kornberg, and Robert L. Sinsheimer), 112

Questions, 120

Answers, 122

---

## Chapter 4  THE STRUCTURES OF RNAs, 129

14  A simplified procedure for the simultaneous isolation of 4S and 5S RNA (H. I. Robins and I. D. Raacke), 132

15  Ribonucleic acid from Escherichia coli; III. The influence of ionic strength and temperature on hydrodynamic and optical properties (R. A. Cox and U. Z. Littauer), 136

16  Molecular weight estimation and separation of ribonucleic acid by electrophoresis in agarose-acrylamide composite gels (Andrew C. Peacock and C. Wesley Dingman), 145

17  A quantitative assay for DNA-RNA hybrids with DNA immobilized on a membrane (David Gillespie and S. Spiegelman), 153

18  Nucleotide sequence of KB cell 5S RNA (Bernard G. Forget and Sherman M. Weissman), 163

Questions, 170

Answers, 174

---

## Chapter 5  GENE TRANSCRIPTION IN VIVO, 185

19  An unstable intermediate carrying information from genes to ribosomes for protein synthesis (S. Brenner, F. Jacob, and M. Meselson), 188

20  Evidence for a nonrandom reading of the genome (T. Kano-Sueoka and S. Spiegelman), 195

21 Gene-specific messenger RNA: isolation by the deletion method
(Ekkehard K. F. Bautz and Eugene Reilly), 201

22 Asymmetric distribution of the transcribing regions on the complementary strands
of coliphage λ DNA (Karol Taylor, Zdenka Hradecna, and Waclaw Szybalski), 205

23 Isolation of the λ phage repressor (Mark Ptashne), 212

Questions, 219

Answers, 222

## Chapter 6  SYNTHESES OF RNAs IN VITRO, 231

24 Deoxyribonucleic acid-directed synthesis of ribonucleic acid by an enzyme from
Escherichia coli (Michael Chamberlin and Paul Berg), 234

25 The role of DNA in RNA synthesis, IX. Nucleoside triphosphate termini in RNA
polymerase products (Umadas Maitra and Jerard Hurwitz), 246

26 Factor stimulating transcription by RNA polymerase (Richard R. Burgess, Andrew A.
Travers, John J. Dunn, and Ekkehard K. F. Bautz), 253

27 A factor that stimulates RNA synthesis by purified RNA polymerase
(J. Davison, L. M. Pilarski, and H. Echols), 261

28 The synthesis of a self-propagating and infectious nucleic acid with a purified enzyme
(S. Spiegelman, I. Haruna, I. B. Holland, G. Beaudreau, and D. Mills), 267

29 The 3'-terminus and the replication of phage RNA
(Ulrich Rensing and J. T. August), 274

Questions, 281

Answers, 283

CHAPTER 1

# THE STRUCTURE OF DEOXYRIBONUCLEIC ACID

Not only is the structure of DNA familiar to almost everyone now, but it is also so real that it is difficult to imagine that when it was proposed by Watson and Crick in 1953, "it was little more than a hypothetical construct, too pretty to be wrong."[1] Its dimensions were, to be sure, taken from x-ray diffraction patterns and thus were correct; but when it comes to structure, x-ray crystallography can only suggest, not prove, and the patterns of DNA certainly allowed more than a single interpretation.

The structure of DNA as proposed by Watson and Crick, which is now known to be correct, beyond any shadow of a doubt, consists of two antiparallel polydeoxyribonucleotide strands wound around each other to yield a helix with a diameter of 20 Å and a repeat of 34 Å occurring every ten nucleotides. The phosphates are on the outside, and the bases are toward the inside. The most characteristic feature of the structure is that the two strands are joined by hydrogen bonds between two complementary pairs of bases, namely adenine and thymine, and guanine and cytosine. There are absolutely no constraints on the nucleotide sequence of one strand, but the sequence of the second strand must be strictly complementary to the first.

To prove the structure it was necessary to show that it is possible to separate DNA into two strands, each possessing half the molecular weight of the native molecule, and a complementary base composition and sequence. This was accomplished by either "melting" the helix by heat or separating the strands by alkali, and then showing that the molecular weight had been halved. It was by no means a straightforward operation, not only because most DNA preparations are "nicked" and therefore yield more than two pieces upon melting, but also because the determination of molecular weight by hydrodynamic methods of a molecule as large as DNA, with as extreme a shape, posed quite a number of problems. I think these problems can best be appreciated by anyone who has contemplated a pot of boiled spaghetti and tried to determine the meanderings of an individual thread.

Eventually, however, accurate molecular weights were determined by means of sedimentation analysis coupled with viscometry, light scattering, band sedimentation in a gradient of CsCl, and directly, by measuring the length of a molecule on an electron micrograph. Halving of the molecular weight upon denaturation could be demonstrated satisfactorily on a number of occasions.

The specific complementarity of the two strands was most convincingly demonstrated by the specific reassociation of homologous strands during annealing, that is, the very slow cooling of a solution, as opposed to the lack of reaction between unrelated strands. This method is the basis for a general analysis for base sequence homology by hybridization between a known and an unknown strand, which has had widespread application. The ultimate proof of the complementarity of the strands came with the direct analysis of the separated strands.

The sense in which we have been talking about the structure of "DNA" is, strictly speaking, that of a segment of a double-stranded DNA molecule. It has become apparent that DNA molecules do not have a singular structure, but come in great variety. In addition to the classical double-stranded rod or filament, there are double-stranded circular DNA molecules, which can be twisted and super-coiled. Furthermore, native DNA need not always be double-stranded, for a number of viruses have single-stranded, circular DNA as their genetic material.

The following papers illustrate the methods used to establish the structure of DNA as we now know it, and some of the practical problems past and present.

**REFERENCE**

1. Watson, J. D.: The double helix; a personal account of the discovery of the structure of DNA, New York, 1968, Atheneum Publishers.

# 1 MOLECULAR STRUCTURE OF NUCLEIC ACIDS; A STRUCTURE FOR DEOXYRIBOSE NUCLEIC ACID

*J. D. Watson*
*F. H. C. Crick*
*Medical Research Council Unit for the Study of the Molecular Structure of Biological Systems*
*Cavendish Physical Laboratory*
*Cambridge University*
*Cambridge, England*

We wish to suggest a structure for the salt of deoxyribose nucleic acid (D.N.A.). This structure has novel features which are of considerable biological interest.

A structure for nucleic acid has already been proposed by Pauling and Corey[1]. They kindly made their manuscript available to us in advance of publication. Their model consists of three intertwined chains, with the phosphates near the fibre axis, and the bases on the outside. In our opinion, this structure is unsatisfactory for two reasons: (1) We believe that the material which gives the X-ray diagrams is the salt, not the free acid. Without the acidic hydrogen atoms it is not clear what forces would hold the structure together, especially as the negatively charged phosphates near the axis will repel each other. (2) Some of the van der Waals distances appear to be too small.

Another three-chain structure has also been suggested by Fraser (in the press). In his model the phosphates are on the outside and the bases on the inside, linked together by hydrogen bonds. This structure as described is rather ill-defined, and for this reason we shall not comment on it.

We wish to put forward a radically different structure for the salt of deoxyribose nucleic acid. This structure has two helical chains each coiled round the same axis (see diagram). We have made the usual chemical assumptions, namely, that each chain consists of phosphate diester groups joining β-D-deoxyribofuranose residues with 3′,5′ linkages. The two chains (but not their bases) are related by a dyad perpendicular to the fibre axis. Both chains follow right-handed helices, but owing to the dyad the sequences of the atoms in the two chains run in opposite directions. Each chain loosely resembles Furberg's[2] model No. 1; that is, the bases are on the inside of the helix and the phosphates on the outside. The configuration of the sugar and the atoms near it is close to Furberg's 'standard configuration', the sugar being roughly perpendicular to the attached base. There is a residue on each chain every 3.4 A. in the z-direction. We have assumed an angle

---

From Nature 171:737-738, 1953. Reprinted with permission.

We are much indebted to Dr. Jerry Donohue for constant advice and criticism, especially on interatomic distances. We have also been stimulated by a knowledge of the general nature of the unpublished experimental results and ideas of Dr. M. H. F. Wilkins, Dr. R. E. Franklin and their co-workers at King's College, London. One of us (J. D. W.) has been aided by a fellowship from the National Foundation for Infantile Paralysis.

**Fig. 1.** This figure is purely diagrammatic. The two ribbons symbolize the two phosphate-sugar chains, and the horizontal rods the pairs of bases holding the chains together. The vertical line marks the fibre axis.

of 36° between adjacent residues in the same chain, so that the structure repeats after 10 residues on each chain, that is, after 34 A. The distance of a phosphorus atom from the fibre axis is 10 A. As the phosphates are on the outside, cations have easy access to them.

The structure is an open one, and its water content is rather high. At lower water contents we would expect the bases to tilt so that the structure could become more compact.

The novel feature of the structure is the manner in which the two chains are held together by the purine and pyrimidine bases. The planes of the bases are perpendicular to the fibre axis. They are joined together in pairs, a single base from one chain being hydrogen-bonded to a single base from the other chain, so that the two lie side by side with identical z-co-ordinates. One of the pair must be a purine and the other a pyrimidine for bonding to occur. The hydrogen bonds are made as follows: purine position 1 to pyrimidine position 1; purine position 6 to pyrimidine position 6.

If it is assumed that the bases only occur in the structure in the most plausible tautomeric forms (that is, with the keto rather than the enol configurations) it is found that only specific pairs of bases can bond together. These pairs are: adenine (purine) with thymine (pyrimidine), and guanine (purine) with cytosine (pyrimidine).

In other words, if an adenine forms one member of a pair, on either chain, then on these assumptions the other member must be thymine; similarly for guanine and cytosine. The sequence of bases on a single chain does not appear to be restricted in any way. However, if only specific pairs of bases can be formed, it follows that if the sequence of bases on one chain is given, then the sequence on the other chain is automatically determined.

It has been found experimentally[3,4] that the ratio of the amounts of adenine to thymine, and the ratio of guanine to cytosine, are always very close to unity for deoxyribose nucleic acid.

It is probably impossible to build this structure with a ribose sugar in place of the deoxyribose, as the extra oxygen atom would make too close a van der Waals contact.

The previously published X-ray data[5,6] on deoxyribose nucleic acid are insufficient for a rigorous test of our structure. So far as we can tell, it is roughly compatible with the experimental data, but it must be regarded as unproved until it has been checked against more exact results. Some of these are given in the following communications. We were not aware of the details of the results presented there when we devised our structure, which rests mainly though not entirely on published experimental data and stereochemical arguments.

It has not escaped our notice that the specific pairing we have postulated immediately suggests a possible copying mechanism for the genetic material.

Full details of the structure, including the conditions assumed in building it, together with a set of co-ordinates for the atoms, will be published elsewhere.

**REFERENCES**

1. Pauling, L., and Corey, R. B., Nature **171**:346 (1953); Proc. U.S. Nat. Acad. Sci. **39**:84 (1953).
2. Furberg, S., Acta Chem. Scand. **6**:634 (1952).
3. Chargaff, E., for references see Zamenhof, S., Brawerman, G., and Chargaff, E., Biochim. et Biophys. Acta **9**:402 (1952).
4. Wyatt, G. R., J. Gen. Physiol. **36**:201 (1952).
5. Astbury, W. T., Symp. Soc. Exp. Biol. 1, Nucleic Acid, 66 (Camb. Univ. Press, 1947).
6. Wilkins, M. H. F., and Randall, J. T., Biochim. et Biophys. Acta **10**:192 (1953).

# 2 SEPARATION OF MICROBIAL DEOXYRIBONUCLEIC ACIDS INTO COMPLEMENTARY STRANDS

*Rivka Rudner*
*John D. Karkas*
*Erwin Chargaff*
*Department of Biological Sciences*
*Hunter College and*
*Cell Chemistry Laboratory*
*Department of Biochemistry*
*College of Physicians and Surgeons*
*Columbia University*
*New York, New York*

*Abstract.* DNA preparations from seven bacterial species and from the *E. coli* phage T4 can, after denaturation with alkali, be separated chromatographically into two distinct components (L and H) through intermittent gradient elution from methylated albumin kieselguhr columns. The direct chemical analysis of the L and H fractions isolated from DNA specimens of the AT type shows them to exhibit a high degree of complementarity; but despite a bias in the distribution of purines and pyrimidines, either fraction contains equimolar quantities of 6-amino and of 6-keto nucleotides. In the L and H components derived from DNA of the equimolar and GC types, the distribution bias appears limited to guanine and cytosine. It is suggested that the L and H fractions represent the complementary DNA strands.

As we reported recently,[1-3] denatured DNA of *Bacillus subtilis* can be separated by a technique of intermittent gradient elution from a column of methylated albumin kieselguhr (MAK), into two fractions, designated, by virtue of their buoyant densities, as L (light) and H (heavy). Taking into account some of the requirements that genuine strands deriving from a native DNA duplex must fulfill,[4] we concluded (on the basis of evidence bearing on the biological activity, nucleotide composition, and temperature-absorbance behavior) that the two fractions isolated from *B. subtilis* DNA represented indeed such complementary strands. They could be annealed together easily, thereby regaining considerable secondary structure as shown by transforming activity and hypochromicity; they were in their nucleotide composition strictly complementary so that the adenine content of fraction L was equivalent to the thymine content of fraction H, the guanine of L to the cytosine of H, etc.; the greater abundance of purines in fraction L was matched by that of pyrimidines in fraction H. As would be expected of genuine strands,[4] the molar sums of A + T and of G + C did not vary in the two fractions[5] and were identical with those found in the native DNA, and this was, therefore, also true of the dissymmetry ratio A + T/G + C; but the preparations also exhibited an unexpected regularity, which would not have been predicted for a single strand, namely, the equality of the 6-amino and the 6-keto nucleotides, A + C and G + T.

We were naturally interested in the generality of these observations and in their biological meaning with respect to template functions, etc. While the latter problem will be taken up in a subsequent paper, we present here evidence that denatured DNA from seven microbial species and from bacteriophage T4, representing DNA varieties of the AT, GC, and equimolar types, can be separated on MAK columns into two distinct and reproducible chromatographic fractions, many of which can be shown by direct analysis to be strictly complementary.

---

From Proceedings of the National Academy of Sciences, U.S.A. 63:152-159, 1969. Reprinted with permission.

This work has been supported by research funds of the City University of New York, by grant E-54 of the American Cancer Society, and by USPHS grants GM 7191 and (in the final stage) GM 16059. The excellent technical assistance of Mrs. V. Remeza is gratefully acknowledged.

## Materials and methods

*Strains used.* We are grateful to Dr. D. Dubnau for *Bacillus megaterium*, to Dr. S. M. Friedman for the thermophile *Bacillus stearothermophilus*, strain 2184, and to Dr. E. Shahn for bacteriophage T4, strain $B_{01}{}^r$. The strains of *Bacillus subtilis* have been described previously.[6] Most of the following strains were used in previous work of this laboratory:[7,8] *Salmonella typhimurium*, strain LT-2; *Escherichia coli*, strains HfrH and W 1157; *Serratia marcescens*; and *Proteus vulgaris*.

*Isolation and denaturation of DNA.* The DNA preparations from the genus *Bacillus* were isolated in the usual manner.[6] From all other organisms DNA was isolated by a procedure[9] in which the cell wall was degraded by pronase instead of lysozyme. DNA of bacteriophage T4 was prepared by phenol extraction[10] and, before denaturation by alkali, subjected to shear by stirring the solution (20-40 μg DNA/ml) for 10 min in a Waring Blendor at approximately 5000 rpm. All DNA preparations were denatured by alkali[6] and their solutions were subjected to dialysis against 0.7 $M$ NaCl.[1] All saline solutions were made 0.05 $M$ with respect to phosphate buffer of pH 6.7.

*Strand separation.* The application of an intermittent gradient elution technique to chromatography on MAK columns has been described previously.[1] Routinely, 2-4 mg of denatured DNA were applied to a standard MAK column and eluted by means of a linear salt gradient of 0.7-1.5 $M$ NaCl, a total of 500 ml of eluent being employed. Individual fractions (0.25-0.75 mg DNA) were rechromatographed on small MAK columns containing about one quarter of the usual quantity of adsorbent; intermittent gradient elution was then performed with 200-250 ml of total eluent. Before being thus recycled, the DNA fractions were dialyzed in the cold against 1-liter volumes, first of 0.3 and then of 0.7 $M$ NaCl.

*Transformation assays.* The procedures have been described before.[6] The recipient strain of *B. subtilis* employed was Mu8u5u16 (ade⁻, leu⁻, met⁻).

*Base composition.* The acid hydrolysis and the paper chromatography of the hydrolysates were performed as before.[3]

## Results and discussion

*Rechromatography of B. subtilis DNA fractions.* It was desirable to ascertain the homogeneity of the two fractions from denatured *B. subtilis* DNA described in our previous communication.[1] The eluates of the fractions collected from a standard-size MAK column (Fig. 1A) were rechromatographed separately on small MAK columns, again with the use of intermittent gradient elution. As can be seen in Figure 1B, the repeated chromatography revealed no gross inhomogeneity of the initial L and H fractions, the contamination amounting mostly to about 5 per cent and never to more than 10 per cent. A third chromatography (Fig. 1C) yielded essentially homogeneous fractions. One may conclude that the fractionation observed was not due to artifacts produced by the intermittent elution procedure.

The partial restoration of the transforming

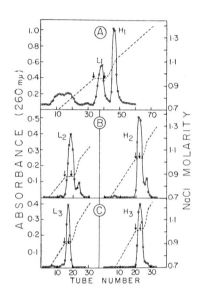

**Fig. 1.** Separation and purification of fractions of denatured *B. subtilis* DNA by intermittent gradient elution from MAK columns.

**A,** Initial separation: elution of alkali-denatured DNA of strain W23 from standard MAK column (total volume of eluent, 500 ml). Recovery (as percentage of input DNA): total, 85; fraction L, 38; fraction H, 36.

**B,** First rechromatography of separated L and H components on small MAK columns. Fraction L: tubes 37-39 (see **A**) were pooled and dialyzed, as described in the text; 13.4 absorbance units (260 mμ) were subjected to elution with a total volume of 250 ml; recovery (as percentage of input): major peak, 72; minor peak, 8. Fraction H: tubes 46 and 47 (see **A**); 10 absorbance units, eluent volume 250 ml; recovery: major peak, 95%; minor peak, 5%.

**C,** Second rechromatography of L and H components on small MAK columns, eluent volumes as in **B**. Fraction L: tubes 16-21 (see **B**); 7.3 absorbance units; recovery, 82%. Fraction H: tubes 20-25 (see **B**); 9.5 absorbance units; recovery, 75%.

The elution gradients are shown, with the first arrow indicating when the gradient was disconnected and the second arrow showing when the gradient was reconnected.

## TABLE 1
*Restoration of transforming activity by renaturation of initial and rechromatographed fractions of B. subtilis DNA*

| Preparation no. | Chromatographic passages (no.) | Strands subjected to renaturation | Proportion of initial transforming activity (%) | | |
|---|---|---|---|---|---|
| | | | Ade | Leu | Met |
| 1 | 0 | $L_0:H_0$ | 55.4 | 42.6 | 34.2 |
| 2 | 1 | $L_1$ | 2.2 | 1.2 | 1.3 |
| 3 | 2 | $L_2$ | 2.0 | 1.1 | 0.8 |
| 4 | 3 | $L_3$ | 0.7 | 0.4 | 0.4 |
| 5 | 1 | $H_1$ | 27.8 | 7.8 | 7.4 |
| 6 | 2 | $H_2$ | 12.1 | 4.6 | 5.0 |
| 7 | 3 | $H_3$ | 6.2 | 2.9 | 3.3 |
| 2 + 5 | 1 | $L_1:H_1$ | 55.4 | 47.2 | 37.2 |
| 3 + 6 | 2 | $L_2:H_2$ | 56.1 | 41.7 | 46.9 |
| 4 + 7 | 3 | $L_3:H_3$ | 41.0 | 34.6 | 33.1 |

The transforming activity of the untreated intact DNA was taken as 100%. Preparation no. 1 represents the alkali-denatured DNA subjected directly to annealing without chromatography. The transforming activities (as per cent of initial) for the adenine marker shown by the separated strands before renaturation were, respectively, for $L_1$, $L_2$, $L_3$: 2.2, 2.1, 0.7, and for $H_1$, $H_2$, $H_3$: 1.6, 1.1, 0.6. They were even lower for the other markers.

## TABLE 2
*The composition of native DNA and of separated DNA fractions from different microbial species*

| Prep. no. | Microbial species | Preparation | Composition (mole %) | | | | Molar ratios | | |
|---|---|---|---|---|---|---|---|---|---|
| | | | A | G | C | T | Pu/Py | A+T/G+C | 6-Am/6-K |
| 1 | P. vulgaris | L | 31.7 | 19.8 | 18.3 | 30.2 | 1.06 | 1.62 | 1.00 |
| | | H | 30.0 | 18.7 | 19.8 | 31.5 | 0.95 | 1.60 | 0.99 |
| | | L + H | 30.9 | 19.2 | 19.1 | 30.9 | 1.00 | 1.61 | 1.00 |
| | | Native | 30.8 | 19.3 | 19.5 | 30.4 | 1.00 | 1.58 | 1.01 |
| 2 | B. megaterium | L | 32.3 | 20.6 | 18.1 | 29.0 | 1.12 | 1.58 | 1.02 |
| | | H | 29.1 | 18.1 | 20.9 | 31.9 | 0.89 | 1.56 | 1.00 |
| | | L + H | 30.7 | 19.3 | 19.5 | 30.5 | 1.00 | 1.58 | 1.01 |
| | | Native | 31.1 | 19.7 | 18.9 | 30.3 | 1.03 | 1.59 | 1.00 |
| 3 | B. subtilis* | L | 30.1 | 23.7 | 20.1 | 26.1 | 1.16 | 1.28 | 1.01 |
| | | H | 27.3 | 19.6 | 23.2 | 29.8 | 0.88 | 1.33 | 1.02 |
| | | L + H | 28.6 | 21.7 | 21.7 | 28.0 | 1.01 | 1.30 | 1.01 |
| | | Native | 28.2 | 21.9 | 21.6 | 28.3 | 1.00 | 1.30 | 0.99 |
| 4 | B. stearo-thermophilus | L | 28.7 | 23.4 | 21.1 | 26.8 | 1.09 | 1.25 | 0.99 |
| | | H | 26.9 | 21.3 | 23.3 | 28.5 | 0.93 | 1.24 | 1.01 |
| | | L + H | 27.8 | 22.4 | 22.2 | 27.7 | 1.01 | 1.24 | 1.00 |
| | | Native | 27.6 | 22.6 | 22.7 | 27.1 | 1.01 | 1.21 | 1.01 |
| 5 | E. coli | L | 24.4 | 27.3 | 24.3 | 24.0 | 1.07 | 0.94 | 0.95 |
| | | H | 23.8 | 24.9 | 27.0 | 24.3 | 0.95 | 0.93 | 1.03 |
| | | L + H | 24.1 | 26.1 | 25.7 | 24.2 | 1.01 | 0.93 | 0.99 |
| | | Native† | 24.6 | 25.5 | 25.6 | 24.3 | 1.00 | 0.96 | 1.01 |
| 6 | S. typhimurium | L | 24.7 | 27.5 | 24.2 | 23.6 | 1.09 | 0.93 | 0.96 |
| | | H | 23.9 | 24.7 | 27.6 | 23.8 | 0.95 | 0.91 | 1.06 |
| | | L + H | 24.3 | 26.1 | 25.9 | 23.7 | 1.02 | 0.92 | 1.01 |
| | | Native | 24.0 | 26.4 | 25.9 | 23.7 | 1.02 | 0.91 | 1.00 |
| 7 | S. marcescens | L | 20.1 | 31.1 | 28.2 | 20.6 | 1.05 | 0.69 | 0.93 |
| | | H | 20.9 | 28.0 | 30.2 | 20.9 | 0.96 | 0.72 | 1.04 |
| | | L + H | 20.5 | 29.6 | 29.2 | 20.7 | 1.00 | 0.70 | 0.99 |
| | | Native | 20.3 | 29.4 | 29.7 | 20.6 | 0.99 | 0.69 | 1.00 |

*Taken from a previous report.[3]
†Composition of DNA of *E. coli* HfrH taken from a previous paper.[8]

activity, especially of that for adenine, observed previously upon the renaturation of the H fraction by itself,[1] was very much diminished, though not abolished entirely, through the repeated passage of the material through MAK. This reduced reactivation upon self-annealing is not due to degradation during purification, as shown by the high recovery of transforming activity when the purified complementary strands are renatured together. Representative results are shown in Table 1. If the restoration of biological activity by the renaturation of separate strands reflects the extent of their contamination, it can be estimated roughly that the twice-recycled L fractions were 95-100 per cent pure and the similarly treated H fractions 85-90 per cent.

**Strand separation in various microbial DNA specimens.** The DNA preparations were, after denaturation by alkali, subjected to chromatography on MAK columns. The selection of elution diagrams in Figure 2 demonstrates the resolution of all specimens into two distinct fractions designated as L and H.[11] The elution profiles were quite reproducible and resembled those obtained previously with denatured *B. subtilis* DNA (Fig. 1A; also Fig. 1B and 1C in ref. 1). The recoveries from the columns were relatively high, sometimes amounting to 90-100 per cent of the input.[12]

**Base composition of L and H fractions.** Since it is now obvious that the elution procedure used here can serve to produce two distinct fractions from many different DNA preparations applied, in the denatured state, to MAK columns, one may inquire whether these chromatographic fractions can be regarded as complementary strands. Data on the direct analysis of the L and H fractions into which the DNA from seven bacterial species was separated are assembled in Table 2 and compared in each case with the composition of the intact DNA of the organism. This table also lists the characteristic molar ratios.

For an evaluation of the results, it will be best to consider first the DNA specimens belonging to the AT type (preps. 1-4). Especially in the DNA derived from the genus *Bacillus* (preps. 2-4), there can be little doubt that the L and H fractions fulfill the analytical requirements of completely complementary strands. Though the L fractions are invariably richer in

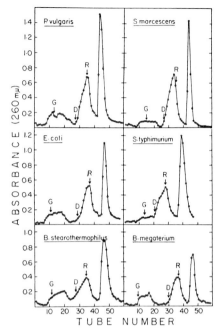

**Fig. 2.** Profiles, by intermittent gradient elution from standard MAK columns, of alkali-denatured DNA of organisms listed below. G, points at which the gradient (0.7-1.5 M NaCl) was started; D, those at which it was disconnected; and R, those at which it was reconnected. The recoveries of separated fractions are given as percentage of input: *P. vulgaris:* total, 100; L, 42; H, 51. *S. marcescens:* total, 88; L, 50; H, 34. *E. coli:* total, 84; L, 40; H, 36. *S. typhimurium:* total, 94; L, 40; H, 50. *B. stearothermophilus:* total, 83; L, 29; H, 37. *B. megaterium:* total, 79; L, 33; H, 30.

purines and the H fractions richer in pyrimidines and though A and T are not paired in the same fraction nor G and C, the sums of A + T and of G + C are the same in both fractions and identical with those found in the intact DNA of the species. Moreover, there is a high degree of complementarity of the L and H fractions derived from the same DNA: the relationships that could be called the complementarity ratios, namely, the expressions $A_L/T_H$, $G_L/C_H$, $C_L/G_H$, $T_L/A_H$, $Pu_L/Py_H$, and $Py_L/Pu_H$, are all very near unity. In fact, the mean of all these ratios for all fractions listed in Table 2 is 1.00 with a standard deviation of 0.016. In a subsequent paper we shall show that the DNA of phage T4, whose separation into two fractions will be mentioned below, also belongs

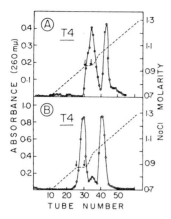

**Fig. 3.** Separation of fractions of sheared and denatured DNA of phage T4 by intermittent gradient elution from MAK columns. Elution gradients and their interruption indicated as in Fig. 1. Recovery (as percentage of input). **A**, Total, 67; L, 40; H, 27. **B**, Total, 88; L, 40; H, 41; minor peak (mixture of L and H), 7.

to the DNA group yielding recognizably complementary strands.

Such complementarity is, perhaps, less obvious, though still arguable, in the case of the DNA specimens belonging to the equimolar and GC types (preps. 5-7, Table 2). In these instances, the A and T contents of the separated fractions are not widely different, and one could almost consider these bases as being paired in the same fraction; but the G and C contents vary significantly in each fraction and show full complementarity with their counterparts. For instance, in *E. coli* DNA (no. 5), whereas $A_L/T_L = 1.02$ and $A_H/T_H = 0.98$, the ratios of $G_L/C_L$ and of $G_H/C_H$ are 1.12 and 0.92, respectively; but the complementarity ratio $G_L/C_H$ is 1.01 and $C_L/G_H$ is 0.98. These observations on partial base-pairing are reminiscent of previous findings[8,14,15] concerning the composition of rapidly labeled RNA in bacteria. For example, the composition of pulse-labeled RNA produced, during an 80-second pulse of $P^{32}$-orthophosphate, by a nonsynchronous culture of *E. coli* was found (unpublished experiments) to contain (as mole %): A, 23.5; G, 27.3; C, 25.8; U, 23.4; this is quite similar to what would be expected of a transcript of the H fraction of the DNA (cf. no. 5, Table 2).[16]

Our previous work on the complementary strands of *B. subtilis* DNA suggested an additional, entirely unexpected regularity, namely, the equality in either strand of 6-amino and 6-keto nucleotides (A + C = G + T).[2,3] This relationship, which would normally have been regarded merely as the consequence of base-pairing in a DNA duplex and would not have been predicted as a likely property of a single strand, is shown here to apply to all strand specimens isolated from denatured DNA of the AT type (Table 2, preps. 1-4). It cannot yet be said to be established for the DNA specimens from the equimolar and GC types (nos. 5-7).

*Preliminary observations on strand separation of denatured DNA of bacteriophage T4.* In our previous communication on strand separation of *B. subtilis* DNA,[1] we mentioned the possibility that RNA fragments adhering to the strands might have contributed to their column separation. It was therefore decided to carry out a few experiments with a DNA specimen free of RNA, namely, that of phage T4. When the denaturation product of the intact preparation was subjected to fractionation on a MAK column, no elution took place even at very high molarities of eluent or at a high ratio of DNA to adsorbent. Very good elution profiles, with a relatively high recovery, were, however, obtained with phage DNA that was broken by shear before being denatured. Although direct comparison of our preparations with those previously studied[17,18] is difficult, it is not unlikely that they contained a high proportion of quarter-length and smaller fragments, i.e., of a range of mol wt $15\text{-}30 \times 10^6$. Figure 3 illustrates two experiments on the separation of denatured T4 DNA into two well-defined fractions which, as will be reported in a subsequent paper, can be shown by transcription analysis with RNA polymerase to exhibit a fully complementary base composition. The minor peak seen in Figure 3B between the L and H components had a composition corresponding to a mixture of the two.

One may conclude that the separation of DNA strands on MAK does not depend upon contamination with RNA fragments, but does depend upon size of the DNA. We shall, on a later occasion, report observations on the temperature-absorbance profiles of the L and H components treated together under conditions of annealing.

### Concluding remarks

There are instances in the literature in which the separation of denatured DNA into two strands by means of density-gradient centrifugation or MAK chromatography has been reported. Without claim to completeness we may mention observations on a *B. megaterium* phage,[19-21] *B. subtilis* phages,[22,23] and a *B. stearothermophilus* phage,[24] as well as reports on "satellite" DNA specimens from mouse and guinea pig cells.[25,26] In all these cases the separated chains varied considerably, and sometimes vastly, in their nucleotide composition; and it is presumably for this reason that they were much more easily separable than the preparations described here. Almost all strand preparations described in the literature share with the ones discussed in the present communication a high degree in complementarity between the L and H chains; but only a few of the phage DNA specimens[23,27] can be said to approach the equality in either strand of the 6-amino and 6-keto nucleotides that we have stressed above.

It may be of interest to note that the DNA specimens that gave the best separation in our experiments (nos. 2-4 in Table 2) came from strains that are known to be hosts for phages yielding DNA readily separable into complementary strands, as has been mentioned. Whether this is more than a coincidence cannot be said, nor whether it is significant that these strains are also spore-forming.

As is shown in Table 2, in all preparations examined by us the H fractions were richer in cytosine than the L fractions, and in the AT types (nos. 1-4) they were also richer in thymine. Whether the cytosine moieties are really more frequently clustered in the H strands, as has been suggested recently,[28] cannot yet be decided. The investigation of the distribution of pyrimidine isostichs in native microbial DNA preparations, at any rate, has shown polycytidylic acid runs of length 4 or 5 to be comparatively rare.[7] The extension of such studies to separated chains is being pursued at present.

If one considers that the microbial DNA preparations studied here may have existed in the cell as macromolecules of a molecular weight range of $10^8$ to $10^9$, or even higher, it stands to reason that the specimens subjected to separation must represent fragments of such giant structures, so that the preparations of strands obtained should be regarded as families rather than as molecular individualities. Despite this stricture, it is unlikely that much mixing can have taken place, fragments of the original L strand being collected together with fraction H and vice versa: both the results of the transformation tests[1] and the constancy of the dissymmetry ratio A + T/G + C (Table 2) speak against this being a frequent occurrence.

### REFERENCES AND NOTES

1. Rudner, R., J. D. Karkas, and E. Chargaff, these Proceedings **60**:630 (1968).
2. Karkas, J. D., R. Rudner, and E. Chargaff, these Proceedings **60**:915 (1968).
3. Rudner, R., J. D. Karkas, and E. Chargaff, these Proceedings **60**:921 (1968).
4. Chargaff, E., Progr. Nucleic Acid Res. Mol. Biol. **8**:297 (1968).
5. The abbreviations A, G, C, T, and U refer to adenine, guanine, cytosine, thymine, and uracil, respectively. Pu designates purines, Py pyrimidines, 6-Am the sum of 6-amino nucleotides (A + C), and 6-K 6-keto nucleotides (G + T).
6. Rudner, R., H. J. Lin, S. E. Hoffmann, and E. Chargaff, Biochim. Biophys. Acta **149**:199 (1967).
7. Rudner, R., H. S. Shapiro, and E. Chargaff, Biochim. Biophys. Acta **129**:85 (1966).
8. Rudner, R., B. Prokop-Schneider, and E. Chargaff, Nature **203**:479 (1964).
9. Thomas, C. A., Jr., K. I. Berns, and T. J. Kelly, Jr., in Procedures in Nucleic Acid Research, ed. G. L. Cantoni and D. R. Davies (New York and London: Harper & Row, 1966), p. 535.
10. Grossman, L., S. S. Levine, and W. S. Allison, J. Mol. Biol. **3**:47 (1961).
11. It should be stressed that the designation of the two elution peaks in Figure 2 as L (light) and H (heavy) was chosen in conformity to our previous observations on *B. subtilis* DNA,[1] though the densities of the preparations under discussion here have not yet been determined.
12. Denatured DNA preparations of *P. vulgaris, B. stearothermophilus,* and *S. typhimurium* yield slightly higher proportions of the H than of the L component. All other DNA preparations discussed in this paper show a consistent deficit in the recovery of the H fraction, as pointed out before for *B. subtilis*.[1] The ratio of L to H in the recovered material varies slightly for any given DNA specimen and may depend upon the efficacy

of the alkali treatment and, perhaps, also of the column. Occasionally, incompletely denatured DNA will give an elution profile of the L peak that is skewed, or shows a shoulder, on the right side; sometimes a minor peak appears between the L and H components (cf. Fig. 3). A partial separation of denatured DNA of *B. stearothermophilus* by elution from a MAK column with a conventional salt gradient has been reported previously.[13]

13. Ageno, M., E. Dore, C. Frontali, M. Arcà, L. Frontali, and G. Tecce, J. Mol. Biol. **15**:555 (1966).
14. Gray, E. D., A. M. Haywood, and E. Chargaff, Biochim. Biophys. Acta **87**:397 (1964).
15. Rudner, R., E. Rejman, and E. Chargaff, these Proceedings **54**:904 (1965).
16. Similarly composed specimens of rapidly labeled RNA are also encountered at certain stages of synchronized cultures of *E. coli*.[8,15] This is consistent with the suggestion that the H strand, richer in pyrimidines or at least in cytosine, is transcribed preferentially so as to yield RNA with a high G/C ratio.[2]
17. Hershey, A. D., and E. Burgi, J. Mol. Biol. **2**:143 (1960).
18. Burgi, E., and A. D. Hershey, J. Mol. Biol. **3**:458 (1961).
19. Cordes, S., H. T. Epstein, and J. Marmur, Nature **191**:1097 (1961).
20. Tocchini-Valentini, G. P., M. Stodolsky, A. Aurisicchio, M. Sarnat, F. Graziosi, S. B. Weiss, and E. P. Geiduschek, these Proceedings **50**:935 (1963).
21. Aurisicchio, S., E. Dore, C. Frontali, F. Gaeta, and G. Toschi, Biochim. Biophys. Acta **80**:514 (1964).
22. Thomas, C. A., Jr., and L. A. MacHattie, Ann. Rev. Biochem. **36**:485 (1967).
23. Riva, S., M. Polsinelli, and A. Falaschi, J. Mol. Biol. **35**:347 (1968).
24. Saunders, G. F., and L. L. Campbell, Biochemistry **4**:2836 (1965).
25. Flamm, W. G., M. McCallum, and P. M. B. Walker, these Proceedings **57**:1729 (1967).
26. Corneo, G., E. Ginelli, C. Soave, and G. Bernardi, Biochemistry **7**:4373 (1968).
27. Marmur, J., and S. Cordes, in Informational Macromolecules (New York: Academic Press, 1963), p. 79.
28. Kubinski, H., Z. Opara-Kubinska, and W. Szybalski, J. Mol. Biol. **20**:313 (1966).

# 3 STRAND SEPARATION AND SPECIFIC RECOMBINATION IN DEOXYRIBONUCLEIC ACIDS: PHYSICAL CHEMICAL STUDIES

P. Doty
J. Marmur
J. Eigner
C. Schildkraut*
Conant Laboratory
Department of Chemistry
Harvard University
Cambridge, Massachusetts

The separation and reunification of the complementary molecular strands of DNA, so clearly indicated by the restoration of biological activity,[1] can be demonstrated by physical chemical techniques as well. Such studies permit a more quantitative description of the phenomenon, make possible the inclusion of DNA samples that do not participate in bacterial transformation and lead to a better insight into the controlling features of the reaction. This, in turn, should provide a better basis for understanding the possibilities and limitations that DNA has *in vivo* for the macromolecular reactions that are the counterpart of its genetic function.

In summarizing here our current work in this direction we begin by showing how strand separation and recombination is reflected in three physical chemical methods in the order in which we took them up. To proceed from these to more quantitative studies it became necessary to find reliable, routine means for determining the molecular weight of DNA in the various forms with which we were confronted, so that the complicating problems of aggregation and depolymerization could be assessed and minimized. Success along these lines made it possible to follow the molecular weight changes accompanying the processes being studied and to examine the effect of molecular weight thereon. In the final section we present the results of a study of recombination between strands which differ either in isotopic label or in species of origin. In both cases "hybrid" or "heterozygous" reformed molecules can be demonstrated.

**Observation of strand recombination**

*Absorbance-temperature curves.* It has been established that when DNA solutions are slowly heated a dramatic macromolecular change occurs in a very restricted temperature range. The change is a cooperative melting out of the one-dimensional helical structure yielding disorganized, coiled polynucleotide chains. At 0.2 molar sodium ions the midpoint of this transition, $T_m$, lies in the interval of 80° to 100° depending on the guanine-cytosine content of the DNA.[2] The change, which involves a 40 per cent increase in absorbance, can be easily and accurately followed by measuring the absorbance at 260 m$\mu$ as a function of temperature.[3] When the solution is cooled to 25° the absorbance decreases until it is about 12 per cent above that of the original solution at room temperature. This appears to be due to the formation of short, imperfect, intrachain helical regions in which a major portion of the bases are paired. Upon reheating such solutions, in the case of calf thymus and other mammalian DNA, the absorbance increases gradually without a region of rapid rise since short, imperfect helices have a broad range of transition temperatures.[3]

When DNA of *Diplococcus pneumoniae* was thermally denatured and then reheated it was

---

From Proceedings of the National Academy of Sciences, U.S.A. 46:461-476, 1960. Reprinted with permission.

The authors would like to thank Miss D. Lane for valuable assistance and Professor A. Peterlin, Professor J. D. Watson, and Dr. N. Sueoka for helpful advice and discussions. We are very grateful to Professor C. E. Hall of the Massachusetts Institute of Technology for the electron micrographs and to Drs. S. E. Luria and M. Demerec, for supplying us with strains of *Shigella dysenteriae* and *Salmonella typhimurium* (LT-2). This work was supported by a grant from the United States Public Health Service (C-2170).

*Predoctoral Fellow of the National Science Foundation.

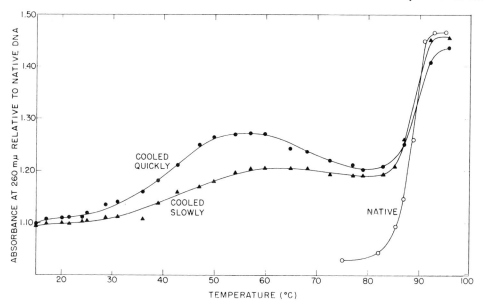

**Fig. 1.** Thermal transitions of native, slowly cooled and quickly cooled *D. pneumoniae* DNA. *D. pneumoniae* DNA was heated at 100° for 10 min in standard saline-citrate. Hot concentrated saline-citrate was then added to a final concentration of 0.27 M NaCl plus 0.02 M Na citrate and subdivided into two portions, one cooled quickly in ice-water, the other cooled slowly. The native DNA was an aliquot of the same sample. The absorbance at 260 mµ, corrected for thermal expansion, was recorded after each solution was exposed for 15 min to the indicated temperature. The ordinate gives this reading relative to the absorbance of the native sample at 25°.

found, in contrast to the general behavior just described, that a distinct maximum occurred in the middle of the temperature range; beyond this there was a sharp rise coincident with the latter half of the original curve for native DNA. This behavior suggested that a typical helix-coil transition was taking place involving the denaturation of long, perfect helices containing about half of the bases in the DNA sample.

The meaning of this became more clear when the absorbance-temperature curves of quickly and slowly cooled samples were measured. Such results are shown in Figure 1. The rate of cooling of these samples is the same as that described in reference 1. The absorbance of the slowly cooled sample is seen to rise to a plateau and then continue to the abrupt region coincident with the thermal denaturation of the native DNA. This sample had retained a large degree of its original transforming activity as described in the foregoing paper.[1] From this it is concluded that the slowly cooled sample contained substantial amounts of the Watson-Crick helix. The curve for the quickly cooled sample, which had very low transforming activity, can now be understood. At room temperature it is devoid of long, complementary helical regions, but during the gradual heating period short, imperfect helical regions melt out, releasing chains that then have time to develop long, complementary helices: these then melt in the vicinity of the characteristic temperature, $T_m$. Thus the display of the sharply melting region by the quickly cooled sample is seen to be an artifact in that this is not evidence that long, complementary helices exist in the same DNA sample at room temperature. This is consistent with its near absence of biological activity. We propose to call this form of DNA *denatured* and refer to the slowly cooled DNA with substantial amounts of reformed complementary helical regions as *renatured*.

**Density gradient ultracentrifugation.** When a very dilute solution of DNA (∼2 γ/ml) in about 7.8 molal cesium chloride is centrifuged at high speed in an analytical ultracentrifuge for 30-40 hours it is found that a density gradient has been set up and the DNA has migrated to a band at a position corresponding to its effective

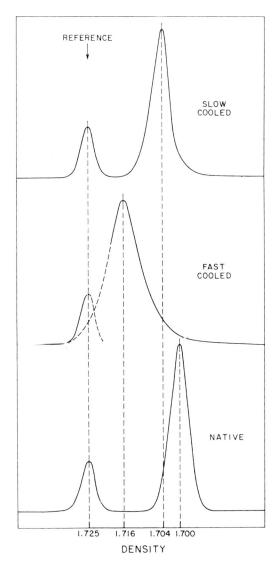

**Fig. 2.** Molecular reconstitution in *D. pneumoniae* DNA. Equilibrium concentration distributions of DNA samples banded in CsCl gradient. See text for a description of samples. Centrifugation at 44,770 rpm. The ordinate represents the DNA concentration as a function of the distance from the axis of rotation. Density increases towards the left. DNA at 20 μg/ml in standard saline-citrate was heated to 100° for 10 minutes. The slow-cooled sample was taken slowly from 90° to room temperature.

density. This is the basis of a valuable new technique[4] which permits the very accurate assessment of the density of DNA and, through the band profile, reflects in a combined fashion the molecular weight and the density heterogeneity of the DNA sample.[5] It has recently been shown that the density of DNA varies linearly with the guanine-cytosine content[6,7], and that the distribution of guanine-cytosine among the molecules making up a DNA sample is relatively narrow, particularly for bacterial DNA.[6-8] It has been also shown that the density of DNA increases about 0.016 units upon denaturation.

In this situation it was obvious that the density of denatured and renatured pneumococcal DNA should be determined because our view of renaturation, brought about by slowly cooling thermally denatured DNA, predicts that such material should have a density close to the original native DNA. The photometric traces of the ultraviolet photographs of the appropriate density gradient ultracentrifugations are shown in Figure 2. At the bottom of the figure one sees the band for native pneumococcal DNA with a density of 1.700 together with a small band of DNA from *Pseudomonas aeruginosa*, which is present in each case to provide a reference density. The trace for the quickly cooled, denatured pneumococcal DNA is shown in the middle, exhibiting the expected higher density of 1.716. At the top is seen the tracing for the renatured DNA: it has a density of 1.704. Thus the renatured DNA is found to have a density close to that of native DNA, supporting the view that it consists predominately of re-formed, complementary helices of the same character, and density, as the native DNA.

Although the molecular weights of these three forms will be discussed at a later point it may be of interest to mention here that the pronounced widening of the band for the denatured DNA reflects an approximately fivefold reduction in molecular weight; hence only a part of this can be due to strand separation alone. The near equivalence of the profile for the renatured DNA to the native DNA involves, therefore, not only a recombination of two strands but, under the conditions employed, some aggregation which by chance increases the molecular weight to about that of the original native DNA.

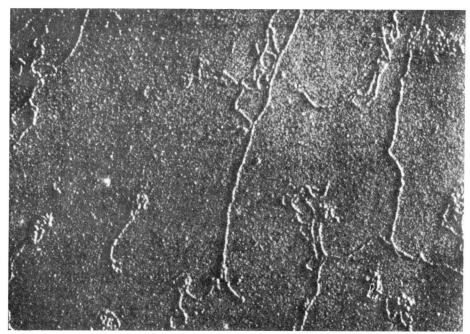

**Fig. 3.** Electron micrograph of renatured *D. pneumoniae* DNA. Magnification 95,000 ×. Shadow casting is from the right at a shadow to height ratio of 10:1. See text for heating and cooling conditions.

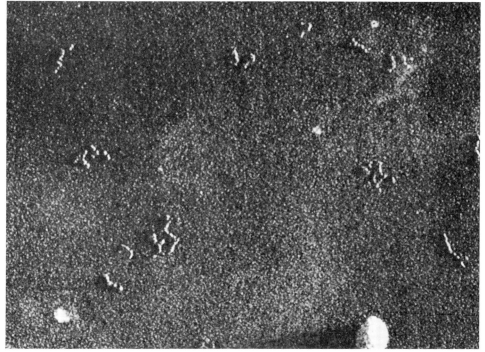

**Fig. 4.** Electron micrograph of denatured *D. pneumoniae* DNA. Magnification 95,000 ×. The polystyrene reference sphere at the lower right of the micrograph has a diameter of 880 Å. Shadow casting is from the right at a shadow to height ratio of 10:1. See text for heating and cooling conditions.

*Electron microscopy.* The evidence presented in the two previous sections clearly predicts that electron microscopy should reveal in the renatured DNA the long, cylindrical threads of 20 A. diameter that are characteristic of native DNA: these should be absent in the denatured (quickly cooled) DNA. Accordingly, pneumococcal DNA was heated in saline-citrate at 100° for 10 minutes, diluted with more concentrated saline-citrate to a final concentration of 18 γ/ml in 0.30 $M$ NaCl + 0.03 $M$ Na citrate, and cooled in two parts, one quickly and one slowly. Both solutions were then dialysed into 0.05 $M$ ammonium carbonate plus 0.1 $M$ ammonium acetate. Thereafter Professor C. E. Hall sprayed the solutions on freshly cleaved mica and, proceeding in a manner previously described[9,10] obtained electron micrographs shown in Figure 3 (slowly cooled) and Figure 4 (quickly cooled). The former shows the structure characteristic of native DNA. The only difference between this and micrographs of native DNA is the more frequent occurrence of irregular patches at the ends of cylindrical threads. These are probably regions of denatured DNA arising from incomplete recombination or the inequality in length of the two strands that are paired. The micrograph for the quickly cooled sample (Fig. 4) shows only irregularly coiled molecules with clustered regions. These results are very similar to those found for ribonucleic acid.[11] Thus the identification of renatured DNA as being for the most part similar to native DNA and denatured DNA as being an irregular chain with considerable base pairing in short regions seems to be complete.

*Dependence of molecular recombination on concentration.* In the foregoing report[1] it is shown that the transforming activity of renatured pneumococcal DNA increases with the concentration at which it is slowly cooled (see Fig. 4 of ref. 1). This suggests that the extent of molecular recombination is quite low at concentrations considerably below those employed in the experiments described in the foregoing sections, that is, ~20 γ/ml. In order to find out if the biological activity reflected the state of the DNA sample itself we heated and slowly cooled a pneumococcal DNA sample in the usual manner at a concentration of 1 γ/ml. The density of this sample was found to be 1.715. This value matches that of quickly cooled pneumococcal DNA (1.716) prepared either at the usual concentration, 20 γ/ml, or at this very low concentration, 1 γ/ml. Thus cooling at what we have described as a slow rate, that is from 90 to 60° in 30 minutes, does not lead to substantial recombination of strands when the concentration is quite low.

With the physical exhibition of the molecular recombination, as well as the biological, shown to be quite concentration dependent, it seemed permissible to attempt to account for the concentration dependence observed (Fig. 4 of ref. 1) by assuming a bimolecular reaction, $A+A \rightleftarrows A_2$, which reaches effective equilibrium at a particular temperature that lies within the region where slow cooling takes place. If α represents the mole fraction of strands that are combined in the helical form, the equilibrium constant $K = \alpha/[(1-\alpha)^2 c]$ where $c$ represents the DNA concentration. This can be applied to the data in Figure 4 of reference 1 by assuming that α is equal to the fractional activity of the renatured DNA. By choosing one point to permit an evaluation of the product $Kc$, α can be calculated as a function of $c$. The dashed line drawn in Figure 4 of reference 1 is the result calculated in this way. Since it fits the experimental data there is reason to conclude that the molecular recombination is indeed bimolecular, as the character of the Watson-Crick structure for DNA would demand. Moreover, the success of this simple approach suggests that properly designed experiments which yield $K$ as a function of temperature should lead to the determination of the heat and entropy of the recombination as well as the free energy change involved in this process at the temperature of cellular division.

*Dependence of molecular recombination on the source of DNA.* With the view that molecular recombination in DNA consists simply of two complementary strands coming together with the fraction of strands complexed under control of an equilibrium constant, it follows that the manifestation of molecular recombination depends critically on the concentration of complementary strands. Now, the concentration of complementary strands is not identical with the concentration of DNA, when DNA from different sources is considered. Take the extreme cases, of typical bacterial DNA and the DNA of mammalian cells. Here the DNA content per cell differs by the order of a thousand fold. If we take the average molecular weight to be the same in both cases and if it is assumed that each DNA molecule in a given cell is different, it follows that at the same weight concentration, the concentration of complementary pairs will be a thousand fold greater for the bacterial than for the animal DNA. From this it follows that molecular recombination will not be expected for animal DNA under the conditions that are just sufficient to permit its occurrence in bacterial DNA.

The expectation that animal DNA will not show complementary reformation when bacterial DNA will do so has been borne out in two kinds of experiments.[12] In one set the absorbance temperature profiles for a number of denatured DNA samples have been determined. The majority of bacterial DNA samples showed some of the character seen in Figure 1, that is, a sharply rising region at high temperature approximately coincident with the curve for native DNA. By contrast calf thymus and salmon sperm DNA showed essentially a continuous and gradual rise with temperature. This shows that on the time scale of the heating curve there was some molecular reconstitution in most of the bacterial samples but not in the cellular DNA.

In another experiment samples of thermally denatured DNA were cooled quickly and then brought to 80° and the absorbance followed as a function of time in 0.30 $M$ NaCl plus 0.03 $M$ sodium citrate. The pneumococcal DNA fell from 1.40 times the absorbance of the native DNA at 25° to 1.08 whereas the calf thymus showed no drop at all. This again demonstrates the inability of calf thymus DNA to reform the helical configuration under the conditions which do permit pneumococcal DNA to do so. Consequently, the great difference in concentration of complementary strands in the denatured DNA appears to offer the explanation.

## The macromolecular properties of native, denatured, and renatured DNA

A means of making reasonably accurate molecular weight assignments to DNA in each of the forms under study is necessary in order to examine quantitatively the phenomenon of molecular recombination in DNA and to enable the complicating features of aggregation and depolymerization to be understood and eliminated in so far as possible. Although light scattering measurements were the first method to satisfactorily establish the molecular weight and shape of DNA in solution,[13, 14] its accuracy and applicability in routine use, particularly for molecular weights greater than 5 million, has been found wanting elsewhere[15] as well as in this Laboratory. Consequently, a systematic investigation of sedimentation constants, evaluated by extrapolation to zero concentration, and intrinsic viscosities evaluated by extrapolation to zero gradients, has been undertaken[16] and molecular weights have been derived by a careful application of the Mandelkern-Flory equation:[17]

$$M = \left[ \frac{s^\circ [\eta]^{1/3} \eta_0 N}{\beta(1 - \bar{v}\rho)} \right]^{3/2}$$

In this equation, $s^\circ$ = the sedimentation constant at zero concentration, $[\eta]$ = the intrinsic viscosity at zero gradient, $\eta_0$ = the solvent viscosity, $N = 6.03 \times 10^{23}$, $(1-\bar{v}\rho)$ = the buoyancy factor and $\beta$ = a constant which has a value which ranges from $2 \times 10^6$ to about $4 \times 10^6$ as the permeability of a coiled molecule or the axial ratio of a rigid ellipsoid increase. For most flexible polymers the value of $\beta$ has been found to be about $2.6 \times 10^6$.

*Native DNA.* An investigation[18] of DNA samples in which the molecular weight had been varied by ultrasonic radiation and the molecular weights measured by light scattering showed $\beta$ to increase monotonically from 2.56 at 300,000 to 3.29 at 8,000,000. Other light scattering experiments, and an independent study[19, 20] in which molecular weights were determined by sedimentation-diffusion, have shown good agreement with this result. Such an empirical evaluation of $\beta$, based as it is on absolute methods, permits us to use the Mandelkern-Flory equation to derive weight average molecular weights from sedimentation-viscosity data.

For a number of samples $s^\circ$ and $[\eta]$ have been measured in a solvent of lower ionic strength, 0.011 $M$ in Na$^+$ ions, as well as in the more common solvent that is 0.195 $M$ in Na$^+$ ions. The former solvent contains 0.0025 $M$ Na$_2$HPO$_4$, 0.0050 $M$ NaH$_2$PO$_4$ and 0.001 $M$ sodium ethylenediaminetetraacetate and has a

### TABLE 1
*Values of $s°$ and $[\eta]$ for selected molecular weights of native DNA*

| Molecular weight | $s°_{20,w}$ in HMP | $[\eta]$ in HM°P | $s°_{20,w}$ in SSC | $[\eta]$ in SSC |
|---|---|---|---|---|
| 1,000,000 | 8.8 S | 9.1 dl/gm | 10.0 S | 8.3 dl/gm |
| 2,000,000 | 11.5 | 21.0 | −13.1 | 18.7 |
| 4,000,000 | 15.4 | 45.5 | 17.5 | 38.0 |
| 10,000,000 | 22.9 | 120 | 26.4 | 82.5 |
| 16,000,000 | 28.2 | 190 | 32.8 | 119 |

### TABLE 2
*Constants for the empirical sedimentation and viscosity relations for denatured DNA in phosphate-versene solution 0.011 M in $Na^+$ (HMP)*

| Source of DNA | Ks | as | K | a |
|---|---|---|---|---|
| D. pneumoniae | $0.054_5$ | $0.35_4$ | $3.4_6 \times 10^{-5}$ | $0.93_3$ |
| E. coli K-12 | $0.055_8$ | $0.36_1$ | $3.1_1 \times 10^{-5}$ | $0.91_2$ |

pH of 6.8. For convenience this is referred to as HMP (hundredth molar phosphate). The more common solvent consists of 0.15 M NaCl and 0.015 M $Na_3(C_6H_5O_7)$ with a pH of about 7. This will be referred to as SSC (standard saline-citrate). On a double logarithmic plot of $s°$ against $M$ the data in these two solvents fall on two parallel lines exhibiting slight upward curvature. Similarly, log $[\eta]$ against log $M$ yields lines with downward curvature. Table 1 summarizes the smoothed $s°_{20,w}$ and $[\eta]$ values in these two solvents for selected molecular weights.

*Denatured DNA and the problem of aggregation.* When DNA is thermally denatured and cooled, a large fraction of the bases become paired.[3] While most of this pair formation can occur within each chain, some interchain bonding can be expected and this would increase with DNA concentration and molecular weight. Consequently, the determination of the molecular weight of denatured DNA requires proof that aggregation has been eliminated. Early light scattering studies showed no significant change in molecular weight on thermal denaturation,[21] and thereby suggested that the two strands had not separated. Other reports claimed a decrease to one-half the molecular weight but they were not convincing.[22] Moreover, the $s°$ and $[\eta]$ values for such denatured DNA did not yield a molecular weight in agreement with the light scattering values by a wide margin. Upon further investigation of this dilemma we have been able to show that significant aggregation occurs upon cooling denatured DNA in SSC at concentrations in excess of 40-100 γ/ml. Since light scattering and viscosity measurements require concentrations in this range or higher we were able to conclude that aggregation had occurred in most samples previously studied.

In order to eliminate this aggregation in the useful range of concentration we have gone to a solvent of lower ionic strength, HMP. Although there is some reformation of base pairs (about 25 per cent) in this solvent at room temperature no aggregation is evident by any of the methods employed, and a satisfactory dependence of $s°$ and $[\eta]$ on molecular weight has been established by using the Mandelkern-Flory equation with a value of $2.6 \times 10^6$ for β. The results of such studies can be fitted with the usual empirical relations: $s° = K_s M^{as}$ and $[\eta] = KM^a$.

Now it has been shown that the guanine-cytosine bond is stronger than the adenine-thymine bond[2,8] and as a result it is to be expected that the amount of base pairs in denatured DNA will increase with the guanine content. Indeed this has been shown to be the case with RNA samples of different composition.[3] Consequently it is possible that the dependence of $s°$ and $[\eta]$ on molecular weight for denatured DNA will depend on the composition of the DNA. This has been found to be the case and it is illustrated in Table 2, where the constants for the empirical equations are listed for denatured DNA (quickly cooled)

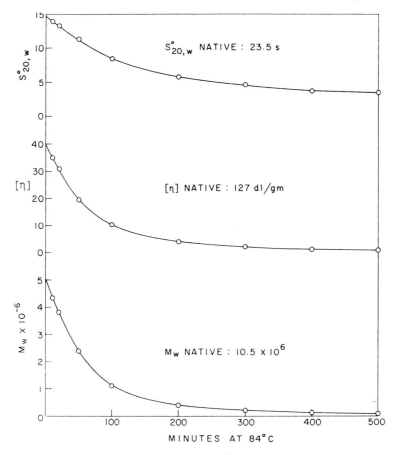

**Fig. 5.** Thermal degradation of *E. coli* DNA in HMP at 84°. Molecular weights derived from Mandelkern-Flory equation[17] using $\beta = 2.6 \times 10^6$.

from *D. pneumoniae* and *Escherichia coli* (K-12).

These results indicate a rather highly swollen chain configuration for DNA, despite some base pair interaction. For a given chain length the molecular dimensions that can be deduced from this are in the range expected for typical polyelectrolytes. These dimensions are much larger than those for RNA of the same chain length and in the same solvent: this is probably due to the additional contraction of the RNA coil due to additional hydrogen bonding made possible by the 2-OH of the ribose ring.

With these results at hand it is possible to assess the molecular weight as a function of time of heating at a given temperature and to demonstrate whether or not strand separation has occurred at the early stages of heating.

*Strand separation and depolymerization.* In the low ionic strength solvent (HMP) the midpoint of the thermal denaturation profile, $T_m$, for *D. pneumoniae* is shifted downward from 85° in SSC to 64° and the transition is complete at 70°. For *E. coli* DNA the corresponding values are 5° higher. A temperature 15° above the $T_m$ is well beyond the melting-out region for DNA in this solvent, indeed by about the same amount as 100° is for the higher ionic strength solvent (SSC). Consequently 79° and 84° were chosen as the temperatures at which to follow the thermal degradation of these two DNA samples. Aliquots were removed periodically from a large amount of the solution that had been brought to this temperature, and cooled quickly. The $s°$ and $[\eta]$ were measured at 25° and the molecular weight calculated as indicated above. The results for *E. coli* DNA are shown in Figure 5. It is seen that the molecular weight falls continuously but rather slowly through the period of exposure to elevated temperature. The extrapolation back to zero time is obvious and the molecular weight value there is 5.0 million. The molecular weight of the native DNA was 10.5 million ($s^o_{20,w}$ = 23.5 $S$ and $[\eta]$ = 126). In this case we have, therefore, a clear example of the molecular weight decreasing by a factor of two upon thermal denaturation, and with aggregation eliminated and depolymerization taken into account it can be said that strand separation did occur in the very early stages of the exposure to the elevated temperature.

The behavior of pneumococcal DNA in a similar experiment was roughly the same except that in this case the molecular weight fell from 8.2 million for the native DNA to 2.9 million for the denatured DNA at zero time. This change by a factor of somewhat more than 2, in this case 2.83, probably reflects an occasional enzymatically induced single chain scission in the original DNA preparation.

From the data obtained during the first 100 minutes of the above experiments the rate of thermal depolymerization of the polynucleotide chains of DNA can be evaluated. In practical units this can be expressed as 0.165 scission per 10 million molecular weight per 10 minutes at 79°. Similar experiments at 100° yields 1.41 in the same units. These and other results lead to a value of about 20 kcal for the activation energy for hydrolytic scission of individual strands of DNA, a value similar to that found in a quite different study of ribonucleic acid.[23] With this evaluation of the rate of bond scission it is possible to assess the extent of fragmentation induced by the exposure to 100° for 10 minutes followed by fast or slow cooling. For fast cooling the value stated above is the answer desired since the units have been chosen for convenience in this situation. Since pneumococcal DNA is often in the range of 10 million molecular weight each native molecule will receive on the average between one and two scissons as a result of the heating to 100°. Slow cooling will involve a longer exposure to temperatures where degradation is significant: in the case of our particular cooling rate the depolymerization would be about doubled.

*Renatured DNA.* Renatured DNA was prepared by heating pneumococcal DNA of molecular weight 8.2 million at 100° for 10 minutes in standard saline-citrate at a concentration of 20 γ/ml and cooling slowly in double the saline-citrate concentration. In standard saline-citrate this material had a $S^o_{20,w}$ of 23.5 $S$ and an $[\eta]$ of 21.6 from which a molecular weight of 6.0 million was deduced. An aliquot of the same heated solution, quickly cooled, had a molecular weight of 2.0 million. When dialyzed into the lower ionic strength solvent (HMP), the renatured DNA had a $s^o_{20,w}$ of 15.4 $S$ and an $[\eta]$ of 43. A molecular weight of 4.0 million was obtained using these results. The denatured DNA showed no change in molecular weight when similarly transferred to the lower ionic strength solvent.

From these results it appears that a small amount of aggregation occurs when renaturation is carried out at 20 γ/ml in 0.3 $M$ NaCl plus 0.03 $M$ Na citrate and that the molecular weight falls by about the amount expected from the depolymerization rates reported above. In the lower ionic strength solvent, where the aggregation appears to have been eliminated, it is interesting to note that the $s^o$ and $[\eta]$ values are essentially equal to those given in Table 1 for 4 million molecular weight native DNA. This supplies further evidence of the close structural similarity of the renatured to the native DNA.

The results have an important bearing on the transforming activity that can be expected for renatured pneumococcal DNA. The loss in activity to be expected from the molecular weight reduction can be obtained directly from an earlier study[24] of the molecular weight dependence of transforming activity. For a reduction of 10 to 4 million the activity should fall to 70 per cent of its original value. However, the activity would be expected to be lowered still further due to the fact that the complementary strands that recombine will generally be of different length as a result of the mild thermal depolymerization. If there was no selection according to chain length in recombination there would be a loss of about 50 per cent of the nucleotides from helical regions due to this effect and a drop to 35 per

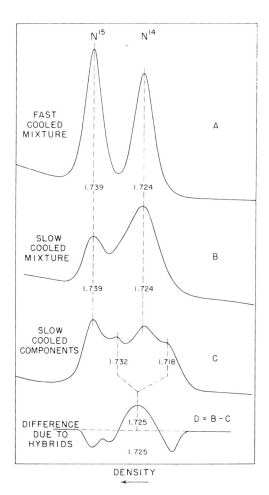

**Fig. 6.** Hybrid formation in *E. coli* B DNA. Equilibrium concentration distributions of DNA samples banded in CsCl gradient. See text for a description of samples. Centrifugation at 44,770 rpm. The ordinate represents the concentration of DNA in the centrifuge cell. The area of each band is proportional to the amount of DNA it contains. $N^{15}$ and $N^{14}$ labeled *E. coli* B DNA were prepared by growing wild-type cells in a synthetic medium containing $N^{15}H_4Cl$ or $N^{14}H_4Cl$, respectively, as the sole nitrogen source. The DNA was isolated by a procedure to be published (J. Marmur, in preparation).

cent would be expected for the maximum activity. However, there is probably some preference for complementary chains of more nearly equal molecular weight to recombine, and the evidence from the absorbance-temperature profile and DNA density in CsCl indicates that about 75 per cent of the nucleotides do reform in the helical configuration. Consequently the maximum biological activity of renatured transforming DNA under the conditions employed may be placed at about 50 per cent. The highest observed[1] has been approximately 50 per cent.

### Recombination of DNA strands of different density and from different species

In this final section we return to the technique of density gradient ultracentrifugation and ask if by its use recombination between strands of different density can be observed. Two cases are of interest. In one we mix two DNA samples from the same species, one of which is denser by virtue of having $N^{15}$ instead of $N^{14}$. In the other we mix $N^{15}$ *E. coli* DNA with $N^{14}$ DNA from other bacteria to see if renatured molecules of intermediate density can be observed. Such "heterozygous" molecules may be expected from closely related bacteria but not from distantly related ones.

From what has been stated in the previous section concerning aggregation and strands of unequal length in the renatured molecules, it is not surprising to find that the optimal resolution of hybrid molecules has proved somewhat elusive. Thus, while maximum helix development in the renatured sample is favored by high DNA concentrations, so is aggregation. Obviously aggregation of renatured molecules of different density will smear out the hybrid band one seeks to resolve. Likewise the inequality of chain lengths in the hybrid molecules will spread out their densities with a corresponding broadening of the hybrid band.

Against this background we present the results obtained with a mixture of $N^{14}$ and $N^{15}$ *E. coli* DNA each at a concentration of 5 γ/ml. This concentration is low enough to prevent aggregation but at the price of only limited recombination to the helical form. The results are collected in Figure 6. At the top *(A)* is shown the tracing for the quickly cooled, denatured DNA. The bands are seen to be completely separated without any hybrid formation. Thus, the reforming of base pairs has been entirely intrachain.

If an aliquot of the same heated mixture is slowly cooled, the results are as shown in the second tracing (B). Here the peaks have remained at the same densities as in the denatured DNA but about one-third of the total DNA has shifted to a lighter density distributed about the value of 1.724. While this is indicative of hybrid formation it is far from proof since the densities of the peaks are essentially those of denatured $N^{14}$ and $N^{15}$ DNA and do not approach that of renatured material which should be about 0.008-0.012 density units lighter than the denatured materials.

In this situation it is necessary to provide a control consisting of the mixture of the two separately cooled DNA samples, that is, samples that have had the same thermal history as the slowly cooled mixture (B) but have been prevented from forming hybrids. The tracing of this control is shown in the Figure as (C). This provides a key to our interpretation since peaks corresponding to the renatured molecules are clearly evident at the right of each of the denatured peaks.

If the control C is now subtracted from the trace for the slowly cooled mixture (B), the migration of material to the density corresponding to renatured hybrids should be evident. This is shown at the bottom as (D). A shift of material from extremities to the density of the hybrid is clearly evident. About 20 per cent of the total DNA is involved: this is about that expected from (C) since there it can be seen that about 40 per cent of the material is in the bands corresponding to the renatured form.

By using 4 times the concentrations employed in the above experiment the denatured species can be essentially eliminated but the additional spreading introduced by aggregation results in only a single band of somewhat irregular shape, and a proper correction cannot be made.

Despite this difficulty we have made some preliminary observations on mixtures of DNA from two different species. In these cases the peak densities appear to be sufficient for interpretation.

The results of Baron et al.[25] showing that *E. coli*, *Shigella* and *Salmonella* are closely related genetically, as well as the report of the similarity of their base compositions[26] prompted an experiment to see whether $N^{15}$ labeled DNA strands from *E. coli* would recombine with $N^{14}$ DNA strands from the other enteric organisms.

$N^{15}$ *E. coli* B DNA and $N^{14}$ *Shigella dysenteriae* (obtained from Dr. S. E. Luria) DNA were mixed, heated as usual and slowly cooled. The densities of the native samples were 1.725 and 1.710 respectively. If we assume 50 per cent of the resultant mixture is composed of hybrid molecules in which helical regions have formed to the extent of 75 per cent of the native helical content we would expect that, as was the case for *E. coli* and *D. pneumoniae*, the density of the resultant hybrid band should not be the mean of the native values but 0.004 density units heavier. A band centered at a density of 1.722 was indeed found and thus hybrid, or heterozygous molecules appear to have formed.

A similar prediction can be made for $N^{15}$ *E. coli* B DNA and $N^{14}$ *Salmonella typhimurium*(LT-2) (obtained from Dr. M. Demerec) since the base compositions are the same. However, a density of 1.729 was observed. Thus it appears that heterozygous renatured molecules have not formed. Since the observed density is midway between that predicted for denatured molecules and renatured, homogenous molecules, it is possible that occasional heterozygous regions have combined preventing a full development of homogeneous renatured molecules. Perhaps under optimal conditions interspecies molecular hybridization of the DNA would be observed in this case as well.

In another experiment, $N^{14}$ *D. pneumoniae* DNA was heated and slowly cooled with $N^{14}$ *Serratia marcescens* DNA. The density of the native samples are 1.700 and 1.717 respectively. The resulting trace showed two clearly separated bands of reformed *D. pneumoniae* of density 1.704 and reformed *Serratia* of density 1.721. There was no evidence of a component of intermediate density indicating that neither aggregation nor interspecies hybridization had occurred.

This serves as an ideal control to show that molecules having approximately the same density difference (0.017 density unit) but not the genetic similarity discussed above, show a strong preference for renaturing only with members of their homologous species. Only in the case where similarity in base composition has been correlated with genetic interaction has it been possible to demonstrate the appearance of heterozygous molecules.

*Discussion.* Our aim in this paper has been to present a series of preliminary reports of closely related work on strand separation and molecular recombination in DNA. Consequently, the discussion will be postponed until the fuller accounts are published. Nevertheless, perhaps two points may be emphasized briefly.

The demonstration that strand separation actually occurs in a matter of a few minutes or less eliminates some of the objections[27] pertaining to the unwinding of DNA that have been put forward as a criticism of the Watson-Crick mechanism[28] for DNA duplication. What has been observed here favors the more recent calculations of the time required for this process.[29,30] Also, although it is not as yet apparent whether the reformation of denatured DNA during slow cooling has a biological counterpart, the demonstration that it does occur suggests a mechanism for genetic exchange in closely related DNA molecules.

The second point is the recognition that the routine reduction of DNA into single strands and their specific reformation into essentially native molecules provides a means of creating entirely new DNA molecules. The study of the progeny of such heterozygous molecules will be of obvious interest. A particular instance is that of a heterozygous molecule each strand of which contains a different marker. The progeny from cells having a single nucleus may be pure clones, mixed clones or transformants with markers normally linked. Any of these results would, of course, illuminate the molecular aspects of the replication act. If linkage can be found, not only will crossing-over be demonstrated but the possibility of making *in vitro* new forms of viable DNA not previously existent will be assured. It seems likely that heterozygous DNA molecules will find other uses as well. For example, active hybrid molecules containing a greater variety of alterations in one strand. New possibilities of testing whether or not homologous ribonucleic acid has sequences in common with DNA can now be explored. The formation of molecular hybrids between closely related organisms should be a useful tool for plotting homologies in base sequences where no genetic exchanges have yet been demonstrated.

*Summary.* When solutions of bacterial DNA are denatured by heating and then cooled, two different molecular states can be obtained in essentially pure form depending on the choice of conditions, that is, rate of cooling, DNA concentration, and ionic strength. One state corresponding to fast cooling consists of single-stranded DNA having about half the molecular weight of the original DNA. The other state corresponding to slow cooling consists of recombined strands united by complementary base pairing over most of their length. This form has as much as 50 per cent of its original transforming activity and is called renatured. The quickly cooled, single-stranded form is essentially inactive and is called denatured. These two forms are clearly identified by differences in (1) absorbance-temperature curves, (2) density, (3) appearance in electron micrographs and (4) hydrodynamic properties. The recombination depends on concentration in the manner expected for an equilibrium between two independent strands and a bimolecular complex.

Molecular weight determinations of the native and renatured forms have been based on an extension of an earlier calibration of sedimentation and intrinsic viscosity in terms of light scattering measurements. For the denatured form molecular weights were determined by the use of the Mandelkern-Flory relation. By these means thermal degradation was accurately assessed. To avoid aggregation it was necessary to substantially lower the cation concentration over that previously used. In this way denatured DNA has been characterized as a single-chain, unaggregated polymer and strand separation demonstrated.

Density gradient experiments on $N^{14}$ and $N^{15}$ *E. coli* DNA have shown the existence of hybrids in the DNA renatured from the mixture. Similarly, hybrids have been shown to form between the strands of bacteria that are closely related genetically. Thus the possibility of forming by renaturation heterozygous DNA molecules with different genetic markers or chemical modifications in the two strands seems assured.

## REFERENCES

1. Marmur, J., and D. Lane, these Proceedings **46**:453 (1960).
2. Marmur, J., and P. Doty, Nature **183**:1427 (1959).
3. Doty, P., H. Boedtker, J. R. Fresco, R. Haselkorn, and M. Litt, these Proceedings **45**:482 (1959).
4. Meselson, M., F. W. Stahl, and J. Vinograd, these Proceedings **43**:581 (1957).
5. Sueoka, N., these Proceedings **45**:1480 (1959).
6. Sueoka, N., J. Marmur, and P. Doty, Nature **183**:1429 (1959).
7. Rolfe, R., and M. Meselson, these Proceedings **45**:1039 (1959).
8. Doty, P., J. Marmur, and N. Sueoka, Brookhaven Symposia in Biology **12**:1 (1959).
9. Hall, C. E., and M. Litt, J. Biophys. and Biochem. Cytology **4**:1 (1958).
10. Hall, C. E., and P. Doty, J. Am. Chem. Soc. **80**:1269 (1958).
11. Hall, C. E., and H. Boedtker (unpublished results).
12. Wright, G., senior thesis, Harvard University, 1960.
13. Doty, P., and B. H. Bunce, J. Am. Chem. Soc. **74**:5029 (1952).
14. Reichmann, M. E., S. A. Rice, C. A. Thomas, and P. Doty, J. Am. Chem. Soc. **76**:3407 (1954).
15. Butler, J. A. V., D. J. R. Laurence, A. B. Robins, and K. V. Shooter, Proc. Roy. Soc. **A250**:1 (1959).
16. Eigner, J., doctoral thesis, Harvard University, 1960.
17. Mandelkern, L., H. A. Sheraga, W. R. Krigbaum, and P. J. Flory, J. Chem. Phys. **20**:1392 (1952).
18. Doty, P., B. B. McGill, and S. A. Rice, these Proceedings **44**:432 (1958).
19. Kawade, Y., and I. Watanabe, Biochim. Biophys. Acta **19**:513 (1956).
20. Iso, K., and Watanabe, I., J. Chem. Soc. Japan, Pure Chem. Section **78**:1268 (1957), in Japanese.
21. Rice, S. A., and P. Doty, J. Am. Chem. Soc. **79**:3937 (1957).
22. Alexander, P., and K. A. Stacey, Biochem. J. **60**:194 (1955).
23. Bacher, J. R., and W. Kauzmann, J. Amer. Chem. Soc. **74**:3779 (1952).
24. Litt, M., J. Marmur, H. Ephrussi-Taylor, and P. Doty, these Proceedings **44**:144 (1958).
25. Baron, L. S., W. F. Carey, and W. M. Spilman, these Proceedings **45**:1752 (1959).
26. Lee, K. Y., R. Wahl, and E. Barbu, Ann. Inst. Pasteur **91**:212 (1956).
27. Delbrück, M., and G. S. Stent, in The Chemical Basis of Heredity, ed. W. D. McElroy and B. Glass (Baltimore: Johns Hopkins Press, 1957), p. 699.
28. Watson, J. D., and F. H. C. Crick, Nature **171**:964 (1953).
29. Levinthal, C., and H. R. Crane, these Proceedings **42**:436 (1956).
30. Longuet-Higgins, H. C., and B. H. Zimm, J. Mol. Biol. (in press).

# 4 STUDIES ON HYBRID MOLECULES OF NUCLEIC ACIDS;
## I. DNA-DNA HYBRIDS ON NITROCELLULOSE FILTERS

J. Legault-Démare
B. Desseaux
T. Heyman
S. Séror
G. P. Ress
Institut du Radium
Radiobiologie
Orsay, France

The technique of Gillespie and Spiegelman (1965) for the detection of DNA-RNA hybrids is based on the observation that nitrocellulose is able to adsorb denatured DNA, although RNA and native DNA are not retained (Nygaard and Hall, 1964).

The same principle may be applied to the study of DNA-DNA homology if only one of the two DNA's to be compared is immobilized. This requirement is fulfilled in the technique of Denhardt (1966) which makes use of a mixture of high polymers to prevent the adsorption of the DNA to be tested on the excess free sites of the nitrocellulose filters carrying the reference DNA. In the technique of Warnaar and Cohen (1966), hybridization is carried out as for DNA-RNA hybrids, and the non-hybridized DNA is subsequently eluted, leaving the hybrid molecules adsorbed on the membrane.

In this paper, we describe a technique of formation of DNA-DNA hybrids on membranes, based upon the observation that in the presence of moderate amounts of dimethyl sulfoxide (DMSO), denatured DNA is not retained by the nitrocellulose, although it is still able to renature or to hybridize with homologous DNA.

## Materials and methods

*Membranes.* Membranfilter Gesellschaft, Göttingen, Germany, type MF 50, diameter 25 mm.

*Dimethyl sulfoxide.* Purissimum Fluka, Switzerland.

*DNA* was extracted from T2 phage by the phenol technique of Mandell and Hershey (1960). $^3$H and $^{32}$P labels were introduced in the form of thymidine and of ortho phosphate respectively during phage growth.

*Fixation of DNA on membranes.* $^3$H labeled, heat denatured T2 DNA in 6×SSC was slowly filtered through the membranes, which were subsequently washed and dried according to Gillespie et al. (1965). The amount of DNA retained on the filters was usually 85 to 90% of the input, and the experiments described in this paper were done with membranes loaded with about 20 µg of DNA ($10^4$ counts/min under our counting conditions).

*Melting curves.* The denaturation-renaturation profiles were measured in a Beckman DU Spectrophotometer equipped with dual thermospacers. A special lid permitted the introduction of a thermoelectric couple inside a reference cuvette in the cuvette holder, for the accurate measure of temperature.

After heating well above the point of complete denaturation, the renaturation curves were obtained during slow cooling of the circulating fluid (100 to 30°C in 3 hours). The concentration of DNA in these experiments was about 15 µg/ml (O.D.$_{260}$ = 0.300), and the concentration of DMSO varied from 0 to 40% (v/v).

*Counting of radioactivity.* The dried membranes were placed in scintillation vials and counted in a Packard TriCarb spectrometer; the cross-contamination of $^{32}$P in the $^3$H canal was of 0.66%.

## Results

### 1. Fixation of DNA on membranes in the presence of DMSO

The membranes were washed with 50 ml of 2×SSC or 6×SSC containing 0 to 40% DMSO (v/v). 40 µg of labeled denatured T2 DNA in 5 ml of the same mixture were slowly filtered through the washed membranes. The filters were finally washed with 100 ml of the same mixture, dried, and the radioactivity was determined.

Fig. 1 shows that in the presence of 20% DMSO, the fixation of the DNA to the filters was almost completely abolished. However, a small amount of radioactivity (about 0.4%) was

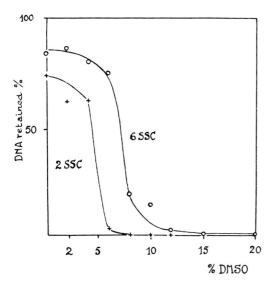

**Fig. 1.** Filtration of denatured T2 DNA through nitrocellulose filters in the presence of different concentrations of DMSO. See details in the text.

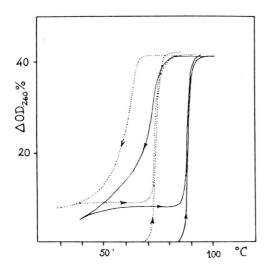

**Fig. 2.** Denaturation-renaturation profiles of T2 DNA. Concentration: 15 μg/ml, in 2×SSC (full line) or in 2×SSC containing 30% DMSO (dotted line).

retained, even in the presence of 40% DMSO. This contamination was fairly constant, and was not modified when the DNA was submitted to repeated pronase and phenol treatments; however, it was reduced when the solution of denatured DNA in 2×SSC-30% DMSO was filtered several times through nitrocellulose membranes. The significance of this phenomenon has not been investigated.

### 2. Denaturation and renaturation of DNA in the presence of DMSO

A number of substances are known to lower the $T_M$ of DNA (Helmkamp and Ts'o, 1961; Marmur and Ts'o, 1961; Hanlon, 1966).

Fig. 2 presents two of the denaturation-renaturation profiles obtained in the experimental conditions described under "Methods". It can be seen that in the presence of 30% DMSO, the $T_M$ of T2 DNA was in fact lowered by 14°C, but that the renaturation curve was displaced by the same value. Although in these conditions complete renaturation was not achieved, reheating of the renatured DNA showed that the renatured portion had the same $T_M$ than native T2 DNA.

These results allowed us to conclude that hybridization as well as renaturation should be possible in the presence of DMSO.

### 3. DNA-DNA hybridization

The following experimental conditions were found to give the best results in studying DNA-DNA hybridization; the subscript letters refer to the comments.

*Technique.* The membranes, loaded with $^3H$ labeled reference $DNA_{(a)}$ following Gillespie et al. (1965), and the same number of blank filters, were soaked for a few minutes in cold 2×SSC containing 30% DMSO (v/v)$_{(b)}$.

Each labeled membrane was transferred with one blank filter into a small plastic vial (diameter 30 mm) containing 1 to 5 μg $_{(c)}$ of denatured $^{32}P$ input DNA in 4 ml of 2×SSC-30% DMSO$_{(d)}$.

The vials were gently agitated in a thermostatic bath during the desired time$_{(e)}$.

After incubation, the vials were cooled in ice, and the membranes were washed on a Büchner funnel, on one side with 50 ml of 2×SSC containing 30% DMSO, on the other side with 100 ml of 2×SSC.

# TABLE 1
*Hybridization $^3H$ T2 DNA $\times$ $^{32}P$ T2 DNA in different conditions*

Solvent: 5 × 2 means 5 ml of 2×SSC. Incubation time for all experiments: 16 h. Samples from II-9 to III-8 were shaken during the incubation. Input DNA in sample I-2 was submitted to shearing in a VirTis homogenizer, 15 min at 10,000 rpm at 0°. I, II and III correspond to three batches of loaded filters. Control filters were counted either directly or after incubation in 2×SSC 16 hours at 60°. The following values were obtained: Batch II: 17.8, 14.8, 17.7, 15.5 µg; mean value: 16.4 ± 0.8. (incubated). Batch III: 23.0, 19.6 (incubated); 17.6, 22.6 (not incubated).

| N° | DNA on the filter µg | Input DNA µg/ml | Solvent Volume ml. SSC | DMSO % | Temperat. °C | Hybrid % | Yield | Blank % of input |
|---|---|---|---|---|---|---|---|---|
| I- 1 | 27.7 | 2.0 | 5 × 2 | 20 | 60 | 3.1 | 8.6 | 0.23 |
| 2 | 25.8 | 2.0 | 5 × 2 | 20 | 60 | 1.9 | 4.9 | 0.65 |
| II- 1 | 15.8 | 0.33 | 3 × 1 | 30 | 30 | 0.45 | 6.8 | 0.30 |
| 2 | 16.6 | ,, | ,, | ,, | ,, | 0.46 | 7.6 | |
| 3 | 16.5 | ,, | ,, | ,, | 40 | 0.38 | 6.2 | |
| 4 | 16.3 | ,, | ,, | ,, | ,, | 0.40 | 6.5 | |
| 5 | 16.3 | ,, | ,, | ,, | 50 | 0.48 | 7.8 | |
| 6 | 14.3 | ,, | ,, | ,, | ,, | 0.50 | 7.3 | |
| 7 | 11.0 | ,, | ,, | ,, | 60 | 0.70 | 7.7 | |
| 8 | 11.5 | ,, | ,, | ,, | ,, | 0.64 | 7.2 | |
| 9 | 15.9 | 0.5 | 4 × 2 | 30 | 50 | 2.3 | 18. | 0.22 |
| 10 | 15.0 | ,, | ,, | ,, | ,, | 2.1 | 16. | 0.25 |
| 11 | 17.3 | ,, | ,, | ,, | ,, | 3.2 | 27. | 0.33 |
| 12 | 15.4 | ,, | ,, | ,, | ,, | 4.0 | 31. | 0.27 |
| III- 1 | 18.1 | 0.5 | 4 × 2 | 30 | 30 | 2.13 | 19.3 | 0.30 |
| 2 | 18.4 | ,, | ,, | ,, | ,, | 1.80 | 16.5 | 0.30 |
| 3 | 15.6 | ,, | ,, | ,, | 40 | 3.45 | 27. | 0.17 |
| 4 | 16.8 | ,, | ,, | ,, | ,, | 2.90 | 24. | 0.18 |
| 5 | 20.4 | ,, | ,, | ,, | 50 | 3.0 | 30. | 0.32 |
| 6 | 18.7 | ,, | ,, | ,, | 60 | 2.06 | 19.3 | 0.11 |
| 7 | 15.8 | ,, | ,, | ,, | ,, | 2.21 | 17.5 | 0.18 |
| 8 | 17.0 | ,, | ,, | ,, | ,, | 2.74 | 23.2 | 0.23 |

The membranes were dried under an infrared lamp, and the radioactivity was determined.

## Comments

(a) The term "reference DNA" applies to the DNA fixed to the membranes, and "input DNA" to the DNA introduced into the hybridization medium.

(b) Dry filters introduced directly into the reaction mixture yielded higher blanks.

(c) The main factor to be controlled is the concentration of denatured input DNA in the solution. The best results were obtained with 0.5 to 2.0 µg/ml. Higher concentrations gave higher blanks, owing to the adsorption to the filters of a small percentage of the input DNA.

(d) Although 10 to 15% DMSO were enough to eliminate the adsorption of DNA on the membranes during a filtration, it was necessary to use 20 to 30% of this compound to prevent blank contamination during incubation.

On the other hand, this amount of DMSO in 2×SSC did not alter the filters even at 60°C, and did not remove the DNA which had been previously fixed. The figures listed in column 2,

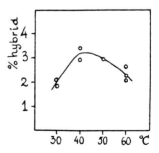

**Fig. 3.** Variation of the extent of hybridization in function of temperature. Membranes carrying an average of 18 µg of $^3H$ T2 DNA were incubated for 16 hours in 2×SSC, 30% DMSO at the indicated temperatures, in the presence of 2 µg of $^{32}P$ T2 DNA.

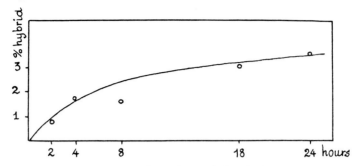

**Fig. 4.** Kinetics of DNA-DNA hybridization. Conditions as in Fig. 3; the temperature was maintained at 50° for the desired time.

Table I, show that for a given batch of loaded membranes, the amount of reference DNA remaining on the filters after incubation lay in the range of 85 to 90% of the controls, except in 1×SSC, 30% DMSO, 60°C, where it was of only 70%.

Owing to the variations observed between simultaneously prepared filters, it was considered necessary to use labelled reference DNA to secure a reasonable precision in the results of the experiments.

(e) The yield and the % hybrid were significantly enhanced when the vials were gently shaken during the incubation: in Table I, samples II-11 and 12, which were shaken, gave results 1.8 times higher than samples II-9 and 10, which were not.

*Expression of the results.* The ratio

$$\frac{\mu g \text{ input DNA hybridized}}{\mu g \text{ DNA on the filter}} \times 100$$

is designated by *% hybrid,* and the *yield* of the reaction is defined by the ratio

$$\frac{\mu g \text{ input DNA hybridized}}{\mu g \text{ total input DNA}} \times 100.$$

*Results.* This technique was tested with the system T2 $^3$H DNA (reference DNA) against T2 $^{32}$P DNA (input DNA). Table I and figs. 3 to 5 show the results obtained in different experiments. The optimum temperature was found to be 40 to 50°C in 2×SSC, 30% DMSO (fig. 3). In fig. 5, the results of a competition experiment are presented, showing a normal competition between cold T2 DNA and $^{32}$P T2 DNA, and no competition between cold thymus DNA and $^{32}$P T2 DNA.

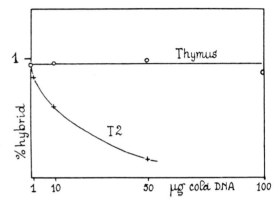

**Fig. 5.** Competition experiment. Membranes carrying an average of 20 µg of $^3$H T2 DNA were incubated for 16 hours at 50°C in 3 ml of 2×SSC, 30% DMSO, containing 1 µg of $^{32}$P T2 DNA and increasing amounts of denatured cold competitor DNA.

## Discussion

The technique presented here permits hybridization of analogous DNA's at relatively low temperatures, in the presence of a denaturing agent. This technique is analogous to the procedure of Denhardt (1966), but eliminates the necessity of a pretreatment of the membranes.

In accordance with the properties of denatured DNA and of nitrocellulose, Denhardt found 80% of the input DNA bound to blank filters if the pretreatment was omitted. Thus, in the technique of Warnaar et al (1966), as no preincubation is used, the same amount of input DNA must be lost by adsorption to the filters. The remaining 20% input DNA may then yield renatured molecules, either in the solution, or with the adsorbed 80%; it may also eventually form hybrid molecules with the reference DNA. Apparently, this technique takes advantage of the drying procedure of Gillespie *et al.*, which results in a very strong linkage between nitrocellulose and denatured reference DNA. Consequently, the small percentage of input DNA actually hybridized is not removed by the washing procedure at low ionic strength, which eliminates the input DNA directly attached to the filters.

In the procedure presented here, as in the technique of Denhardt, the input DNA can only either renature or hybridize with reference DNA; however, the yields obtained by the three techniques are of the same order of magnitude.

In addition, competitive hybridizations between homologous DNA's were recently obtained (Hotta *et al.*, 1966), although the input DNA adsorbed on the filters was not eliminated at all, even by the unexpected RNAse treatment proposed by the authors.

All these data lead to the conclusion that the so-called DNA-DNA hybridization techniques on membranes are based on one of the multiple reactions which may occur reversibly when free denatured DNA, adsorbed denatured DNA and excess nitrocellulose are annealed together and submitted to various empirical treatments. One must be very careful in interpreting the results of such experiments.

**REFERENCES**

1. D. T. Denhardt, Biochem. Biophys. Res. Commun. **23**:641 (1966).
2. D. Gillespie and S. Spiegelman, J. Mol. Biol. **12**:829 (1965).
3. S. Hanlon, Biochem. Biophys. Res. Commun. **23**:861 (1966).
4. G. K. Helmkamp and P. O. P. Ts'o, J. Amer. Chem. Soc. **83**:138 (1961).
5. Y. Hotta, M. Ito and H. Stern, Proc. Natl. Acad. Sci. **56**:1184 (1966).
6. J. D. Mandell and A. D. Hershey, Anal. Biochem. **1**:66 (1960).
7. J. Marmur and P. O. P. Ts'o, Biochim. Biophys. Acta **51**:32 (1961).
8. A. P. Nygaard and B. D. Hall, J. Mol. Biol. **9**:125 (1964).
9. S. O. Warnaar and J. A. Cohen, Biochem. Biophys. Res. Commun. **24**:554 (1966).

## QUESTIONS FOR CHAPTER 1

**Paper 1**
1. What, in your view, was the most fruitful aspect of the Watson-Crick structure of DNA in terms of stimulating later research?
2. If possible, try to build a model of DNA, containing at least a complementary pair of dinucleotides, with space-filling atomic models of the Corey-Pauling-Koltun type. From contemplating this model, your experience in building it, and your general knowledge of the subject, try to answer the following questions:
   a. The paper states that it is probably impossible to build a model similar to DNA with a ribose sugar in place of the deoxyribose. Is this correct?
   b. Could you build a similar model with D-arabinose as the sugar?
   c. The five valences of the P are on four sides of the atom. Are these sides equivalent for building the correct sugar-phosphate backbone of the polynucleotide chain?
   d. In order to obtain proper complementarity of the bases in the helical structure, do they have to be in the *anti* or *syn* conformation with respect to the sugar?
   e. Is the hydrogen bonding between complementary bases given in the paper correct?

**Paper 2**
1. a. The paper states that the DNA was prepared in the "usual" manner. Can you think of another generally used method for isolating DNA?
   b. Why are the DNA strands separated by alkali? Write down the reactions.
2. a. What is an MAK column?
   b. On what basis does the MAK column separate the two DNA strands?
   c. If the H strand were broken into five pieces and the L were intact, would this affect the separation? How?
   d. Peak $H_1$ in Fig. 1A was isolated in a continuous gradient. Why was an intermittent gradient used for its further purification in Fig. 1B?
   e. What is the concentration of DNA in tube 47, Fig. 1A?
3. Why is it important to separate and analyze the strands of many different DNAs?
4. Is any kind of DNA suitable for demonstrating the complementarity of the bases in the two strands? Justify your answer with examples.
5. For the determination of the base ratios,
   a. Is it necessary to obtain quantitative recovery of the material put on the column?
   b. Is it necessary to know how much DNA was taken?
   c. What are the necessary and sufficient conditions for obtaining the base ratios from radioactivity determinations?
   d. Is it necessary to know the molecular weight of the DNA?

**Paper 3**
1. On the basis of your knowledge of the factors affecting denaturation of DNA, indicate which of the following you would expect to increase or decrease the $T_m$:
   a. Increased salt concentration of the medium
   b. Slightly acid pH of the medium
   c. Alkaline pH of the medium
   d. Strongly acid pH of the medium
   e. High A-T content of the DNA
2. How would you expect the density of DNA to be changed by the following conditions?
   a. An increase in G-C content of the DNA
   b. An increase in alkalinity to pH 9
   c. Shearing of the DNA
   d. An increase in the concentration of DNA
3. What conclusions can you draw from the following:
   a. A comparison of the tracings of a run in neutral CsCl with one at pH 13 shows no increase in heterogeneity but a fivefold decrease in molecular weight.
   b. A molecular weight determination by ultracentrifugation and viscosity measurements on a given DNA solution yielded an average value of $1 \times 10^6$ when carried out at $0°$ C. When, however, the temperature was raised to $27°$ C, no macromolecular component appeared in the Schlieren diagram.
   c. A sample of DNA exposed to $100°$ C and rapidly cooled in a low salt-containing solution showed a 3.5 increase in molecular weight.
   d. Would the tracing of the rapidly cooled sample above show any significant change in $T_m$?
   e. Do the data say anything about the intactness of the strands of the original DNA sample in 3c?
4. Would you expect the viscosity of a DNA solution:

a. To increase with ionic strength?
b. To increase with molecular weight?
c. To change with denaturation, and which way? Why?
d. To be different for a slowly cooled or a quickly cooled sample?

5. It is stated in the paper that the experimental data indicate that the molecular recombination of denatured *E. coli* DNA is indeed a bimolecular reaction. It is now known, however, that the molecular weight of *E. coli* DNA is $2 \times 10^9$, whereas the molecular weight assumed in this paper (Fig. 5) was $10.5 \times 10^6$. The solution of denatured *E. coli* DNA therefore contained not two, but approximately 400, different molecules. How do you explain the observed second-order kinetics?

## Paper 4

1. a. Can you guess the nature of the bonds that fix denatured DNA to the Millipore filters from the fact that dimethyl sulfoxide, at suitable concentrations, prevents single-stranded DNA from being fixed to the filters?
   b. What is SSC?
   c. What are the concentrations of the various ingredients in 6xSSC?
   d. Why does the amount of fixed DNA increase with the concentration of SSC?
   e. Why is the DNA fixed in high salt concentration more resistant to elution with DMSO?
   f. It is stated in the paper that a small amount of fixed DNA cannot be eluted with DMSO, but can be displaced by single-stranded DNA. Can you give a plausible explanation for the different modes of binding of the bulk DNA and that of the strongly fixed fraction?

2. Why is hybridization aided by the presence of DMSO?

3. a. If you had a series of unknown DNAs to be tested for homology against a known DNA, which one is fixed to the membranes first, and why?
   b. Which one should be labeled and why?
   c. Why does shaking improve the amount of hybridization?
   d. Is it possible to determine the amount of reference DNA fixed to the filters when unlabeled DNA is used?

4. a. In determining the degree of relatedness of two DNAs, the maximum amount of hybrid formed with a reference DNA is usually ascertained. How is such a maximum determined?
   b. Draw a figure of the expected results and label the ordinate and abscissa.
   c. In what way does your curve resemble a common curve used in enzymology?
   d. Can you draw a Lineweaver-Burk plot of the data you used for your figure in 4b?
   e. Can you obtain the maximum amount of hybrid obtainable under a given set of conditions by the above plot, and what would be the advantage of this method?

5. a. Explain how a competition experiment like that in Fig. 5 works.
   b. Is it necessary to have saturated conditions in order to perform a competition experiment?
   c. Which method is most suitable for determining homology between DNAs: the comparison of hybridization maxima or competition?
   d. Suppose you were given 1.0 $O.D._{260m\mu}$ units of an unknown cold DNA and asked to determine if it was of bacterial or animal origin. What would you do?

# ANSWERS FOR CHAPTER 1

**Paper 1**

1. In my view the most fruitful aspect of the Watson-Crick model was that it suggested a precise mechanism for the replication of DNA, and hence for the transmittal of genetic information. The principle of base complementarity, which assured the preservation of the base sequence in a DNA strand, has turned out to be the general principle by which the information stored in nucleic acids is transferred, not only from one generation to the next, but also, in the same generation, into RNA and from there into protein, thus leading to the expression of genetic information in the cell.

2. a. The 2'OH of the ribose does not interfere with the building of a polynucleotide double helix, as has been shown by x-ray analysis of the double-helical RNA of reovirus. The bases may be slightly tilted, causing the chain to be somewhat shortened and adding an extra base pair for each turn of the helix (11 instead of 10), but this is not strictly necessary from a stereochemical point of view.

   b. D-arabinose is identical to D-ribose, except that the 2'OH is on the opposite side. In a polynucleotide chain, in which the sugar is perpendicular to the base, the 2'H of the ribose is right up against the $N^3$ position of the base, and there is therefore no room for the extra oxygen of the inverted 2'OH. D-arabinosyl nucleoside and nucleotides are, in fact, potent inhibitors of nucleic acid metabolism.

   c. The five valences of the P are not exactly equivalent, and it is only when the P atoms have the correct orientation that a proper helical conformation of the backbone can be obtained.

   d. The bases are in the *anti* conformation.

   e. In this first paper Watson and Crick assumed two hydrogen bonds each between A and T and between G and C. It is now known that in the G-C pair there is a third hydrogen bond between the 2-amino of the G and the $2=O$ of C.

**Paper 2**

1. a. Possibly the most widely used method for preparing DNA is that of Marmur (J. Mol. Biol. **3**:208, 1961). Basically, it consists of shaking the DNA-containing crude extract with chloroform-*iso*-amyl alcohol, digesting RNA away by ribonuclease, deproteinizing repeatedly and finally precipitating the DNA with 70% ethanol in the presence of salt.

   b. In strong alkali the bases are ionized, thus destroying the hydrogen bonds. The ionizations of the different bases are as follows:

   $A = T$  $G \equiv C$

2. a. An MAK column is a methylated (bovine serum) albumin-kieselguhr column.

   b. The MAK column operates mainly by forming hydrogen bonds between the protein and the nucleic acid. For nucleic acids having about the same base composition, the number of H bonds formed is roughly proportional to molecular weight. For nucleic acids of the same molecular weight, however, as is the case of the two strands of one DNA molecule, the number of H bonds depends on the base composition, the one with higher $G + C$ content emerging last. This can be seen from Table 2, where with two exceptions, $G + C$ is greater for the H than for the L strand.

   Since the differences between $G + C$ and $A + T$ are quite small in any case, there might be other causes for the separation of the H and L strands. For example, the equality of 6-keto and 6-amino within a single strand indicates that there might be some complementary helix formation within a given strand by some kind of hairpin model. The distribution of the helical regions in the other strand would obviously be quite different, and this would aid in the separation of the two. Double-stranded DNA binds very poorly to MAK columns, as shown by the fact that native DNA elutes right after the low molecular weight tRNA. Furthermore, the conformation of the molecule also influences its behaviour on a MAK column. It may be folded in such a manner as to make certain regions more or less accessible to hydrogen bonding with the protein of the column.

c. Since the MAK column separates molecules primarily on the basis of the number of hydrogen bonds formed with the column, the broken-up H strand might still emerge as a single peak, depending on the distribution of base sequences within the pieces, but it would now emerge quite a distance ahead of the L strand.

d. Under the right conditions any one component will emerge from a chromatographic column in a Gaussian peak, the shape of which will depend on the concentration of the component and its adsorption characteristics. On an ion exchange column the desorption will depend on the concentration and properties of the exchanging ion. Thus, if the salt concentration or pH is just sufficient to start exchange, elution will occur in a very broad, low peak. In order to speed up elution a gradient of salt or pH is used, so that if the rate of elution decreases due to a decrease in the absorbed material, it is compensated for by an increase in eluant concentration, and the rate of desorption thereby stays constant. The material will then emerge in a narrow, non-Gaussian peak. This method has the disadvantage, however, that the increase in eluant not only speeds up the desorption of the tail end of a given peak but also starts the desorption of another material bound slightly more firmly. If the amount of this second material is less than that of the first, the two might emerge in a single, smooth peak. As a matter of fact, depending on the relative concentrations and charges, quite a number of different materials could thus elute in a single peak. The emergence of a single peak in gradient elution ion exchange chromatography is therefore no proof of homogeneity.

The method used in the present paper uses gradient elution to speed up the chromatography. As soon as a peak starts to elute, however, the salt concentration is kept constant in order to permit separation of any separate components that might be present. $H_1$ is isolated in a gradient in order to obtain a peak of high enough concentration for rechromatography. The use of an intermittent gradient for the latter then ensures the isolation of a homogeneous $H_2$ component.

e. The absorption at 260 m$\mu$ for a 0.1% solution of native DNA is 20. According to paper 3, there is a 40% increase in O.D. upon complete denaturation, that is, separation of the strands. The O.D.$_{260}$ of a 0.1% solution of single strands is thus 28.

Tube 47, 1A, has an O.D. of 1.0, which is $1.0/28 = 35.7$ $\mu$g/ml.

3. It is important to analyze the separated strands of a large number of different DNAs, for any molecules of the size of DNAs could show complementarity simply by coincidence. It is particularly important to analyze DNAs of different types, that is, high G-C, low G-C content, etc.

4. Only DNAs with very different base compositions in the two strands are suitable for demonstrating complementarity. If the base composition is equimolar in both strands, the same analytical data would be obtained regardless of the relationship between the two strands.

5. a. It is not necessary to obtain quantitative recovery of the material from the column for base analyses, because the values are either related to the amount of one of the bases (usually adenine, which is taken as equal to 1.00), or the amounts are related to the total amount of bases *recovered* from the column, in which case the sum of the bases is taken as unity.

b. It is not necessary to know how much DNA was taken, for the same reasons related above.

c. If the nucleotides are to be determined by means of radioactivity instead of optical density at 260 m$\mu$, the conditions depend to some extent on the isotope and the precursor used. In any case, however, it is necessary to do the analysis with uniformly labeled DNA so that the radioactivity does, in fact, reflect the amount of a given nucleotide, rather than the metabolic state of the DNA at the time of harvesting. For DNA, in general, it is necessary to grow the cells, or organisms, in the presence of isotope for an integral number of generations, or else for such a large number that the presence of fractional amounts of the initial, unlabeled DNA is no longer noticeable. Another necessary condition is the use of a radioactive precursor that is uniformly incorporated into all four nucleotides, or whose preferential incorporation is known and can be corrected for. If the radioactivity can thus be made to be proportional to the concentration of each nucleotide, it is not necessary to determine the specific activity of any of the nucleotides, nor is it necessary to know the total amount of DNA or the recovery of nucleotides from the column.

34   Molecular biology of DNA and RNA

   d. It is not necessary to know the molecular weight of the DNA in order to calculate the base composition, for the reasons given in 5a and b.

Paper 3

1. a. The $T_m$ of DNA increases with increasing ionic strength. For example, there is an increase of about 40° C in going from a very low salt concentration, just capable of keeping the DNA double-stranded, to 0.1M Na⁺. The stabilization afforded by divalent ions is proportionally even greater than would be expected from their ionic strength.
   b. The amino group of adenine and the $N^1$ of cytosine pick up a proton and thus become ionized at pH 4.1 and 4.6, respectively. Below these pHs, therefore, the hydrogen bonds between the strands would break, but at pHs above those pKs, there would probably be relatively little effect on the $T_m$.
   c. Increasing pH would have relatively little effect on the $T_m$ until above pH 9, when the $N^1$ of guanine and thymine become negatively charged, thus breaking a hydrogen bond and of course decreasing the $T_m$. Above pH 12 the strands separate completely.
   d. In a strongly acid pH the DNA precipitates.
   e. The pair A-T has only two hydrogen bonds, whereas C-G has three; therefore, high A-T content decreases the $T_m$.

2. a. An increase in G-C causes an increase in the density of DNA. As a matter of fact, there is a linear relationship between the two, as follows:
   $$\rho = 1.660 + 0.098(x_G + x_C)$$
   where $x_G$ and $x_C$ are the molar fractions of G and C.
   b. Up to pH 9 it is below the pKs for ionization of the bases, so an increase of pH up to 9 would cause no change in the density.
   c. The density of DNA is independent of molecular weight over a very wide range, so that shearing would not change the density.
   d. An increase in the concentration of DNA would not change the density; it would only increase the height and width of the DNA peak.

3. a. Since the density of DNA is essentially independent of molecular weight, "heterogeneity," that is, non-Gaussian distribution of the material refers to a qualitative heterogeneity to different kinds of DNA with different base compositions. At pH 13 the strands separate, and one would expect to find a twofold drop in molecular weight, as shown by the width of the peak. A fivefold drop means that one of the strands gave rise to two pieces and the other to three, so that there must have been "nicks" in the two chains. Since there was no apparent loss of homogeneity, this must mean that each of the pieces had close to the same base composition.
   b. The preparation must have been grossly contaminated with a nuclease, which digested the DNA at the higher temperature, causing the loss of all macromolecular components.
   c. A denatured and rapidly cooled sample of DNA would be expected to show little renaturation, and should thus show a decrease rather than an increase of molecular weight. For an increase to take place, extensive intra-strand aggregation must have occurred, or else partial renaturations involving a number of strands.
   d. The melting curve of such a sample would be exceedingly broad and most likely would have no determinable $T_m$. It might show a series of steps, corresponding to melting of the short, imperfect helices formed by rapid cooling.
   e. The melting data do not give any information about the intactness of DNA strands, for the $T_m$ are essentially independent of molecular weight.

4. The viscosity of DNA would:
   a. Decrease with ionic strength.
   b. Increase with molecular weight.
   c. Decrease with denaturation, because the molecular weight is diminished.
   d. Probably be higher for a slowly cooled sample of DNA than for a quickly cooled one, unless there was considerable aggregation and concomitant increase in molecular weight in the latter case.

5. The observed second order kinetics for *E. coli* DNA are difficult to explain, unless one assumes that the recombination of the strands was not completely specific, and each of the 400 pieces had approximately the same base composition. It is known now that renaturation under the conditions of this paper is indeed not very specific, as is also indicated by Fig. 1.

Paper 4

1. a. Dimethyl sulfoxide is an avid maker of hydrogen bonds. It is therefore likely that this solvent denatures DNA by making hydrogen bonds with the N--H groups of the bases, thereby breaking the intra-strand hydrogen bonds of the DNA. The binding of the DNA to nitrocellulose must therefore also be by hydrogen bonds.

b. SSC stands for "standard saline-citrate" (see paper 3).
c. Standard SSC contains 0.15M NaCl and 0.015M sodium citrate, hence 6×SSC contains 0.9M NaCl and 0.090M sodium citrate (see paper 3; also Gillespie and Spiegelman, cited in paper 4).
d. Citrate is a chelator of metal ions, especially divalent ones. It therefore removes from the DNA all contaminating metals that hinder the separation of the strands. Thus SSC increases the amount of single-stranded DNA available to bind on the filter.
e. The DNA fixed in high salt concentration makes a larger number of bonds with the filter, and hence requires a higher concentration of DMSO for elution.
f. The fact that the small tightly bound fraction of DNA is not eluted by agents capable of breaking hydrogen bonds suggests that the minor fraction is not bound to the filter by hydrogen bonds. The further statement that this fraction can, however, be eluted by denatured DNA suggests that there may be some more specific adsorption sites on the filters at which the labeled DNA is displaced by unlabeled DNAs.

2. Hybridization is aided by DMSO because at some critical concentration the latter undoes "false hybrids," that is, hybrids in which the two strands are out of phase and held together by a relatively small number of complementary hydrogen bonds or by hydrogen bonds other than the stable Watson-Crick ones. Such "false hybrids" usually preclude the formation of perfectly matched hybrids. DMSO thus works by increasing the number of strands available for true hybrids.

3. a. The DNA available in the largest amount is usually taken as the reference DNA bound to the filter, whereas the unknown, available usually in small amounts, is taken as input. Ideally, of course, each test should involve a reciprocal hybridization.
b. The input DNA must be labeled in order to determine the amount hybridized. (It is easily seen that the reference DNA cannot be labeled if the input DNA is unlabeled). If either the percent hybrid or the yield is to be determined, then the reference DNA should also be labeled, but with a different isotope from that in the input DNA.
c. Because of their extreme length-to-width ratio, all kinds of DNA have an extremely low diffusion coefficient. It is therefore not possible to rely on simple diffusion to replenish the area close to the filter, from which the DNA was removed by the hybridization reaction. Shaking is an efficient way for exposing all of the input DNA to the filter, thereby increasing the actual amount of DNA available for hybridization.
d. If labeled reference DNA is not available, the amount of DNA fixed to the filters can be determined by eluting it with a suitable solvent and reading the optical density. This method, however, leaves much to be desired.

4. a. The most common method for ascertaining the maximum amount of hybrid which a given system is capable of forming is to obtain a classical saturation curve by plotting the amount of hybrid formed against increasing amounts of input DNA.
b. The curve can be constructed by taking the data in Table I and Figs. 3 and 4 as a guide, and filling in a few points. The test of a correct figure is that it gives a reciprocal plot with a positive intercept of the ordinate.
c. The saturation curve for hybrid formation resembles a substrate saturation curve of an enzyme.

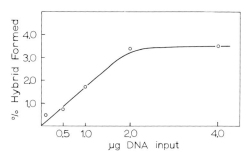

d. A Lineweaver-Burk plot consists of a double reciprocal plot of product versus substrate. For the hybridization reaction this would mean plotting the reciprocal of hybrid formed against the reciprocal of input DNA.

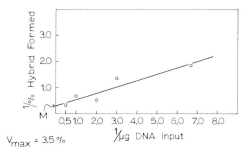

$V_{max} = 3.5\%$

e. A Lineweaver-Burk plot for an enzyme defines the $V_{max}$ of the enzymatic reaction as the point where the line intercepts the ordinate. Likewise, then, the point M in the above graph represents the maximum amount of hybrid obtainable. The advantage of this method lies in the fact that it is not necessary to obtain the entire saturation curve, a number of points at the beginning of the curve being sufficient. In this way one avoids not only having to repeat the experiment if complete saturation was not obtained, but also the possible complications that arise at high concentrations of input DNA.

5. a. In a competition experiment a small amount of labeled input DNA I plus increasing amounts of unlabeled input DNA II are offered to filters containing a reference DNA capable of hybridizing input DNA I at least to some extent. A completely unrelated DNA, like calf thymus DNA in Fig. 5, should give a straight line, whereas in the case when input DNA I and reference are the same, a sharply falling curve, as the one for $T_2$ in Fig. 5, is obtained. DNAs with various degrees of homology with the reference DNA fall between those two extremes.

b. In order to obtain good competition it is necessary to use a large excess of cold DNA. Then, to avoid unwieldy concentrations of DNA, the amount of unlabeled input DNA used should be as small as possible, and always below saturation.

c. Competition hybridization is most suitable for qualitatively determining the degree of homology between a known DNA and an unknown that is suspected to be identical or closely related to it. It is also useful when only a small amount of the unknown labeled DNA is available, provided there is a large amount of cold standard DNA. It should also be pointed out that in competition hybridization it is not necessary to know the exact amount of reference DNA on the filters, although smoother lines will undoubtedly be obtained if percent hybridization, rather than just the number of counts hybridized, is plotted. Competition hybridization has the disadvantage that the results are hard to quantitate unless proper boundaries are established. Therefore, provided enough labeled DNAs are available, it is preferable to determine the maximum hybridization from a Lineweaver-Burk plot if a whole series of DNAs are to be compared.

d. The simplest test to determine whether a given DNA is of animal or bacterial origin, and which does not use up any DNA, is to determine the extent of renaturation under annealing conditions known to allow no renaturation for the animal DNA, but a good amount for the bacterial one. If further definition by hybridization is desired, one could bind the cold DNA to the filters, and determine the extent of hybridization of a number of labeled animal and bacterial DNAs. In the case at hand, however, only 50 $\mu$g (1.0/20) of DNA is available, not enough to prepare the requisite number of filters bearing the optimum amount of DNA (15-20 $\mu$g per filter). The best strategy, then, although not completely satisfactory, would be to divide the DNA in half, and use each for obtaining about three points in competition against a very small amount of highly labeled animal and bacterial DNA, respectively.

# CHAPTER 2

# REPLICATION OF DNA IN VIVO

The most important aspect of the Watson-Crick structure for DNA from a biological point of view was that it clearly predicted the mode of replication of the genetic material, namely, that the two strands would have to separate, and each would serve as the template for a new strand, complementary to the old. A clear demonstration of the semiconservative nature of the replication of DNA would thus not only prove the predicted scheme but also provide solid proof of the correctness of the structure.

A beginning of this demonstration was made by Taylor, who showed by autoradiographic means that chromosomes of *Vicia fava*, allowed one round of replication in tritiated thymidine, showed label in both chromatids, but after a further round in the absence of label, they produced one labeled and one unlabeled descendant. This experiment, while hailed by molecular biologists as proof of the predicted semiconservative replication scheme, did not create any converts among biologists generally. For one thing, autoradiography lacks the precision to be convincing to the uninitiated, and further, the relationship between a chromatid and a double-stranded molecule of DNA was (and is) far from clear.

The semiconservative replication of DNA was definitively demonstrated by Meselson and Stahl, who used density instead of radioactive label, and band centrifugation on CsCl to separate the DNAs of different density. They observed that after growing heavy bacteria in a normal medium for one generation, all the DNA was of hybrid density, whereas after two generations half of the DNA was hybrid, and half was light, exactly as predicted.

The separation of the strands was beautifully demonstrated by a fine resolution autoradiographic study of the replicating *E. coli* chromosome by Cairns, which clearly showed a fork at the point of replication. This study also showed that the bacterial chromosome was circular. The replicating fork has actually been isolated in a number of later experiments, using ingenious combinations of radioactive and density labels.

The enzymes involved in the replication of DNA *in vivo* are not yet known for sure. One problem is that DNA polymerase, the enzyme known to synthesize deoxypolynucleotides *in vitro* (see Chapter 3), works only in one direction, from the $5'$ end to the $3'$, whereas due to the antiparallelism of the two strands, the fork mechanism offhand requires the simultaneous synthesis of two strands in opposite chemical directions. The same seems to be true for a new DNA polymerase associated with membranes, which was found in mutants devoid of the Kornberg polymerase, and which is a better candidate for the "replicase" than the latter.

At present there are various ingenious schemes designed to reconcile the known mechanism of the polymerases with the observation of a replication fork. According to Kornberg, the polymerase starts copying the strand bearing the free $3'$OH group, and when it catches up with the fork it continues copying up the other strand. The new strand is then hydrolyzed at the point of the fork by a "nickase," possibly the DNA polymerase itself, which continues copying the $3'$OH strand by building on the recently freed $3'$OH of the replicated piece. It again copies a piece, catches up with the fork, turns around, and copies the other strand up to the free $5'$ phosphate group of the previously replicated piece. The enzyme polynucleotide ligase (see Chapter 3) then joins the two pieces. Other schemes are basically similar, but dispense with the "nickase," and may have two molecules of polymerases working at the same time, in opposite directions.

The involvement of polynucleotide ligase in replication of DNA *in vivo* has been strongly suggested by the behavior of ligase-less mutants, which do not actually replicate their DNA, but synthesize small pieces of DNA.

The rate of DNA replication in synchronized bacterial cultures has been studied not only by isotopic means but also by determining the rate of replication of genetic markers, as measured by the increase of the corresponding enzymes or by transformation assays. All three methods give the same results. The replication of DNA always starts at the same point on the chromosome.

Despite a wealth of information, there are, however, quite a number of unanswered questions. It is still difficult, for example, to picture how the strands separate to allow replication, especially when one considers that the DNA is many times the length of the bacterium, so that it must exist in a very tightly packed state. Furthermore, what triggers the start of replication is still completely unknown, except that protein synthesis is involved at some stage; for no DNA synthesis can take place if inhibitors of protein synthesis are added, or if the bacterium is deprived of amino acids at some critical stage prior to the start of the replication. It is also tied in with cell division and, hence, with the growth of membranes and cell walls, although the critical factors are, again, unknown. Recently, however, mutants have been isolated in which DNA replication and cell division have been uncoupled, which should be very useful for the study of the factors controlling DNA synthesis. An amusing by-product of these studies was the production of metabolically active "mini" cells entirely devoid of DNA.

The synthesis of another category of DNAs, that of viruses, has also been studied in great detail. The replication of viral DNAs differs from that of their hosts in that the DNA is not attached to a membrane. Furthermore, some viral DNAs are single-stranded, and their mode of replication is of special interest. They form a double-stranded replicative intermediate on which the single-stranded viral DNA is synthesized, but the complete details are not yet known.

The synthesis of DNA in higher organisms, on the other hand, has not been studied in anything like the detail of that of bacteria and their viruses. It is known, however, that both the nuclear and the organelle DNAs replicate in a semiconservative manner, but the enzymology is still obscure although mammalian DNA polymerases have been described.

The papers in this chapter offer only a very brief sampling of the enormous number of pertinent publications. The student who would like to follow up the subject should read papers cited in the bibliographies.

# 5 THE REPLICATION OF DNA IN ESCHERICHIA COLI

*Matthew Meselson**
*Franklin W. Stahl*
*Gates and Crellin Laboratories of Chemistry† and*
*Norman W. Church Laboratory of Chemical Biology*
*California Institute of Technology*
*Pasadena, California*

## Introduction

Studies of bacterial transformation and bacteriophage infection[1-5] strongly indicate that deoxyribonucleic acid (DNA) can carry and transmit hereditary information and can direct its own replication. Hypotheses for the mechanism of DNA replication differ in the predictions they make concerning the distribution among progeny molecules of atoms derived from parental molecules.[6]

Radioisotopic labels have been employed in experiments bearing on the distribution of parental atoms among progeny molecules in several organisms.[6-9] We anticipated that a label which imparts to the DNA molecule an increased density might permit an analysis of this distribution by sedimentation techniques. To this end, a method was developed for the detection of small density differences among macromolecules.[10] By use of this method, we have observed the distribution of the heavy nitrogen isotope $N^{15}$ among molecules of DNA following the transfer of a uniformly $N^{15}$-labeled, exponentially growing bacterial population to a growth medium containing the ordinary nitrogen isotope $N^{14}$.

## Density-gradient centrifugation

A small amount of DNA in a concentrated solution of cesium chloride is centrifuged until equilibrium is closely approached. The opposing processes of sedimentation and diffusion have then produced a stable concentration gradient of the cesium chloride. The concentration and pressure gradients result in a continuous increase of density along the direction of centrifugal force. The macromolecules of DNA present in this density gradient are driven by the centrifugal field into the region where the solution density is equal to their own buoyant density.[11] This concentrating tendency is opposed by diffusion, with the result that at equilibrium a single species of DNA is distributed over a band whose width is inversely related to the molecular weight of that species (Fig. 1).

If several different density species of DNA are present, each will form a band at the position where the density of the CsCl solution is equal to the buoyant density of that species. In this way DNA labeled with heavy nitrogen ($N^{15}$) may be resolved from unlabeled DNA. Figure 2 shows the two bands formed as a result of centrifuging a mixture of approximately equal amounts of $N^{14}$ and $N^{15}$ *Escherichia coli* DNA.

In this paper reference will be made to the apparent molecular weight of DNA samples determined by means of density-gradient centrifugation. A discussion has been given[10] of the considerations upon which such determinations are based, as well as of several possible sources of error.[12]

*Experimental.* *Escherichia coli* B was grown at 36° C. with aeration in a glucose salts medium containing ammonium chloride as the sole nitrogen source.[13] The growth of the bacterial population was followed by microscopic cell counts and by colony assays (Fig. 3).

Bacteria uniformly labeled with $N^{15}$ were prepared by growing washed cells for 14 generations (to a titer of $2 \times 10^8$ ml) in medium containing 100 μg/ml of $N^{15}H_4Cl$ of 96.5 per cent isotopic purity. An abrupt change to $N^{14}$ medium was then accomplished by adding to the growing culture a tenfold excess of $N^{14}H_4Cl$, along with ribosides of adenine and uracil in experiment 1 and ribosides of adenine, guanine, uracil, and cytosine in experiment 2, to give a concentration of 10 μg/ml of each riboside. During subsequent

---

From Proceedings of the National Academy of Sciences, U.S.A. 44:671-682, 1958. Reprinted with permission.

Aided by grants from the National Foundation for Infantile Paralysis and the National Institutes of Health.

*Present address: Harvard University, Cambridge, Massachusetts.

†Contribution No. 2344.

40  Molecular biology of DNA and RNA

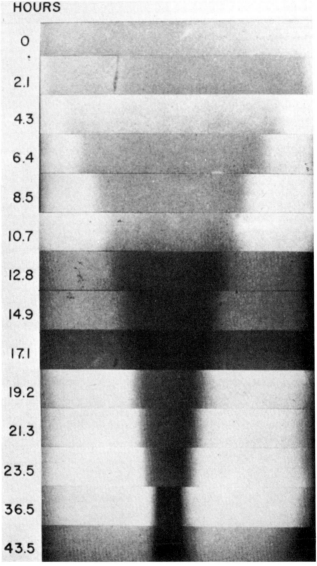

**Fig. 1.** Ultraviolet absorption photographs showing successive stages in the banding of DNA from *E. coli*. An aliquot of bacterial lysate containing approximately $10^8$ lysed cells was centrifuged at 31,410 rpm in a CsCl solution as described in the text. Distance from the axis of rotation increases toward the right. The number beside each photograph gives the time elapsed after reaching 31,410 rpm.

growth the bacterial titer was kept between 1 and 2 × $10^8$/ml by appropriate additions of fresh $N^{14}$ medium containing ribosides.

Samples containing about 4 × $10^9$ bacteria were withdrawn from the culture just before the addition of $N^{14}$ and afterward at intervals for several generations. Each sample was immediately chilled and centrifuged in the cold for 5 minutes at 1,800 × *g*. After resuspension in 0.40 ml of a cold solution 0.01 *M* in NaCl and 0.01 *M* in ethylenediaminetetra-acetate (EDTA) at pH 6, the cells were lysed by the addition of 0.10 ml of 15 per cent sodium dodecyl sulfate and stored in the cold.

For density-gradient centrifugation, 0.010 ml of the dodecyl sulfate lysate was added to 0.70 ml of CsCl solution buffered at pH 8.5 with 0.01 *M* tris(hydroxymethyl)aminomethane. The density of the resulting solution was 1.71 gm cm$^{-3}$. This was

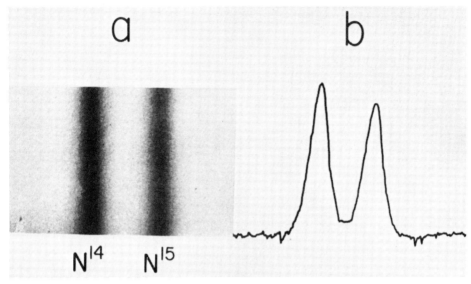

**Fig. 2. a,** The resolution of N¹⁴ DNA from N¹⁵ DNA by density-gradient centrifugation. A mixture of N¹⁴ and N¹⁵ bacterial lysates, each containing about 10⁸ lysed cells, was centrifuged in CsCl solution as described in the text. The photograph was taken after 24 hours of centrifugation at 44,770 rpm. **b,** A microdensitometer tracing showing the DNA distribution in the region of the two bands of Fig. 2a. The separation between the peaks corresponds to a difference in buoyant density of 0.014 gm cm⁻³.

centrifuged at 140,000×g. (44,770 rpm) in a Spinco model E ultracentrifuge at 25° for 20 hours, at which time the DNA had essentially attained sedimentation equilibrium. Bands of DNA were then found in the region of density 1.71 gm cm⁻³, well isolated from all other macromolecular components of the bacterial lysate. Ultraviolet absorption photographs taken during the course of each centrifugation were scanned with a recording microdensitometer (Fig. 4).

The buoyant density of a DNA molecule may be expected to vary directly with the fraction of N¹⁵ label it contains. The density gradient is constant in the region between fully labeled and unlabeled DNA bands. Therefore, the degree of labeling of a partially labeled species of DNA may be determined directly from the relative position of its band between the band of fully labeled DNA and the band of unlabeled DNA. The error in this procedure for the determination of the degree of labeling is estimated to be about 2 per cent.

*Results.* Figure 4 shows the results of density-gradient centrifugation of lysates of bacteria sampled at various times after the addition of an excess of N¹⁴-containing substrates to a growing N¹⁵-labeled culture.

It may be seen in Figure 4 that, until one generation time has elapsed, half-labeled molecules accumulate, while fully labeled DNA is depleted. One generation time after the addition of N¹⁴, these half-labeled or "hybrid" molecules alone are observed. Subsequently, only half-labeled DNA and completely unlabeled DNA are found. When two generation times have elapsed after the addition of N¹⁴, half-labeled and unlabeled DNA are present in equal amounts.

*Discussion.* These results permit the following conclusions to be drawn regarding DNA replication under the conditions of the present experiment.

1. *The nitrogen of a DNA molecule is divided equally between two subunits which remain intact through many generations.*

The observation that parental nitrogen is found only in half-labeled molecules at all times after the passage of one generation time demonstrates the existence in each DNA molecule of two subunits containing equal amounts of nitrogen. The finding that at the second generation half-labeled and unlabeled molecules are found in equal amounts shows that the number of surviving parental subunits is twice the number of parent molecules initially present. That is, the subunits are conserved.

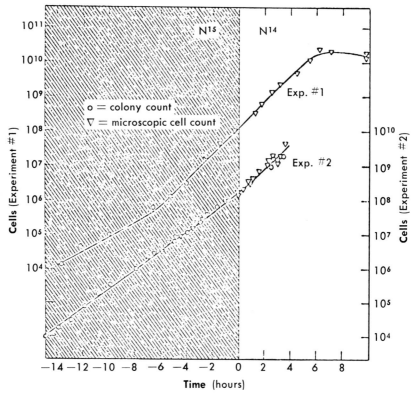

**Fig. 3.** Growth of bacterial populations first in $N^{15}$ and then in $N^{14}$ medium. The values on the ordinates give the actual titers of the cultures up to the time of addition of $N^{14}$. Thereafter, during the period when samples were being withdrawn for density-gradient centrifugation, the actual titer was kept between 1 and $2 \times 10^8$ by additions of fresh medium. The values on the ordinates during this later period have been corrected for the withdrawals and additions. During the period of sampling for density-gradient centrifugation, the generation time was 0.81 hours in Experiment 1 and 0.85 hours in Experiment 2.

2. *Following replication, each daughter molecule has received one parental subunit.*

The finding that all DNA molecules are half-labeled one generation time after the addition of $N^{14}$ shows that each daughter molecule receives one parental subunit.[14] If the parental subunits had segregated in any other way among the daughter molecules, there would have been found at the first generation some fully labeled and some unlabeled DNA molecules, representing those daughters which received two or no parental subunits, respectively.

3. *The replicative act results in a molecular doubling.*

This statement is a corollary of conclusions 1 and 2 above, according to which each parent molecule passes on two subunits to progeny molecules and each progeny molecule receives just one parental subunit. It follows that each single molecular reproductive act results in a doubling of the number of molecules entering into that act.

The above conclusions are represented schematically in Figure 5.

**The Watson-Crick model**

A molecular structure for DNA has been proposed by Watson and Crick.[15] It has undergone preliminary refinement[16] without alteration of its main features and is supported by physical and chemical studies.[17] The structure consists of two polynucleotide chains wound helically about a common axis. The nitrogen base (adenine, guanine, thymine, or cytosine) at each level on one chain is

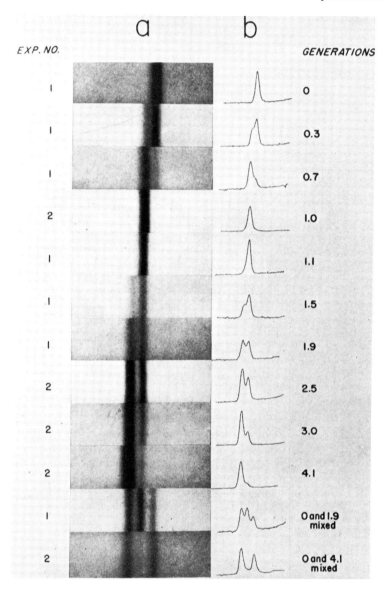

**Fig. 4. a,** Ultraviolet absorption photographs showing DNA bands resulting from density-gradient centrifugation of lysates of bacteria sampled at various times after the addition of an excess of $N^{14}$ substrates to a growing $N^{15}$-labeled culture. Each photograph was taken after 20 hours of centrifugation at 44,770 rpm under the conditions described in the text. The density of the CsCl solution increases to the right. Regions of equal density occupy the same horizontal position on each photograph. The time of sampling is measured from the time of the addition of $N^{14}$ in units of the generation time. The generation times for Experiments 1 and 2 were estimated from the measurements of bacterial growth presented in Fig. 3. **b,** Microdensitometer tracings of the DNA bands shown in the adjacent photographs. The microdensitometer pen displacement above the base line is directly proportional to the concentration of DNA. The degree of labeling of a species of DNA corresponds to the relative position of its band between the bands of fully labeled and unlabeled DNA shown in the lowermost frame, which serves as a density reference. A test of the conclusion that the DNA in the band of intermediate density is just half-labeled is provided by the frame showing the mixture of generations 0 and 1.9. When allowance is made for the relative amounts of DNA in the three peaks, the peak of intermediate density is found to be centered at $50 \pm 2$ per cent of the distance between the $N^{14}$ and $N^{15}$ peaks.

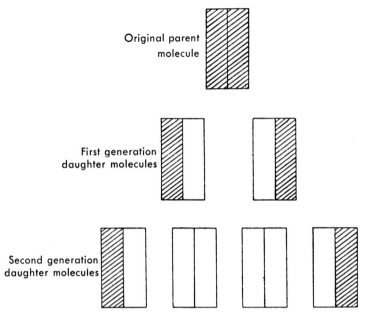

**Fig. 5.** Schematic representation of the conclusions drawn in the text from the data presented in Fig. 4. The nitrogen of each DNA molecule is divided equally between two subunits. Following duplication, each daughter molecule receives one of these. The subunits are conserved through successive duplications.

hydrogen-bonded to the base at the same level on the other chain. Structural requirements allow the occurrence of only the hydrogen-bonded base pairs adenine-thymine and guanine-cytosine, resulting in a detailed complementariness between the two chains. This suggested to Watson and Crick[19] a definite and structurally plausible hypothesis for the duplication of the DNA molecule. According to this idea, the two chains separate, exposing the hydrogen-bonding sites of the bases. Then, in accord with the base-pairing restrictions, each chain serves as a template for the synthesis of its complement. Accordingly, each daughter molecule contains one of the parental chains paired with a newly synthesized chain (Fig. 6).

The results of the present experiment are in exact accord with the expectations of the Watson-Crick model for DNA duplication. However, it must be emphasized that it has not been shown that the molecular subunits found in the present experiment are single polynucleotide chains or even that the DNA molecules studied here correspond to single DNA molecules possessing the structure proposed by Watson and Crick. However, some information has been obtained about the molecules and their subunits; it is summarized below.

The DNA molecules derived from E. coli by detergent-induced lysis have a buoyant density in CsCl of 1.71 gm cm$^{-3}$, in the region of densities found for T2 and T4 bacteriophage DNA, and for purified calf-thymus and salmon-sperm DNA. A highly viscous and elastic solution of $N^{14}$ DNA was prepared from a dodecyl sulfate lysate of E. coli by the method of Simmons[19] followed by deproteinization with chloroform. Further purification was accomplished by two cycles of preparative density-gradient centrifugation in CsCl solution. This purified bacterial DNA was found to have the same buoyant density and apparent molecular weight, $7 \times 10^6$, as the DNA of the whole bacterial lysates (Figs. 7, 8).

**Heat denaturation**

It has been found that DNA from E. coli differs importantly from purified salmon-sperm DNA in its behavior upon heat denaturation.

Exposure to elevated temperatures is known to bring about an abrupt collapse of the relatively rigid and extended native DNA molecule and to make available for acid-base

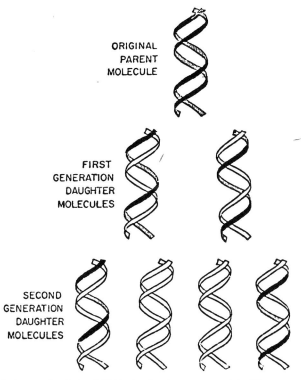

**Fig. 6.** Illustration of the mechanism of DNA duplication proposed by Watson and Crick. Each daughter molecule contains one of the parental chains (*black*) paired with one new chain (*white*). Upon continued duplication, the two original parent chains remain intact, so that there will always be found two molecules each with one parental chain.

titration a large fraction of the functional groups presumed to be blocked by hydrogen-bond formation in the native structure.[19,20,21,22] Rice and Doty[22] have reported that this collapse is not accompanied by a reduction in molecular weight as determined from light-scattering. These findings are corroborated by density-gradient centrifugation of salmon-sperm DNA.[23] When this material is kept at 100° for 30 minutes either under the conditions employed by Rice and Doty or in the CsCl centrifuging medium, there results a density increase of 0.014 gm cm$^{-3}$ with no change in apparent molecular weight. The same results are obtained if the salmon-sperm DNA is pre-treated at pH 6 with EDTA and sodium dodecyl sulfate. Along with the density increase, heating brings about a sharp reduction in the time required for band formation in the CsCl gradient. In the absence of an increase in molecular weight, the decrease in banding time must be ascribed[10] to an increase in the diffusion coefficient, indicating an extensive collapse of the native structure.

The decrease in banding time and a density increase close to that found upon heating salmon-sperm DNA are observed (Fig. 9, *A*) when a bacterial lysate containing uniformly labeled $N^{15}$ or $N^{14}$ *E. coli* DNA is kept at 100° C. for 30 minutes in the CsCl centrifuging medium; but the apparent molecular weight of the heated bacterial DNA is reduced to approximately half that of the unheated material.

Half-labeled DNA contained in a detergent lysate of $N^{15}$ *E. coli* cells grown for one generation in $N^{14}$ medium was heated at 100° C. for 30 minutes in the CsCl centrifuging medium. This treatment results in the loss of the original half-labeled material and in the appearance in equal amounts of two new density species, each with approximately half the initial apparent molecular weight (Fig. 9, *B*). The density difference between the two species is 0.015 gm cm$^{-3}$, close to the

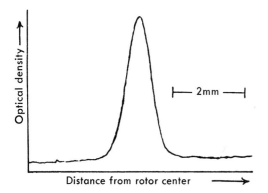

**Fig. 7.** Microdensitometer tracing of an ultraviolet absorption photograph showing the optical density in the region of a band of $N^{14}$ *E. coli* DNA at equilibrium. About 2 µg of DNA purified as described in the text was centrifuged at 31,410 rpm at 25° in 7.75 molal CsCl at pH 8.4. The density gradient is essentially constant over the region of the band and is 0.057 gm/cm⁴. The position of the maximum indicates a buoyant density of 1.71 gm cm⁻³. In this tracing the optical density above the base line is directly proportional to the concentration of DNA in the rotating centrifuge cell. The concentration of DNA at the maximum is about 50 µg/ml.

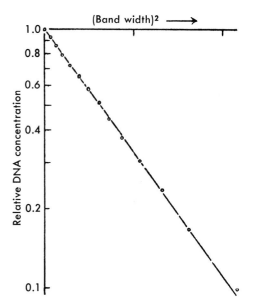

**Fig. 8.** The square of the width of the band of Fig. 7 plotted against the logarithm of the relative concentration of DNA. The divisions along the abscissa set off intervals of 1 mm². In the absence of density heterogeneity, the slope at any point of such a plot is directly proportional to the weight average molecular weight of the DNA located at the corresponding position in the band. Linearity of this plot indicates monodispersity of the banded DNA. The value of the slope corresponds to an apparent molecular weight for the Cs · DNA salt of $9.4 \times 10^6$, corresponding to a molecular weight of $7.1 \times 10^6$ for the sodium salt.

increment produced by the $N^{15}$ labeling of the unheated DNA.

This behavior suggests that heating the hybrid molecule brings about the dissociation of the $N^{15}$-containing subunit from the $N^{14}$ subunit. This possibility was tested by a density-gradient examination of a mixture of heated $N^{15}$ DNA and heated $N^{14}$ DNA (Fig. 9, C). The close resemblance between the products of heating hybrid DNA (Fig. 9, B) and the mixture of products obtained from heating $N^{14}$ and $N^{15}$ DNA separately (Fig. 9, C) leads to the conclusion that the two molecular subunits have indeed dissociated upon heating. Since the apparent molecular weight of the subunits so obtained is found to be close to half that of the intact molecule, it may be further concluded that the subunits of the DNA molecule which are conserved at duplication are single, continuous structures. The scheme for DNA duplication proposed by Delbrück[24] is thereby ruled out.

To recapitulate, both salmon-sperm and *E. coli* DNA heated under similar conditions collapse and undergo a similar density increase, but the salmon DNA retains its initial molecular weight, while the bacterial DNA dissociates into the two subunits which are conserved during duplication. These findings allow two different interpretations. On the one hand, if we assume that salmon DNA contains subunits analogous to those found in *E. coli* DNA, then we must suppose that the subunits of salmon DNA are bound together more tightly than those of the bacterial DNA. On the other hand, if we assume that the molecules of salmon DNA do not contain these subunits, then we must concede that the bacterial DNA molecule is a more complex structure than is the molecule of salmon DNA. The latter interpretation challenges the sufficiency of the Watson-Crick DNA model to explain the observed distribution of parental nitrogen atoms among progeny molecules.

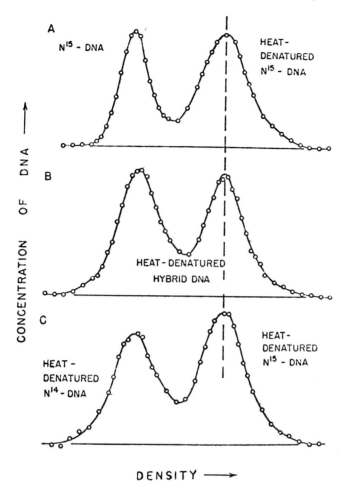

**Fig. 9.** The dissociation of the subunits of *E. coli* DNA upon heat denaturation. Each smooth curve connects points obtained by microdensitometry of an ultraviolet absorption photograph taken after 20 hours of centrifugation in CsCl solution at 44,770 rpm. The base line density has been removed by subtraction. **A,** A mixture of heated and unheated $N^{15}$ bacterial lysates. Heated lysate alone gives one band in the position indicated. Unheated lysate was added to this experiment for comparison. Heating has brought about a density increase of 0.016 gm cm$^{-3}$ and a reduction of about half in the apparent molecular weight of the DNA. **B,** Heated lysate of $N^{15}$ bacteria grown for one generation in $N^{14}$ growth medium. Before heat denaturation, the hybrid DNA contained in this lysate forms only one band, as may be seen in Fig. 4. **C,** A mixture of heated $N^{14}$ and heated $N^{15}$ bacterial lysates. The density difference is 0.015 gm cm$^{-3}$.

## Conclusion

The structure for DNA proposed by Watson and Crick brought forth a number of proposals as to how such a molecule might replicate. These proposals[6] make specific predictions concerning the distribution of parental atoms among progeny molecules. The results presented here give a detailed answer to the question of this distribution and simultaneously direct our attention to other problems whose solution must be the next step in progress toward a complete understanding of the molecular basis of DNA duplication. What are the molecular structures of the subunits of *E. coli* DNA which are passed on intact to each daughter molecule? What is the relationship of these subunits to each other in a DNA molecule? What is the mechanism of the synthesis and dissociation of the subunits in vivo?

## Summary

By means of density-gradient centrifugation, we have observed the distribution of $N^{15}$ among molecules of bacterial DNA following the transfer of a uniformly $N^{15}$-substituted exponentially growing *E. coli* population to $N^{14}$ medium. We find that the nitrogen of a DNA molecule is divided equally between two physically continuous subunits; that, following duplication, each daughter molecule receives one of these; and that the subunits are conserved through many duplications.

## REFERENCES AND NOTES

1. R. D. Hotchkiss, in The Nucleic Acids, ed. E. Chargaff and J. N. Davidson (New York: Academic Press, 1955), p. 435; and in Enzymes: Units of Biological Structure and Function, ed. O. H. Gaebler (New York: Academic Press, 1956), p. 119.
2. S. H. Goodgal and R. M. Herriott, in The Chemical Basis of Heredity, ed. W. D. McElroy and B. Glass (Baltimore: Johns Hopkins Press, 1957), p. 336.
3. S. Zamenhof, in The Chemical Basis of Heredity, ed. W. D. McElroy and B. Glass (Baltimore: Johns Hopkins Press, 1957), p. 351.
4. A. D. Hershey and M. Chase, J. Gen. Physiol. 36:39, 1952.
5. A. D. Hershey, Virology 1:108, 1955; 4:237, 1957.
6. M. Delbrück and G. S. Stent, in The Chemical Basis of Heredity, ed. W. D. McElroy and B. Glass (Baltimore: Johns Hopkins Press, 1957), p. 699.
7. C. Levinthal, these Proceedings 42:394, 1956.
8. J. H. Taylor, P. S. Woods, and W. L. Huges, these Proceedings 43:122, 1957.
9. R. B. Painter, F. Forro, Jr., and W. L. Hughes, Nature 181:328, 1958.
10. M. S. Meselson, F. W. Stahl, and J. Vinograd, these Proceedings 43:581, 1957.
11. The buoyant density of a molecule is the density of the solution at the position in the centrifuge cell where the sum of the forces acting on the molecule is zero.
12. Our attention has been called by Professor H. K. Schachman to a source of error in apparent molecular weights determined by density-gradient centrifugation which was not discussed by Meselson, Stahl, and Vinograd. In evaluating the dependence of the free energy of the DNA component upon the concentration of CsCl, the effect of solvation was neglected. It can be shown that solvation may introduce an error into the apparent molecular weight if either CsCl or water is bound preferentially. A method for estimating the error due to such selective solvation will be presented elsewhere.
13. In addition to $NH_4Cl$, this medium consists of $0.049\ M\ Na_2HPO_4$, $0.022\ M\ KH_2PO_4$, $0.05\ M$ NaCl, $0.01\ M$ glucose, $10^{-3}\ M\ MgSO_4$, and $3 \times 10^{-6}\ M\ FeCl_3$.
14. This result also shows that the generation time is very nearly the same for all DNA molecules in the population. This raises the questions of whether in any one nucleus all DNA molecules are controlled by the same clock and, if so, whether this clock regulates nuclear and cellular division as well.
15. F. H. C. Crick and J. D. Watson, Proc. Roy. Soc. London, A 223:80, 1954.
16. R. Langridge, W. E. Seeds, H. R. Wilson, C. W. Hooper, M. H. F. Wilkins, and L. D. Hamilton, J. Biophys. and Biochem. Cytol. 3:767, 1957.
17. For reviews see D. O. Jordan, in The Nucelic Acids, ed. E. Chargaff and J. D. Davidson (New York: Academic Press, 1955), 1, 447; and F. H. C. Crick, in The Chemical Basis of Heredity, ed. W. D. McElroy and B. Glass (Baltimore: Johns Hopkins Press, 1957), p. 532.
18. J. D. Watson and F. H. C. Crick, Nature 171:964, 1953.
19. C. E. Hall and M. Litt, J. Biophys. and Biochem. Cytol. 4:1, 1958.
20. R. Thomas, Biochim. et Biophys. Acta 14:231, 1954.
21. P. D. Lawley, Biochim. et Biophys. Acta 21:481, 1956.
22. S. A. Rice and P. Doty, J. Am. Chem. Soc. 79:3937, 1957.
23. Kindly supplied by Dr. Michael Litt. The preparation of this DNA is described by Hall and Litt (J. Biophys. and Biochem. Cytol. 4:1, 1958).
24. M. Delbrück, these Proceedings 40:783, 1955.

# 6 THE CHROMOSOME OF ESCHERICHIA COLI

*John Cairns*
*Cold Spring Harbor Laboratory of Quantitative Biology*
*Cold Spring Harbor, New York*

## Introduction

Autoradiography of *Escherichia coli*, labeled with tritiated thymidine and lysed with duponol, has shown that the bacterial chromosome comprises a single piece of DNA which is probably duplicated at a single growing point (Cairns, 1963). Further, it seemed likely from the variety of structures seen that this DNA is in the form of a circle while it is being replicated, even though no intact replicating circles had, at that time, been found.

There is no immediate prospect of proving that the bacterial chromosome is simply a continuous DNA double helix; the existence, for example, of protein linkers scattered along the chromosome (Freese, 1958) could be disproved only by a degree of purification of the intact chromosome that, at the moment, is technically impossible. So, rather than attempt any further purification, an effort was made to extract the chromosome as an intact but replicating circle by lysing labeled bacteria with lysozyme instead of duponol; extraction with lysozyme, it was thought, might leave the DNA complexed with basic proteins and polyamines and therefore less liable to breakage by turbulence (Kaiser, Tabor and Tabor, 1963) or by tritium decay (Hershey, unpublished). Whether or not this reasoning is correct, intact and replicating circles have now been found.

## Results

*E. coli* K12 #3000 Hfr thy⁻ was labeled by growth for about two generations in medium containing $H^3$-thymidine (10 C/mM), as described previously (Cairns, 1963), and was lysed at a concentration of $10^4$/ml by incubation at 37°C, in the usual dialysis chamber, in a medium containing 1.5 M sucrose, 0.01 M KCN, 0.01 M EDTA (pH 8), 5 μg/ml calf thymus DNA, and 200 μg/ml lysozyme. Following incubation for 6 hr, the lysed bacteria were dialyzed against repeated changes of 0.005 M EDTA for 18 hr at room temperature. Finally the membranes (VM Millipore Filters) were subjected to autoradiography.

This procedure displays up to 1% of the chromosomes as more or less tangled circles which, when fully extended, have a circumference of 1100-1400 μ. Usually these circles are seen to be engaged in duplication. Here, one such example will be described in considerable detail. First, however, it is simplest to describe the model of chromosome duplication to which it gives rise.

Figure 1 shows diagrammatically two rounds of duplication of a circular chromosome, following the introduction of a labeled precursor at some arbitrary point in the cycle. The diagram is based on the assumption that duplication always starts at the same point (in this case, at 12 o'clock) and always advances in the same direction (in this case, counter-clockwise). To make the diagram in a sense complete, some provision must be made for free rotation of the unduplicated part of the circle with respect to the rest, so that the parental double helix can unwind as it is duplicated; this provision, which we may noncommittally refer to as a swivel, has been placed at the junction of starting and finishing point and is itself marked as being duplicated just before the completed daughter chromosomes separate and begin the next round of duplication. Aside from any question of the swivel, we see that each daughter chromosome (and two of the four granddaughters) shows the stage in the duplication cycle at which label was originally introduced.

Figure 2 shows the autoradiograph of an unbroken replicating circular chromosome that is almost entirely untangled. For the purpose of grain counts and length measurements it was devided, according to the inset diagram, into

---

From Cold Spring Harbor Symposium on Quantitative Biology 28:43-46, 1963. Reprinted with permission.

I am greatly indebted to V. J. Paral and R. Westen, of the Australian National University, who brought their technical skill to bear on the problem of photographing the autoradiographs of DNA; without their help such pictures could not have been taken.

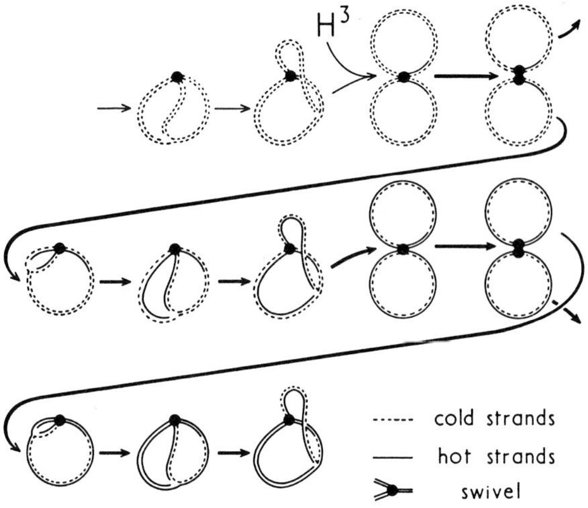

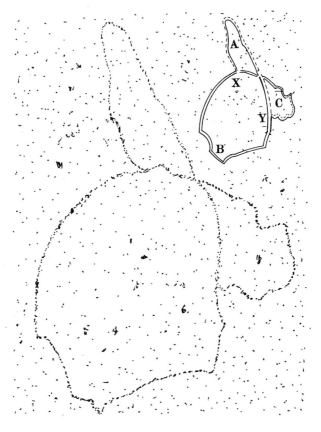

**Fig. 1.** A diagrammatic representation of the replication of a circle, based on the assumption that each round of replication begins at the same place and proceeds in the same direction.

**Fig. 2.** Autoradiograph of the chromosome of *E. coli* K12 Hfr, labeled with tritiated thymidine for two generations and extracted with lysozyme. Exposure time two months. The scale shows 100 μ. Inset, the same structure is shown diagrammatically and divided into three sections (A, B, and C) that arise at the two forks (X and Y).

TABLE 1

*Grain counts and length measurements for the three sections that unite the forks, X and Y, of Fig. 2*

| Section | Grains | Length ($\mu$) | Grains/$\mu$ |
|---|---|---|---|
| A | 714 | 670 | 1.1 |
| B | 1298 | 680 | 1.9 |
| C  Y to C | 213 | 215 | 1.0 |
| C to X | 359 | 205 | 1.8 |

three sections (A, B, and C) which all meet at two forks (X and Y). The grain counts and length measurements are given in Table 1.

We see here a predominately half-hot chromosome that has completed about two-thirds of the process of duplication. Part of the still-unduplicated section is half-hot (from Y to C) and part is hot-hot (from C to X); as shown in Fig. 1, this situation arises when the moment of introduction of the label does not coincide with the start of a round of replication. And it is for this reason that one can, in this instance, identify one of the forks (X) as the starting and finishing point of duplication and the other (Y) as the growing point. Thus the history of this chromosome is taken to be as pictured in Fig. 1.

Discounting the excess due to replication, the total length of the chromosome seen here is $1100\,\mu$ ($420\,\mu$ plus $670\text{-}680\,\mu$) or about 22 times the length of T2 DNA—i.e., equivalent to about $2.8 \times 10^9$ daltons of DNA. This value is slightly higher than that reported earlier (Cairns, 1963) and agrees well with the maximum value of 23 T2-equivalents, obtained when the reported total DNA content of 32 T2-equivalents (Hershey and Melechen, 1957) is multiplied by $ln\,2$ to correct for continuous duplication.

The process of duplication portrayed here is, in most respects, merely the physical embodiment of the conclusions of others. For it has become clear from quite unrelated experiments that the bacterial chromosome is duplicated at a single growing point (Bonhoeffer, 1963) which, at least in *E. coli* Hfr and in certain strains of *B. subtilis*, always starts at the same place and moves in the same direction (Nagata, 1963; Yoshikawa and Sueoka, 1963).

More problematical than the process of DNA synthesis itself, about which there is such satisfactory agreement, are the processes occurring between one round of duplication and the next. These have been represented diagrammatically in Fig. 1 as a separate stage in the duplication of the chromosome during which the swivel is supposed to be duplicated. Presumably, it is at this time that RNA and protein synthesis become obligatory (Maaløe and Hanawalt, 1961), that the Hfr chromosome becomes available for transfer during mating (Bouck and Adelberg, 1963), and that thymineless death may be consummated.

At first sight it seemed surprising to find that the chromosome is physically in the form of a circle even while it is being replicated, for this arrangement demands that somewhere in the cirlce there must be something that acts as a swivel. However, in view of the apparent importance of the structure that unites the ends of the chromosome and so completes the circle, one must now consider the possibility that the structure actively drives DNA replication by rotating one end of the chromosome relative to the other; in this way, single-stranded DNA might be continually produced at the replicating fork to act as primer for the polymerase. In short, it now seems conceivable that rapid DNA synthesis is possible only for circles.

**REFERENCES**

1. Bonhoeffer, F. 1963. Personal communication.
2. Bouck, N., and E. A. Adelberg. 1963. The relationship between DNA synthesis and conjugation in *Escherichia coli*. Biochem. Biophys. Res. Commun. 11:24-27.
3. Cairns, J. 1963. The bacterial chromosome and its manner of replication as seen by autoradiography. J. Mol. Biol. 6:208-213.
4. Freese, E. 1958. The arrangement of DNA in the chromosome. Cold Spring Harbor Symp. Quant. Biol. 23:13-18.
5. Hershey, A. D., and N. E. Melechen. 1957. Synthesis of phage-precursor nucleic acid in the presence of chloramphenicol. Virology 3:207-236.
6. Kaiser, D., H. Tabor, and C. W. Tabor. 1963. Spermine protection of coliphage λ DNA against breakage by hydrodynamic shear. J. Mol. Biol. 6:141-147.
7. Maaløe, O., and P. C. Hanawalt. 1961. Thymine deficiency and the normal DNA replication cycle, I. J. Mol. Biol. 3:144-145.
8. Nagata, T. 1963. The molecular synchrony and

sequential replication of DNA in *Escherichia coli*. Proc. Natl. Acad. Sci. **49:**551-559.
9. Yoshikawa, H., and N. Sueoka. 1963. Sequential replication of *Bacillus subtilis* chromosome. 1. Comparison of marker frequencies in exponential and stationary growth phases. Proc. Natl. Acad. Sci. **49:**559-566.

## Discussion

**Butler:** I should like to mention an idea put forward at a meeting of the British Biophysical Society December 1962 in a paper by Godson, Barr, and myself (see accompanying Fig. 1). The DNA polymerase is pictured as a disc with two holes or slots, through each of which one of the strands of the DNA passes. As it passes up the primer, this disc is pictured as rotating relative to it and so separating and unwinding the strands. The energy required for this could probably be provided by the energy of condensation of the triphosphates in the condensation process (about 8 kcals per nucleotide pair). The two new double fibers of DNA would initially be loosely wound round each other, as is often seen in chromosomes, but

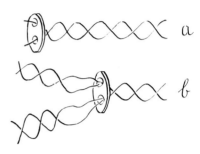

Fig. 1

thermal agitation in the cell would tend to unwind them. The idea suggested by Dr. Cairns that the primer strand itself may rotate would get over the need of the two new fibers of DNA to be wound round each other.

# 7 ISOLATION OF THE GROWING POINT IN THE BACTERIAL CHROMOSOME

*Philip C. Hanawalt*
*Dan S. Ray*
Biophysics Laboratory
Stanford University
Stanford, California

Current ideas and evidence concerning the replication of the bacterial chromosome can be summarized as follows: (1) The chromosome consists of one piece of *double-stranded* DNA[1] (mol wt $2 \times 10^9$) which may exist as a closed circle.[1,2] (2) It replicates *sequentially* from a starting point[3,4] by forming a single growing point at a fork which moves along the structure.[1,5] (3) Protein and/or RNA synthesis is required to initiate replication of the chromosome but the cycle can then be completed under conditions of protein synthesis inhibition.[6-8] (4) Replication is *semiconservative* and involves the separation of parental DNA strands "near" the growing point as complementary daughter strands are formed.[9-12]

In the autoradiographic analysis by Cairns,[1] the entire replicating chromosome was observed, but the linear grain density in the photographic emulsion (approximately one exposed grain per micron) limited the resolution of the growing point region to that of a $2 \times 10^6$ molecular weight segment of DNA. Preparatory to molecular characterization of the growing point region we have developed a procedure for the isolation of chromosome fragments containing this region, and we have defined some of the necessary conditions for the observation of these partially replicated DNA units.

## Materials and methods

The thymine-requiring *E. coli* strain TAU-bar[13] was cultured aerobically in a glucose-salts synthetic medium at 37°C with required supplements as previously described.[6,13] Exponential growth (mean generation time, 40 min) was maintained by periodic dilution into prewarmed medium to keep the cell concentration between $5 \times 10^7$ and $2 \times 10^8$ cells/ml.

From Proceedings of the National Academy of Sciences, U.S.A. **52**:125-132, 1964. Reprinted with permission.

This work was supported by a grant GM 09901 from the U.S. Public Health Service. We are indebted to Dr. V. Bode for supplying $H^3$-labeled λ phage DNA.

The bacterial DNA was uniformly prelabeled with tritium by growth for 8-12 generations in medium containing $H^3$-methyl-thymine (New England Nuclear, Boston) at 0.85 µg/ml and a specific activity of 380 mc/mM.

Media changes were accomplished by the rapid filtration technique previously described.[6]

Density labeling involved the substitution of 5-bromouracil (5 BU) (2 µg/ml in the growth medium in place of thymine. At 37°C the DNA synthesis *rate* in the presence of 5 BU was about half that for cultures growing with thymine.

$P^{32}$-pulse labeling of DNA involved the growth of 100-ml cultures in the minimal medium with the phosphate concentration reduced to $10^{-4}$ M and the subsequent addition of 6 mc $P^{32}$-orthophosphate (Radiochemical Centre, Amersham, England).

*Preparation of cell lysates.* Isotope incorporation and growth were stopped abruptly by the dilution of a 100-ml culture with an equal volume of ice-cold buffer (containing 0.1 M NaCl, 0.01 M EDTA, 0.01 M Tris (Sigma-121), and 0.01 M KCN at pH 8) abbreviated NET-CN. The cells were harvested by collection on a 9-cm diameter membrane filter (Schleicher & Schuell, Grade A coarse)[6] and rinsed with 100 ml cold NET-CN before resuspension in a 4-ml volume of NET-CN containing 1.5 M sucrose. The concentrated cell suspension was placed in dialysis tubing, and 0.1 ml of a 4 µg/ml solution of freshly prepared egg white lysozyme (Worthington, 2X recrystallized) was added. A 30-min incubation at 37°C with the dialysis tube suspended in NET-CN plus 1.5 M sucrose resulted in protoplasting of the cells. Attempts to use the Duponol procedure of Cairns[1] were unsuccessful for the subsequent isolation of newly replicated material in a CsCl density gradient. Up to this point in the procedure no appreciable shearing of the DNA should have occurred because the chromosomes were still contained within cell membranes.

In successive dialysis steps the sucrose concentration was reduced from 1.5 M to 0.75 M to 0, resulting in gradual lysis of the protoplasts. The dialysis tube was opened and the lysate was gently poured directly onto solid cesium chloride (Harshaw, optical grade) in a weighed beaker so that the density could be adjusted to 1.72 gm/cc without sampling. The pH of the cesium chloride solution was adjusted to 10.5 by the addition of 0.5 ml 0.5 M $K_2HPO_4$, pH 10.5, a step which greatly improved the yield of growing point

54   Molecular biology of DNA and RNA

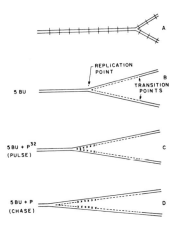

**Fig. 1.** Schematic representation of a segment of the bacterial chromosome containing the growing point. **A,** Fragmentation of the DNA during extraction. Single strands shown as solid lines. Short cross lines indicate lateral scissions and would be expected also for the three lower diagrams. **B,** The same chromosomal segment after a period of density labeling with 5 BU. The dashed lines are 5 BU containing strands. **C,** Same segment after short pulse incorporation of $P^{32}$ to label growing point fragment during growth with 5 BU. $P^{32}$ containing regions indicated by x's. **D,** Transfer of $P^{32}$ containing regions into hybrid fragments during subsequent growth with 5 BU and unlabeled phosphate.

regions, presumably by aiding in the high salt deproteinization of these regions. Use of alkaline CsCl was suggested to us by the paper of Vinograd et al.[14] The pH 10.5 is below that which would be expected to partially denature 5 BU hybrid DNA in CsCl.[12]

*Density gradient equilibrium sedimentation.* Four milliliters of the alkaline CsCl solution was slowly poured into a cellulose tube, layered with mineral oil, and subjected to 48 hr at 37,000 rpm and 20° C in the SW39 rotor of a Spinco Model L ultracentrifuge. After deceleration, drops were collected through a pinhole punched in the tube bottom, and one or two drop fractions were diluted with 0.5 ml 0.01 $M$ NaCl, 0.001 $M$ EDTA, 0.001 $M$ Tris pH 8. For counting, 0.1-ml aliquots were added to 0.2 ml 1.0 $M$ KOH and allowed to stand at room temperature overnight to hydrolyze RNA. Salmon-sperm DNA (10 $\mu$g) was added to each sample as carrier before the addition of 5 ml ice-cold 5% trichloroacetic acid. Samples were then collected by suction filtration on 23-mm diameter Millipore filters (Type HA, 0.45 $\mu$ pore size) and rinsed twice with 5 ml distilled water. After drying under a heat lamp the filters were placed in glass counting vials containing 5 ml toluene, 18 mg PPO, and 0.45 mg dimethyl-POPOP. The Packard TriCarb scintillation spectrometer was adjusted for two channel simultaneous assay of tritium and $P^{32}$. Overlap corrections were made by reference to freshly prepared standards.

*Design of the experiments.* In the random fragmentation of the bacterial chromosome during isolation, only one (if any!) of the isolated fragments would be expected to contain the fork, as illustrated in Figure 1$A$. When bacteria are transferred to a medium containing a density label (e.g., 5 BU) and the DNA is then extracted at an arbitrary time later, two classes of chromosomal fragment would be expected to have densities intermediate between normal (thymine-containing) and hybrid (5 BU replacement of thymine sites in one strand).[15] As shown in Figure 1$B$, these are the fragments containing *transition points* (i.e., the position of the growing point at the instant of transfer to 5 BU growth medium), and the *replication point* as caught at the instant growth was stopped. Fragments containing the transition points have previously been demonstrated.[5,15] As far as the isolation of DNA is concerned, the transition point material should be sufficiently distant from the growing point region to behave as the bulk of the chromosomal fragments.[5] To distinguish between transition point regions and the growing point regions it is necessary to demonstrate that the latter, but not the former, become hybrid within one generation period. Figures 1$C$ and $D$ illustrate a pulse and chase experiment to accomplish this.

## Results

Figure 2 shows the result of a $P^{32}$ pulse and chase experiment to demonstrate the transfer of intermediate density material to the hybrid region. The bacterial culture was transferred to 5 BU medium 30 min before the addition of $P^{32}$. A rebanding of the intermediate density region from the chase (Fig. 3) shows that the $P^{32}$ activity has been removed from that region but that some of the tritium activity remains, as expected for the fragments containing transition points. In Figure 4$A$, the rebanded intermediate density region from the pulse exhibits both $P^{32}$ and $H^3$ activity in the intermediate region, but when an equal aliquot is subjected to further fragmentation by sonication (Fig. 4$B$), both $P^{32}$ and $H^3$ activity are resolved in broader bands at the hybrid and normal densities. Note that no $P^{32}$ appears in the normal band.

To improve the yield of DNA in a subsequent experiment, a portion of the cell lysate was treated with the enzymes lipase, trypsin, and chymotrypsin before the addition of

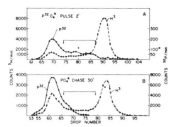

**Fig. 2.** Density distributions of DNA fragments in alkaline CsCl following bacterial growth with 5 BU. A 30-min period of 5 BU incorporation preceded the addition of $P^{32}$ so that the replication point but not the transition point would be labeled during the $P^{32}$ pulse. **A,** After 2′ pulse of $P^{32}$ as indicated in Fig. 1C. **B,** After 30-min chase with unlabeled phosphate as shown in Fig. 1D. •----•, tritium activity, from uniform prelabeling of DNA (see *Methods*). Peak on the right is at the position of normal unreplicated DNA. •—•, $P^{32}$ activity. Peak on the left is at the position of hybrid, completely replicated DNA fragments. Intermediate density fractions for rebanding (Figs. 3 and 4) are indicated by bracketed arrows.

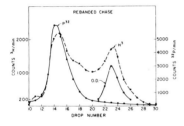

**Fig. 3.** Rebanding of intermediate density fractions from Fig. 2B. Normal unlabeled TAU-bar DNA added as optical density (O.D.) marker. O.D. scale not shown.

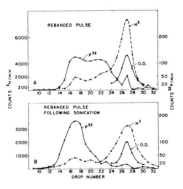

**Fig. 4.** Rebanding of intermediate density fractions from Fig. 2A. Normal unlabeled TAU-bar DNA added as density marker as in Fig. 3. **A,** No further treatment. **B,** Following sonication for 30 sec under conditions previously described.[15]

alkaline CsCl. Equal portions of the lysate with and without such treatment are compared in Figure 5. Note that the enzyme treatment doubles the yield of normal and hybrid fragments but that it more than triples the yield of partially replicated material.

In a third preparation (not shown), the lysate was held at 60°C for 30 min before adding the alkaline CsCl. This treatment also doubled the yield but did not change the *relative* yield of intermediate density material.

It might be predicted that the fork in the replicating DNA would be particularly sensitive to shear and that this would explain, in part, why it has not previously been isolated in density gradients. An experiment to test this is illustrated in Figure 6. The relatively mild shearing in a vortex mixer removes essentially all of the $P^{32}$ activity from the intermediate density region while leaving some $H^3$ activity, as expected if replication points are more sensitive to shear than transition points. To determine the size of the DNA fragment which is being observed in these experiments and the extent of fragmentation by the vortex mixer, the sucrose gradient procedure of Burgi and Hershey[16] was used as shown in Figure 7. The molecular weight distribution observed in Figure 7A represents a *lower* limit for this isolation procedure, since drop collecting from the preparative CsCl gradient probably shears the DNA also. Yet our rebanding experiments show that at least some fragments containing replication points survive the drop-collecting procedure. The DNA preparation as shown in Figure 7A is very heterogeneous (molecular weight range roughly $3 \times 10^7$ to $20 \times 10^7$) but the mean molecular weight is $10^8$, considerably higher than that usually obtained from bacteria.[5,18] Further fragmentation in the vortex mixer reduces this to a more homogeneous preparation in the molecular weight range of roughly $3 \times 10^7$ to $6 \times 10^7$, which is in the range generally obtained[5,18] (Fig. 7B).

*Resistance to isolation of newly replicated DNA.* The importance of quantitating recovery of DNA in studies on its physical state *in vivo* has recently been emphasized by Cohen.[19] Experimental evidence that newly replicated DNA differs in state from the bulk of cellular DNA has come from the studies of Goldstein and Brown[20] in which pulse-labeled bacterial

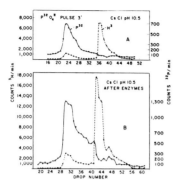

**Fig. 5.** Effect of enzyme treatment on recovery of pulse-labeled DNA in the cesium chloride gradient. Culture treatment same as for Fig. 2A, but 3-min pulse of $P^{32}$ was used. O----O, tritium activity; O—O, $P^{32}$ activity. **A,** Same isolation procedure as for Figs. 2A and B. **B,** Lysate subjected to 37°C incubation with lipase [(wheat germ) B grade, Calbiochem]. 50 γ/ml for 10 min, then trypsin 20 γ/ml and α-chymotrypsin 20 γ/ml for 15 min before adding to CsCl.

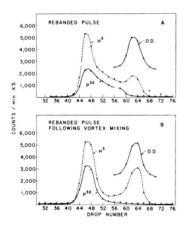

**Fig. 6.** Rebanding of intermediate density fractions 36-40 inclusive from Fig. 5B. Unlabeled TAU-bar DNA added as density marker. **A,** No further treatment. **B,** Following 5 min in Cyclomixer (Clay-Adams, N. Y).

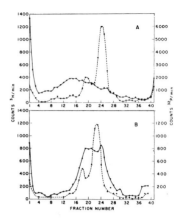

**Fig. 7.** Molecular weight distribution of DNA isolated by our method and the effect of mild shearing on this distribution. $P^{32}$-labeled DNA was isolated from the hybrid band in a preparation such as shown in Fig. 5B. A 0.03-ml (less than 5 γ/ml DNA) volume was added with a 0.07-ml volume containing $H^3$-DNA (phage λ DNA at 50 γ/ml) to the top of a 5-20% linear sucrose gradient. Fractions were collected after 4-hr centrifugation at 28 K and assayed for $H^3$ and $P^{32}$ activity. (See *Methods* and refs. 16 and 17.) •----•, tritium activity, λ DNA marker (31 × $10^6$ MW); •—•, $P^{32}$ activity. The small $H^3$ peak probably represents a dimer of λ DNA molecules, as calculated from its position relative to the main $H^3$ peak. Such dimers have been reported by Hershey et al.[17] and have been shown to occur at high DNA concentrations. The small peak was not observed in a sucrose gradient run in which the λ DNA concentration was reduced by a factor of two (Ray and Hanawalt, to be published). Velocity sedimentation studies on this particular λ DNA preparation by Dr. William Studier indicated that no fragments of the whole λ DNA were present. The high concentration of the λ DNA required a correction in the observed sedimentation rate. **A,** Unsheared preparation. **B,** Following 5 min on Cyclomixer (see text).

DNA was found to be selectively resistant to release from aggregates by sonication. Also, it has been demonstrated that newly synthesized DNA remains in the interface after a chloroform-octanol extraction which releases 90 per cent of cellular DNA from a rabbit kidney cell preparation.[21]

We have observed a similar resistance to isolation of newly replicated DNA from bacterial preparations. In Figure 5, the lipase and proteolytic enzyme treatment increased the yield of DNA from 22 to 40 per cent while also increasing the *proportional* yield of pulse-labeled intermediate density DNA in the cesium chloride gradient. (The unrecovered DNA is found in the meniscus at the end of a density gradient sedimentation run, presumably because it is still bound to protein.)

The deproteinization of a lysate (before addition to alkaline cesium chloride) by shaking 5 min on the vortex mixer with an equal volume of chloroform-octanol 9:1 (v/v)[22] was observed to lose $P^{32}$ pulse-labeled (i.e., newly replicated) DNA from the aqueous phase selectively, even though over-all yields as high as 90 per cent of the total DNA could be obtained. The treatment of such a lysate with lipase, trypsin, and chymotrypsin (as for Fig. 5) before shaking with chloroform-octanol gave a 6 per cent increase in yield of uniformly labeled $H^3$-DNA in the aqueous phase, but a 38 per cent increase in recovery of $P^{32}$ pulse-labeled DNA. In another experiment, the yield of alkaline resistant $P^{32}$ activity in the aqueous phase was improved from 32 to 50% when the lysate was treated with the enzymes prior to chloroform-octanol extraction. Thus far, it has not been possible to obtain a quantitative recovery of newly replicated DNA. The detergent and phenol methods for deproteinization have also been found to select against newly replicated DNA.

## Discussion

We have shown that $P^{32}$ pulse-labeled DNA during density labeling with 5 BU appears in the intermediate density region between normal and hybrid densities and that it leaves this region (presumably by becoming completely hybrid upon chasing with unlabeled phosphate). It could be argued that the lighter-than-hybrid density is due to protein binding to newly replicated hybrid segments and that this protein is released as the growing point moves beyond these segments. We consider this explanation unlikely for a number of reasons. In $P^{32}$ pulse-labeling experiments in which no density label was used (not reported in this paper), the newly synthesized DNA appeared in a symmetrical band at the same density as the bulk of the isolated DNA. In Figure 5, we show that the relative yield of intermediate density DNA is *increased* following proteolytic enzyme treatment of the cell lysate. A comparison of Figures 6A and B shows that the effect of mild shearing on the intermediate density material is to remove the $P^{32}$ counts to the hybrid band but simultaneously to remove tritium activity (i.e., unreplicated fragments) to the normal band. Thus, the shearing involves a separation into normal and hybrid density fragments as expected if the intermediate density region contained the fork in partially replicated DNA units.

The isolation of partially replicated DNA units is just a first step in the physical characterization of the growing point region in the bacterial chromosome. However, in reporting the isolation of such fragments we can define two important problems which may explain why they have not previously been observed.

(1) The growing point is very sensitive to shear, evidently more so than the bulk of the DNA. Any isolation procedure which results in shearing to the 30 million molecular weight range will probably preclude the observation of partially replicated DNA fragments. Thus, shearing of the bacterial chromosome during isolation is not random with respect to the replication point.

(2) The growing point region is more resistant to isolation than the bulk of the DNA and will be selectively lost at the meniscus in cesium chloride density gradients. The use of alkaline cesium chloride and our pretreatment of lysates with lipase and proteolytic enzymes to disaggregate DNA from complexes has been shown to help in recovery of this material. (The specific and independent functions of the lipase and/or proteolytic enzymes in this disaggregation have not been ascertained.)

The nature of the two problems above poses a dilemma. If one uses a very gentle method for isolation and deproteinization of the DNA (e.g., the detergent method of Cairns[1]), then the newly replicated DNA will not be recovered in a cesium chloride gradient. On the other hand, vigorous shaking and extraction with chloroform-octanol may improve the yield of newly replicated material but will certainly shear all of it at the replication point.

We cannot estimate the size of the partially replicated fragments from the relative amount of $P^{32}$ in hybrid and intermediate regions following a pulse of known duration, because of the difference in sensitivity to shear and resistance to isolation of these fragments. However, the mere existence of such fragments rules out that replication is an all-or-none event within defined regions along the chromosome and that breakage necessarily occurs at the ends of such regions. The resistance to isolation of partially replicated DNA regions might have been predicted since the polymerase (or polymerases) engaged there, or perhaps structural proteins involved in the "zipper," should require more stringent deproteinization procedures than for the bulk of the chromosomal DNA.

Our studies have little bearing on the reported observation of prereplicative states of DNA in bacteria by Rolfe,[23] Rosenberg and Cavalieri,[24] and Lark.[25] We have not obtained any detectable amounts of DNA at positions corresponding to denatured DNA components in $P^{32}$ pulse labeling experiments in which no density label was used. It is possible that a denatured portion of the chromosome ahead of the growing point would have a chance to renature in the course of our lysate preparation, as it is perhaps equally possible that the DNA strands might continue to unwind without synthesis at the growing point in other isolation procedures. Continuing studies are designed to examine these possibilities.

## Summary

A thymine-requiring bacterium was transferred to a medium containing the thymine analogue, 5-bromouracil, as a density label for newly replicated DNA. Following a gentle lysis procedure, the density distribution of the DNA molecules was examined in cesium chloride density gradients. A small fraction of the DNA molecules were isolated at densities intermediate between normal (unreplicated) and hybrid (5-bromouracil in one strand), and these were characterized as containing transition points (i.e., the replication point at instant of transfer to 5-bromouracil medium) or replication points (as caught at the instant growth was stopped). The isolation of replication points was specifically shown by $P^{32}$ pulse-labeling during 5-bromouracil incorporation and the demonstration that the intermediate density $P^{32}$ activity could be "chased" into the hybrid density region by subsequent growth with unlabeled phosphate. The extracted DNA was shown to have a mean molecular weight of $10^8$, and mild shear which reduced this to the 3 to 6 $\times 10^7$ range sheared all replication point containing molecules into hybrid and normal density fragments. The resistance to isolation and the sensitivity to shear of partially replicated DNA molecules is discussed.

## REFERENCES

1. Cairns, J., J. Mol. Biol. **6**:208 (1963).
2. Jacob, F., and E. L. Wollman, Symp. Soc. Exptl. Biol. **12**:75 (1958).
3. Nagata, T., these Proceedings **49**:551 (1963).
4. Yoshikawa, H., and N. Sueoka, these Proceedings **49**:559 (1963).
5. Bonhoeffer, F., and A. Gierer, J. Mol. Biol. **7**:534 (1963).
6. Maaløe, O., and P. C. Hanawalt, J. Mol. Biol. **3**:144 (1961).
7. Hanawalt, P. C., O. Maaløe, D. J. Cummings, and M. Schaecter, J. Mol. Biol. **3**:156 (1961).
8. Lark, K. G., T. Repko, and E. J. Hoffman, Biochim. Biophys. Acta **76**:9 (1963).
9. Watson, J. D., and F. H. C. Crick, Nature **171**:737 (1953).
10. Delbrück, M., and G. Stent, in The Chemical Basis of Heredity, ed. W. D. McElroy and B. Glass (Baltimore: Johns Hopkins Press, 1957).
11. Meselson, M., and F. W. Stahl, these Proceedings **44**:671 (1958).
12. Baldwin, R. L., and E. M. Shooter, J. Mol. Biol. **7**:511 (1963).
13. Hanawalt, P. C., Nature **198**:286 (1963).
14. Vinograd, J., J. Morris, N. Davidson, and W. F. Dove, Jr., these Proceedings **49**:12 (1963).
15. Pettijohn, D. E., and P. C. Hanawalt, J. Mol. Biol. **8**:170 (1964).
16. Burgi, E., and A. D. Hershey, Biophys J. **3**:309 (1963).
17. Hershey, A. D., E. Burgi, and L. Ingraham, these Proceedings **49**:748 (1963).
18. Marmur, J., J. Mol. Biol. **3**:208 (1961).
19. Cohen, S. S., J. Cellular Comp. Physiol., Suppl. 1, **62**:43 (1963).
20. Goldstein, A., and B. J. Brown, Biochim. Biophys. Acta **53**:19 (1961).
21. Ben-Porat, T., A. Steere, and A. S. Kaplan, Biochim. Biophys. Acta **61**:150 (1962).
22. Sevag, M. G., D. B. Lackman, and J. Smolens, J. Biol. Chem. **124**:425 (1938).
23. Rolfe, R., these Proceedings **49**:386 (1963).
24. Rosenberg, B. H., and L. F. Cavalieri, these Proceedings **50**:826 (1964).
25. Lark, K., in Molecular Genetics, ed. J. H. Taylor (New York: Academic Press, 1963), Part I, p. 153.

# 8 MECHANISM OF DNA CHAIN GROWTH, I. POSSIBLE DISCONTINUITY AND UNUSUAL SECONDARY STRUCTURE OF NEWLY SYNTHESIZED CHAINS

*Reiji Okazaki*
*Tuneko Okazaki*
*Kiwako Sakabe*
*Kazunori Sugimoto*
*Akio Sugino*
*Institute of Molecular Biology and Department of Chemistry*
*Faculty of Science*
*Nagoya University*
*Nagoya, Japan*

---

*In vivo* studies[1-7] of chromosome replication have led to the inference that both daughter strands of chromosomal DNA grow continuously, the direction of synthesis being 3′ to 5′ on one strand and 5′ to 3′ on the other (Fig. 1A). No enzymatic mechanism for the biosynthesis of deoxypolynucleotide in the 3′ to 5′ direction has been demonstrated, although 5′ to 3′ *in vitro* synthesis of DNA is accomplished by DNA polymerase.[8] If discontinuous synthesis of DNA could occur *in vivo* (Fig. 1B-D), short stretches could be synthesized by a reaction in the 5′ to 3′ direction and subsequently connected to the growing polynucleotide chain by formation of phosphodiester linkages.

It is possible to distinguish between continuous and discontinuous chain growth by elucidating the structure of the most recently replicated portion of the chromosome, that is, that portion selectively labeled by an extremely short radioactive pulse. If the chromosome replicates discontinuously by one of the mechanisms shown in Figures 1B, C, or D, a large portion of the radioactive label would be found in unconnected short chains which can be isolated, after denaturation, from the large DNA chains derived from the other portion of the chromosome. No such difference in the molecular size between the pulse-labeled and bulk DNA would be expected from a mechanism of continuous synthesis (Fig. 1A).

Our results to be described here, together with those reported previously,[9] indicate that in a variety of bacterial systems and in one bacteriophage system most of the recently synthesized portion of the chromosome can be obtained after denaturation as small DNA molecules with a sedimentation coefficient of about 10S. This supports the prediction of those mechanisms by which two daughter strands are synthesized in a discontinuous fashion (Fig. 1C or D). It is also shown that the secondary structure of the chromosomal region containing these newly synthesized chains may differ from that of ordinary double-stranded DNA.

## Materials and methods

Organisms used were as follows: *Escherichia coli* strains B, 15T⁻, W3110, and 1100 (endonuclease I-deficient strain provided by Dr. H. Hoffman-Berling), *Bacillus subtilis* strain SB 19, and bacteriophages T4 (wild-type) and δA (provided by Dr. I. Watanabe).[10]

*Reagents.* The following commercial products were used: $H^3$-thymidine and $C^{14}$-thymidine (New England Nuclear); crystalline pancreatic DNase and RNase and egg-white lysozyme (Worthington); Pronase P (Kaken Chemical). *E. coli* exonuclease I was a gift of Dr. I. R. Lehman. *B. subtilis* nuclease was fraction I-A described previously.[11] Bacterial α-amylase was provided by Dr. F. Fukumoto. Hydroxylapatite was prepared according to Miyazawa and Thomas.[12] $C^{14}$-*E. coli* DNA used as standard substrate for DNase was prepared as described previously.[11] DNA from phase δA was obtained by phenol extraction.[10]

---

From Proceedings of the National Academy of Sciences, U.S.A. 59:598-605, 1968. Reprinted with permission.

We are grateful to Dr. K. Gordon Lark and Dr. Rollin D. Hotchkiss for helpful discussions during preparation of the manuscript.

Supported by the Research Fund of the Ministry of Education of Japan and by a grant from the Jane Coffin Childs Memorial Fund for Medical Research. This work was presented at the Symposium on Nucleic Acid Synthesis, Tokyo, March 1967 (Okazaki, R., T. Okazaki, K. Sakabe, and K. Sugimoto, Jap. J. Med. Sci. Biol. **20**:255 (1967)), and at the Seventh International Congress of Biochemistry, Tokyo, August 1967 (Okazaki, R., and K. Sakabe, Abstract B-10 (International Union of Biochemistry, 1967)).

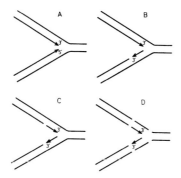

**Fig. 1.** Models for the possible structure and reaction in the replicating region of DNA.

*Culture media.* Medium A: glucose salt medium containing 0.1 M potassium phosphate buffer, pH 7.3, 1 mM $MgSO_4$, 0.02 M $(NH_4)_2SO_4$, 0.002 mM $Fe(NH_4)(SO_4)_2$ and 1% glucose; medium B: medium A supplemented with 0.5% casamino acids, 0.01% cysteine and DL-tryptophan, and $1.2 \times 10^{-5}$ M thymidine; medium C: M9 synthetic medium supplemented with 0.5% casamino acids. Media A, B, and C were used for experiments with E. coli B, E. coli 15T⁻, and T4 phage-infected E. coli B, respectively. Medium B containing no thymidine was used for E. coli W3110 and 1100.

*Pulse labeling.* To pulse-label the bacteria with no thymine requirement or T4 phage-infected cells, $H^3$-thymidine (14 mc/μmole) was added to the stirred culture ($5 \times 10^8$ cells/ml) at 20° (at 30° with B. subtilis to a concentration of $10^{-7}$ M. After allowing the cells to incorporate $H^3$-thymidine for a desired time, the culture was poured onto crushed ice and KCN (to 0.02 M), and the cells were collected by centrifugation at 0°. To pulse-label E. coli 15T⁻, cells grown in medium containing thymidine were precipitated and resuspended in a small volume of fresh medium at 0° containing no thymidine. The cell suspension was poured into a larger volume of stirred medium at 20°. $H^3$-thymidine ($10^{-7}$ M) was added and the reaction was stopped by KCN and ice.

*Extraction of DNA.* (a) *Extraction of native DNA by the Thomas procedure*[13] (Figs. 3, 4, 7, and 8; Table 1). This was carried out as described previously[9] except that in some experiments sodium dodecyl sulfate (SDS) treatment was at 37° and the DNA solution was concentrated by filtration through a collodion membrane. DNA from 1 ml of culture was finally obtained in a volume of 0.5-1 ml. In E. coli B, recovery of DNA labeled by various lengths of pulse was greater than 90%. With E. coli 15T⁻, recovery varied from 30 to 60% but no systematic difference was found between the pulse- and uniformly labeled DNA's in parallel experiments.

(b) *Extraction of denatured DNA by NaOH-EDTA* (Figs. 2, 5, and 6). The cells were suspended in ice-cold 0.1 N NaOH containing 0.01 M ethylenediaminetetraacetic acid (EDTA) at a concentration of $5 \times 10^9$ cells/ml. The suspension was incubated at 37° for 20 min with occasional gentle stirring with a glass rod, and the insoluble material was removed by low-speed centrifugation. More than 80% of E. coli DNA and 50-80% of pulse-labeled DNA from T4 phage-infected cells were recovered by this procedure.

*Denaturation of DNA.* DNA extracted in the native state was denatured by incubation in 0.1 N NaOH containing 1 mM EDTA at room temperature for 20 min.

*Zone sedimentation in sucrose gradients.* (a) *Alkaline sucrose gradient.* Either the SW25.1 or SW25.3 rotor of a Spinco L or L2 centrifuge was used. With the SW25.1 rotor, 1 ml of DNA sample in 0.1 N NaOH containing 0.01 M EDTA was layered on a 29-ml 5-20% linear sucrose gradient containing 0.1 N NaOH, 0.9 M NaCl, and 1 mM EDTA. With the SW25.3 rotor, the volumes of the sample and gradient were 0.3 and 16 ml respectively. Five mμmoles of DNA from bacteriophage δA was added to each sample as internal reference. After centrifugation, fractions were collected from the bottom of the tube. Radioactive DNA in each fraction was counted in a Tri-Carb liquid scintillation spectrometer after repeated precipitation with cold 5% trichloroacetic acid (TCA) and solubilization by 5% TCA at 90°. Distribution of δA DNA among fractions was determined by assaying aliquots for infectivity in E. coli protoplasts.[10] Distance of sedimentation was shown relative to the distance from the meniscus to the band of δA DNA. Sedimentation coefficients were calculated from the value of 19S for this marker DNA, obtained by boundary sedimentation in 0.1 N NaOH-0.9 M NaCl.

(b) *Neutral sucrose gradient.* Centrifugation was carried out in the SW25.1 rotor, layering 1 ml of DNA sample over a 29-ml 5-20% sucrose gradient, pH 7.0, containing 0.15 M NaCl, 0.015 M sodium citrate, and 1 mM EDTA.

Recovery of DNA from alkaline and neutral sucrose gradients was more than 90%.

*Other methods.* Chromatography of DNA on hydroxylapatite was carried out according to Bernardi.[14] Recovery of DNA from the column was 60-65%. Formation of acid-soluble product by enzymatic degradation of labeled DNA was measured as described by Lehman.[15]

## Results

***Nature of the replicating region as revealed by alkaline sucrose gradient sedimentation.*** To facilitate labeling of a small portion near the growing end of the daughter strands, all the

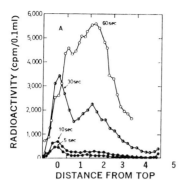 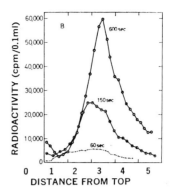

Fig. 2. Alkaline sucrose gradient sedimentation of pulse-labeled DNA from *E. coli* B. Cells were grown at 37° to a titer of 3 × 10⁸ cells/ml and then at 20° to 5 × 10⁸ cells/ml and the 10-ml culture was pulse-labeled with $10^{-7}$ $M$ H³-thymidine at 20° for the indicated time. DNA was extracted by NaOH-EDTA treatment and sedimented in the SW25.3 rotor for 10 hr at 22,500 rpm and 4°. Distance from top is relative to that of infective DNA from phage δA (19$S$, reference).

pulse-labeling experiments with *E. coli* (normal or T4 phage-infected) were carried out at 20°. The rate of macromolecular synthesis at 20° is estimated to be about one sixth of the rate at 37°, since at 20° the generation time (and doubling time of DNA) of *E. coli* is about 3 hours and the lysis by T4 phage occurs about 140 minutes after infection.

In the experiment presented in Figure 2, growing cells of *E. coli* B were exposed to H³-thymidine for various times. DNA was extracted in the denatured state by the NaOH-EDTA treatment and sedimented in alkaline sucrose gradients. Infectious DNA from phage δA used as internal reference had a sedimentation coefficient of 19$S$ in 0.1 $N$ NaOH-0.9 $M$ NaCl. Most of the radioactivity incorporated into DNA during the five-second pulse was recovered in a distinct component with an average sedimentation rate of 11$S$. Some radioactivity was found in material sedimenting at faster rates. Increasing the pulse time to 10 or 30 seconds increased the radioactivity in the "11$S$ component" as well as the radioactivity in the fast-sedimenting DNA. Further increasing the pulse time resulted in large increases of the radioactivity in the fast-sedimenting DNA with little or no increase in the radioactive "11$S$ component." The presence of the latter was obscure after the 150- or 600-second labeling because of the possible trailing of the high molecular DNA containing a large amount of radioactivity. The average sedimentation rate of the fast-sedimenting component increased

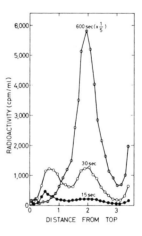

Fig. 3. Alkaline sucrose gradient sedimentation of pulse-labeled DNA from *E. coli* B. A 25-ml culture was pulse-labeled as in Fig. 2. DNA was extracted by the Thomas method. An aliquot was denatured in alkali and sedimented in the SW25.1 rotor for 15 hr at 20,500 rpm and 8°.

gradually and was about 50$S$ after the ten-minute pulse.

Essentially the same result was obtained by using the Thomas method for DNA extraction (Fig. 3).

Similar results were also obtained with other *E. coli* strains, i.e., *E. coli* 15T⁻, W3110, and 1100 (endonuclease I-deficient) (Figs. 4 and 5), and *B. subtilis* strain SB 19. The initial label of H³-thymidine always appeared in the DNA component with an average sedimentation rate of 10-11$S$.

That the "11S component" is really DNA was substantiated by several facts. It is degraded by the action of pancreatic DNase or by *E. coli* exonuclease I at the same rate as the denatured *E. coli* DNA routinely used as standard DNase substrate.[11] It is also completely degraded by *B. subtilis* nuclease[11] but not by alkali, pancreatic RNase, or bacterial α-amylase.

Figure 6 shows a result obtained with T4 phage-infected *E. coli* B. Cells were pulse-labeled after 70 minutes of infection at 20°, when phage DNA is being synthesized actively. After a two-second pulse the radioactive label incorporated was recovered almost exclusively in DNA component with a sedimentation coefficient of 9S. After a longer period of labeling, the radioactivity was found also in faster-sedimenting material. The radioactivity in the "9S component" increased quickly and reached a maximum in about 30 seconds, whereas the radioactivity in the fast-sedimenting component increased almost linearly and in two minutes attained a level ten times higher than the radioactivity in the "9S component." The sedimentation rate of the fast component increased gradually as in growing bacterial cells. The average rate was about 40S after the two-minute pulse. In other experiments average rates of 45 and 50S were obtained for five- and ten-minute pulse DNA, respectively.

In these experiments the pulse labeling was stopped by KCN and ice, cells were precipitated, and denatured DNA was obtained by either (a) extraction by the Thomas method followed by alkali denaturation, or (b) extrac-

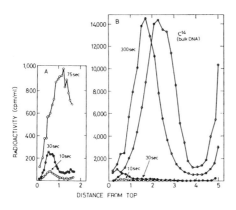

**Fig. 4.** Alkaline sucrose gradient sedimentation of pulse-labeled DNA from *E. coli* 15T⁻. A 5-ml culture was pulse-labeled at 20° for the indicated period. To the 300-sec sample, a small amount of culture uniformly labeled by C¹⁴-thymidine was added before DNA extraction by the Thomas method. Sedimentation was carried out in the SW25.1 rotor at 8° and 20,500 rpm for (**A**) 25 hr or (**B**) 10 hr.

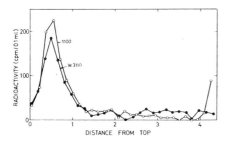

**Fig. 5.** Alkaline sucrose gradient sedimentation of a 10-sec pulse DNA of *E. coli* W3110 and 1100. Experiments were carried out as in Fig. 2.

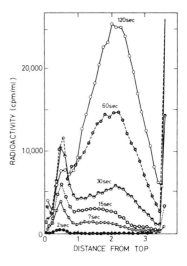

**Fig. 6.** Alkaline sucrose gradient sedimentation of pulse-labeled DNA from T4 phage-infected *E. coli* B. Cells grown at 37° to 5 × 10⁸ cells/ml were suspended in M9 medium containing no glucose at 10⁹ cells/ml and incubated for 15 min at 37°. Following addition of DL-tryptophan (40 μg/ml), the cells were infected with T4 phage (MOI = 10). After 5 min at 37°, the culture was cooled to 20° and an equal volume of M9 medium containing twice as much glucose and casamino acids as medium C was added. After incubation with stirring at 20° for 70 min, the 10-ml culture was pulse-labeled with H³-thymidine for the indicated time. DNA was extracted by NaOH-EDTA treatment and sedimented in the SW25.1 rotor for 15 hr at 20,500 rpm and 8°.

tion with NaOH-EDTA. The following changes in these procedures did not alter the essential feature of the results: (1) omission of the phenol step from *(a)*, (2) addition of a pretreatment with lysozyme to *(b)*, (3) directly adding NaOH-EDTA to the culture with or without prior addition of KCN and ice in *(b)*, (4) denaturation with formamide in *(a)*, and (5) extraction by the method of Nomura et al.[16] followed by alkali denaturation.

*Secondary structure of the replicating region.* Pulse-labeled DNA, isolated by the Thomas procedure but not subjected to denaturation, was analyzed by sedimentation in neutral sucrose gradients. A result obtained with *E. coli* B is shown in Figure 7. While most of the DNA isolated from the cells labeled with $H^3$-thymidine for ten minutes sedimented at a rate faster than δA DNA, having a sedimentation coefficient of 29S in 0.5 $M$ NaCl, pH 7.0, a considerable fraction of 15-second pulse DNA was recovered in a band sedimenting at a much slower rate. It was shown in other experiments that the fraction of the radioactivity found in the slowly sedimenting band decreased with increasing pulse time.

On the other hand, a large fraction of the DNA labeled by a short pulse was found to be susceptible to degradation by *E. coli* exonuclease I, which specifically hydrolyzes single-stranded DNA[17] (Table 1). Approximately the same fraction of the labeled DNA was eluted from hydroxylapatite at the relatively low phosphate concentration expected for single-stranded DNA and was found to be completely susceptible to the action of exonuclease I (Fig. 8 and Table 1). The susceptibility of unfractionated pulse DNA to exonuclease I and the fraction eluted from hydroxylapatite at low phosphate concentrations decrease with increasing pulse time (Table 1 and Fig. 8). Furthermore, the slowly sedimenting component of pulse DNA recovered from the neutral sucrose gradient was shown to be highly susceptible to exonuclease I, while the fast-sedimenting component had a low susceptibility to the enzyme (Table 1).

Thus an appreciable fraction of the newly synthesized material as isolated appears to be single-stranded, and this fraction is sedimented slowly in the neutral sucrose gradient.

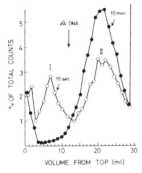

**Fig. 7.** Neutral sucrose gradient sedimentation of pulse-labeled DNA from *E. coli* B. Native DNA samples of Fig. 3 were sedimented for 15 hr at 20,500 rpm and 8°.

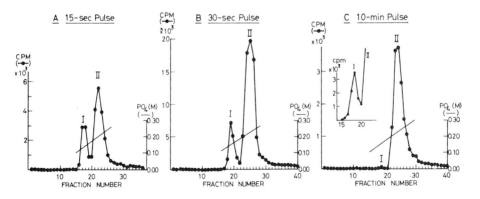

**Fig. 8.** Hydroxylapatite chromatography of pulse-labeled DNA of *E. coli* B. The native DNA samples of Fig. 3 were dialyzed against 0.01 $M$ potassium phosphate buffer, pH 6.8. Elution was achieved with a linear 0.01-0.7 $M$ gradient of the same buffer (total vol 140 ml). Fractions of 2.5 ml were collected at 30-min intervals.

## TABLE 1
*Susceptibility of pulse-labeled DNA to E. coli exonuclease I prior to denaturation treatment*

| | Extent of degration by exonuclease I (%) | | | | |
|---|---|---|---|---|---|
| | | Hydroxylapatite fraction | | Neutral sucrose gradient fraction | |
| Pulse time | Unfractionated | I | II | I | II |
| 5 Sec | 45 | | | | |
| 10 Sec | 32* | | | 77* | 24* |
| 15 Sec | 30 | 96 | 24 | | |
| 30 Sec | 24, 24* | 96, 88* | 18, 15* | | |
| 10 Min | 4, 0* | 78 | 2 | | |

E. coli B was pulse-labeled as in Fig. 2. Extraction and fractionation of labeled DNA were carried out as in Figs. 3, 7, and 8. SDS treatment for DNA extraction was 37°* or at 60°. Hydroxylapatite fractions I and II are shown in Fig. 8, and neutral sucrose gradient fractions I and II in Fig. 7.

For susceptibility to exonuclease, the 150-µliter reaction mixture, containing 10 µmoles glycine-KOH buffer, pH 9.2, 1 µmole $MgCl_2$, 0.15 µmole 2-mercaptoethanol, 60-µliter DNA sample (300-12,000 cpm), and 3 units of E. coli exonuclease I (DEAE-cellulose fraction), was incubated at 37°. After 60 min, 3 units of enzyme were added to the mixture and the incubation was continued for another 60 min. Acid-soluble and insoluble counts were determined at 0, 60, and 120 min. More than 85% of the radioactive DNA degraded during the 120-min period was already acid soluble at 60 min.

## Discussion

Average chain growth rate of E. coli chromosome is estimated to be about 400 nucleotides per second at 20°. Therefore, a 5 second pulse would label the stretches of about 2000 nucleotides, or a 0.05 per cent portion of the whole chromosome. Our experiments show that the portion of the chromosome, labeled by such a short radioactive pulse is separable in alkali from the bulk of chromosomal DNA as small molecules. Observations described in this and a previous paper[9] indicate that this represents an intermediary state in the formation of chromosomal DNA. This result conforms to the prediction from the replication mechanisms by which two daughter strands are synthesized discontinuously (Figs. 1C and D). The replication mechanism by which only one of the two daughter strands is synthesized discontinuously (Fig. 1B) is less likely, because virtually all the label is recovered in the slowly sedimenting component after the very short pulse. The sedimentation coefficient of the initially labeled material is 10-11S in various bacterial systems and 8-9S in the T4 phage system, suggesting that the length of the "unit" may be 1,000-2,000 nucleotides. This corresponds to the dimension of cistron.

Figure 9 illustrates a possible structure of the daughter strands in the vicinity of the growing end. "Units" synthesized at the growing point would be joined together by phosphodiester bonds to form longer strands located in the non-terminal position. The number of "units" and of chains with intermediate lengths would be determined by the relative rates of synthesis and of joining.

An alternative interpretation of our results is that artificial breaks may be introduced selectively in the newly replicated region during DNA extraction. This possibility, which in any case suggests selective weakness in the newly replicated region, is diminished by the fact that similar results are obtained using different methods in a number of different systems (including an endonuclease I-deficient E. coli strain).

Our results on native DNA do not distinguish clearly between the two mechanisms for discontinuous chain growth shown in Figures 1C and D. Although a fraction of the pulse-labeled DNA sediments at a much slower rate than the bulk of DNA in the neutral sucrose gradient, this material proved to be single-stranded. The remaining portion, which is in a duplex form, is not separated from the

**Fig. 9.** Schematic illustration of a possible structure of the daughter strand in the vicinity of the growing end.

bulk DNA by sedimentation. The fact that an appreciable fraction of the pulse DNA is isolated in the single-stranded form would imply either that most of the newly formed "units" exist as single strands in the cell or that the secondary structure of the replicating region containing these "units" is abnormally unstable. It may represent a unique state during replication or might indicate functioning of the newly synthesized "units" or the complementary portions of the parental strands as templates for RNA synthesis.

Our hypothesis of discontinuous DNA chain growth is encouraged by the discovery of polynucleotide-joining enzyme (ligase) in normal and T4 phage-infected *E. coli*.[18-21] The enzyme is encoded in one of the T4 genes previously implicated as a structural gene controlling DNA synthesis.[22] It has been used in *in vitro* synthesis of biologically active circular DNA in conjunction with DNA polymerase.[23-24] The synthesis and joining of the "units" assumed in our hypothesis may be carried out by DNA polymerase and polynucleotide ligase, respectively. A similar idea has recently been suggested by Kornberg and co-workers.[24,25] Further support for such hypotheses will await proof of the following: (1) the "units" are synthesized in the cell only by a reaction in the $5'$ to $3'$ direction; (2) the "units" are joined in the cell by the ligase reaction.

*Note added in proof:* Recent studies indicate that cells infected with temperature-sensitive mutants of T4 phage defective in ligase accumulate a large amount of the newly synthesized short DNA chains at 42°.

## REFERENCES AND NOTES

1. Cairns, J., J. Mol. Biol. **6**:208 (1963).
2. Cairns, J., in Cold Spring Harbor Symposia on Quantitative Biology, vol. 28 (1963), p. 43.
3. Nagata, T., these Proceedings **49**:551 (1963).
4. Yoshikawa, H., and N. Sueoka, these Proceedings **49**:559 (1963).
5. *Ibid.* **49**:806 (1963).
6. Bonhoeffer, F. B., and A. Gierer, J. Mol. Biol. **7**:534 (1963).
7. Lark, K. G., T. Repko, and E. J. Hoffman, Biochim. Biophys. Acta **76**:9 (1963).
8. Mitra, S., and A. Kornberg, J. Gen. Physiol. **49**:59 (1966).
9. Sakabe, K., and R. Okazaki, Biochim. Biophys. Acta **129**:651 (1966).
10. A filamentous bacteriophage specific to male strains of *E. coli*. The DNA extracted from this phage with phenol has a single-stranded circular structure and is infective to *E. coli* protoplasts (Okazaki, R., M. Morimyo, and K. Sugimoto, in preparation).
11. Okazaki, R., T. Okazaki, and K. Sakabe, Biochim. Biophys. Res. Commun. **22**:611 (1966).
12. Miyazawa, Y., and C. A. Thomas, Jr., J. Mol. Biol. **11**:223 (1965).
13. Thomas, C. A., Jr., K. I. Berns, and T. J. Kelley, Jr., in Procedures in Nucleic Acid Research (New York: Harper and Row, 1966), p. 535.
14. Bernardi, Nature **22**:779 (1965).
15. Lehman, I. R., J. Biol. Chem. **235**:1479 (1960).
16. Nomura, M., K. Matsubara, K. Okamoto, and R. Fujimura, J. Mol. Biol. **5**:535 (1962).
17. Lehman, I. R., and A. L. Nussbaum, J. Biol. Chem. **239**:2628 (1964).
18. Gellert, M., these Proceedings **57**:148 (1967).
19. Olivera, B. M., and I. R. Lehman, these Proceedings **57**:1426 (1967).
20. Gefter, M. L., A. Becker and J. Hurwitz, these Proceedings **58**:241 (1967).
21. Weiss, B., and C. C. Richardson, these Proceedings **57**:1021 (1967).
22. Fareed, G. C., and C. C. Richardson, these Proceedings **58**:665 (1967).
23. Goulian, M., and A. Kornberg, these Proceedings **58**:1723 (1967).
24. Goulian, M., A. Kornberg, and R. L. Sinsheimer, these Proceedings **58**:2321 (1967).
25. Mitra, S., P. Richard, R. B. Inman, L. L. Bertsch, and A. Kornberg, J. Mol. Biol. **24**:429 (1967).

# 9 ROLE OF POLYNUCLEOTIDE LIGASE IN T4 DNA REPLICATION

*Jean Newman*
*Philip C. Hanawalt*
*Department of Biological Sciences*
*Stanford University*
*Stanford, California*

Sakabe & Okazaki (1966) have reported that the newly replicated DNA in *Escherichia coli* 15T⁻ exists in relatively small 7 s single-stranded fragments and that these fragments are subsequently converted into material of high molecular weight. A similar result was found for *E. coli* strain B and for strain B infected with bacteriophage T4 (Okazaki, Okazaki, Sakabe & Sugimoto, 1967; Okazaki, Okazaki, Sakabe, Sugimoto & Sugino, 1968). It was postulated that semiconservative replication proceeds by the synthesis of short single-stranded fragments 1000 to 2000 nucleotides in length, and that these fragments are later joined end-to-end to the older contiguous strands of DNA. It seemed appropriate to implicate the joining enzyme, polynucleotide ligase, in this latter step. Such ligases have been reported in *E. coli* (Gellert, 1967; Olivera & Lehman, 1967) and are known to catalyze the joining of

From Journal of Molecular Biology 35:639-642, 1968. Reprinted with permission.

Our research was supported by the award of a U.S. Public Health Service grant GM 09901 to one of us (P. H.) and a predoctoral fellowship to the other (J. N.).

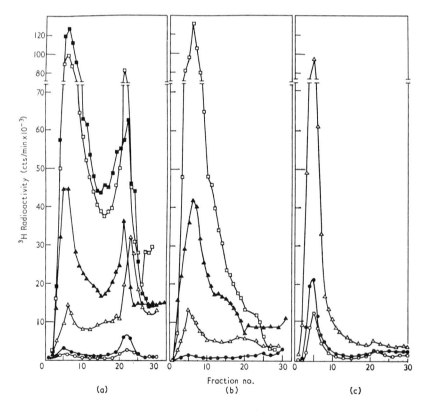

**Fig. 1.** [³H]Thymidine pulse experiments as described in the text. **a,** *ts* A80-infected cells pulse labeled at 43°C. **b,** *ts* A80-infected cells pulsed at 23°C. **c,** T4r⁺-infected cells pulsed at 43°C. Pulse times: — o — o — 5-sec; — ● — ● — 10-sec; — △ — △ — 30-sec; — ▲ — ▲ — 60-sec; — □ — □ — 120-sec; — ■ — ■ — 180-sec.

preformed polynucleotides through a 3'-5' phosphodiester bond. A ligase with similar properties is induced upon infection of E. coli with phage T4 (Weiss & Richardson, 1967; Becker, Lyn, Gefter & Hurwitz, 1967). Our experiments confirm the findings of Okazaki et al. (1967, 1968) on phage T4-infected E. coli and show that polynucleotide ligase is indeed involved in joining these small newly replicated DNA fragments to larger DNA segments.

A conditional lethal mutant of phage T4, ts A80, was obtained through the kindness of Charles Richardson. This mutant is known to direct the synthesis of a ligase that is functional at 25°C but inactive at 37°C (Fareed & Richardson, 1967). Exponentially growing cultures of E. coli B3/r were infected at a cell concentration of $2 \times 10^8$/ml. with ts A80 or with the wild-type T4$r^+$ at a multiplicity of 10. This leads to 99% infection within 20 minutes at 23°C. Pulse experiments indicate increased rates of DNA synthesis after 20 minutes at 23°C, showing that T4 DNA replication is in progress. At this time the culture was concentrated 20-fold by low-speed centrifugation. The concentrated cells were incubated for four minutes at either 23 or 43°C prior to the administration of a pulse label of [$^3$H]TdR (62.5μc of [$^3$H]TdR at 11c/m-mole to 1 ml. of concentrated cells). Incorporation of label was stopped and cell lysis effected by making the suspension of 1% Sarcosyl, 0.2 N-NaOH, 0.01 M-NaN$_3$ and 0.01 M-EDTA. Lysates were mixed by Vortex for two minutes and then layered on a 5 to 20% linear alkaline (0.1 N-NaOH) sucrose density-gradient over a 62% sucrose shelf (Smith & Hanawalt, 1967). Centrifugation was carried out in the SW25 rotor of a model L centrifuge at 24,000 rev./min for 16 hours at 8°C. Fractions were collected and cold 5% tricholoroacetic acid precipitates were trapped on Millipore HA filters for assay by liquid-scintillation counting.

Figure 1 indicates the distribution of pulse-label in different size classes of DNA as synthesized in the wild-type phage at 43°C and in the mutant at 23°C and at 43°C. It is evident that low molecular weight DNA fragments accumulate at the restrictive temperature in cells infected with the mutant ts A80, but not with the wild type, T4$r^+$. This strongly suggests that DNA is synthesized in short pieces and

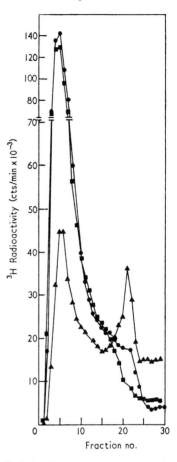

**Fig. 2.** Pulse-chase experiment, as described in the text. — ▲ — ▲ —, 60-sec pulse; — ■ — ■ —, 3-min chase at 23°C; — ● — ● —, 3-min chase at 43°C.

that ligase function is normally required for their conversion to high molecular weight DNA.

The fact that there is a slow conversion to high molecular weight material in the mutant at 43°C could be attributed to the normal function of the bacterial ligase, which might also act on replicating T4 DNA. It would seem likely that the T4-induced ligase if formed in order to accommodate the increased number of DNA growing points in the phage-infected cell.

For longer pulse times in the mutant at 43°C there is proportionately less accumulation of small fragments. It is possible that there has been some degradation of newly synthesized DNA that has not yet been joined to bulk DNA. Label uptake curves indicate that some breakdown does occur at 43°C in ts A80-

infected cells. Degradation of newly replicated DNA has been reported by Hosoda (1967) in *am* H39 (a T4 ligase amber mutant) infection of the non-permissive host and attributed to enzymes responsible for degradation of host DNA (Hosoda, 1967). Selective breakdown of pulse-labeled DNA following inhibition of DNA synthesis has also previously been reported for uninfected bacteria (Hanawalt & Brempelis, 1967).

T4 DNA synthesis is markedly reduced in *ts* A80-infected cells upon incubation at the restrictive temperature; although after only two minutes at $43°C$, no difference between wild type and the mutant is evident. After four minutes, the rate is depressed to 60% of that observed after two minutes at $43°C$, and after eight minutes the rate is further depressed to 32%. However, pulse labeling after only two minutes at the restrictive temperature yielded results qualitatively similar to those reported here.

Preparative CsCl density-gradient equilibrium sedimentation of the low molecular weight labeled material with an added $^{32}P$-labeled *E. coli* DNA density marker, verified that all of the $^3H$ label was in material of the expected density for T4 phage DNA.

Pulse labeling (60 seconds) of the mutant was followed with cold thymidine chases at $23°C$ and at $43°C$ by the addition of $10^4$-fold increased concentration of thymidine for 1.5 or 3 minutes. The results of the three-minute chase are seen in Figure 2. Although the results are complicated by additional incorporation of label during the chase, it is clear that pulse label is moving into the bulk DNA region at both temperatures. A similar result was seen in the 1.5-minute chase although the chase was not as complete. This result strongly implies that the short fragments are indeed precursors for material of high molecular weight material rather than a degradative product.

It is unlikely that the observed fragments are the result of another mode of DNA replication such as non-conservative repair, since repaired regions are much smaller in size and newly repaired regions are not released from the larger fragments upon denaturation (Pettijohn & Hanawalt, 1964).

Our results are in accord with the model proposed by Okazaki, that newly synthesized daughter DNA strands exist in short fragments that may be later "stitched" together. The polynucleotide ligase is specifically implicated in the stitching step in our experiments. Studies in progress are designed to determine whether one or both of the daughter strands are synthesized in this discontinuous fashion.

We have learned that R. Okazaki has performed similar experiments with essentially identical results (Note in Proof, Okazaki *et al.*, 1968).

**REFERENCES**

1. Becker, A., Lyn, G., Gefter, M. & Hurwitz, J. (1967). Proc. Nat. Acad. Sci., Wash. **58**:1996.
2. Fareed, G. & Richardson, C. C. (1967). Proc. Nat. Acad. Sci., Wash. **58**:665.
3. Gellert, M. (1967). Proc. Nat. Acad. Sci., Wash. **57**:148.
4. Hanawalt, P. & Brempelis, I. (1967). Proc. 7th Int. Congr. Biochem. p. 650. In the press.
5. Hosoda, J. (1967). Biochem. Biophys. Res. Comm. **27**:294.
6. Okazaki, R., Okazaki, T., Sakabe, K. & Sugimoto, K. (1967). Japanese J. Med. Sci. Biol. **20**:255.
7. Okazaki, R., Okazaki, T., Sakabe, K., Sugimoto, K. & Sugino, A. (1968). Proc. Nat. Acad. Sci., Wash. **59**:598.
8. Olivera, B. M. & Lehman, I. R. (1967). Proc. Nat. Acad. Sci., Wash. **57**:1426.
9. Pettijohn, P. & Hanawalt, P. (1964). J. Mol. Biol. **8**:170.
10. Sakabe, K. & Okazaki, R. (1966). Biochim. biophys. Acta **129**:651.
11. Smith, D. & Hanawalt, P. (1967). Biochim. biophys. Acta **149**:519.
12. Weiss, B. & Richardson, C. C. (1967). Proc. Nat. Acad. Sci., Wash. **57**:1021.

# QUESTIONS FOR CHAPTER 2

## Paper 5

1. The *E. coli* was first grown in $^{15}$N medium, and then transferred to $^{14}$N to observe the replication. Is this order critical, and why or why not?
2. Suppose one wanted to increase the quantitative discrimination of the experiment by the use of a radioactive label to monitor the amount of DNA in each density species. Would one label the $^{15}$N or the $^{14}$N? Why?
3. Could the same experiment be carried out with sheared DNA?
4. The experiments were carried out in an analytical ultracentrifuge.
   a. Could one employ label instead of UV to visualize the bands of DNA in this instrument?
   b. Could one carry out the experiments in a preparative centrifuge with a swinging bucket rotor? With an angle rotor?
   c. Could one carry out the experiment as described, but with ordinary optics?
   d. What modifications in the experiment would have to be made if no ultraviolet optics and/or ultraviolet-sensitive film were available?
   e. Where are the other components of the bacterial lysate, cell wall, membranes, proteins, RNA, and small molecules?
5. a. It is stated that the curves in Fig. 9, A, show that the heat-denatured DNA has twice the molecular weight of the native one. What is the basis for this statement?
   b. Draw the separate peaks of $^{14}$N and $^{15}$N in Fig. 9, C.
   c. Assuming that the concentration of native DNA in Fig. 9, A, is 1, what is the relative amount of heat-denatured DNA?
   d. It can be seen from Fig. 9 that the density of native $^{15}$N DNA is exactly the same as that of heat-denatured $^{14}$N DNA. What is the significance of this?

## Paper 6

1. It is stated in the paper that the bacteria were lysed in the usual dialysis chamber. What is the purpose of dialyzing the bacteria?
2. a. How was the radioautograph in Fig. 2 obtained?
   b. What order of magnitude must the concentration of the bacteria be in order to obtain a single chromosome like the one in Fig. 2, without interference from other chromosomes?
   c. How was the autoradiograph magnified?
   d. On what basis did the author decide which parts of the chromosome had one strand, and which had both strands labeled?
3. The paper states that the chromosome seen in Fig. 2 measures 1100 $\mu$. How is that length accurately determined?
4. In the bacterium, is the chromosome present in its extended state, as shown in Figs. 1 and 2?
5. Keeping in mind that the chromosome is a double-stranded helix, which of the loops A, B, and C (Fig. 2) have to move in order to unravel the strands?

## Paper 7

1. a. Why was bromouracil instead of $^{15}$N used as a density label?
   b. What are the limitations to the use of bromouracil?
   c. Would the experiment work equally well if $^{14}$C-thymine were used instead of $^{3}$H-thymine?
   d. Can you suggest why the Duponol procedure of Cairns (paper 6) was unsuccessful for the subsequent isolation of growing points?
   e. Why was CsCl at pH 10.5 used for the density centrifugations?
   f. Why was the CsCl-DNA solution layered with mineral oil before centrifugation?
2. It is suggested in the paper that the growing point would be particularly difficult to deproteinize. Why?
3. a. What would Fig. 2, A, look like if the fragmentation lines shown in Fig. 1 would go right through the growing point?
   b. Explain why, on rebanding, the intermediate-density fractions of Fig. 2, A, yield peaks both in the normal DNA position and in the density-labeled one (Figs. 3 and 4, A).
   c. After completion of the centrifugation, the gradients are dripped out through a pinhole in the bottom of the tubes. Conceivably, the DNA could be sheared in passing through the narrow orifice. If it were sheared, would that affect the results as shown in Fig. 2? Fig. 3? Fig. 4?
4. It is stated that the growing point is more resistant to isolation than the bulk of the DNA and will be selectively lost at the meniscus of the CsCl gradient centrifugation. It is also accumulated at the interphase when the lysate is extracted with octanol-chloroform. What do these observations mean?

69

5. It is stated that no denatured DNA has been found which could correspond to the separated strands ahead of the growing point. How big would this region have to be to stand a chance to be seen in experiments of this type?

### Paper 8

1. a. It is said that the DNA solution was concentrated by filtration through a collodion membrane. What is the common name of this procedure, and how does it work?
   b. What is the purpose of running alkaline sucrose density gradients? Why not CsCl gradients?
   c. Phage DNA was used as a marker, and its distribution was determined by assaying for infectivity of *E. coli* spheroplasts. Why was such a relatively laborious procedure chosen? Why was not an isotopically labeled DNA used?
2. a. From the data given in the paper, how many base pairs were synthesized by one *E. coli* cell in 90 minutes at 20° C? At 37°?
   b. What is the doubling time of DNA of *E. coli* in $T_4$ phage-infected cells?
3. a. In order to obtain the desired results, is it absolutely necessary to extract the DNA in the denatured state?
   b. What is the advantage of extracting DNA in the denatured state?
   c. Assuming a 90% yield, what was the amount of DNA put on the gradient in Fig. 2, A?
4. Assuming that the "11S" component is saturated with label at 100 sec, how big is it?

### Paper 9

1. a In the lysing mixture, which of the components actually effects the lysis?
   b. Why did the authors use a "shelf" of 62% sucrose under their alkaline sucrose gradients?
   c. What order of magnitude of total DNA was used for each gradient?
   d. Which tube numbers in Fig. 1 correspond to the small pieces, and why?
2. Can you offer any explanation for the fact that in the previous paper (8) there always remained the intermediate peak of small pieces, whereas in Fig. 1, the peak of small material diminishes after 120 sec?
3. a. Why is the incorporation into the phage DNA peak much greater in Fig. 1, *C*, than in Fig. 1, *A*?
   b. Why is the incorporation into the phage DNA peak at 23° C about the same as at 43°?
4. a. Draw the curve that would result if the material used for Fig. 1, *A*, 60 sec pulse were anlayzed on a CsCl gradient.
   b. Draw a figure of the same material on an alkaline CsCl gradient.
   c. Is there any point in running a CsCl gradient in this experiment?
5. a. What is the basis for the statement in the text that more incorporation of isotope occurred during the 3 min chase?
   b. The small pieces take only about 120 sec to synthesize, as can be seen from Fig. 1, *A*; yet it is stated that after 3 min the chase was not complete. What is the explanation of this discrepancy?
   c. What is the size of the small pieces synthesized in this paper?

# ANSWERS FOR CHAPTER 2

**Paper 5**

1. It is critical to grow the *E. coli* first in $^{15}$N medium for two reasons. First, $^{15}$N is an unnatural isotope, and the culture shows a lag when first transferred into heavy medium; it is thus better to allow it to adapt slowly by growing it for many generations in $^{15}$N. Second, in order to obtain isotopically clean DNAs, it is necessary to change the cellular pools of precursors as well as the external medium. This is easily done in going from $^{15}$N to $^{14}$N by adding bases and nucleosides instead of $NH_4Cl$, but it would be very difficult and unnecessarily expensive to obtain the $^{15}$N compounds necessary for carrying out the experiment in the opposite direction.

2. In order to increase the quantitative discrimination of the experiment, one would want to use radioactivity instead of UV for measuring the amount of DNA present, because the former is much more sensitive and would permit the use of much smaller amounts of DNA. Since both density species of DNA have to be visualized, both would have to be labeled. This can be done without interfering with the density label by adding $^{32}$P to both media. If, on the other hand, one wished to use radioactive label in order to reinforce the density label and to provide for easy quantitation of the amount of DNA synthesized after the transfer, then the label would have to be added to the $^{14}$N medium; for if it were added to the $^{15}$N, there would be no change in the pattern after the first round of replication, since hybrid $^{15}$N/$^{14}$N molecules remain present throughout.

3. The density of DNA is essentially independent of molecular weight, so that shearing would have no effect on the experiment. As a matter of fact, in the light of present knowledge, the DNA used in the Meselson-Stahl experiment, which had a given molecular weight of $7 \times 10^6$, was extensively broken up.

4. a. In the analytical ultracentrifuge the sample is in a metal cell with a quartz window to permit observation of the boundaries during centrifugation by a complicated optical system. There is no provision for determining the radioactivity of the sedimenting molecules.

   b. Yes, one could carry out the experiment in a preparative ultracentrifuge. Swinging bucket rotors have traditionally been used for gradient analysis, but recently it has been shown that it is also possible to use ordinary fixed-angle rotors for this purpose. The latter even have the advantage of having a higher capacity for the amount of DNA. Other recent improvements in the technique include the use of preformed gradients and greater speeds, both of which cut the time for reaching the equilibrium position by severalfold.

   c. No, one could not carry out the experiment as described with ordinary optics. These generally take advantage of the differences in refractive index between the solution and the solvent, and the concentrations of DNA used are too small to be visible by these means.

   d. If no ultraviolet optics or UV-sensitive film to record the bands as they sediment were available, the experiment would have to be carried out in a preparative ultracentrifuge, for in this case the gradient is dripped out of the tube through a pinhole punched in the bottom, and the UV adsorption of the collected fractions can then be determined in a spectrophotometer.

   e. As stated in the paper, the middle of the tube, where the DNA is located, has a density of about 1.7. The density of RNA is greater than 1.8, so it would be at the bottom of the tube, whereas protein, with a density of about 1.2, would form a pellicle at the top. The cell walls in the lysed bacteria would be largely intact and would therefore not be in solution; due to their size, they would probably pellet. The membranes, on the other hand, are dissolved by the sodium dodecyl sulfate, and would float. The small molecules, due to their high diffusion coefficient, would be distributed throughout the cell, even though the free bases and their derivatives have densities similar to those of DNA, which would make them peak in the DNA region. Fortunately, their concentrations are too small to interfere with the optical density measurements.

5. a. If $\rho_0$ is the midpoint of the buoyant density distribution of a given molecular species, the molecules form a Gaussian distribution around it. The exact shape of the distribution depends to a large extent on the diffusion coefficient of the molecules. The variance $\sigma$ (the width of the Guassian peak at half height) is related to the molecular weight (the lower the molecular weight, the higher

the diffusion) as follows:

$$\sigma^2 = \frac{RT}{M(d\rho/dr)\bar{v}\omega^2 r_0}$$

where R is the gas constant, T the absolute temperature, $\rho$ the buoyant density of the molecule, r the absolute density, $d\rho/dr$ the density gradient, $\bar{v}$ the partial specific volume, $\omega$ the angular velocity of the centrifuge, and $r_0$ the density at the center of the Gaussian distribution.

When two molecules of the same type are compared, then the equation becomes simply

$$\frac{\sigma^2}{\sigma_1^2} = \frac{M_1}{M}$$

In Fig. 9, A, the ratio of the $\sigma^2$ is 2.55. Since the size of the Gaussian peak for the denatured species has to be corrected for the 40% increase in O.D. upon denaturation, the ratio is indeed close to 2, indicating a halving of the molecular weight upon separation of the strands.

b. In the overlap region between the two peaks in Fig. 9, C, the optical density is the sum of the contributions of the two peaks. The outline for each peak, therefore, is considerably lower than the composite outline, as shown in part B of the figure below; the

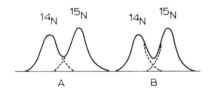

outline shown in part A is often seen, but it is incorrect. If Gaussian peaks can be expected, the correct outline of each component can be obtained by drawing a mirror image of the outer edge. If the sums of the contributions of the two peaks should not correspond to the experimental composite outline, the chances are that a third component is present.

c. The area of a Gaussian peak can be approximately determined by multiplying its height, h, by its width at h/2. Other methods generally applicable to all kinds of peaks are to overlay the drawing with graph paper and count the squares; to cut out the peaks, weigh them, and compare them with a standard; or to measure the area with a planimeter. In the case at hand, the peak of native DNA measures 3.53 cm$^2$, and that of denatured DNA, 6.23 cm/$^2$. Since, however, the optical density of DNA increases 40% upon denaturation, there is 4.82/3.53 or 1,365 times more DNA in the denatured peak than in the native one.

d. The fact that the density of the native $^{15}$N DNA is exactly the same as that of denatured $^{14}$N DNA is purely coincidental.

Paper 6
1. It had been found in a previous paper (Cairns, J. Mol. Biol. **4**;407 [1962]) that if the bacteria are lysed in a dialysis chamber of a certain construction (a cylinder, closed on one side with a Millipore filter against the lysing solution), the DNA liberated from the bacteria is adsorbed onto the membrane in an extended configuration, making it possible to remove it for autoradiography without exposing it to any handling that might shear it. By dialyzing the bacteria against the Duponol, they lyse slowly.

2. a. The bacteria were lysed in chambers as described above. The Millipore membranes with their adsorbed DNA were fixed onto microscope slides and overlaid with a radiation-sensitive strip film. After an exposure of two months, the films were developed, and examined under the microscope. Of the many molecules on the membranes, the best were selected and photographed through the microscope.

   b. The chromosome has a diameter of about 350 $\mu$. In order to allow room around it, let us assume that each chromosome occupies a square of 500 $\times$ 500 $\mu$, or 0.25 mm$^2$. The membrane has a diameter of 20 mm, or a surface of 62.8 mm$^2$, which could accommodates 250 chromosomes. The minimum number of bacteria would therefore also be 250, or, since the volume of the chamber is 157 mm$^2$, 0.8 $\times$ 10$^3$/ml. Actually, the number should probably be greater, since not all the liberated chromosomes will reach the membrane. The concentration used was, in fact, 10$^4$ bacteria per milliliter.

   c. The autoradiograph was magnified by a light microscope.

   d. The author counted the density of photographic grains, and on the basis of these average numbers decided which parts were labeled in both strands, and which only in one.

3. The length of the chromosome was accurately determined by measuring a blowup of known magnification. For example, a slide of the autoradiograph might be projected on a wall and the contour measured with a flexible tape measure;

or a large print might be obtained, and the contour traced and measured with an appropriate instrument; or else the contour might be cut out so as to obtain a fine ribbon that can be stretched out and measured accurately. The same methods are used to measure electron micrographs of DNA for molecular weight determinations.

4. The chromosome is stated to have a contour length of 1100 $\mu$, which corresponds to a diameter of 350 $\mu$. An *E. coli* spheroplast measures only about 10 $\mu$ in diameter, and only one fifth of it is occupied by DNA. The chromosome inside the bacterium must therefore exist in a highly convoluted and tightly packed state.

5. Make a double helix from two twisted pieces of string. You will notice that it can be unraveled by grasping the ends and unwinding; it can also be unraveled, however, by firmly grasping the ends, applying a finger to the fork, and pressing downward. In this way, unwinding is accomplished not by moving the strands around each other but by rotating the helix. Given the high viscosity of the nucleoplasm, the extreme length of the DNA and the close proximity of the DNA fibers to each other, it is extremely unlikely that unwinding takes place by moving the strands relative to each other, if for no other reasons than that the operation becomes increasingly more difficult as the distance from the ends increases. It is much more likely that a force is applied to the fork (possibly the replicating enzyme itself) and the double helix rotates as it is replicated. This corresponds to loop B (Fig. 2). If the above experiment is repeated with the ends of the twisted string fastened together so as to form a circle, one finds that as one presses the finger down the fork, the helix becomes progressively more tightly twisted until it becomes impossible to proceed. This is the reason for Cairns' statement that a "swivel" was needed for the replication of a circle. A single polynucleotide strand differs from a string, however, in that it can rotate freely around any of its P-O single bonds. Therefore, as long as one of the strands of the double helix is "nicked," the helix becomes freely movable around a P-O bond, and no "swivel" is necessary. It is now known, in fact, that a single-strand scission precedes the replication of double-stranded circular DNA.

## Paper 7

1. a. To obtain enough of a density difference with $^{15}N$, it is necessary to grow the bacteria on $^{15}NH_4Cl$ as the only N source. This precludes the use of auxotrophs in the experiments. It also precludes obtaining a sudden difference in density by merely adding $^{15}NH_4Cl$ to the medium. It is always necessary to centrifuge the cells and resuspend them in heavy medium. Bromouracil, on the other hand, offers all the advantages and the flexibility of a radioactively labeled compound.

   b. The only disadvantage of bromouracil is that it is mutagenic, and more so with some strains than with others. Occasionally this complicates its use.

   c. $^3H$-thymine, instead of $^{14}C$-thymine, is used as the "base" label because the activity curves of $^3H$ and $^{32}P$ do not overlap. It is, of course, possible to do the same experiment with $^{14}C$, but the $^{14}C$ counts must be corrected for overlapping $^{32}P$ counts. Such corrections are not only extremely time-consuming but they also lessen the confidence in the counts, especially when an isotope of short half-life like $^{32}P$ (14 days) is involved.

   d. The procedure described by Cairns yields DNA adsorbed onto the dialysis membranes, which would drastically curtail the yield of DNA in solution. Perhaps the growing points have some characteristics which make them adsorb preferentially to the membranes.

   e. The high pH, together with the high salt concentration of the CsCl, effects an additional deproteinization, thereby cleaning up the gradient and increasing the yield, especially of the growing point. Yet, at pH 10.5, the pH is not high enough to cause separation of the strands.

   f. The smallest tube for swinging buckets holds 4.5 ml, which is much more than is needed to form a CsCl gradient. The tube is therefore filled with mineral oil, which is very inert, does not mix with water, and will not penetrate the solution during the centrifugation. If the tube is not completely full, it will collapse in the vacuum during the centrifugation, and ruin the run.

2. It seems, offhand, that the replicating enzyme would sit in the growing fork. This is the assumption of the authors. However, now that it is likely that the replication of DNA proceeds by a much more complicated mechanism, it seems much less certain that the protein that sticks so stubbornly to the growing point is DNA polymerase.

3. a. If fragmentation would go right through the growing point, there would be no $^{32}P$ on the heavy side of the $^3H$ peak.

   b. In any kind of separation, the true position of a material is given by the middle of the peak. The position of the sides depends mostly on the amount of material, although the diffusion coefficient also plays a role in determining the width of a peak. At any rate, the tails of the peak may be quite far from the center. Now, when the tails of two peaks overlap, the counts, or whatever is used in the quantitation, add up and often give the appearance of a third component between the two. This can be resolved by rerunning the intermediate material. If a third component is present it will run true; if, however, the material constituted the tails of other peaks, it will move to the center position of the parent peak, and now, because there is much less material, the peaks will be less broad, and as a consequence, no intermediate material will appear in the second run. In Fig. 2, A, for example, there appears to be some $^{32}$P-labeled material of intermediate density, as well as a little $^{32}$P-labeled material of normal density. both of which, according to Fig. 1, originate from the growing region. A rerun (Fig. 4, A) gives the same picture. In Fig. 2, B, on the other hand, the material has been chased with cold P in the presence of bromouracil and should therefore contain no $^{32}$P in either the intermediate or the normal density region. Figure 2, B, is unclear, but a rerun of the intermediate region (Fig. 3) clearly shows that all the $^{32}$P is in the heavy peak.

c. Shearing of the DNA after the centrifugation clearly would not affect Fig. 2, because the separation has already taken place, and shearing would not affect either the counts or the optical density of any fraction. Shearing might have affected Fig. 3, because although it would be expected not to affect the density of the main peaks, it might reduce the amount of intermediate material; but then the results of Fig. 4, A, show that such has not been the case. Fig. 4, B, again, would not be affected, for the sonication did precisely what shearing would have done.

4. These observations mean that the growing point is more strongly associated not only with protein but with lipid as well.

5. It can be stated, as a general rule for most separation procedures, that a minor component has to be present at a concentration of 10% of the major one (5% at the outmost) in order to be detected. Given the molecular weight of $2 \times 10^9$ for the DNA of *E. coli*, 5% would represent a piece of $10^8$. This is far too big, for even if there is a single-stranded region, it is hard to imagine that it would encompass more than 1,000 to 2,000 base pairs, corresponding to $6.4 \times 10^6$ and $12.8 \times 10^6$, respectively, or 0.032% to 0.064%.

Paper 8

1. a. Ultrafiltration is the forced passage of a solution through a collodion membrane, which allows water and small molecules to pass, but retains macromolecules. Ultrafiltration thus results in the concentration of a solution with respect to macromolecules, while there is no change in the concentration of small solutes.

   b. The purpose of running alkaline sucrose gradients is to separate small single strands, which might be annealed to a long strand, from the large pieces. An alkaline CsCl gradient, although it would also denature the DNA, would separate out the small pieces only if they had a strikingly different base composition, for density is independent of molecular weight.

   c. It would have been perfectly feasible to use a $^{32}$P-labeled phage DNA as a marker. The reasons for not doing so must be due to the particular setup of the laboratory or to some other, trivial reason.

2. a. One base pair has an average weight of 640 daltons, and each *E. coli* cell has on the average two chromosomes weighing $4 \times 10^9$ daltons. Each cell, therefore, has $6.25 \times 10^6$ base pairs. At 20° the generation time of this particular strain is 180 min, that is, in that time the DNA doubles. Thus $6.25 \times 10^6$ base pairs are synthesized in 180 min, *$3.125 \times 10^6$ base pairs* in 90 min. At 37° C the rate of DNA synthesis is six times faster; thus *$1.875 \times 10^7$ base pairs* are synthesized in 90 min.

   b. T$_4$ phage infection stops endogenous DNA synthesis.

3. a. It is stated in the paper (p. 63) that the results were not changed by extracting the DNA first and then denaturing it.

   b. Because the alkaline pH decreases aggregation and aids in the deproteinization, the yield is probably better if the DNA is extracted in the denatured state.

   c. As was stated above, each bacterium contains $4 \times 10^9$ daltons of DNA; $6.023 \times 10^{23}$ bacteria, therefore, contain $4 \times 10^9$ gm of DNA. For the experiment in Fig. 2, 10 ml containing $5 \times 10^8$ cells/ml were used, which would contain $3.32 \times 10^{-6}$ gm or 3.32 μg of DNA. A yield of 90% of that would be 2.99 μg.

4. From the results in 2a it can be calculated that at 20°C an *E. coli* cell synthesized $5.785 \times 10^2$

base pairs/sec. In the present case only a single strand is being laid down, but according to the model of this paper, the two strands are being synthesized by different polymerase molecules, and therefore the rate of synthesis of the single strand should be the same as that of the total DNA, that is, $5.785 \times 10^2$ bases per sec. In 100 sec, then, there would be 57,850 bases, weighing $57,850 \times 320 = $ *$1.85 \times 10^7$ daltons.* This is surprisingly large, but since the bacteria used in this study were not defieient in polynucleotide ligase, it is impossible to say whether the original pieces laid down were that large (see paper 9). Furthermore, the 11S piece may be completed in considerably less time than the 100 sec assumed here, which would make it smaller.

## Paper 9

1. a. Sarcosyl is a detergent and is the effective lysing agent in the lysing mixture.
   b. A "shelf" of concentrated sucrose is commonly used with sucrose gradients to prevent the pelleting of large molecular weight material, and the disturbances resulting therefrom. A pellet in the gradient tube interferes with the analysis of the gradient, for even if it does not clog up the pinhole through which the gradient is withdrawn, it might leach small amounts of material into all the fractions as they come in contact with the pellet. Furthermore, a pellet does not give a peak, and is therefore not analyzed with the gradient. If, on the other hand, there is a "shelf" or "cushion" of concentrated sucrose, either the heavy material will not penetrate the density barrier and will remain as a stationary peak in the interface, or if it penetrates the sucrose, its sedimentation will be sufficiently slowed to prevent pelleting.
   c. The culture contains $2 \times 10^8$ cells, but just before use it is concentrated 20 times and 1 ml is taken, that is, $4 \times 10^9$ cells. Each cell is infected with 10 $T_4$ phages. The cells are harvested "after $T_4$ replication is in progress." For the sake of calculation, let us then assume that 20% of the final yield has been synthesized; assuming a burst size of 100, each cell should contain 20% of 1,000, or 200 $T_4$ DNA equivalents. (This may be an overestimate, for not all phages might produce progeny at this high multiplicity of infection.) The total amount of DNA is thus:

$$\frac{4 \times 10^9}{0.6023 \times 10^{24}} [4 \times 10^9 + (200 \times 1.48 \times 10^8)] \text{ gm} = 2.2 \times 10^{-6} \text{ gm or } 2.2 \text{ μg}$$

where $0.6023 \times 10^{24}$ is Avogadro's number

N, $1.48 \times 10^8$ is the molecular weight of the $T_4$ DNA, and $4 \times 10^9$ is the total amount of host DNA per cell as well as the number of cells.
   d. The small material is in tubes 21 to 25. Since the gradient is dripped out through the bottom, tube 1 is the heaviest and tube 30 is the lightest.
2. The decrease in the amount of label in the small materials peak in Fig. 1, A, is due to the fact, stated in the paper, that the mutant $T_4$ grows well at the restrictive temperature for the first few minutes, but that after 2 min (120 sec) the rate of incorporation drops markedly.
3. a. The difference in counts in the phage DNA in Fig. 1, A and C, is largely due to the fact that there is no peak of small materials in Fig. 1, C, and that the main peak is much narrower in Fig. 1, C. When all the counts are added over the whole gradient, the differences are within the limits of error. For example, for the 30 sec pulse, there are 245,000 cpm for Fig. 1, A, and 330,000 cpm for Fig. 1, C.
   b. Although it is true that usually the incorporation increases sharply with temperature, in this case 23° is a permissive temperature, whereas 43° is not only above the optimum but is a restrictive temperature. When the counts are added over the whole gradient, there are in fact somewhat more counts in Fig. 1, A, than in 1, B. Otherwise the equality of the incorporation at the same temperatures is largely coincidental and of no profound significance.
4. a. From the calculations in question 1c, it is apparent that there is about seven to eight times more phage than host DNA. In the CsCl gradient, therefore, there would be one large peak at the density of the phage DNA, and a very low one at the density of the host.
   b. In alkaline CsCl the results would be the same, except that both peaks would be displaced 0.016 unit of density toward the heavier side, due to the separation of the strands.
   c. As stated in the paper, it was important to show that the small pieces had the same density as the phage DNA, for this indicates that they are precursors of the bulk DNA.
5. a. The basis for the statement in the text that more incorporation occurred during the chase is that the increase after the chase in the main peak of Fig. 2 is much greater than the sum of the two peaks seen after the 60 sec pulse.
   b. The necessary and sufficient test for a precursor-product relationship is that the presumptive intermediate is chased as fast as

it is synthesized. Since the small material behaves like a precursor, it is surprising that it takes so long to be chased completely into the main product. A possible explanation might be that there is a great excess of label available, and even a $10^4$ dilution still leaves a significant amount to be incorporated. It may also be that some of the labeled precursors are in a metabolic compartment and are therefore not as efficiently diluted as one would expect.

c. There are no data given in the paper that would permit one to calculate the size of the small precursor piece. It is stated in the beginning of the paper that the precursors are 1,000 to 2,000 nucleotides long.

CHAPTER 3

# SYNTHESIS OF DNA IN VITRO

The synthesis of DNA *in vitro* is best discussed in terms of the enzymes known to synthesize and modify deoxypolynucleotides. The term "DNA" in this context means a piece of deoxyribonucleic acid, rather than a molecule of DNA, and "synthesis of DNA *in vitro*" should not be taken to imply that entire molecules of DNA—that is, entire genomes—can be synthesized in the test tube, although a very simplified genome, that of the single-stranded, circular virus $\phi$X174, has been thus made.

The first enzyme capable of synthesizing a piece of DNA was DNA polymerase, discovered by Kornberg in 1957. The enzyme needs both a template and a primer, and synthesizes antiparallel, complementary chains building onto the 3'OH group of the primer, according to the overall reaction:

$$DNA_n + dXTP \underset{}{\overset{Mg^{++}}{\rightleftharpoons}} DNA_{n+1} + PP_i$$

The enzyme is capable of very extensive net synthesis of DNA, and the product is all double-stranded, antiparallel, and strictly complementary. Furthermore, the new DNA does have the characteristics of the template as far as base composition and nearest neighbor frequency is concerned. Because of this, it was assumed in the early years that DNA polymerase was *the* DNA replicating enzyme. As work progressed, however, this became very much less certain. For one thing, despite enormous efforts, it proved impossible to synthesize biologically active DNA, that is, transforming DNA, although it was possible to use the enzyme to restore biological activity previously lost by partial nuclease digestion. Furthermore, the synthetic DNA differed from natural DNA in that it was capable of instant renaturation after melting. This indicated that the strands were not separate, but that single strands were bent back on themselves. Electron microscopy of the synthetic product showed that, in fact, the material consisted of many branches formed by hairpin turns. Natural DNA, by contrast, was completely unbranched.

For a while, therefore, DNA polymerase was relegated to the secondary status of a "repair enzyme," for despite intensive efforts, no other replicase turned up until recently. Success was at hand when some polymerase-less mutants were discovered that managed nevertheless to replicate their DNA at a normal rate. These mutants possess a DNA polymerase activity in the membrane fractions. This activity is capable of synthesizing DNA at near *in vivo* rates but continues for a short time only. However, even though DNA polymerase is not the replicase, it is by far the best-known enzyme involved in DNA metabolism.

DNA polymerase has been very thoroughly purified and intensively studied over the years by Kornberg and his associates. The picture which has emerged is that it is a very complicated enzyme, capable of three other activities besides its synthetic one: a pyrophosphate exchange, a pyrophosphorolysis of the DNA from the 3'OH end, and a simple hydrolysis of the DNA. The reactions involved are as follows:

1. $dXTP + DNA_n + PP_i^* \rightleftharpoons dXTP^* + DNA_n + PP_i$
2. $DNA_n + PP_i^* \rightleftharpoons DNA_{n-1} + dXTP^*$
3. $DNA + H_2O \longrightarrow DNA_{n-1} + dXMP$

The hydrolysis (3) can proceed both from the 3' and the 5' ends. The latter can be made the exclusive reaction by phosphorylating the 3'OH.

The enzyme is apparently formed by a single polypeptide chain with a molecular weight of 109,000, and it possesses only one active center. Within this center, however, there are several active sites in which the various reactions take place: a template site, a primer site, a primer terminus site, a triphosphate site, and a $5'$ phosphate site.

The enzyme binds to the DNA exclusively through the primer terminus site, and the prime requirement for binding at this site is a $3'$OH group in the "ribose" conformation. The number of polymerase molecules (determined through a radioactive Hg combined with an unessential SH group) bound to a DNA molecule thus provides an accurate count of the number of ends or "nicks" possessing a $3'$OH group. This site also binds deoxymonophosphates, and the binding of DNA is competitively inhibited by all four deoxymonophosphates. It is in this site that hydrolysis of the primer from the $3'$OH end takes place, for this hydrolysis, too, is inhibited by dXMP.

The triphosphate site is adjacent to the primer terminus site, and it is the synthetic site. It "sees" the triphosphate part of the substrate, being essentially insensitive to the sugar and the base. The question thus arises how the insertion of the correct nucleoside triphosphate is ensured. The best guess is that the enzyme responds allosterically to some feature of a base pair, rather than to an individual nucleotide. When the correct base pair has been formed, it could trigger some change in conformation which signals the formation of a phosphodiester bond.

The pyrophosphorolysis of the primer from the $3'$OH end is essentially a reversal of the synthetic step and also takes place in the triphosphate site. The pyrophosphate exchange, on the other hand, does not involve an enzyme-triphosphate complex, and must therefore take place in a transition state when the $3'$OH of the primer is in the process of attacking a triphosphate not yet settled in the triphosphate site.

The $5'$ phosphate site is above the triphosphate site, as can be demonstrated by the fact that when a deoxypolynucleotide blocked in the $3'$OH position and terminated by a $5'$ triphosphate, instead of a phosphate, is hydrolyzed by the polymerase, the triphosphate end "slips down" into the triphosphate site, and the first hydrolysis product, instead of being a monophosphate, is a dinucleotide.

It is obvious from these properties that a completely intact double helix cannot be replicated. In fact, $T_7$ DNA, which is a linear double helix when prepared with extreme care, is not a substrate for the polymerase. Because of its requirement for a $3'$OH primer, the polymerase cannot start a strand *de novo*. Therefore, the original picture of the replication of a Watson and Crick helix, that is, the separation of two intact strands and the simultaneous replication of each, cannot be true at the enzymatic level.

The DNA polymerase of *E. coli* is by far the most thoroughly studied, but there are other polymerases. For example, the T-even viruses make their own polymerases, and the one induced by $T_4$ has been isolated and well studied. It differs from the *E. coli* polymerase in only one way: it cannot hydrolyze the primer in the $3' \to 5'$ direction. Because of this, it cannot utilize "nicks" in the double-stranded structure to start replication, for it cannot hydrolyze the 5-P-containing nucleotide to make room for a triphosphate.

When polymerase is incubated with nucleoside triphosphates in the absence of template and primer, it synthesizes certain polymers after a long lag. Thus, in the presence of dATP, for example, poly dA is made; if dATP and TTP are present, a poly dAT is made, in which the A and T residues are strictly alternating; and in the presence of dGTP and dCTP, the resulting polymer has one strand of poly dG and the other of poly dC. All these synthetic DNAs are efficient templates for the production of more polymer.

When these polymers first appeared, it was hypothesized that a small oligonucleotide was synthesized first, and that it subsequently grew by a mechanism of slippage: one strand slips with respect to the other, leaving at each end a single-stranded stretch which is then "repaired" by the polymerase, and so on. This mechanism has now been proved, for Khorana and his group have shown that any deoxyoligonucleotide would give rise to a polymer showing a strict repetition of the nucleotide sequence of the oligonucleotide, provided it was double-stranded and the sequence was repeated twice.

Polymers containing all the triplets constituting the genetic code were thus prepared.

Macromolecules, like cells and organisms, are not merely replicated, but between birth and death they have a life history, in the course of which they are extensively modified. DNA is no exception, and in addition to the replicase and polymerases, there are many enzymes that modify existing DNA molecules. These enzymes may be grouped into two classes—those that modify the ends of the chain, and those that modify the bases.

Among the enzymes of the first group there are phosphatase III, which specifically removes a phosphate from the $3'$OH, and polynucleotide kinase, which specifically phosphorylates a $5'$OH end group according to the reaction:

OH-Xp...pX + ATP → P-O-Xp.....pX + ADP

The enzyme polynucleotide ligase joins a $5'$ phosphate of one chain to the $3'$OH of another, if they are apposed in a double-stranded structure. The enzyme does not join two single-stranded polynucleotides. The *E. coli* enzyme uses DPN (nicotinamide adenine dinucleotide) as a coenzyme, and the reaction takes place in different steps:

1. N-P-O-P-A + Ligase ⇌ Ligase.AMP + NMN
2. Ligase.AMP + P-O-XpXp...X ⇌
   Ligase + AMP-O-P-XpXp...X
3. pXpX...pX-OH + AMP-O-P-O-P-XpXp....pX →
   pXpX...pXpXpXp....pX + AMP

The T-even phages induce their own ligase in *E. coli*; the $T_4$ ligase differs from the *E. coli* one in that it utilizes ATP instead of DPN as a cofactor.

In mammalian cells there are a number of other end-modifying enzymes, nucleotidyl transferases, which transfer both deoxynucleotides and ribonucleotides to the ends of DNA, and synthesize various oligonucleotides or homopolymers on the ends. The function of these enzymes is not known, and it is not known whether they have the same specificity *in vivo*, but they are useful adjuncts to our arsenal of isolated enzymes.

Among the second class of modifying enzymes, those modifying the bases, there are various phage glucosylating enzymes, which add glucose residues (from UDP-glucose) to the hydroxymethylcytosine residues of the phages.

The most important and ubiquitous enzymes, however, are the methylating enzymes, which transfer methyl groups from methionine, through S-adenosylmethionine, to various bases, according to the overall reaction:

S-adenosylmethionine + R →
$\qquad$ R-CH$_3$ + Ado + Homocysteine

There are many different enzymes with different specificities, some being specific for purines, or for pyrimidines, and others, for only a single base. One of the outstanding characteristics of methylating enzymes is that they are strictly species-specific, that is, they will methylate bases in different positions, depending on the origin of the DNA. For example, *E. coli* methylating enzymes will not use natural *E. coli* DNA as a substrate because all the positions that can be methylated have already been methylated *in vivo*; it will, however, add further methyl groups to DNA from another organism, and the DNA from *E. coli* can likewise be further methylated by a heterologous enzyme. Since only about one in a thousand or more bases in methylated, and because of the species specificity, the methylating enzymes are the most specific enzymes known. They must not only be capable of recognizing a rather large stretch of base sequences—for a more limited sequence would be repeated many more times in a DNA—but they must also be capable of recognizing tertiary structure.

The exact function of the methylating enzymes is not known, for undermethylated DNA seems to be biologically active, but they must be involved in species specificity.

Given the existence of all these modifying enzymes, the picture that emerges is far from the previously held notion of the extreme stability of the genetic material. In addition to an enzymatically complicated mechanism of replication, very different from the popularly pictured "zipper" mechanism, the DNA appears to be in a constant state of flux, being constantly worked over by a variety of enzymes capable of repairing local damage imparted by degradative enzymes, radiation, and other damaging factors in the environment.

# 10 ENZYMATIC SYNTHESIS OF DEOXYRIBONUCLEIC ACID;
# I. PREPARATION OF SUBSTRATES AND PARTIAL PURIFICATION OF AN ENZYME FROM ESCHERICHIA COLI

I. R. Lehman*
Maurice J. Bessman**
Ernest S. Simms
Arthur Kornberg
Department of Microbiology
Washington University School of Medicine
St. Louis, Missouri

In considering how a complex polynucleotide such as DNA[1] is assembled by a cell, the authors were guided by the known enzymatic mechanisms for the synthesis of the simplest of the nucleotide derivatives, the coenzymes. The latter, whether composed of an adenosine, uridine, guanosine, or cytidine nucleotide, are formed by a nucleotidyl transfer from a nucleoside triphosphate to the phosphate ester which provides the coenzymatically active portion of the molecule (1, 2). This condensation, which has been regarded as a nucleophilic attack (3) on the innermost or nucleotidyl phosphorus of the nucleoside triphosphate, results in the attachment of the nucleotidyl unit to the attacking group and in the elimination of inorganic pyrophosphate. By analogy, the development of a DNA chain might entail a similar condensation, in this case between a deoxynucleoside triphosphate with the hydroxyl group of the deoxyribose carbon-3 of another deoxynucleotide. Alternative possibilities involving other activated forms of the nucleotide (as, for example, nucleoside diphosphates which have proved reactive in the enzymatic synthesis of ribonucleic acid (4)) were not excluded.

Earlier reports (1, 2, 5-7) briefly described an enzyme system in extracts of *Escherichia coli* which catalyzes the incorporation of deoxyribonucleotides into DNA. Purification of this enzyme led to the demonstration that all four of the naturally occurring deoxynucleotides, in the form of triphosphates, are required. In addition, polymerized DNA and $Mg^{++}$ were found to be indispensable for the reaction. Deoxynucleoside diphosphates are inert; and as a further indication of the specificity of the enzyme for the triphosphates, the synthesis of DNA is accompanied by a release of inorganic pyrophosphate, and reversal of the reaction is specific for inorganic pyrophosphate.

These considerations have led to a provisional formulation of the reaction as follows:

$$\begin{bmatrix} dTP^*PP \\ dGP^*PP \\ dCP^*PP \\ dAP^*PP \end{bmatrix} + DNA \rightleftharpoons DNA - \begin{bmatrix} dTP^* \\ dGP^* \\ dCP^* \\ dAP^* \end{bmatrix} + 4(n)PP$$

The purpose of this report is to describe in detail the methods for the partial purification and assay of the enzyme from *E. coli* and for the preparation of the substrates for the reaction. In order to facilitate reference in this report, the enzyme responsible for deoxyribonucleotide incorporation is designated as "polymerase." The succeeding report will present evidence for the net synthesis of the DNA and other general properties of the system.

## Experimental

### Preparation of substrates

*Preparation of $P^{32}$-labeled deoxyribonucleotides.* DNA was isolated from *E. coli* cells grown to a limit on $P^{32}$-orthophosphate. The DNA was then degraded

---

From Journal of Biological Chemistry **233**:163-170, 1958. Reprinted with permission.

We gratefully acknowledge the grants from the Public Health Service and the National Science Foundation which made this work possible.

*Fellow of the American Cancer Society.

**Fellow of the National Cancer Institute, Public Health Service.

[1] The following abbreviations are used: ATP, adenosine triphosphate; dATP or dAPPP, deoxyadenosine triphosphate; dCTP or dCPPP, deoxycytidine triphosphate; dGTP or dGPPP, deoxyguanosine triphosphate; DNA, deoxyribonucleic acid; DNase, deoxyribonuclease; P*, $P^{32}$-labeled phosphate; $P_i$, inorganic orthophosphate; Tris, tris(hydroxymethyl)aminomethane; dTTP, dTPPP, thymidine triphosphate.

to 5'-mononucleotides, which were separated by anion exchange chromatography. 100 ml. of glycerol-lactate medium (8) containing $8 \times 10^{-4}$ M phosphate with a specific radioactivity of about 0.3 mc. per µmole was inoculated with 0.1 ml. of a 7 hour broth culture of *E. coli* strain B. After 18 hours' growth at 37°, the cells were harvested and washed twice with 10 ml. portions of a solution containing 0.5 per cent each of KCl and NaCl. The packed cells were suspended in 6 ml. of alcohol-ether (3:1), incubated at 37° for 30 minutes, centrifuged, and resuspended in 6 ml. of the alcohol-ether solution. After centrifugation, the cells were dried over KOH *in vacuo*. The dry powder was then suspended in 3.0 ml. of 1 N NaOH, and left for 15 hours at 37°. The turbid, viscous solution was chilled to 0° and treated with 1.0 ml. of 5 N perchloric acid. The precipitate, collected by centrifugation, was suspended in 1.75 ml. of water and dissolved by the addition of 0.25 ml. of 1 N NaOH. This solution contained approximately 5 µmoles of purine deoxynucleotides, as judged by deoxypentose estimation, and from 90 to 100 per cent of the deoxypentose of the original alkaline digest. The DNA was precipitated from solution by the addition of 0.23 ml. of 5 N perchloric acid, washed with 2 ml. of cold water, and then redissolved in 2 ml. of 0.12 N NaOH. This reprecipitation was repeated until the radioactivity in the perchloric acid supernatant solution was less than 0.5 per cent of that in the dissolved precipitate. Usually two to three reprecipitations were sufficient. The precipitate was suspended finally in 1.0 ml. of water, the pH adjusted to approximately 7.5 with 1 N NaOH, and the volume brought to 2.0 ml.

Digestion with pancreatic DNase was carried out at 37° in a mixture containing 80 µmoles of Tris buffer, at pH 7.5, 10 µmoles of $MgCl_2$, and 300 µg. of crystalline pancreatic DNase (Worthington Biochemical Corporation) in a final volume of 2.25 ml. Aliquots removed at intervals from the digestion mixture were precipitated with an equal volume of 1 N perchloric acid in the presence of thymus DNA, which was added as carrier. 90 to 100 per cent of the radioactivity was rendered acid-soluble after approximately 3 hours.

The polydeoxyribonucleotides of the DNase digest were degraded to mononucleotides by a snake venom phosphodiesterase purified free of mononucleotidase (9). The incubation mixture (10 ml.) consisted of the DNase digest, 180 µmoles of magnesium acetate, 600 µmoles of glycine buffer (at pH 8.5), and 50 units of snake venom phosphodiesterase (an amount which releases mononucleotides from a pancreatic DNase digest of thymus DNA at a rate of 50 µmoles per hour). Formation of deoxyribonucleotide was measured in aliquots by the release of inorganic orthophosphate upon incubation with purified 5'-nucleotidase (10). Conversion of the DNase digest to deoxyribonucleotides was usually complete within 3 hours. The reaction mixture was heated in a boiling water bath for 2 minutes; a flocculent precipitate was separated by centrifugation, washed with 10 ml. of water, and then discarded. The supernatant fluid and wash were combined and adsorbed on a column of Dowex 1 (chloride form, 2 per cent cross-linked, $6.0 \times 1.0$ cm.$^2$). The elution rate was 0.5 ml. per minute. Deoxycytidylate was eluted between 12 and 15 resin bed volumes of 0.002 N HCl, after which the eluant was changed to 0.01 N HCl. Deoxyadenylate appeared in the next 5 to 6 resin bed volumes. The eluant was changed to 0.01 N HCl containing 0.05 M KCl, and then deoxyguanylate and thymidylate were eluted together in approximately 3 resin bed volumes. The tubes containing the latter fractions were pooled, neutralized, and reapplied to a Dowex 1 column (chloride form, 2 per cent cross-linked, $10.0 \times 1.0$ cm.$^2$). Deoxyguanylate was eluted between 15 and 20 resin bed volumes of 0.01 N HCl, and thymidylate was eluted between 30 and 36 resin bed volumes. Approximately 1.5 µmoles of each deoxyribonucleotide, with a specific radioactivity of about 0.3 mc. per µmole, were obtained under these conditions.

*Enzymatic synthesis of deoxynucleoside 5'-triphosphates.* The $P^{32}$-labeled deoxyribonucleotides isolated as described above and the unlabeled nucleotides obtained from the California Foundation for Biochemical Research were converted to the triphosphates by enzymes partially purified from *E. coli*. The preparation and assay of these kinases, and the synthesis and isolation of the triphosphates are described below.

*1. Assay of deoxynucleotide kinase.* This assay measures the conversion of a $P^{32}$-labeled deoxynucleotide that is susceptible to the action of semen phosphatase to a form that is resistant to the phosphatase. Adsorption on Norit is used to distinguish between the phosphate that is liberated from the nucleoside and that which is bound to it. The assay consists of three stages (Scheme I).

In a control incubation, from which enzyme was omitted in Stage I, no radioactivity was found adsorbed to the Norit, whereas with an excess of the deoxynucleotide kinase, all the radioactivity was accounted for in the Norit precipitate. A unit of enzyme was defined as the amount that converts 100 mµmoles of deoxynucleotide to a phosphatase-resistant form in 1 hour under these assay conditions. The reaction rate was proportional to the amount of enzyme added; with 0.005, 0.01, and 0.02 ml. of streptomycin fraction (Table I), there were obtained 3.92, 4.33, and 4.38 units per ml., respectively.

*2. Purification of deoxynucleotide kinase.* An enzyme preparation (partially purified from extracts of *E. coli*) catalyzes the phosphorylation, by ATP, of deoxyadenylate, deoxyguanylate, deoxycytidylate, and thymidylate to yield the corresponding triphosphates (Table I). The purification procedures were carried out at 0° to 4°. Bacterial extract (60 ml. of

## SCHEME I

In Stage I, the incubation mixture (0.25 ml.) contained 2 μmoles of ATP, 4 μmoles of $MgCl_2$, 16 mμmoles of $P^{32}$-labeled deoxynucleotide (0.5 to 1 × $10^5$ c.p.m. per μmole), 0.05 to 0.2 unit of enzyme, and 10 μmoles of Tris buffer (pH 7.5). After incubation at 37° for 20 minutes, 0.5 ml. of water was added, the tube was immersed in boiling water for 2 minutes, and then was chilled in an ice bath. In Stage II, 4 μmoles of $MgCl_2$ (0.04 ml.), 100 μmoles of sodium acetate buffer, at pH 5.0, (0.1 ml.), 0.2 μmole of unlabeled deoxynucleotide (0.02 ml.), and 50 units of human semen phosphatase (0.02 ml.) (11) were added to the mixture, which was then incubated at 37°; after 15 minutes it was chilled. In Stage III, 0.1 ml. of cold 2 N HCl and 0.15 ml. of a Norit suspension (20 per cent packed volume) were added and shaken for 2 to 3 minutes with the incubation mixture. The Norit precipitate was collected by centrifugation, washed three times with 2.5 ml. portions of cold water, and suspended in 0.5 ml. of 50 per cent ethanol containing 0.3 ml. of concentrated ammonium hydroxide per 100 ml. The entire suspension was plated and counted.

| Stage | Reagent | Reaction | |
|---|---|---|---|
| I | Deoxynucleotide kinase | Deoxynucleoside-P* $\xrightarrow{ATP}$ Deoxynucleoside-P*-P (P) | |
| | | ↓ | ↓ |
| II | Phosphatase (monoesterase) in excess | Deoxynucleoside + $P_i$* | no change |
| | | ↓ | ↓ |
| III | Norit, in excess | Radioactivity unadsorbed | Radioactivity adsorbed |

## TABLE I
*Purification of deoxynucleotide kinase*

| Fraction | Units total | Protein mg./ml. | Specific activity* units/mg. protein |
|---|---|---|---|
| Extract (Fraction I) | 918 | 14.8 | 1.0 |
| Streptomycin fraction | 1210 | 3.4 | 3.1 |
| Calcium phosphate gel fraction | 836 | 1.6 | 4.8 |
| Alumina Cγ gel eluate | 633 | 1.8 | 6.8 |

*Using thymidine 5'-phosphate as a substrate.

Fraction I as described later for "polymerase") was diluted with 30 ml. of glycylglycine buffer (0.05 M, pH 7), and 27 ml. of 5 per cent streptomycin sulfate[2] was added slowly as the solution was stirred. The suspension was allowed to stand for 5 minutes and then was centrifuged for 10 minutes at 10,000 × g. The supernatant solution (streptomycin fraction, 115 ml.) was next treated with calcium phosphate gel (12) as follows: 66 ml. of gel (15 mg. of solids per ml.) was centrifuged and the supernatant fluid was discarded. The streptomycin fraction (110 ml.) was mixed with the packed gel; after 5 minutes, the suspension was centrifuged, and the supernatant solution was collected (calcium phosphate gel fraction, 113 ml.). This fraction (108 ml.) was added to the packed residue from 45 ml. of alumina Cγ gel (13) (15 mg. of solids per ml.). The gel was thoroughly dispersed, and, after 5 minutes, was collected by centrifugation. The enzyme was then eluted from the gel with 54 ml. of potassium phosphate buffer (0.066 M, pH 7.4) (alumina Cγ gel fraction, 52 ml.).

The alumina Cγ fraction showed no significant loss of activity over a 4-month period when stored at $-10°$. The specific activity (units per mg. of protein) of a typical enzyme preparation, using equivalent levels of deoxyadenylate, deoxyguanylate, deoxycytidylate, and thymidylate as substances, was 8.8, 7.1, 4.0, and 6.8, respectively. Although this preparation served as a kinase for all the deoxynucleotides, specific kinases for each of the deoxynucleotides can be obtained by further purification.[3]

3. *Enzymatic preparation of the triphosphates: dGTP, dCTP or dTTP.* The reaction mixture (90 ml.) consisted of the following: Tris buffer, at pH 7.5, 3 mmoles; $MgCl_2$, 300 μmoles; cysteine, 50 μmoles; adenosine diphosphate, 35 μmoles; acetyl phosphate, 175 μmoles; a deoxynucleoside 5'-phosphate, 35 μmoles; acetokinase, 5 units (14); and the deoxynucleotide kinase (alumina Cγ gel fraction), 300 units.

---

[2] We are extremely grateful to Merck and Company, Inc., for generous gifts of this substance.

[3] Further purification procedures developed by Dr. Jerard Hurwitz have shown distinct kinases for deoxyguanylate, deoxycytidylate, and thymidylate. The presence in *E. coli* of a kinase for deoxyadenylate (very likely adenylate kinase) which is distinguishable from these other kinases has been suggested (30). Heating of the alumina Cγ gel eluate at 100° for 3 minutes at pH 7.4 permitted the survival of some kinase activity for deoxyadenylate, whereas that for the other deoxynucleotides was completely destroyed. Phosphorylation of deoxynucleotides by enzyme preparations from animal (31, 32) and other bacterial cells (33) has been reported.

After incubation at 37° for 2 hours, the mixture was heated for 2 minutes in a boiling water bath, and was then immediately chilled; the precipitated proteins were removed by filtration.

*dATP*: The incubation mixture was the same as that described above, except that glycylglycine buffer, at pH 7.4, was used in place of Tris.

Although the chromatographic procedures to be described effectively separated the excess ATP in the reaction mixture from dGTP, dCTP and dTTP, they did not resolve it from dATP. ATP was therefore destroyed in the dATP mixture before chromatography.

4. *Selective destruction of ATP.* ATP may be selectively degraded in the presence of deoxyribonucleotides by periodate and then alkaline treatment, by Whitfeld's procedure for oligoribonucleotides (15). According to this author, the procedure should result first in an oxidation of the ribose and then a cleavage of the adenine from the nucleotide. The dATP reaction mixture containing 35 μmoles of dATP was treated with 175 μmoles of sodium metaperiodate at 25°. Spectrophotometric measurement of periodate reduction (16) indicated that under these conditions 90 per cent of the theoretical amount was consumed within 4 minutes and that the reaction was essentially complete within 8 minutes. After incubation for 30 minutes, the excess periodate was destroyed by adding 200 μmoles of glucose and incubating the mixture for 30 minutes at 25°. The pH was then adjusted to 10 with glycine buffer (1 M, at pH 10.2), and the solution was incubated for 12 to 16 hours at 37°. Prior to treatment with alkali, it appeared that about 10 per cent of the ATP remained, as judged by hexokinase assay (17); however, the slow rate of reaction with hexokinase as compared with an ATP control (ATP added after the periodate was destroyed by glucose) suggests that it was a periodate-oxidation product of ATP which reacted with hexokinase or an associated kinase. After treatment with alkali, ATP was completely destroyed; less than 1 per cent remained, as determined by the hexokinase assay. Such an exposure of ATP to alkali without prior periodate oxidation resulted in losses of only 5 per cent or less. A more sensitive assay of the removal of ATP involved the use of ATP labeled with $C^{14}$ in the adenine. After periodate and alkaline treatment, the mixture was made 1 N with respect to HCl, and passed over a column of Dowex 50-$H^+$ resin. Only 0.6 per cent of the radioactivity passed through the column, indicating that 99.4 per cent of the ATP had been degraded to adenine, which is quantitatively adsorbed by this resin, whereas ATP is not adsorbed. Comparable results have been obtained for adenosine 5'-phosphate and presumably should be expected for other derivatives of adenosine. When applied to uridine nucleotides, this procedure did not result in a quantitative cleavage of the pyrimidine base from the nucleotide. Determinations by the orcinol method (18) of the DATP isolated after ion exchange chromatography have indicated 0.04 to 0.05 μmole of orcinol-reactive material per μmole of DATP; the significance of this determination is not clear.

5. *Isolation of triphosphates.* The chilled incubation mixtures were adsorbed on Dowex 1 columns (2 per cent cross-linked, chloride form, 10 cm. × 3.8 $cm.^2$), and the individual triphosphates were eluted as symmetrical peaks. All fractions were neutralized or made slightly alkaline with $NH_4OH$ immediately upon collection.

*dATP*: Elution was begun with 0.01 N HCl-0.08 M LiCl. When the adenine resulting from the periodate degradation of ATP had been removed completely, dATP was eluted with 0.01 N HCl-0.2 M LiCl. The fractions between 2 and 4 resin bed volumes represented a yield of dATP of approximately 60 per cent, based on the amount of deoxyadenylate initially added to the reaction mixture.

*dGTP*: Following the elution of ATP from the column with 0.01 N HCl-0.1 M LiCl, the dGTP was eluted with 0.01 N HCl-0.2 M LiCl. This appeared between 11 and 16 resin bed volumes of effluent with a yield of about 75 per cent.

*dCTP*: The column was first washed with 0.01 N HCl-0.05 M LiCl to remove any mono- or diphosphates present in the reaction mixture. dCTP was eluted with 0.01 N HCl-0.08 M LiCl. The fraction between 4 and 12 resin bed volumes of effluent represented a yield of triphosphate of about 60 per cent.

*dTTP*: After removal of ATP from the column, the dTTP was eluted with 0.02 N HCl-0.2 M LiCl, and appeared between 10 and 18 resin bed volumes of eluate, and about 75 per cent of the starting thymidylate was recovered as the triphosphate.

The triphosphates were concentrated by precipitation as the barium salts, then metathesized with Dowex 50-$K^+$ resin. A typical preparation was carried out in the following manner. The ion exchange fractions were pooled (285 ml.). Glycine buffer (1 N, at pH 10.2) was added to a final concentration of 0.01 M, and the solution was adjusted to pH 8.5 by the dropwise addition of 4 N LiOH. 2 ml. of a saturated solution of barium bromide and 285 ml. of cold ethanol were added. After 30 minutes at 0°, the precipitate was collected by centrifugation, and was dried in a vacuum desiccator over KOH for about an hour.

The barium salt was dispersed in about 2 ml. of cold water, and 1 ml. (packed volume) of washed Dowex 50-$K^+$ resin was added. The suspension was shaken for 15 minutes at 0°, then adjusted to a pH of about 6 with 0.5 N HCl, and again shaken for 15 minutes. The suspension was then filtered through a

**TABLE II**
*Analysis of the deoxynucleoside triphosphates*

| Compound | Absorbance ratios* | | Base | Deoxypentose | Acid-labile P† | Total P |
|---|---|---|---|---|---|---|
| | λ250/λ260 | λ280/λ260 | | | | |
| dTTP | 0.64 (0.64) | 0.72 (0.72) | 1 | ‡ | 1.90 | 3.18 |
| dTTP§ | 0.63 | 0.74 | 1 | ‡ | 1.98 | 3.26 |
| dCTP | 0.44 (0.43) | 2.14 (2.12) | 1 | ‡ | 1.92 | 3.20 |
| dATP | 0.77 (0.79) | 0.14 (0.15) | 1 | 0.99 | 1.82 | 3.00 |
| dGTP | 1.14 (1.15) | 0.66 (0.68) | 1 | 1.03 | 1.78 | 2.97 |

*The values cited in the literature for the corresponding 5'-monophosphates (20) are given in parentheses for comparison. dTTP and dCTP were determined at pH 2 and dATP and dGTP at pH 7.

†Hydrolysis was for 10 minutes at 100° in 1 N HCl. In the case of dATP and dGTP, 0.65 μmole of P per μmole of base was subtracted in order to correct for the hydrolysis of the phosphate linked to the deoxyribose; this correction, based on the hydrolysis rate of deoxyadenylate and deoxyguanylate, is only an approximation. Hydrolysis of dATP and dGTP for 30 minutes yielded 93 to 94 per cent of the total P.

‡Deoxypentose determinations of dTTP and dCTP by the cysteine-sulfuric acid method (19) did not yield reproducible values.

§Chemically synthesized.

sintered glass funnel, and the resin was washed with cold water. The combined filtrate and washings were neutralized with 1 N KOH. The recovery of the triphosphates from the column effluents by this procedure ranged from 90 to 100 per cent. Analyses of the four deoxynucleoside triphosphates are presented in Table II.

6. *Chemical synthesis of dTTP.* dTTP was also prepared by chemical synthesis[4] according to the method described by Hall and Khorana (21) for uridine triphosphate. The calcium salt of thymidine 5'-phosphate (1 gm., California Foundation for Biochemical Research) was first converted to the pyridine salt on Dowex 50 pyridine, taken to dryness, and then treated with proportions of phosphoric acid, pyridine, water, and dicyclohexylcarbodiimide, as previously described (21) (Table I). The mono-, di-, tri- and higher polyphosphates of thymidine were separated by ion exchange chromatography with recoveries of 14, 22, 40 and 24 per cent, respectively, and dTTP was finally collected as a barium salt after the chromatogram eluates were adsorbed on and eluted from Norit (Table II).

## Preparation of deoxyribonucleic acids

$P^{32}$-*labeled bacteriophage DNA.* Lysates of T2r+ were prepared in the glycerol-lactate medium (8). 200 ml. of medium containing 0.60 μmole of inorganic orthophosphate per ml. (specific activity, 10 μc. per μmole) were inoculated with 0.1 ml. of an 18 hour culture of *E. coli* strain B and then incubated at 37°. When the culture reached a level of 2 × 10⁸ cells per ml., bacteriophage at a multiplicity of 1 was added. Lysis was usually complete in 8 to 10 hours with titers of about 4 × 10¹⁰ infectious particles per ml. The bacteriophage was purified as described by Herriott (22), with omission of the filtration through Super-cel, and finally taken up in 0.5 ml. of 0.15 M NaCl. In order to disrupt the bacteriophage by osmotic shock (23), 180 mg. of NaCl were dissolved in the suspension, to which 25 ml. of distilled water were added rapidly, with mixing. The mixture stood for 1 hour at 25° and was then centrifuged for 90 minutes at 20,000 × g to remove intact phage and debris. The supernatant solution, designated $P^{32}$-phage DNA, contained 0.16 μmole per ml. of phosphorus, and had a molar extinction coefficient, $E(P)$, on this basis, of 6900 at 260 mμ.

*Calf thymus DNA.* This was prepared by the method of Kay, Simmons and Dounce (24). A solution of 0.5 mg. per ml. had an extinction number at 260 mμ of about 7.0.

## Enzyme assays

*Assay of "polymerase."* This assay measures the conversion of acid-soluble $P^{32}$-labeled deoxynucleoside triphosphates into an acid-insoluble product. The incubation mixture (0.3 ml.) contained 0.02 ml. of glycine buffer (1 M, pH 9.2), 0.02 ml. of MgCl₂ (0.1 M), 0.03 ml. of 2-mercaptoethanol (0.01 M), 0.02 ml. of thymus DNA (0.5 mg. per ml.), 0.01 ml. of dATP (0.5 μmole per ml.), 0.02 ml. of dGTP (0.5 μmole per ml.), 0.01 ml. of dCTP (0.5 μmole per ml.), 0.01 ml. of dTP³²PP[5] (0.5 μmole per ml., 1.5 × 10⁶ c.p.m. per μmole), and 0.005 to 0.05 unit of

---

[4] This preparation was synthesized in the laboratory of Dr. H. G. Khorana, to whom we are indebted for guidance and hospitality.

[5] dCP*PP, dAP*PP, and dTTP labeled in carbon-2 of thymine have also been used in routine assays during the course of enzyme purification, with essentially identical results.

enzyme. Dilutions of the enzyme for assay were made in Tris buffer (0.05 M, pH 7.5) containing 0.1 mg. per ml. of thymus DNA. After incubation at 37° for 30 minutes, the tube was placed in ice, and 0.2 ml. of a cold solution of thymus DNA (2.5 mg. per ml.) was added as carrier. The reaction was stopped, and the DNA was precipitated by the immediate addition of 0.5 ml. of ice-cold 1 N perchloric acid. After 2 to 3 minutes, the precipitate was broken up thoroughly with a snug-fitting glass pestle, 2 ml. of cold distilled water were added, and the precipitate was thoroughly dispersed. After centrifugation for 3 minutes at 10,000 × g, the supernatant fluid was discarded. The precipitate was dissolved in 0.3 ml. of 0.2 N NaOH, the DNA was reprecipitated by the addition of 0.40 ml. of cold 1 N perchloric acid, 2.0 ml. of cold water were added, and the precipitate was thoroughly dispersed. After centrifugation, this precipitate was again dissolved, reprecipitated again and recentrifuged. Finally, the precipitate was dissolved by the addition of 0.2 ml. of 0.1 N NaOH, the entire solution was pipetted into a shallow dish, dried, and the radioactivity measured.

Controls for the crude enzyme fractions (I to III) were incubation mixtures to which the enzyme fraction was added after completion of the incubation period, but just before the perchloric acid was added. With more purified fractions (Fractions IV to VII), an incubation mixture lacking $Mg^{++}$ or one of the deoxynucleoside triphosphates served as well. Precipitates obtained from control incubation mixtures contained less than 0.10 per cent of the total radioactivity added, and, in most assays they contained radioactivity in the order of 2 per cent of the experimental values. A unit of enzyme was defined as the amount causing the incorporation of 10 m$\mu$moles of the labeled deoxynucleotide into the acid-insoluble product during the period of incubation. The specific activity was expressed as units per mg. of protein.

With the exception of Fraction I, proportionality of enzyme addition, with the amount of labeled substrate incorporated into the product, was obtained. With Fraction IV, for example, the addition of 0.5, 1.0, and 1.5 $\mu$g. of the enzyme preparation yielded specific activities of 21.0, 18.7 and 20.5, respectively. Assays of Fraction I are only an approximation, and the levels of activity assayed should not exceed 0.02 unit; even so, augmentation of incorporation by as much as 50 per cent was obtained at times with the addition of ATP (0.0025 M) and the deoxynucleotide kinase (0.7 unit of alumina C$\gamma$ gel eluate fraction).

With respect to the optimal pH for the assay, the rate was most rapid at about 8.7; at pH 6.5, 8.0 and 10.0, the respective rates were 15, 70 and 15 per cent of that observed at pH 8.7. The pH usually achieved in the assay mixture was in the neighborhood of 8.7 to 9.0. It is noteworthy that high concentrations of salt interfere with the assay; for example, NaCl at a final concentration of 0.2 M produced a 97 per cent inhibition. The use of fluoride to inhibit the action of phosphatases in the assay of crude enzyme fractions is limited by the inhibitory action of fluoride on the "polymerase" (0.05 M KF produced a 90 per cent inhibition).

*Assays of DNase.* There is an abundance of DNase activity in cell-free extracts of *E. coli* which degrades both the DNA added initially to the assay mixture and the newly synthesized DNA. As will be described in a later section, there are indications for the existence of at least 3 distinct DNases. One activity (or activities) to be designated tentatively as "DNase A", degrades calf thymus DNA with a pH optimum near 8.5, whereas it does not degrade it at all at pH 10. Although the cleavage of bacteriophage DNA is only one-tenth as rapid, the use of this substrate labeled with $P^{32}$ provided a sensitive assay when this enzyme activity is reduced to trace concentrations. Another activity, designated "DNase B", is characterized by its effective action on the enzymatically synthesized DNA even at pH 10. Immunological and fractionation procedures, to be described later, suggest that there are at least two distinct enzymes in this group. The assays of DNase A on thymus and bacteriophage DNA and of DNase B on enzymatically prepared DNA are described below.

*1. Assay of DNase A on thymus DNA.* This assay depends on the degradation of DNA to acid-soluble fragments which are measured spectrophotometrically. The reaction mixture (0.3 ml) contained 0.1 ml. of thymus DNA (0.5 mg. per ml.), 0.02 ml. of Tris buffer (1 M, pH 7.5), 0.02 ml. of MgCl$_2$ (0.1 M) and about 4 units of enzyme. After incubation for 30 minutes at 37°, 0.2 ml. of a thymus DNA solution (2.5 mg. per ml.) was added as carrier, followed by 0.5 ml. of cold 1 N perchloric acid. After 5 minutes at 0°, the precipitate was removed by centrifugation. The supernatant solution was diluted with an equal volume of distilled water, and the optical density at 260 m$\mu$ was determined. One unit of enzyme was defined as the amount causing the production of 10 m$\mu$moles of acid-soluble DNA polynucleotides in 30 minutes (assuming a molar extinction coefficient, $E(P)$, of 10,000). The assay showed proportionality between the substrate split and the enzyme added at levels of 1 to 8 units. With the addition of 0.02, 0.05, and 0.10 ml. of a 1:10 dilution of an enzyme fraction (the ammonium sulfate fraction preceding "polymerase" Fraction VI), 405, 362, and 394 units per ml., respectively, were obtained. The activity of this enzyme at pH 7.5 is approximately twice that observed at pH 9.

*2. Assay of DNase A on bacteriophage DNA.* This assay depends on the degradation of DNA to acid-soluble fragments which are determined by measurement of radioactivity. The incubation mixture (0.30 ml.) contained 0.02 ml. of glycine buffer (1 M, pH 8.5), 0.01 ml. of $MgCl_2$ (0.10 M), 0.03 ml. of $P^{32}$-phage DNA (0.15 µmole of P per ml.; specific activity $5 \times 10^6$ c.p.m. per µmole of P), and 0.01 to 0.10 unit of enzyme. After incubation for 30 minutes at 37°, 0.20 ml. of thymus DNA (2.5 mg. per ml.) was added as carrier, and then 0.50 ml. of 0.5 N perchloric acid was added. After standing 2 to 3 minutes in ice, the precipitate was removed by centrifugation, and 0.5 ml. of the supernatant fluid was transferred to a shallow dish, and the radioactivity was determined. An incubation mixture from which enzyme was omitted served as a control for nonenzymatic hydrolysis of bacteriophage DNA. The proportionality of the amount of enzyme added to the amount of substrate hydrolyzed was obtained when 2 to 20 per cent of the substrate was converted to acid-soluble material. The unit of enzyme activity is the same as that defined in the thymus DNA assay.

*3. Assay of DNase B.* This assay also measures the release of radioactive acid-soluble fragments from DNA. The incubation mixture (0.3 ml.) contained 0.03 ml. of $P^{32}$-labeled, enzymatically prepared, DNA[6] (0.3 µmole per ml., $2 \times 10^6$ c.p.m. per µmole of DNA nucleotide), 0.03 ml. of glycine buffer (1 M, at pH 9.2), 0.02 ml. of $MgCl_2$ (0.1 M), 0.03 ml. of 2-mercaptoethanol (0.01 M), and about 0.05 unit of enzyme. After incubation of the mixture for 30 minutes at 37°, carrier DNA and cold perchloric acid were added, as in the assay for DNase A. After removal of the precipitate by centrifugation, 0.2 ml. of the supernatant solution was plated, and the radioactivity was determined. The enzyme unit is the same as for DNase A. The proportionality between the amount of substrate hydrolyzed and the amount of enzyme added was observed from 0.05 to 0.25 unit of enzyme. With the addition of 0.02, 0.05, and 0.10 ml.

---

[6] The reaction mixture consisted of 0.15 µmole each of dTTP, dGTP, dATP and dCTP (the latter having a specific radioactivity of 20 µc. per µmole), 0.33 ml. of thymus DNA (2.5 mg. per ml.), glycine buffer, at pH 9.2 (330 µmoles), $MgCl_2$ (330 µmoles) and Fraction VI (20 units) in a final volume of 5 ml. After incubation at 37° for 1 hour, the incubation mixture was chilled and 0.55 ml. of 4 N trichloracetic acid was added, and the suspension was centrifuged. The precipitate was dissolved in 1.5 ml. of cold 0.02 N NaOH, and then was treated with 1.5 ml. of cold 1 N perchloric acid and 2 ml. of water. The precipitate was collected by centrifugation, dissolved, reprecipitated, and finally dissolved in 1.0 ml. of 0.02 N NaOH. 0.1 N HCl was added dropwise to adjust the pH to 7.5, and cold water was added to bring the volume to 3.3 ml. Such a preparation had a specific activity of $2.5 \times 10^6$ c.p.m. per µmole of DNA nucleotide.

of a 1:100 dilution of an enzyme fraction ("polymerase" Fraction VI), 340, 260 and 257 units per ml., respectively, were obtained.

## Other methods

Determinations of phosphate, pentose and protein, procedures for ion exchange chromatography and measurements of radioactivity have been cited previously (25). The efficiency of the gas-flow counter was approximately 50 per cent. Deoxypentose of purine deoxynucleotides was determined by the diphenylamine method of Dische (26).

## Purification of "polymerase"

*Growth and harvest of bacteria.* E. coli strain B, or ML30, was grown in a medium containing 1.1 per cent $K_2HPO_4$, 0.85 per cent $KH_2PO_4$, 0.6 per cent Difco yeast extract, and 1 per cent glucose. Cultures, usually 60 liters, were grown with vigorous aeration in a large growth tank,[7] and were harvested about 2 hours after the end of exponential growth. The cultures were chilled by the addition of ice, the cells collected in a Sharples supercentrifuge, washed once in a Waring Blendor by suspension in 0.5 per cent NaCl-0.5 per cent KCl (3 ml. per gm. of packed cells), centrifuged, and then stored at $-12°$. The yield of packed wet cells was approximately 8 gm. per liter of culture, and cells stored as long as 1 month were used. All subsequent operations were carried out at 0° to 3° unless otherwise specified.

*Preparation of extract.* Cells were suspended in .05 M glycylglycine buffer, at pH 7.0 (4 ml. per gm. of packed cells), and were disrupted by treatment for 15 minutes in a Raytheon 10-KC sonic oscillator. The suspension was centrifuged for 15 minutes at 12,000 $\times$ g, and the slightly turbid supernatant liquid was collected. The protein content was determined and adjusted to a concentration of 20 mg. per ml. by addition of the same glycylglycine buffer (Fraction I) (Table III).

*Streptomycin precipitation.* 8400 ml. of Fraction I, obtained from 2 kilos of packed cells, were treated in the following manner. To a 525 ml. batch were added 525 ml. of Tris buffer (0.05 M, at pH 7.5), then slowly, with stirring, 81 ml. of 5 per cent streptomycin sulfate[2] were added. After 10 minutes, the precipitate was collected by centrifugation at 10,000 $\times$ g. This precipitation with streptomycin was carried out four times on this scale, and the precipitates were collected in the same centrifuge cups. The precipitates, with potassium phosphate buffer (0.05 M, at pH 7.4) added to a total volume of 430 ml., were homogenized in a Waring Blendor for 30 minutes at low speed. This

---

[7] Available from Rinco Instruments Company, Greenville, Illinois.

**TABLE III**
*Purification of "polymerase"*

| Fraction no. | Step | Units per ml. | Units Total | Protein mg./ml. | Specific activity units/mg. protein |
|---|---|---|---|---|---|
| I | Sonic extraction | 2.0 | 16,800 | 20.0 | 0.1 |
| II | Streptomycin | 13.0 | 19,500 | 3.0 | 4.3 |
| III | DNase, dialysis | 12.1 | 18,100 | 1.80 | 6.7 |
| IV | Alumina gel | 15.4 | 12,300 | 0.78 | 19.8 |
| V | Concentration of gel eluate | 110 | 9,900 | 4.90 | 22.4 |
| VI | Ammonium sulfate | 670 | 6,030 | 8.40 | 80.0 |
| VII | Diethylaminoethyl cellulose* | 120 | 3,600 | 0.60 | 200.0† |

*This step was actually carried out many times on a smaller scale (see the text), and these values are calculated for the large-scale procedure.
†In some runs, values as high as 400 have been obtained.

suspension was centrifuged for 2 hours at 78,000 × g in a Spinco model L centrifuge, and the supernatant fluid was collected (Fraction II).

*Deoxyribonuclease digestion.* To 1500 ml. of Fraction II (derived from 8400 ml. of Fraction I) were added 15 ml. of 0.3 M $MgCl_2$ and 1.5 ml. of pancreatic deoxyribonuclease (100 μg. per ml.). This mixture was incubated at 37° for about 5 hours, until 85 to 90 per cent of the ultraviolet-absorbing material was rendered acid-soluble.[8] A considerable amount of protein settled out during the digestion, but it was not removed at this time. The digest was dialyzed for 16 hours against 24 liters of Tris buffer (0.01 M, at pH 7.5), then centrifuged for 5 minutes at 10,000 × g, and the supernatant fluid was collected (Fraction III).

*Alumina Cγ gel adsorption and elution.* Enough aged alumina gel (13) (195 ml. containing 15 mg. dry weight per ml.) was added to 1500 ml. of Fraction III in order to adsorb 90 to 95 per cent of the enzyme. The mixture was stirred for 5 minutes, and then centrifuged. The supernatant fluid was discarded and the gel washed with 400 ml. of potassium phosphate buffer (0.02 M, at pH 7.2). The gel was then eluted twice with 400 ml. portions of potassium phosphate buffer (0.10 M, at pH 7.4) in order to remove the enzyme, and the eluates were combined (Fraction IV).

*Concentration of alumina Cγ eluate.* To 800 ml. of Fraction IV were added 16 ml. of 5 N acetic acid and then 480 gm. of ammonium sulfate. After 10 minutes at 0°, the resulting precipitate was collected by centrifugation (30 minutes, 30,000 × g) and dissolved in 90 ml. of potassium phosphate buffer (0.02 M, at pH 7.2) (Fraction V).

*Ammonium sulfate fractionation.* To 90 ml. of Fraction V were added 9 ml. of potassium phosphate buffer (1 M, at pH 6.5) and 0.90 ml. of a 0.10 M solution of 2-mercaptoethanol. 24.7 gm. of ammonium sulfate were added, and after 10 minutes at 0°, the precipitate was removed by centrifugation at 12,000 × g for 10 minutes.[9] To the supernatant fluid an additional 9.6 gm. of ammonium sulfate were added, and, after 10 minutes at 0°, the precipitate which formed was collected by centrifugation at 12,000 × g for 10 minutes. This precipitate was dissolved in 9 ml. of potassium phosphate buffer (0.02 M, at pH 7.2) (Fraction VI).

*Diethlaminoethyl cellulose fractionation.* A column (11 × 1 cm.) was prepared from diethylaminoethyl cellulose (27) which had previously been equilibrated with $K_2HPO_4$ (0.02 M). 1.2 ml. of Fraction VI was diluted to 8.0 ml. with 0.02 M $K_2HPO_4$ and passed through the column at a rate of 12 ml. per hour. The column was washed with 3.0 ml. of the same buffer, and then eluted (flow rate of 9 ml. per hour) with pH 6.5 potassium phosphate buffers as follows: 8 ml. of 0.05 M, 10 ml. of 0.10 M, 3 ml. of 0.20 M, and finally 4 ml. of 0.20 M. Approximately 60 per cent of the enzyme applied to the adsorbent was obtained in the last elution with 0.20 M buffer (Fraction VII).

Samples of Fraction VII have been further purified by another treatment with diethylaminoethyl cellulose or by treatment with a phosphocellulose adsorbent (27). Specific activities of 250 to 350 have thus been obtained.

### Stability of "polymerase"

With the exception of the alumina gel eluate (Fraction IV), which in some instances lost activity on storage, each of the enzyme fractions has been stored for at least several weeks at −12° without any

---

[8] The optical density at 260 mμ was determined before and after precipitation with an equal volume of cold 1 N perchloric acid.

[9] This precipitate dissolved in 9 ml. of potassium phosphate buffer (0.02 M, at pH 7.2) has been designated AS 1 and used as an antigen in studies cited below.

**TABLE IV**
*DNase activities in "polymerase" fractions*

| "Polymerase" fraction | Ratios of DNase to "polymerase"* | | |
|---|---|---|---|
| | DNase A on thymus DNA | DNase A on phage DNA | DNase B on enzymatic DNA |
| Fraction I | † | † | 34-62 |
| Fraction V | 1 | 0.1 | 1 |
| Ammonium sulfate fractionation of Fraction V: | | | |
| AS 1 | 5 | 0.6 | 6 |
| AS 2 (Fraction VI) | 1 | 0.1 | 1 |
| AS 3 | 2 | 0.3 | 2 |
| Ammonium sulfate fractionation of AS 3: | | | |
| AS 3a | 1 | 0.4 | 2 |
| AS 3b | 26 | 2.1 | 6 |
| Fraction VII | 0.2 | 0.003 | 0.3 |
| Fraction VII‡ | | | 0.06 |

*"Polymerase" was determined as described in the text section on assays. Nucleases were determined with thymus DNA, bacteriophage DNA, or synthetic DNA as substrates. The ratios here express:

$$\frac{\text{Nucleotide rendered acid-soluble, } \mu\text{moles}}{\text{Nucleotide rendered acid-insoluble, } \mu\text{moles}}$$

†No accurate assay was possible because of the high DNA content of this fraction.

‡Tested under conditions which differed from the standard assay (see the text) as follows: potassium phosphate buffer, at pH 7.4, replaced the glycine buffer, and the 2-mercaptoethanol was omitted.

significant loss in activity. When heated at pH 7.2 for 3 minutes at 80°, all the activity was destroyed, and, after 10 minutes at 60°, only a trace (less than 2 per cent) remained. Incubation for 10 minutes at 45° or for 30 minutes at 37° produced losses of 75 and 30 per cent, respectively, whereas similar incubations, but at pH 8.7, resulted in respective losses of 95 and 85 per cent. Since the assay of the enzyme is carried out for 30 minutes at 37° at a pH of about 8.7, it was of interest to determine the stabilizing components in the assay mixture. Among the constituents of the assay mixture only DNA was active in this regard. Complete protection against inactivation of the enzyme during incubation was provided at a level of 17 μg. per ml. Replacement of the DNA by bovine serum albumin (1 mg. per ml.), apurinic acid (100 μg. per ml.) (28), ribonucleic acid from crystalline turnip yellow mosaic virus[10] (68 μg. per ml.), or thymus DNA treated with 1 N NaOH for 15 hours at 37° (37 μg. per ml.), resulted in losses in enzyme activity of 90, 70, 85, and 70 per cent, respectively. Thymus DNA heated for 3 minutes at 100° had about half the stabilizing effect of the untreated DNA.

### Deoxyribonucleases in "polymerase" fractions

During the course of purification of "polymerase" the relative amounts of DNase were considerably reduced but DNase activity was not completely removed even from the best preparation (Table IV).

Results of immunological studies not only supported the indications from fractionation data (Table IV) that DNase A and B were distinguishable, but also suggested that there are two distinct enzymes in the DNase B group.

Using an enzyme fraction rich in DNase A and B (AS 1 (Table IV), collected just prior to Fraction VI) as antigen, rabbit antisera were produced which neutralized 95 per cent of DNase B but only 50 per cent or less of DNase A, even with larger amounts of the sera. In addition, these sera neutralized less than 10 per cent of the DNase B in the adjacent enzyme fraction (AS 2). Inasmuch as the failure of the antisera to neutralize much of the DNase B in fraction AS 2 might have been due to an inhibitory substance in AS 2, equal amounts of fraction AS 1 and AS 2 (in terms of DNase B units) were mixed and then treated with the antisera; approximately 50 per cent of the DNase was neutralized, indicating the absence of such an inhibitor. In the most purified enzyme fractions (Fraction VII refractionated with diethylaminoethyl cellulose), DNase A activity was reduced to levels of 5 per cent, or less than that of "polymerase". DNase B, although persisting to a significant extent, was considerably reduced in activity by modifying the assay conditions of "polymerase" (last line of Table IV).

### Discussion

The extensive purification of the enzyme has not yet resulted in a homogeneous preparation,

---
[10] Gift from Dr. L. A. Heppel.

and it has not removed the last traces of DNase activities. Further enzyme purification is limited by the small yields of the purified fraction (1 kilo of *E. coli* yields less than 10 mg. of the purified enzyme), and the indications are that there are at least three distinct nucleases to be considered. Preliminary studies of enzymes which appear to be comparable to the "polymerase" in extracts of other bacteria indicate lower activities than those found in *E. coli* extracts, and the DNA-synthesizing systems in acetone powder extracts of calf thymus gland of HeLa (tumor) cells appear to be only 1 to 2 per cent as active. It would appear that *E. coli*, with a generation time of only 20 minutes, is likely to prove at least as fertile a source of this enzyme as most other cells available for large-scale work.

In connection with the phosphorylation of the four deoxynucleotides (which commonly occur in DNA) to the triphosphate level by enzymes in *E. coli*, it is of interest to recall the observation (29) that no such kinase activity for deoxyuridine 5'-phosphate was detectable. This finding led to the suggestion that the lack of an enzyme to make deoxyuridine triphosphate might explain the lack of uracil in DNA. Current studies[11] which show that deoxyuridine triphosphate, prepared by chemical deamination of deoxycytidine triphosphate, can replace thymidine triphosphate and can be incorporated into the enzymatically synthesized DNA, furnish additional support for this suggestion.

## Summary

An enzyme which catalyzes the incorporation of deoxyribonucleotides from the triphosphates of deoxyadenosine, deoxyguanosine, deoxycytidine and thymidine into deoxyribonucleic acid has been purified from cell-free extracts of *Escherichia coli* in excess of 2000-fold. The reaction mixture includes polymerized deoxyribonucleic acid and $Mg^{++}$.

The deoxynucleoside triphosphate substrates were synthesized from the deoxynucleotides by kinases partially purified from *Escherichia coli*. Procedures for the preparation of $P^{32}$-labeled deoxynucleotides have also been described.

---

[11] Unpublished results.

## REFERENCES

1. Kornberg, A., Advances in Enzymol. **18**:191 (1957).
2. Kornberg, A. In W. D. McElroy, and B. Glass (Editors), The chemical basis of heredity, Johns Hopkins Press, Baltimore, 1957, p. 579.
3. Koshland, D. E. Jr. In W. D. McElroy, and B. Glass (Editors), The mechanism of enzyme action, Johns Hopkins Press, Baltimore, 1954, p. 608.
4. Grunberg-Manago, M., Ortiz, P. J., and Ochoa, S., Biochim. et Biophys. Acta **20**:269 (1956).
5. Kornberg, A., Lehman, I. R., and Simms, E. S., Federation Proc. **15**:291 (1956).
6. Kornberg, A., Lehman, I. R., and Bessman, M. J., and Simms, E. S., Biochim. et Biophys. Acta **21**:197 (1956).
7. Bessman, M. J., Lehman, I. R., Simms, E. S., and Kornberg, A., Federation Proc. **16**:153 (1957).
8. Hershey, A. D., and Chase, M., J. Gen. Physiol. **36**:39 (1952-53).
9. Sinsheimer, R. L., and Koerner, J. F., J. Biol. Chem. **198**:293 (1952).
10. Heppel, L. A., and Hilmoe, R. J., J. Biol. Chem. **188**:655 (1951).
11. Wittenberg, J., and Kornberg, A., J. Biol. Chem. **202**:431 (1953).
12. Keilin, D., and Hartree, E. F., Proc. Roy. Soc. London, B **124**:397 (1938).
13. Willstätter, R., and Kraut, H., Ber. deut. chem. Ges. **56**:1117 (1923).
14. Rose, I. A., Grunberg-Manago, M., Korey, S. R., and Ochoa, S., J. Biol. Chem. **211**:737 (1954).
15. Whitfeld, P. R., Biochem. J. **58**:390 (1954).
16. MacDonald, N. S., Thompsett, J., and Mead, J. F., Anal. Chem. **21**:315 (1949).
17. Kornberg, A., J. Biol. Chem. **182**:779 (1950).
18. Mejbaum, W., Z. physiol. Chem. Hoppe-Seyler's **258**:117 (1939).
19. Brody, S., Acta Chem. Scand. **7**:502 (1953).
20. Beaven, G. H., Holiday, E. R., and Johnson, E. A. In E. Chargaff, and J. N. Davidson (Editors), The nucleic acids, Vol. I, Academic Press, Inc., New York, 1955, p. 493.
21. Hall, R. H., and Khorana, H. G., J. Am. Chem. Soc. **76**:5056 (1954).
22. Herriott, R. M., and Barlow, J. L., J. Gen. Physiol. **36**:17 (1952-53).
23. Anderson, T. F., Rappaport, C., and Muscatine, N. A., Ann. inst. Pasteur **84**:5 (1953).
24. Kay, E. R. M., Simmons, N. S., and Dounce, A. L., J. Am. Chem. Soc. **74**:1724 (1952).
25. Littauer, U. Z., and Kornberg, A., J. Biol. Chem. **226**:1077 (1957).
26. Dische, Z. In E. Chargaff, and J. N. Davidson (Editors), The nucleic acids, Vol. I, Academic Press, Inc., New York, 1955, p. 285.
27. Peterson, E. A., and Sober, H. A., J. Am. Chem. Soc. **78**:751 (1956).
28. Tamm, C., Hodes, M. E., and Chargaff, E., J. Biol. Chem. **195**:49 (1952).

29. Friedkin, M., and Kornberg, A. In W. D. McElroy, and B. Glass (Editors), The chemical basis of heredity, Johns Hopkins Press, Baltimore, 1957, p. 609.
30. Klenow, H., and Lichter, E., Biochim. et Biophys. Acta **23**:6 (1957).
31. Hecht, L. I., Potter, V. R., and Herbert, E., Biochim. et Biophys. Acta **15**:134 (1954).
32. Sable, H. Z., Wilber, P. B., Cohen, A. E., and Kane, M. R., Biochim. et Biophys. Acta **13**:1956 (1954).
33. Ochoa, S., and Heppel, L. In W. D. McElroy, and B. Glass (Editors), The chemical basis of heredity, Johns Hopkins Press, Baltimore, 1957, p. 615.

# 11 ENZYMATIC SYNTHESIS OF DEOXYRIBONUCLEIC ACID; XXVIII. THE PYROPHOSPHATE EXCHANGE AND PYROPHOSPHOROLYSIS REACTIONS OF DEOXYRIBONUCLEIC ACID POLYMERASE

Murray P. Deutscher*
Arthur Kornberg
Department of Biochemistry
Stanford University School of Medicine
Stanford, California

*Summary.* The pyrophosphate ($PP_i$) exchange reaction catalyzed by *Escherichia coli* DNA polymerase is identical with the polymerization reaction in its requirements for a template, strict specificity in base pairing, and a 3'-hydroxyl-terminated primer strand. However, in contrast to polymerization, appreciable exchange is obtained in the absence of a full complement of deoxyribonucleoside triphosphates. Inhibition of synthesis by $PP_i$ as measured by the difference between $PP_i$ release and nucleotide incorporation from deoxyribonucleoside triphosphates, can be accounted for by $PP_i$ exchange. The $PP_i$ exchange reaction appears to represent the removal by $PP_i$ of the newly incorporated nucleotide prior to its complete stabilization by the entry of the next triphosphate.

The degradation of DNA by $PP_i$ (pyrophosphorolysis) appears to differ from the $PP_i$ exchange reaction in that it is inhibited by deoxyribonucleoside triphosphates, attains a steady state plateau, and has a lower pH optimum and a slower rate. Some of these differences may depend on the requirement that progressive pyrophosphorolysis places on progressive movement of the DNA chain relative to the enzyme.

The results are discussed in relation to the mechanism of the $PP_i$ exchange reaction and to all polymerase functions in a model with a single active center for the enzyme.

From Journal of Biological Chemistry 244:3019-3028, 1969. Reprinted with permission.
This study was supported in part by grants from the National Institutes of Health (United States Public Health Service), the National Science Foundation and the National Aeronautics and Space Administration. The previous paper in this series is Reference 1.
*Postdoctoral Fellow of the American Cancer Society. Present address, Department of Biochemistry, University of Connecticut Health Center, Farmington, Connecticut 06032.

The purified DNA polymerase of *Escherichia coli* catalyzes several different reactions. These include (as shown below) (*a*) polymerization of deoxyribonucleoside triphosphates with the concomitant release of $PP_i$ (2), (*b*) exchange of the $\beta, \gamma$-phosphates of deoxyribonucleoside triphosphates with $PP_i$ (2, 3), (*c*) pyrophosphorolysis of DNA to form deoxyribonucleoside triphosphates (2, 3), and (*d*) hydrolysis of DNA to form deoxyribonucleoside monophosphates (4).

$$XTP + DNA_n \rightleftharpoons DNA_{n+1} + PP_i \quad (a)$$
$$XTP + DNA_n + PP_i^* \rightleftharpoons XTP^* + DNA_n + PP_i \quad (b)$$
$$DNA_n + PP_i \rightleftharpoons DNA_{n-1} + XTP \quad (c)$$
$$DNA + H_2O \rightleftharpoons DNA_{n-1} + XMP \quad (d)$$

These reactions are all carried out by homogeneous preparations of DNA polymerase and thus are not a consequence of the presence of contaminating activities. The ability of DNA polymerase to catalyze all these reactions has enabled us to examine its mechanism of action from several different directions. We expect that an understanding of all the aspects of DNA polymerase activity will contribute to a deeper insight into its mode of action.

In this paper we will describe the properties, relative magnitudes, and relationship of the various reactions catalyzed by DNA polymerase. Particular emphasis will be put on the pyrophosphate exchange reaction and the question of whether an isolatable nucleotidyl-enzyme intermediate is formed during polymerase action. In subsequent papers we will examine more closely the hydrolytic activities associated with DNA polymerase, particularly the ability to degrade DNA chains from the 5' end (5) and the way in which hydrolysis is affected by concomitant synthesis.[1] The latter

[1] R. B. Kelly, M. P. Deutscher, N. R. Cozzarelli, L. R. Lehman, and A. Kornberg, unpublished observation.

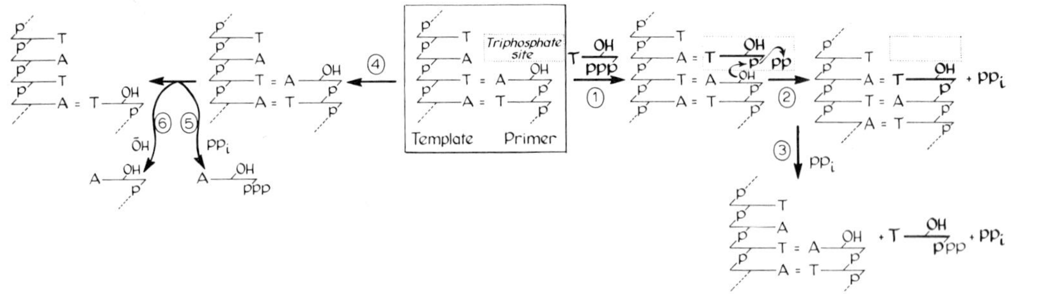

**Scheme 1.** Speculative formulation of the reactions catalyzed by DNA polymerase. See the text for discussion.

study will show the importance of the secondary structure of the DNA within the active site to the catalytic activity of the enzyme.

To facilitate the presentation of our results, these several reactions of the enzyme have been formulated in a simplified and provisional scheme (Scheme 1). In Step 1 the enzyme-DNA complex binds the proper deoxyribonucleoside triphosphate in a manner that results in the covalent linkage of the deoxyribonucleotide to the primer, the release of $PP_i$, and the downward movement of the chain (Step 2) to reopen the triphosphate site of the enzyme. Another consequence of the reaction in Step 1 is that $PP_i$ in the medium can exchange with the $\beta,\gamma$-pyrophosphate of the triphosphate, as in Step 3. There is no evidence that the stage at which $PP_i$ reverses the reaction is a nucleotidyl-enzyme intermediate; rather, the incoming nucleotide appears to have approached the state of covalent attachment to the primer and to have left the nascent triphosphate site of the enzyme in order that the next triphosphate may enter. Degradation of the primer can occur by pyrophosphorolysis (Step 5) of hydrolysis (Step 6). It is not certain what movement, if any, of the primer (substrate) chain (Step 4) is necessary to bring the diester bond which is targeted for attack by $PP_i$ or $OH^-$ within the proper subsite of the enzyme.

A single center for several DNA polymerase functions has been proposed by Beyersmann and Schramm (6) in their recent studies of the mechanism of its action.

### Experimental procedure
*Materials*

Unlabeled deoxyribonucleoside and ribonucleoside triphosphate and *p*-nitrophenyl-5'-deoxyribothymidylate were purchased from Calbiochem. $\alpha,\beta$-dTTP-methylene diphosphonate was prepared by a method to be described (7). 2',3'-Dideoxy-TTP, 2',3'-dideoxy-3'-iodo-TTP, and 2',3'-dideoxy-2',3'-dehydro-TTP were gifts from Dr. Alan F. Russell and Dr. John G. Moffatt. 3'-dTTP and 3'-dATP were gifts from Dr. John Josse. $\alpha$-$^{32}$P-dTTP was prepared as previously described (8) with the use of $\gamma$-$^{32}$P-dATP in the deoxythymidine kinase reaction. $\gamma$-$^{32}$P-dTTP and $\gamma$-$^{32}$P-dATP were prepared as reported earlier (9). $^{32}$P-Orthophosphate was obtained from New England Nuclear. $^{32}$P-Pyrophosphate was prepared by the method of Bergmann (10) and further purified on a Dowex 1-chloride column.

Poly d(A-T),[2] labeled or unlabeled, was prepared as described by Schachman, Adler, Radding, Lehman, and Kornberg (11), except that homogeneous DNA polymerase was used (12). 3'-Phosphate-terminated poly d(A-T) and exonuclease III-treated poly d(A-T) were prepared as reported by Falaschi and Kornberg (13). The oilgomers d(A-T)$_2$ and d(A-T)$_4$ were prepared as described previously (14). d(A-T)$_{10}$, mixed oligo d(A-T), d(A-T)$_{21}$, and circular d(A-T)$_{20}$ were prepared and characterized by Dr. Immo E. Scheffler. $^3$H-*E. coli* DNA was prepared from gently lysed cells (15) by phenol extraction, ribonuclease treatment, and isopropyl alcohol precipitation. Denatured DNA was prepared by heating native DNA in 0.02 M NaCl for 10 min at 90°. Poly dA and poly dT were prepared as described by Riley, Maling, and Chamberlin (16). Oligo dT (chain length about 150) was a gift from Dr. Baldomero M. Olivera and was prepared as described earlier (17). pTpT was a gift from Dr. I. R. Lehman.

---

[2] The abbreviations used are: poly d(A-T), alternating copolymer of deoxyriboadenylate and deoxyribothymidylate, previously designated dAT; d(A-T)$n$, an oligomer of $2n$ alternating residues of deoxyriboadenylate and deoxyribothymidylate; poly dA, polydeoxyriboadenylate; poly dT, polydeoxyribothymidylate; 3'-P, a 3'-phosphate-terminated chain.

Homogeneous DNA polymerase (18,000 poly d(A-T) units per mg) was prepared as described by Jovin, Englund, and Bertsch (12). Pancreatic DNase and micrococcal nuclease were products of Worthington.

## Methods

*Assay of synthesis.* The assay measures the conversion of $\alpha\text{-}^{32}$P-deoxyribonucleoside triphosphate into an acid-insoluble product according to the method of Richardson *et al.* (3). The incubation mixture (0.3 ml) contained 67 mM potassium phosphate buffer, pH 7.4, 6.7 mM $MgCl_2$, 1 mM mercaptoethanol, 7.8 µmoles of poly d(A-T) (expressed as nucleotide), 33 µM dATP, 33 µM $\alpha\text{-}^{32}$P-dTTP, and the indicated amounts of DNA polymerase. The reaction was stopped by chilling and the addition of 0.1 ml of 0.1 M sodium pyrophosphate and 0.5 ml of cold 1 M perchloric acid. The acid-insoluble radioactivity was determined by collection of the precipitate on a glass fiber filter. The filter was washed with 30 ml of 5% trichloracetic acid containing 0.02 M sodium pyrophosphate, followed by 5 ml of ethanol-ether, 1:1. The filters were dried and the radioactivity was determined in a liquid scintillation counter. One unit of DNA polymerase, the conversion of 10 nmoles of nucleotide to an acid-insoluble form, is the same as described previously (3).

*Assay of degradation.* The assay measures the conversion of $^3$H- or $^{32}$P-labeled polynucleotide to an acid-soluble form as described by Lehman and Richardson (4). The incubation mixture was identical with that for synthesis except that labeled polynucleotide was present. The reaction was stopped by chilling and the consecutive additions of 0.1 ml of calf thymus DNA (2.5 mg per ml) and 0.4 ml of 1 M perchloric acid. After 10 min at $0°$, the precipitate was removed by centrifugation and 0.4 ml of the supernatant fluid was added to 0.7 ml of 1.15 M Tris. Dioxane-based scintillation fluid (12) was added (10 ml), and acid-soluble radioactivity was determined in a liquid scintillation counter.

*Assay of $PP_i$ release.* The assay measures the conversion of $\gamma\text{-}^{32}$P-deoxyribonucleoside triphosphate to an acid-soluble, Norit-nonadsorbable form. The incubation mixture was identical with that for synthesis, except that $\gamma\text{-}^{32}$P-triphosphate was present. The reaction was stopped by chilling and the successive addition of 0.1 ml of 0.1 M sodium pyrophosphate, 0.4 ml of 1 M perchloric acid, and 2 drops of 25% (w/v) Norit. After 10 min at $0°$, the Norit was removed by centrifugation and the radioactivity in 0.4 ml of the supernatant fluid was determined as described above for the assay of degradation.

*Assay of $PP_i$ exchange and pyrophosphorolysis.* The assay measures the conversion of $^{32}$P-$PP_i$ to an acid-soluble, Norit-adsorbable form. The incubation mixture was identical with that for synthesis except that 1 mM $^{32}$P-$PP_i$ (0.5 to 1.0 µCi per µmole) was present. No triphosphates were present in the assay for pyrophosphorolysis. The reaction was stopped by chilling and the addition of 1.0 ml of 2 M perchloric acid containing 0.4 M sodium pyrophosphate followed by the addition of 2 drops of 25% Norit. After 10 min at $0°$, the Norit was collected on a glass fiber filter, washed with 50 ml of water, and placed in a planchet, and the radioactivity was determined in a gas flow counter.

## Results

*Identity and magnitude of reactions catalyzed by DNA polymerase.* In order to understand the mechanism of polymerization and the relationships among the various reactions catalyzed by DNA polymerase, the relative magnitudes of these different processes were examined. The reactions were measured under identical conditions with the use of $^3$H-labeled poly d(A-T) and either $^{32}$P-labeled or unlabeled deoxyribonucleoside triphosphates and $PP_i$. In this manner we were able to determine the rate of degradation of the original poly d(A-T) by hydrolysis or pyrophosphorolysis, as well as the rate of polymerization of triphosphates, $PP_i$ release, and $PP_i$ exchange. These reactions were measured in the presence or absence (where applicable) of 1 mM $PP_i$ in order to assess the magnitude of the synthetic reaction under conditions in which $PP_i$ exchange also was occurring. Synthesis was determined by the amount of $\alpha\text{-}^{32}$P-deoxyribonucleoside triphosphate incorporated into polymer as well as by the amount of $^{32}$P-$PP_i$ released from a $\gamma$-labeled triphosphate. The determination of $^{32}$P-$PP_i$ release has the added advantage of measuring total nucleotide incorporation, since any incorporated nucleotide which is subsequently released by degradation of the newly synthesized poly d(A-T) is also detected. That the product released from $\gamma\text{-}^{32}$P-deoxyribonucleoside triphosphate is exclusively $PP_i$ was shown by electrophoresis of the Norit-nonadsorbable product.

The rates of these reactions at pH 7.4 are presented in Table I. In the absence of $PP_i$, synthesis, as measured by nucleotide incorporation, was essentially identical with $PP_i$ release, indicating that little, if any, of the newly synthesized poly d(A-T) was degraded during the 30-min incubation. This conclusion was

## TABLE I
*Relative magnitude of DNA polymerase reactions*

The various reactions were assayed as described in "Methods." Reaction mixtures contained, in 0.3 ml, 67 mM potassium phosphate buffer, pH 7.4; 6.7 mM $MgCl_2$; 1 mM mercaptoethanol; 10 nmoles, as nucleotide, of $^3$H-poly d(A-T) (1,000 cpm per nmole); 0.05 µg of DNA polymerase; and, when present, 33 µM dATP, 33 µM dTTP, and 1 mM $PP_i$. The release of $PP_i$ was measured with $\gamma$-$^{32}$P-dTTP (14,000 cpm per nmole); synthesis was determined with $\alpha$-$^{32}$P-dTTP (9,000 cpm per nmole); and exchange and pyrophosphorolysis were measured, with $^{32}$P-$PP_i$ (1,400 cpm per nmole). Unlabeled triphosphates or $PP_i$ were added as required. Incubations were for 30 min at 37°.

| Reaction measured | Rate of reaction | |
|---|---|---|
| | $-PP_i$ | $+PP_i$ |
| | nmoles/30 min | |
| Poly d(A-T) + dATP + dTTP | | |
| Release of $PP_i$ | 4.22 | 4.38 |
| Synthesis | 4.24 | 2.30 |
| $PP_i$ exchange | | 2.19 |
| Degradation | 2.35 | 1.71 |
| Poly d(A-T) | | |
| Degradation | 1.66 | 1.56 |
| Pyrophosphorolysis | | 0.38 |

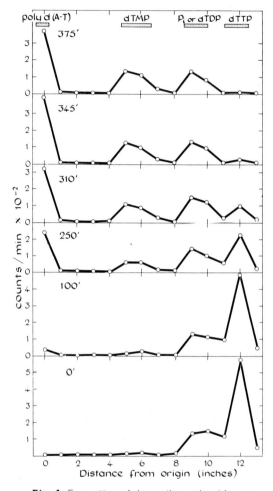

**Fig. 1.** Formation of deoxyribonucleoside mono- and triphosphates during the course of extensive synthesis of poly d(A-T). The reaction mixture (10 ml) contained 60 mM potassium phosphate buffer, pH 7.4, 6 mM $MgCl_2$, 1 mM mercaptoethanol, 0.5 mM dATP, 0.5 mM $^{32}$P-dTTP ($2 \times 10^4$ cpm per n mole), 2.2 µM poly d(A-T), and 9 units of DNA polymerase. Incubation was at 37°. At the indicated times, 2-µl aliquots were withdrawn and added to 20 µl of 10 mM dTMP-10 mM dTTP. Samples of 10 µl were subjected to electrophoresis (5000 volts) in 20 mM sodium citrate, pH 3.5, for 25 min. dTMP and dTTP were located by an ultraviolet lamp. Strips of 1 inch were cut out and the radioactivity was determined. The peak of radioactivity between dTMP and dTTP represents orthophosphate or dTDP contaminating the $^{32}$P-dTTP preparation. The size of this peak remained constant throughout the experiment. The $^{32}$P-poly d(A-T) synthesized during the experiment remained at the origin.

strengthened by an experiment in which $^{32}$P-dTTP was incorporated into poly d(A-T), followed by the addition of a 50-fold excess of unlabeled triphosphates to prevent any further incorporation of label. During a subsequent 30-min incubation less than 2% of the incorporated radioactivity was made acid-soluble. Newly synthesized poly d(A-T) can, however, be hydrolyzed to a greater degree under conditions which favor a manyfold replication of poly d(A-T). In such experiments, only 50 to 60% of the added triphosphates are incorporated into polymer, and the remainder are converted to monophosphates, indicating that they have been polymerized into poly d(A-T) and subsequently degraded. Under these conditions of extensive synthesis, as shown in Fig. 1, mononucleotides are, in fact, formed relatively early in the reaction, and increase in amount continuously. The difference between these results and those shown in Table I is probably due to the fact that conditions for extensive synthesis require only one-tenth the usual starting level of poly d(A-T) and an excess of enzyme. This leads to newly synthesized poly d(A-T) itself being utilized as a template and being rendered susceptible to degradation by exonuclease functions of polymerase. Appreciable degradation of $^3$H-template poly d(A-T) is also seen in Table I, at a rate of about 50% of the poly d(A-T) synthesized. The presence of triphosphates stimulated the degradation of template (2.35 compared to 1.66 nmoles). The stimulating effect of synthesis on the hydrolysis of template will be described in detail in a subsequent paper.[1]

The data in Table I also indicate that $PP_i$ release is not affected by 1 mM $PP_i$. (However, higher levels of $PP_i$ do inhibit; for instance, 2 mM $PP_i$ inhibited about 20%.) When, on the other hand, synthesis was measured by nucleotide incorporation, there was a strong inhibition, amounting to about 50% at 1 mM, $PP_i$ (Table I) and about 85% at 2 mM $PP_i$. The inhibition was found to be first order with respect to $PP_i$, indicating the involvement of only one $PP_i$ molecule in the reaction. The percentage inhibition of synthesis by $PP_i$ under these conditions was found to be greatest at pH 6.5, being approximately 70%, and decreasing to about 20% at pH 9.2. As shown in Table I, the difference between synthesis as measured by nucleotide incorporation and as measured by $PP_i$ release can all be accounted for by $PP_i$ exchange. This result suggests that some intermediate state exists during the polymerization sequence which leads to the release of $PP_i$ from the incoming triphosphate (Step 2 of Scheme 1) and in which there can be attack by $^{32}$P-$PP_i$, resulting in the exchange reaction (Step 3 of Scheme 1). The properties of the exchange reaction and the attempts to define this intermediate state will be discussed more fully below.

An intermediate state might be inferred from studies of $PP_i$ release, polymerization, and $PP_i$ exchange in the presence of a single triphosphate (Table II). With poly d(A-T) and either dATP or dTTP there was relatively little $PP_i$ release, and this value was essentially identical for dATP and dTTP. Although $PP_i$ release from a single triphosphate was low, it was, in fact, approximately 100-fold greater than the incorporation of that nucleotide into poly d(A-T) (200 μμmoles compared to about 2). This result contrasts with that found when both triphosphates were present, and in which

### TABLE II
*DNA polymerase reactions in presence of single deoxyribonucleoside triphosphate*

The various reactions were assayed as described in "Methods" and the legend to Table I, except that either dATP or dTTP was present. Essentially identical results were obtained with each triphosphate; the values shown represent the average of several measurements with each triphosphate alone.

| Reaction measured | Rate of reaction | |
|---|---|---|
| | $-PP_i$ | $+PP_i$ |
| | nmoles/30 min | |
| Release of $PP_i$ | 0.20 | 1.50 |
| Synthesis | 0.002 | |
| $PP_i$ exchange | | 1.47 |

synthesis measured by $PP_i$ release was identical with the process measured by nucleotide incorporation (Table I). The discrepancy between $PP_i$ release from a single triphosphate and incorporation of that nucleotide must be explained by the hydrolytic release of a deoxyribonucleoside monophosphate. Thus, under conditions in which extended synthesis cannot occur, hydrolysis by DNA polymerase of the extended poly d(A-T) chain appears to be extremely active. As will be discussed below, the mechanism of attack in the hydrolytic reaction may be the same as that during pyrophosphorolysis (Steps 5 and 6 of Scheme 1).

Addition of $PP_i$ resulted in a greatly increased rate of $PP_i$ release (Table II). The presence of excess $PP_i$ presumably drives the reaction back, allowing another molecule of triphosphate to react, and thus permits the cycle to be repeated in a type of "shuttle" mechanism. Inorganic tripolyphosphate did not substitute for $PP_i$ in stimulating $PP_i$ release. The stimulation of $^{32}$P-$PP_i$ release by $PP_i$ is identical with $PP_i$ exchange in the presence of a single triphosphate, since the latter is presumably a measure of the same reaction in the reverse direction.

In the absence of triphosphates, DNA polymerase also converts poly d(A-T) to an acid-soluble form. In the presence of $PP_i$ part of the degraded material appears as deoxyribonucleoside triphosphates, by a reversal of polymerization, or pyrophosphorolysis (Table I). With the use of poly d(A-T), the

## TABLE III
*Inhibition of pyrophosphorolysis by deoxyribonucleoside triphosphates*

The production of acid-soluble material from $^3$H-poly d(A-T) (1000 cpm per nmole) and $^3$H-*E. coli* DNA (600 cpm per nmole) was measured as described in "Methods." Reactions were performed at pH 6.5 in 67 mM potassium phosphate buffer with either 10 nmoles of poly d(A-T) or 46 nmoles of DNA and 0.05 μg of DNA polymerase. Where indicated, deoxyribonucleoside triphosphates were present at 33 μM and PP$_i$ at 1 mM. Incubations were at 37° for 30 min.

| Polynucleotide and additions | Nucleotide made acid-soluble |
|---|---|
| | nmole |
| Poly d(A-T) | |
| None | 0.18 |
| dATP, dTTP | 0.30 |
| PP$_i$ | 0.63 |
| dATP, dTTP, PP$_i$ | 0.23 |
| DNA | |
| None | 0.05 |
| dATP, dTTP, dGTP, dCTP | 0.09 |
| PP$_i$ | 0.14 |
| dATP, dTTP, dGTP, dCTP, PP$_i$ | 0.04 |

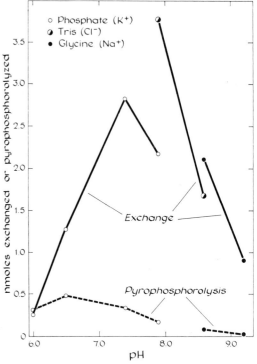

**Fig. 2.** pH optimum for pyrophosphate exchange and pyrophosphorolysis. Exchange and pyrophosphorolysis were measured as described in "Methods," with 1 unit of DNA polymerase. Buffers of the indicated pH were present at 67 mM. pH values for the different buffers were determined at concentrations of 50 mM at 37°. Incubations were for 30 min at 37°.

maximal rate of hydrolysis to form deoxyribonucleoside monophosphates was approximately 5 times more rapid than pyrophosphorolysis under optimal conditions for each reaction. However, owing to limitations on the amount of PP$_i$ that can be added to the reaction mixture, the maximum possible rate of pyrophosphorolysis might not have been attained.

Results similar to those shown in Table I for poly d(A-T) have also been obtained with native and denatured *E. coli* DNA as template, although the rates of the various reactions were only about 2% of those obtained with poly d(A-T) (3).

*Comparison of pyrophosphorolysis and pyrophosphate exchange.* Pyrophosphate exchange into a deoxyribonucleoside triphosphate is several times more rapid than the pyrophosphorolytic release of triphosphates (Table I), but otherwise might appear to resemble the latter. Is pyrophosphate exchange simply some form of triphosphate-stimulated pyrophosphorolysis? On the contrary, triphosphates actually inhibit pyrophosphorolysis. In one experiment $^{32}$P-poly d(A-T) was degraded to mono- and triphosphates at pH 7.4 in the presence of 1 mM PP$_i$. The triphosphates, as determined by Dowex 1 chromatography, represented 15 to 20% of the acid-soluble products. In the presence of dATP and dTTP, the hydrolytic reaction (monophosphate production) was stimulated, but pyrophosphorolysis was inhibited by approximately 50%. In a second experiment $^3$H-poly d(A-T) and $^3$H-*E. coli* DNA were each degraded by DNA polymerase at pH 6.5 in the presence or absence of PP$_i$ (Table III). At this pH, which is optimal for pyrophosphorolysis (see Fig. 2), the major portion of acid-soluble material formed in the presence of PP$_i$ represents triphosphates. The addition of dATP and dTTP in the case of poly d(A-T), and of all four triphosphates in the case of DNA, stimulated the hydrolytic reaction but abolished pyrophosphorolysis. dATP or dTTP, present alone, also inhibited the pyrophosphorolytic reaction with poly d(A-T). Triphosphate inhibition of pyrophosphorolysis has also

Synthesis of DNA in vitro 97

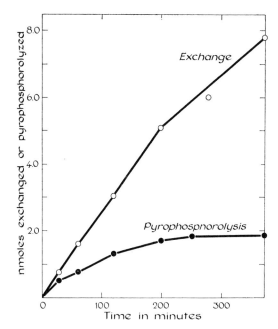

**Fig. 3.** Kinetics of PP$_i$ exchange and pyrophosphorolysis of poly d(A-T). Pyrophosphorolysis and exchange were measured under the usual conditions at pH 6.5, except that 31 nmoles of poly d(A-T) and 1 unit of DNA polymerase were present. Triphosphates (0.33 mM) were present in the exchange assays. During the 6-hour period of the experiment, only 3 nmoles of poly d(A-T) were hydrolyzed by exonuclease, so that saturating levels of poly d(A-T) remained throughout the incubation.

been reported by Beyersmann and Schramm (6). As is shown in Table IV, only triphosphates which can be incorporated are able to inhibit the pyrophosphorolysis of poly d(A-T) when present at the same levels as dATP and dTTP.

Pyrophosphate exchange and pyrophosphorolysis have different pH optima (Fig. 2). Exchange exhibits an optimum at pH 7.4 in potassium phosphate, whereas pyrophosphorolysis is optimal at pH 6.5. These results also show that the rate of pyrophosphate exchange is considerably greater than that of pyrophosphorolysis at every pH value tested except 6.0. Although the two processes may involve the same intermediate state, pyrophosphorolysis also requires progressive movement of the polynucleotide chain relative to the enzyme, and this may slow down the rate.

Whereas pyrophosphorolysis of poly d(A-T) reaches a maximal level and then stops (Fig. 3), exchange proceeds at a nearly linear rate for at least 6 hours. The apparent cessation of pyrophosphorolysis is probably due to the accumulation of deoxyribonucleoside triphosphates, which are then reincorporated into poly d(A-T). The levels of triphosphates present at the plateau would be sufficient to maintain a rate of synthesis equal to that of pyrophosphorolysis. The plateau of pyrophosphorolysis was not due to inactivation of the enzyme, since the addition of dATP and dTTP at 6 hours resulted in the expected increment due to exchange over a 30-min period.

*Effects of nucleoside triphosphates on exchange and pyrophosphorolysis.* Does the strict base specificity which obtains in the synthetic reaction (18) apply to the formation of the intermediate state as well? Various deoxyribonucleoside triphosphates and their analogues were tested for their ability to support exchange in the presence of poly d(A-T), as shown in Table IV. In the absence of triphosphates, the incorporation of $^{32}$P-PP$_i$ into Norit-adsorbable material represents pyrophosphorolysis of the poly d(A-T) chain. As shown in Tables I and II, appreciable exchange occurred in the presence of dATP and dTTP, as well as in the presence of each triphosphate alone, in contrast to synthesis in which a full complement of triphosphates was required for extended polymerization. Results similar to these were observed early in the course of study of DNA polymerase (2), with DNA as the substrate. The lower exchange with dATP or dTTP alone was not due to a suboptimal triphosphate concentration, since doubling the dATP level had no effect on the rate of exchange. Presumably, the lower exchange in the single triphosphate (amounting to about 50% of the value with two triphosphates present) is due to the fact that with the alternating poly d(A-T) copolymer, at any given instant, half the enzyme molecules would be bound to poly d(A-T) chains requiring the triphosphate that was absent. These complexes would be abortive for exchange, and would account for the 50% diminution of rate.

The results in Table IV also show that with poly d(A-T) as template there is no exchange into dGTP or dCTP, alone or in combination. Furthermore, these incorrect triphosphates do not inhibit exchange into dATP and dTTP, and also have no effect on pyrophosphorolysis of

TABLE IV
*Pyrophosphate exchange with different nucleoside triphosphates in presence of poly d(A-T)*

Pyrophosphate exchange reactions were performed under the usual conditions of assay (see "Methods"). Nucleoside triphosphates were present at 33 μM. Incubations were at 37° for 30 min with 0.05 μg of DNA polymerase.

| Nucleoside triphosphates present | $^{32}$P-PP$_i$ converted to Norit-adsorbable form |
|---|---|
| | nmoles |
| None | 0.34 |
| Standard deoxyribonucleoside 5'-triphosphates | |
| dATP, dTTP | 2.57 |
| dATP | 1.44 |
| dTTP | 1.50 |
| dGTP | 0.33 |
| dCTP | 0.33 |
| dGTP, dCTP | 0.34 |
| dATP, dTTP, dGTP, dCTP | 2.50 |
| Ribonucleoside triphosphates | |
| ATP | 0.31 |
| UTP | 0.38 |
| Deoxyribonucleoside 3'-triphosphates | |
| 3'-dATP, 3'-dTTP | 0.36 |
| 3'-dATP, 3'-dTTP + dATP, dTTP | 2.62 |
| Other 5'-triphosphate analogues | |
| 2',3'-Dideoxy-3'-iodo-TTP | 0.29 |
| 2',3'-Dideoxy-3'-iodo-TTP + dATP | 0.85 |
| 2',3'-Dideoxy-TTP | 0.11 |
| 2',3'-Dideoxy-TTP + dATP | 0.14 |
| 2',3'-Dideoxy-2',3'-dehydro-TTP | 0.09 |
| 2',3'-Dideoxy-2',3'-dehydro-TTP + dATP | 0.15 |
| α,β-dTTP-methylene diphosphonate | 0.31 |
| α,β-dTTP-methylene diphosphonate + dATP | 1.46 |

poly d(A-T). When dGTP and dCTP are present at 10-fold higher concentrations they do inhibit exchange into the correct triphosphate, by 42% and 17%, respectively. However, under these conditions there is still no exchange into dGTP or dCTP, and no inhibition of pyrophosphorolysis. Similarly, there is no exchange with ATP or UTP, and these triphosphates also do not inhibit pyrophosphorolysis of poly d(A-T). Identical results were also obtained with the 3'-triphosphates of dATP and dTTP. These compounds did not support exchange and did not inhibit exchange with dATP and dTTP, nor did they have any effect on pyrophosphorolysis. These results indicate that the base specificity of DNA polymerase extends to the intermediate stage represented by exchange, and that at usual levels the incorrect triphosphates do not inhibit exchange or pyrophosphorolysis.

We then sought to examine pyrophosphate exchange with a series of analogues of dTTP which would be expected to have base pairing properties identical with those of dTTP. These analogues included three that were modified in the sugar residue, 2',3'-dideoxy-TTP, 2',3'-dideoxy-2',3'-dehydro-TTP, and 2',3'-dideoxy-3'-iodo-TTP, and one which contained a methylene bridge between the α and β phosphates and thereby could not be activated (α,β-dTTP-methylene diphosphonate). None of these analogues supported pyrophosphate exchange, but of even more significance was the finding that in the case of 2',3'-dideoxy-TTP and 2',3'-dideoxy-2',3'-dehydro-TTP less $^{32}$P-PP$_i$ was incorporated into a Norit-adsorbable form than in the absence of triphosphates. Thus, these two analogues inhibited pyrophosphorolysis by as much as 70%. They also abolished exchange into dATP, whereas the phosphonate had no effect and 3'-iodo-dTTP inhibited exchange by only about 30%. Since the phosphonate did not act as an inhibitor of either pyrophosphorolysis or exchange, these results suggest that incorporation of the 2',3'-dideoxy-TTP and 2',3'-dideoxy-2',3'-dehydro-TTP is necessary for their inhibitory action. Preliminary results[3] with $^3$H-2',3'-dideoxy-TTP indicate that this analogue is, in fact, incorporated on the end of a poly d(A-T) chain, but that it is removed very slowly by exonuclease action. Since this analogue inhibits pyrophosphorolysis, it is also not removed readily from the end of a poly d(A-T) chain by PP$_i$. The significance of these results in relation to the nature of the activated intermediate will be discussed more fully below.

*Effects of polynucleotides on exchange and pyrophosphorolysis.* Very small oligonucleotides, such as d(A-T)$_2$ and d(A-T)$_4$, supported little exchange or pyrophosphorolysis under standard conditions (Table V). Similar results have previously been observed for these oligonucleotides with respect to their ability to act as primers for synthesis (14). In contrast,

---

[3] M. R. Atkinson, M. P. Deutscher, A. Kornberg, A. F. Russell, and J. G. Moffatt, unpublished observation.

d(A-T)$_{10}$ and a pancreatic DNase digest of poly d(A-T) were as effective as poly d(A-T) for both exchange and pyrophosphorolysis. However, one difference has been observed between the abilities of poly d(A-T) and d(A-T)$_{10}$ to serve as substrates for these reactions. With poly d(A-T), the level of polymer required to produce maximal rates was the same for the exchange reaction and for pyrophosphorolysis. With d(A-T)$_{10}$, however, the concentrations required for maximal pyrophosphorolysis (as in Table V) were approximately 500-fold higher than those required for optimal PP$_i$ exchange. This result may be due to the necessity for polymerase to find repeatedly new chains after pyrophosphorolysis of a short oligomer, such as d(A-T)$_{10}$, whereas this would not be required with a polymeric substrate. The length of the chain, once greater than a minimum value, is not important for the pyrophosphate exchange since this reaction involves only the addition and removal of the last entering nucleotide, but would not affect the chain length of the substrate.

3'-P-Poly d(A-T) and 3'-P-DNA were both essentially inactive as substrates for both exchange and pyrophosphorolysis (Table V). These results are consistent with earlier observations that 3'-P groups on DNA inhibit synthesis (19). Removal of the 3'-P groups from poly d(A-T) with exonuclease III restored full activity in both reactions. These results emphasize the importance of the 3'-hydroxyl group for DNA polymerase action, including the step represented by PP$_i$ exchange. These conclusions will be discussed more fully below.

Poly dA and poly dT, likewise, did not support exchange or pyrophosphorolysis when present alone (Table V). Similarly, these polymers did not act as templates for polymerization even when present at 5 times the level used in these experiments. However, when both polymers were present in a reaction mixture, presumably forming the double-stranded poly dA-poly dT homopolymer pair, there was appreciable exchange, pyrophosphorolysis, and synthesis. Furthermore, the exchange into either dATP or dTTP, each added individually, was similar.

Compared to the synthetic polynucleotides, native *E. coli* DNA was a relatively poor substrate for the exchange and pyrophosphoro-

## TABLE V
*Pyrophosphate exchange and pyrophosphorolysis with different polynucleotides*

Pyrophosphate exchange and pyrophosphorolysis were measured as described in "Methods" with the following amounts of polynucleotide substrates (nanomoles of nucleotide per 0.3 ml): poly d(A-T), 7.8; d(A-T)$_2$, 7.9; d(A-T)$_4$, 4.0; d(A-T)$_{10}$, 8.7; pancreatic DNase-treated poly d(A-T), 7.0; 3'-P-poly d(A-T), 7.2; exonuclease III-treated 3'-P-poly d(A-T), 7.3; poly dA, 3.6; poly dT, 3.5; poly dA-poly dT, 3.6:3.5; and *E. coli* DNA (in all experiments), 46 nmoles. Exonuclease III-treated DNA, degraded 30%, was also present at a level which corresponded to 46 nmoles prior to the degradation. DNA polymerase, 1 unit, was present in all experiments except in those with DNA, which contained 5 units. Incubations were for 30 min at 37°.

| | $^{32}$P-PP$_i$ converted to Norit-adsorbable form | |
|---|---|---|
| Polynucleotide present | Exchange | Pyrophos-phorolysis |
| | nmoles | |
| d(A-T)$_n$ series | | |
| Poly d(A-T) | 2.79 | 0.34 |
| d(A-T)$_2$ | 0.05 | 0.01 |
| d(A-T)$_4$ | 0.09 | 0.05 |
| d(A-T)$_{10}$ | 2.17 | 0.35 |
| Poly d(A-T), treated with DNase | 2.46 | 0.31 |
| 3'-P-poly d(A-T) | 0.12 | 0.03 |
| 3'-P-poly d(A-T), treated with exonuclease III | 2.29 | 0.30 |
| Poly dA-poly dT series | | |
| Poly dA | <0.01 | <0.01 |
| Poly dT | <0.01 | <0.01 |
| Poly dA-poly dT | 1.00 | 0.15 |
| *E. coli* DNA Series | | |
| Native DNA | 0.25 | 0.05 |
| 3'-P-DNA | 0.02 | <0.01 |
| Native DNA, treated with DNase | 2.45 | 0.16 |
| Native DNA, treated with exonuclease III | 1.42 | 0.07 |

lytic reactions, a result which is in agreement with those obtained earlier for the polymerization reaction (3). However, introduction of single-strand nicks into the DNA by pancreatic DNase, an enzyme which produces 3'-hydroxyl groups, resulted in a 10-fold increase in the rate of PP$_i$ exchange and a 3-fold increase in the rate of pyrophosphorolysis, as well as in the well-known effect of increasing synthesis (20). This stimulation by DNase occurred under conditions in which the native DNA had been present in amounts that appeared to saturate the enzyme. Thus, it seems that reaction rates at single-strand breaks are considerably greater than at ends of DNA chains. Another possiblity is that a large proportion of DNA polymerase molecules actually bind to sites on the DNA which do not result in active catalysis. The addition of more DNA does not increase reaction rates, since both active and inactive sites for DNA polymerase are simultaneously added. However, introduction of 3'-hydroxyl groups presumably increases the number of catalytically active binding sites that are suitable for polymerase action.

Treatment of DNA with exonuclease III, which is known to increase rates of polymerization by allowing repair synthesis to occur (21), also led to greatly increased rates of $PP_i$ exchange; however, pyrophosphorolysis was only slightly stimulated (Table V).

*Importance of 3'-hydroxyl-terminated primer in $PP_i$ exchange reaction.* The similarities between the $PP_i$ exchange, $PP_i$ release, and polymerization reactions are all consistent with an intermediate state which is actually part of the sequence leading to polymerization. The question then arises as to the nature of this intermediate state, as well as to the requirements for its formation. Although DNA polymerase is able to bind triphosphate in the absence of a DNA (7), it catalyzes neither $PP_i$ exchange, $PP_i$ release, nor incorporation of a nucleotide onto the enzyme in the absence of a polynucleotide. This is true even with stoichiometric amounts of enzyme. Furthermore, as shown in Table V, the presence of a template or a primer strand alone also is not sufficient for activation, since neither poly dA nor poly dT supports exchange when present at usual levels. Addition of both a template and a primer strand, as in poly dA-poly dT, permits appreciable exchange to occur. In addition to a requirement for both a template and a primer strand, activation of a triphosphate also requires the presence of a 3'-hydroxyl group. For instance, 3'-P-terminated polynucleotides do not support exchange (Table V).

The importance of 3'-hydroxyl groups is further emphasized by two additional experiments. In one, $d(A-T)_{20}$ circles prepared with the *E. coli* polynucleotide joining enzyme (22) were used as substrates for DNA polymerase.

**TABLE VI**
*Pyrophosphate exchange and pyrophosphorolysis with $d(A-T)_{20}$ circles*

Pyrophosphate exchange and pyrophosphorolysis were measured under the usual conditions with 0.6 nmole of circular $d(A-T)_{20}$ or linear $d(A-T)_{21}$. Incubations were for 30 min at 37° with 1 unit of DNA polymerase. Pancreatic DNase, when used, was present during the reaction.

| Polynucleotide present | $^{32}P\text{-}PP_i$ converted to Norit-adsorbable form | |
|---|---|---|
| | Exchange | Pyrophosphorolysis |
| | nmoles | |
| Linear $d(A-T)_{21}$ | 2.69 | 0.13 |
| Circular $d(A-T)_{20}$ | 0.06 | <0.01 |
| Circular $d(A-T)_{20}$ + DNase, 17 ng per ml | 0.64 | |
| Circular $d(A-T)_{20}$ + DNase, 170 ng per ml | 1.73 | |

**TABLE VII**
*Comparison of poly dA-poly dT and poly dA-oligo dT as substrates for $PP_i$ exchange*

Pyrophosphate exchange was measured under the usual conditions as described in "Methods." Poly dA, 3.6 nmoles; poly dT, 3.6 nmoles; and oligo dT, 3.5 nmoles, were present as indicated. Triphosphates were present at 33 μM. Incubations were at 37° for 30 min with 2 units of DNA polymerase.

| Polynucleotide and triphosphate present | $^{32}P\text{-}PP_i$ converted to Norit-adsorbable form |
|---|---|
| | nmoles |
| Poly dA-poly dT | |
| dATP + dTTP | 0.95 |
| dATP | 0.35 |
| dTTP | 0.61 |
| None | 0.17 |
| Poly dA-oligo dT | |
| dATP + dTTP | 1.42 |
| dATP | 0.11 |
| dTTP | 1.42 |
| None | 0.10 |

The data in Table VI reveal that these circles, which have many of the properties of linear poly d(A-T) but lack ends, show no significant support of $PP_i$ exchange or pyrophosphorolysis compared to linear poly d(A-T) molecules of the same chain length. The absence of activity with $d(A-T)_{20}$ circles was not due to modification of the polynucleotide during production of

## TABLE VIII
*Kinetic parameters for pyrophosphate as substrate and inhibitor of DNA polymerase*

The various reactions were assayed at pH 7.4 as described in "Methods." For determination of the apparent $K_m$ values in exchange and pyrophosphorolysis, the $^{32}$P-PP$_i$ concentration was varied from 0.08 to 2 mM; for determination of the $K_i$ values for PP$_i$ as an inhibitor of polymerization, the concentrations of dATP and dTTP or of poly d(A-T) were varied in the presence or absence of 1 mM PP$_i$. Incubations were carried out for 30 min at 37° with 1 unit of DNA polymerase for the exchange and pyrophosphorolysis measurements and 0.2 unit for the polymerization measurements.

| PP$_i$ reactions | $K_m$ | $K_i$ |
|---|---|---|
| | mM | mM |
| PP$_i$ as substrate | | |
|   Exchange | 0.6 | |
|   Pyrophosphorolysis | 0.6 | |
| PP$_i$ as inhibitor of polymerization | | |
|   With respect to poly d(A-T) | | 0.7 |
|   With respect to triphosphates | | 0.7 |

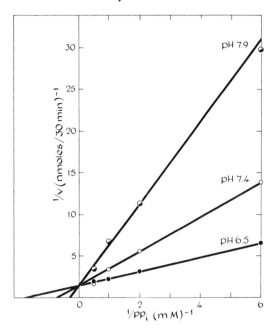

**Fig. 4.** Competition between PP$_i$ and OH$^-$ in the degradative reactions of DNA polymerase. Pyrophosphorolysis was measured as described in "Methods" in 67 mM potassium phosphate buffer at pH 6.5, 7.4, or 7.9 in the presence of 10 nmoles of poly d(A-T) and 1 unit of DNA polymerase. Incubations were for 30 min at 37°.

the circles. This was shown by the restoration of the ability of poly d(A-T) to act as a substrate for exchange when pancreatic DNase was included in the reaction mixture.

In a second experiment, the importance of 3'-hydroxyl groups was shown by a comparison of the exchange dependent on poly dA-poly dT with that found using identical amounts of poly dA-oligo dT. The important difference between these two substrates is that poly dA-oligo dT contains a relatively large number of T ends relative to A ends. As shown in Table VII, when poly dA-poly dT was used as substrate, the rates of exchange into dATP and DTTP were very similar. However, with poly dA-oligo dT virtually all the exchange was into dTTP. The results implicate the 3'-hydroxyl ends of the oligo dT chains and, thus, show a direct involvement of the 3'-hydroxyl group on the primer strand in the PP$_i$ exchange reaction. The simplest interpretation of these results is that the intermediate state implicated in PP$_i$ exchange represents covalent attachment of the incoming nucleotide to the 3'-hydroxyl terminus of the primer. This conclusion will be discussed more fully below.

### Evidence for one active center for all DNA polymerase functions.
Since DNA polymerase catalyzes several different reactions, the question arises as to how these various reactions are related, and whether any evidence exists for their occurrence at the same active center. Several lines of evidence suggest that all polymerase reactions occur within a limited area on the enzyme. First of all, the apparent $K_m$ values for poly d(A-T) in all the reactions of polymerase were very similar, being approximately 4 to 10 μM (expressed as nucleotide). Furthermore, as shown in Table VIII, the apparent $K_m$ values for PP$_i$ as a substrate for exchange and pyrophosphorolysis, as well as its $K_i$ values as a noncompetitive inhibitor of synthesis with respect to poly d(A-T) and triphosphates, are essentially all identical.

It was reported previously that very small oligonucleotides were hydrolyzed by the exonuclease function of DNA polymerase (4). However, small oligonucleotides are not substrates for synthesis, PP$_i$ exchange, or pyrophosphorolysis. If all polymerase functions do occur at the same site, it would be expected that an oligonucleotide which was a substrate for exonuclease, but not for synthesis, would nevertheless be an inhibitor of polymerization. Such inhibition was, in fact, observed with the dinucleotide pTpT. This compound, although not a substrate for polymerization, was a competitive inhibitor of synthesis with respect to poly d(A-T). We observed a $K_i$ value of approximately 1 mM, which is similar to its apparent $K_m$ of 0.34 mM as a substrate for exonuclease action from the $3'$ end (4).

A further correlation among the reactions of DNA polymerase is seen in the reciprocal relationship between hydrolysis and pyrophosphorolysis. Hydrolysis of poly d(A-T) is optimal at pH 9.2 and is approximately 150 times more rapid than pyrophosphorolysis at this pH. Pyrophosphorolysis, on the other hand, increases with decreasing pH, and is optimal at pH 6.5 (Fig. 2). At pH 6.0, pyrophosphorolysis is actually 5 times more active than hydrolysis. Competition between hydrolysis and pyrophosphorolysis was examined by measuring the latter reaction as a function of $PP_i$ concentration at three pH values, 6.5, 7.4, and 7.9. The double reciprocal plots shown in Fig. 4 are consistent with a competition between $PP_i$ and $OH^-$ for the same active site. Results similar to these have recently been reported by Beyersmann and Schramm (6).

## Discussion

In addition to polymerization of nucleotides to form DNA, *E. coli* DNA polymerase also catalyzes several other reactions which appear to be closely related to its synthetic function. These include degradation of DNA by hydrolysis and pyrophosphorolysis, and an exchange reaction between pyrophosphate and deoxyribonucleoside triphosphates. The fact that DNA polymerase catalyzes an exchange reaction seems particularly relevant for an understanding of the mechanism of polymerization, since it suggests that at some stage during the course of polymerization an intermediate is formed which can react with pyrophosphate to regenerate triphosphates. The properties of the exchange reaction in regard to the strict base specificity for activation and to the requirement for a template strand and a $3'$-hydroxyl-terminated primer strand all indicate that the intermediate involved in $PP_i$ exchange is also part of the polymerization sequence, and, in fact, probably represents attack by $PP_i$ on the nucleotide when it is forming a diester bond with the primer. Furthermore, the inhibition by $PP_i$ of polymerization as measured by nucleotide incorporation compared to its determination by $PP_i$ release from a triphosphate can all be accounted for by exchange. This again indicates a close relationship between polymerization and exchange and suggests that in the presence of $PP_i$ the two reactions compete.

We have been unable to find any evidence for the existence of a nucleotidyl-enzyme intermediate. Rather, our data are most consistent with covalent attachment of the incoming nucleotide to the primer. This newly incorporated nucleotide can be attacked by pyrophosphate to regenerate triphosphate or it can be made resistant to $PP_i$ attack by the entry of the next required triphosphate.

Thus, the observation that exchange occurs even in the presence of a full complement of triphosphates suggests that the release of $PP_i$ from the triphosphate and its replacement by $^{32}P-PP_i$ occurs more rapidly than the stabilization of the last nucleotide by the subsequent entry of a new triphosphate. Although only limited synthesis occurs in the presence of a single triphosphate, appreciable exchange can take place since $PP_i$ can remove the last nucleotide, another triphosphate can enter, and the cycle can continue. The observation that there is considerable monophosphate production in the presence of only a single triphosphate suggests that the newly incorporated nucleotide can also be attacked by hydroxide ion. However, since this is not seen when a full complement of triphosphates is present, the attack by hydroxide ion may be relatively slow compared to the stabilization of the terminal nucleotide.

The covalent attachment of the incoming nucleotide to the primer strand as the intermediate involved in $PP_i$ exchange is strongly suggested by studies with analogues of dTTP. Two of these, $2',3'$-dideoxy-TTP and $2',3'$-dideoxy-$2',3'$-dehydro-TTP, inhibited exchange into dATP. This result can be explained most easily by assuming that the analogues are incorporated into the primer, but since they lack a 3'-hydroxyl group they cannot bind properly to the primer site and subsequent exchange with dATP cannot occur. Preliminary evidence indicates that 2',3'-dideoxy-TTP is, in fact, incorporated into the primer.[3] Since these analogues do not support exchange themselves, and also inhibit pyrophosphorolysis, it appears that once incorporated, their removal from the primer by $PP_i$ or, as preliminary evidence suggests, by hydroxide ion, is extremely slow.[3] The observation that 2',3'-dideoxy-TTP is incorporated strongly suggests that an enzyme-nucleotide intermediate is not involved in $PP_i$ exchange. If such an intermediate participated in polymerization, it should also form during the incorporation of the analogue, and lead to an exchange reaction. The absence of exchange with 2',3'-dideoxy-TTP, thus, argues against the existence of an enzyme-nucleotide intermediate. Another analogue, $\alpha,\beta$-dTTP methylene diphosphonate, which cannot be added to the primer, also fails to inhibit exchange into dATP and has no effect on the pyrophosphorolysis of poly d(A-T).

Degradation of DNA by pyrophosphorolysis involves attack by pyrophosphate on the 3'-terminal nucleotide and, in this respect, clearly resembles $PP_i$ exchange. Because of this similarity, it seems plausible that the terminal nucleotide would occupy a position similar to that of the newly incorporated nucleotide involved in $PP_i$ exchange. $PP_i$ exchange appears to differ from pyrophosphorolysis in that the latter reaches an apparent limit, is inhibited by triphosphates, exhibits a different pH optimum, and is relatively slow. The limit reached in pyrophosphorolysis may be ascribed to a steady state between synthesis and breakdown, the triphosphate inhibition may be explained by masking of termini undergoing breakdown, and the lower pH optimum for pyrophosphorolysis as compared to $PP_i$ exchange may reflect the higher pH optimum for the synthetic component of the exchange reaction. What cannot be readily accounted for and requires additional study is the very much slower rate of pyrophosphorolysis. Perhaps the chain terminus attacked in $PP_i$ exchange, having just been generated by the attachment of a nucleotide, may be in a favorable transition state not attained as readily in the pyrophosphorolytic reaction.

The fact that DNA polymerase catalyzes both synthetic and degradative reactions suggests that, relative to a DNA chain, it is capable of progressive movement in both directions. The mechanisms of this movement, and whether the enzyme dissociates from the DNA after each addition or removal of a nucleotide unit, have not been determined. What effect, if any, triphosphates have on this process also remains to be elucidated. These questions about DNA polymerase are also relevant to other enzymes which have the dual ability to synthesize and degrade polymers.

Since DNA polymerase catalyzes several reactions, the question naturally arises as to how these various reactions are related, and whether they occur at the same active center. The close relationship between synthesis, $PP_i$ exchange, and pyrophosphorolysis, and their likely occurrence within the same site, have already been discussed. Similarly, it is not difficult to imagine that the hydrolytic activity of DNA polymerase also occurs within the same site, except that a hydroxide ion would substitute for $PP_i$ resulting in the release of a monophosphate rather than a triphosphate. The conclusion is strengthened by the observation (23) that DNA polymerase contains only a single site for binding polynucleotides. Furthermore, the dinucleotide pTpT which is a substrate for the hydrolytic activity (4), but which is inactive in synthesis, exchange, or pyrophosphorolysis, is, nevertheless, a competitive inhibitor of poly d(A-T) in the synthetic reaction, again suggesting a common site. The fact that hydroxide ion competes with $PP_i$ in the pyrophosphorolytic reaction (see Fig. 4) is also consistent with a common site of action for all polymerase reactions, as shown in Scheme 1.

Several additional observations may serve to support the simple model presented in Scheme 1. (a) DNA polymerase, modified by an acylating agent, retains its ability to hydrolyze DNA but becomes essentially inactive for synthesis (1). Since the ability of the modified enzyme to bind a deoxynucleoside triphosphate has been impaired (1), it may be argued that the triphosphate site is defective but that the primer site, where hydrolysis occurs, remains relatively intact. (b) Although hydroxide ion and $PP_i$ compete with each other, substrates such as pTpT and p-nitrophenyldeoxythymidylate are hydrolyzed but not pyrophosphorolyzed.[1] This may be explained by assuming that pyrophosphorolysis is the reversal of synthesis and that microscopic reversibility requires the formation of a base pair with a template which is lacking in these two substrates. (c) 3'-P-Poly d(A-T) can be hydrolyzed from its 5' terminus (5), but neither this polymer nor 3'-P-DNA is a substrate for pyrophosphorolysis (see Table V).

---

[4] N. R. Cozzarelli, R. B. Kelly, and A. Kornberg, unpublished observation.

These and related observations (5)[4] can be interpreted by assuming that there is a distinctive site for degradation from the 5' terminus which supports hydrolysis but not pyrophosphorolysis.

The studies described here and those to be reported from this laboratory are all in keeping with a number of sites within a single active center of the enzyme. Scheme 1 serves only as first step in formulating a detailed description of the mechanism of action of DNA polymerase.

## REFERENCES

1. Jovin, T. M., Englund, P. T., and Kornberg, A., J. Biol. Chem. **244**:3009 (1969).
2. Bessman, M. J., Lehman, I. R., Simms, E. S., and Kornberg, A., J. Biol. Chem. **233**:171 (1958).
3. Richardson, C. C., Schildkraut, C. L., Aposhian, H. V., and Kornberg, A., J. Biol. Chem., **239**:222 (1964).
4. Lehman, I. R., and Richardson, C. C., J. Biol. Chem. **239**:233 (1964).
5. Deutscher, M. P., and Kornberg, A., J. Biol. Chem. **244**:3029 (1969).
6. Beyersmann, D., and Schramm, G., Biochem. Biophys. Acta **159**:64 (1968).
7. Englund, P. T., Huberman, J. A., Jovin, T. M., and Kornberg, A., J. Biol. Chem. **244**:3038 (1969).
8. Okazaki, R., and Kornberg, A., J. Biol. Chem. **239**:269 (1964).
9. Mitra, S., Reichard, P., Inman, R. B., Bertsch, L. L., and Kornberg, A., J. Mol. Biol. **24**:429 (1967).
10. Bergmann, F. H., in S. P. Colowick and N. O. Kaplan (Editors), Methods in enzymology, Vol. 5, Academic Press, New York, 1962, p. 708.
11. Schachman, H. K., Adler, J., Radding, C. M., Lehman, I. R., and Kornberg, A., J. Biol. Chem. **235**:3242 (1960).
12. Jovin, T. M., Englund, P. T., and Bertsch, L. L., J. Biol. Chem. **244**:2996 (1969).
13. Falaschi, A., and Kornberg, A., J. Biol. Chem. **241**:1478 (1966).
14. Kornberg, A., Bertsch, L. L., Jackson, J. F., and Khorana, H. G., Proc. Nat. Acad. Sci. U. S. A. **51**:315 (1964).
15. Godson, G. N., and Sinsheimer, R. L., Biochim. Biophys. Acta **149**:476 (1967).
16. Riley, M., Maling, B., and Chamberlin, M. J., J. Mol. Biol. **20**:359 (1966).
17. Olivera, B. M., and Lehman, I. R., Proc. Nat. Acad. Sci. U. S. A. **57**:1426 (1967).
18. Trautner, T. A., Swartz M. N., and Kornberg, A., Proc. Nat. Acad. Sci. U. S. A. **48**:449 (1962).
19. Richardson, C. C., Schildkraut, C. L., and Kornberg, A., Cold Spring Harbor Symp. Quant. Biol. **28**:9 (1963).
20. Okazaki, T., and Kornberg, A., J. Biol. Chem. **239**:259 (1964).
21. Richardson, C. C., Inman, R. B., and Kornberg, A., J. Mol. Biol. **9**:46 (1964).
22. Olivera, B. M., Scheffler, I. E., and Lehman, I. R., J. Mol. Biol., in press.
23. Englund, P. T., Kelly, R. B., and Kornberg, A., J. Biol. Chem. **244**:3045 (1969).

# 12 ENZYMATIC JOINING OF POLYNUCLEOTIDES, V. A DNA-ADENYLATE INTERMEDIATE IN THE POLYNUCLEOTIDE-JOINING REACTION

*Baldomero M. Olivera**
*Zach W. Hall***
*I. R. Lehman*
Department of Biochemistry
Stanford University School of Medicine
Palo Alto, California

In the reaction catalyzed by the polynucleotide-joining enzyme from *Escherichia coli*, phosphodiester bonds are synthesized between the 3'-hydroxyl and 5'-phosphoryl termini of properly aligned DNA chains coupled to the cleavage of the pyrophosphate bond of DPN.[1-4] The first step in this over-all reaction occurs in the absence of DNA and consists of the formation of a covalently linked enzyme-adenylate (E-AMP) intermediate and simultaneous release of NMN[5,6] (Fig. 1).

This paper presents evidence for a second intermediate in the joining reaction. This intermediate is formed by reaction of E-AMP with a DNA chain to generate a new pyrophosphate bond linking the 5'-phosphoryl terminus of the DNA and the phosphoryl group of the AMP. In the final step of the joining reaction we presume that the DNA phosphate in the pyrophosphate bond of the DNA-adenylate is attacked by the 3'-hydroxyl group of the neighboring chain, displacing the activating AMP group and effecting the synthesis of the phosphodiester bond (Fig. 1).

## Experimental procedure
### Materials

Unlabeled nucleotides were purchased from Calbiochem. $H^3$-labeled ATP and AMP were obtained from Schwarz BioResearch. $\gamma$-$P^{32}$-ATP was prepared by the method of Glynn and Chappell.[7] DPN was obtained from the Sigma Chemical Co. $H^3$-adenine-labeled DPN (spec. act. 1.5-2.0 $\times$ $10^3$ cpm/μμmole) was synthesized from $H^3$-ATP and NMN with the hog liver DPN pyrophosphorylase[8] and purified by chromatography on DEAE-Sephadex. $d(pC)_3$ was prepared according to Khorana, Turner, and Vizsolyi.[9] Phage λ DNA was isolated from the purified phage by phenol extraction as described by Kaiser and Hogness.[10]

Single-strand scissions with 5'-phosphoryl and 3'-hydroxyl termini were introduced into λ DNA by treatment with pancreatic DNase. The reaction mixture (4.0 ml) contained 0.1 M Tris-HCl, pH 8.0, 10mM $MgCl_2$, λ DNA, $A_{260}$ = 4.0, and 0.1 μg of pancreatic DNase. After incubation at 30° for 30 min, 1 M EDTA, pH 8.7, was added to a final concentration of 20 mM. The solution was extracted 3 times with 2-ml aliquots of phenol (equilibrated with 0.3 M Tris-HCl, pH 8.6) and dialyzed (3 changes) against 1-liter portions of 10 mM Tris-HCl, pH 8.0, 10 mM EDTA, then against 1 liter of 10 mM Tris-HCl, pH 8.0. This preparation of DNA contained an average of 70 single-strand scissions per molecule as judged by the ability of the DNA to accept $P^{32}$ from $\gamma$-$P^{32}$-ATP in the presence of polynucleotide kinase after treatment of the DNA with *E. coli* alkaline phosphatase at 65°.[11]

Single-stranded DNA chains with $H^3$-riboadenylate

---

From Proceedings of the National Academy of Sciences, U.S.A. 61:237-244, 1968. Reprinted with permission.

We wish to acknowledge the expert help of Mrs. Janice R. Chien with many of the experiments. We are very grateful to Dr. F. N. Hayes for his gift of calf thymus deoxynucleotidyl transferase, to Dr. Margaret Lieb for her preparation of the DPN pyrophosphorylase and to Dr. John Moffatt for his help in the synthesis of poly dT-adenylate.

The abbreviations used are: AMP and ATP, adenosine 5'-mono- and triphosphate; CMP and CDP, cytidine 5'-mono- and diphosphate; DPN, diphosphopyridine nucleotide; NMN, nicotinamide mononucleotide; $d(pC)_3$, a trinucleotide composed of deoxycytidylate residues terminated by a 5'-phosphate group; $d(pT)_3$ or pTpTpT, a trinucleotide of deoxythymidylate residues terminated by 5'-phosphate; poly dT, a homopolymer of deoxythymidylate residues; poly dA, a homopolymer of deoxyadenylate; AppTpTpT, AMP bound in pyrophosphate linkage to the 5'-phosphate of pTpTpT; poly dT-adenylate, AMP in pyrophosphate linkage to the 5'-phosphate of poly dT; DEAE, diethylaminoethyl; EDTA, ethylenediaminetetraacetate; and Tris, tris(hydroxymethyl) aminomethane.

This work was supported in part by research grants from the National Institutes of Health, USPHS.

*Postdoctoral fellow of the Damon Runyon Memorial Fund for Cancer Research. Present address: Department of Biochemistry, University of the Philippines, College of Medicine, Manila, Philippines.

**Postdoctoral fellow of the USPHS. Present address: Department of Neurobiology, Harvard Medical School, Boston, Massachusetts.

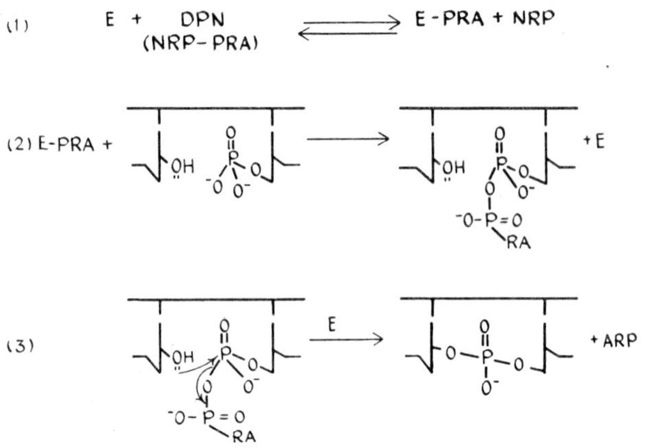

**Fig. 1.** Postulated mechanism of the reaction catalyzed by the *E. coli*-joining enzyme. DPN is written as NRP-PRA to emphasize the pyrophosphate bond linking the nicotinamide mononucleotide (NRP) and adenylic acid (PRA) moieties of the DPN molecule. The designation E-PRA for enzyme-adenylate is not meant to imply that linkage of AMP to the enzyme is necessarily through the phosphate group.

at their 3' termini were synthesized as described by Richardson and Kornberg.[12]

The DNA-adenylate intermediate isolated by CsCl density gradient centrifugation was purified by filtration through Sephadex G-25. $H^3$-containing fractions banding at the density of λ DNA (Fig. 2) were pooled and applied to a column (20 × 1 cm) of Sephadex G-25 equilibrated with 1 mM Tris-HCl, pH 8.0–1 mM EDTA–10 mM KCl. The DNA-bound AMP appeared in the void volume and was well separated from the bulk of the $H^3$, presumably unreacted DPN; it was concentrated approximately tenfold by evaporation under reduced pressure.

Poly dT-adenylate was prepared by condensing $H^3$-AMP with $p^{32}$TpTpT by the morpholidate method of Moffatt and Khorana;[13] deoxythymidylate residues were then added to the 3'-hydroxyl end of the d(pT)$_3$ moiety by using the $H^3$-App$^{32}$TpTpT as an initiator in a reaction catalyzed by calf thymus deoxynucleotidyl transferase.[14] The average chain length was 100 deoxythymidylate residues. The $P^{32}$ in the product was acid-precipitable and was insusceptible to alkaline phosphatase except after being heated at 100° in 1 N HCl for 15 min, or after treatment with *E. coli* exonuclease I and snake venom phosphodiesterase. Details of the preparation and characterization of the poly dT-adenylate will be published elsewhere.

Pancreatic DNase, *E. coli* alkaline phosphatase, micrococcal nuclease, spleen, and snake venom phosphodiesterases were purchased from the Worthington Biochemical Co. *E. coli* exonuclease I (DEAE-cellulose fraction) was purified by the method of Lehman and Nussbaum.[15] *E. coli* polynucleotide-joining enzyme (fraction V) was prepared and assayed as described previously;[1] it was further purified by gradient chromatography on DEAE-Sephadex.[16]

## Methods

Density gradient sedimentation was carried out at 25° in the International B60 centrifuge with the SB 405 rotor. To 2.7 gm of solution to be analyzed (containing 30 μmoles of 1 M Tris-HCl, pH 8.6, and 30 μmoles of EDTA, pH 8.7) were added 3.5 gm CsCl, and the solution was centrifuged for 30 hr at 53,000 rpm. The bottom of the tube was then punctured and 15-drop fractions (usually a total of 16) were collected. Acid-insoluble radioactivity was determined as described elsewhere.[17]

Paper electrophoresis of nucleotides was performed at 20° in 0.015 M sodium citrate buffer, pH 5.5, at a potential of 5000 volts for 30-75 min. Descending paper chromatography on Whatman 3 MM paper was carried out with the 1-propanol-ammonia-water system of Hanes and Isherwood.[18] After electrophoresis of chromatography, the paper was cut into strips and the radioactivity determined. Measurements of radioactivity were made using a Nuclear-Chicago model 724 liquid scintillation counter.

## Results

### Isolation of DNA-adenylate

When joining enzyme, $H^3$-labeled DPN, and phage λ DNA containing multiple single-strand scissions were briefly incubated (5 min) at 0°, a small peak of acid-precipitable $H^3$ with the buoyant density of λ DNA was identified by density gradient centrifugation (Fig. 2). This

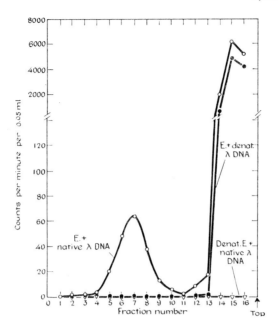

**Fig. 2.** Demonstration of DNA-adenylate by CsCl density gradient centrifugation. A reaction mixture (0.1 ml) containing 25 mM Tris-HCl, pH 8.0, 10 mM MgCl$_2$, 2.5 mM EDTA, 12.5 μg of bovine plasma albumin, 70 units of joining enzyme (E), and 0.68 μM H$^3$-DPN (1.8 × 10$^3$ cpm/μμmole) was incubated for 5 min at 30°. Pancreatic DNase-treated λ DNA (0.1 ml) was added, and the mixture incubated for 5 min at 0°. More DNA (0.1 ml) and H$^3$-DPN (34 μμmoles) was added, the solution mixed rapidly at 0°, and the reaction immediately terminated by the addition of 0.02 ml of 0.66 M glycine-NaOH, pH 10.2, containing 0.33 M EDTA. Two control reaction mixtures were prepared and treated as above, except that in one the enzyme had been heated to 100° for 2 min and in the second the λ DNA had been heated to 100° for 2 min. CsCl density gradient centrifugation and determination of acid-insoluble H$^3$ were then performed as described in *Methods*.

DNA-adenylate peak amounted to approximately 1 per cent of the H$^3$ found at the top of the gradient where E-AMP would be expected to band. Assay of the fractions from the CsCl gradient showed that joining-enzyme activity was exclusively at the top of the gradient; less than 0.01 per cent could be detected at the density of λ DNA. DNA-adenylate did not appear when denatured DNA was used in place of native λ DNA. Similarly, termination of the reaction after mixing H$^3$-DPN and enzyme, but before the addition of DNA, prevented the accumulation of H$^3$ in the position of the gradient occupied by λ DNA (Table 1). In both of these cases, despite the absence of DNA-adenylate, E-AMP was formed, as judged by the appearance of high levels in acid-insoluble radioactivity at the top of the density gradient. When joining enzyme which had been denatured by heating for two minutes at 100° was used, acid-insoluble H$^3$ could not be detected at any point in the gradient.

Radioactivity found in the position of λ DNA appeared within 0.5 minute of incubation at 0° and then fell precipitously with further incubation. On the other hand, the amount of radioactivity in the E-AMP fraction remained relatively constant and even increased slightly with longer periods of incubation (Table 1). After incubation for five minutes at 30°, there was no detectable DNA-adenylate peak.

When the number of single-strand breaks in the DNA was raised by about ten-fold the yield of DNA-adenylate was increased. Under these conditions, the DPN concentration became limiting and the levels of both DNA-adenylate and E-AMP fell upon incubation; however, the ratio of the two remained relatively constant (14.7% at 0.5 min and 6.7% at 5 min).

**TABLE 1**
*Effect of incubation conditions on yield of DNA-adenylate*

| Time of incubation at 0° (min) | Units of enzyme | $H^3$ in E-AMP (cpm) | $H^3$ in DNA-adenylate (cpm) | Yield (%) |
|---|---|---|---|---|
| 0 | 250 | 30,300 | <10 | <0.03 |
| 0.5 | 290 | 12,200 | 970 | 7.9 |
| 5.0 | 145 | 25,300 | 406 | 1.6 |
| 5.0 | 250 | 38,000 | 250 | 0.7 |

The reaction mixtures (0.1 ml) for the samples incubated for 5 min contained 25 mM Tris-HCl, pH 8.0, 10 mM $MgCl_2$, 2.5 mM EDTA, 1 µM $H^3$-DPN (1.8 × $10^3$ cpm/µµmole), 12.5 µg of bovine plasma albumin, and the indicated amounts of enzyme. After 5 min at 30° the mixtures were chilled to 0° and 0.2 ml of pancreatic DNase-treated λ DNA (prepared as described in *Methods*) was added. After 5 min the reactions were terminated by adding glycine-EDTA as described in the legend to Fig. 2. The "0-min" reaction mixture was prepared in the same way except that the glycine-EDTA was added before the λ DNA. The composition of the reaction mixture for the sample incubated for 0.5 min was the same as that described above, except that it was scaled up to 0.3 ml. After the 5-min preincubation period at 30°, the mixture was chilled to 0° and 0.5 ml of λ DNA was added. After 0.5 min the reaction was terminated. All of the samples were then subjected to CsCl density gradient centrifugation as described in *Methods*. An aliquot (0.1 ml) of each fraction was used to determine acid-insoluble $H^3$.

**TABLE 2**
*Release of AMP from DNA-adenylate in the absence of DPN*

| Units of enzyme | "Native" DNA-adenylate | Heat-denatured DNA-adenylate |
|---|---|---|
| | (% of $H^3$ made acid-soluble) | |
| 6.0 | 100 | 35* |
| 0.2 | 60 | – |
| 0.06 | 32 | 2 |

Reaction mixtures (0.1 ml) contained 10 mM Tris-HCl, pH 8.0, 2 mM $MgCl_2$, 1 mM EDTA, 5.0 µg of bovine plasma albumin, the indicated amounts of joining enzyme and either native DNA-adenylate (2 µg of DNA, 20 cpm) or DNA-adenylate denatured by heating at 100° for 2 min. After incubation at 37° for 30 min, the reaction mixtures were heated at 100° for 2 min and acid-insoluble $H^3$ was determined. A minimum of 400 counts over background was recorded for the sample in which all the AMP was retained in the DNA-adenylate.
*The acid-soluble $H^3$ formed in this sample was not free AMP (see text).

### Evidence that DNA-adenylate is an intermediate in the reaction

The isolated DNA-adenylate displayed the properties expected of an intermediate in the joining reaction. Thus, the $H^3$-AMP was quantitatively released into an acid-soluble form upon treatment of the DNA-adenylate with joining enzyme in the absence of added DPN (Table 2). The liberated $H^3$ was identified chromatographically as 5'-AMP; after treatment with phosphatase it cochromatographed with adenosine. When the DNA-adenylate was heat-denatured before incubation with joining enzyme, approximately one third of the AMP was converted to an acid-soluble form (Table 2). However, after treatment with phosphatase and chromatography, essentially all (>90%) of the $H^3$ remained at the origin, well separated from the 5'-AMP and adenosine markers and at a position where oligonucleotide material would be anticipated in the solvent system used. The release of AMP linked to oligonucleotide from heat-denatured DNA-adenylate is therefore most probably due to nuclease contamination of the joining enzyme preparation, which became significant at the relatively high levels of enzyme used in this experiment.

### Evidence that AMP is linked to the 5'-terminus of DNA

The two most probable sites in DNA to which AMP may be linked are at the 5'-phosphoryl and the 3'-hydroxyl termini. E. coli exonuclease I should be capable of distinguishing between these two possibilities. This enzyme degrades single-stranded DNA sequentially from the 3'-hydroxyl end, and produces 5'-mononucleotides but leaves the 5'-terminal dinucleotide intact.[15] Thus, if the AMP were linked to the 5'-phosphoryl terminus of the DNA, digestion by the exonuclease I should yield a trinucleotide in which the AMP is linked to the terminal dinucleotide through a pyrophosphate bond. On the other hand, if the AMP

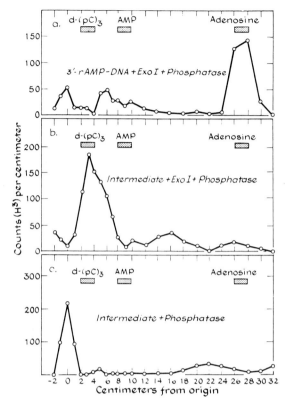

**Fig. 3.** Paper electrophoresis of DNA-adenylate intermediate and DNA chains with riboadenylate at their 3'-termini (3'-rAMP-DNA) after treatment with exonuclease I (Exo I) and phosphatase. Three reaction mixtures (0.2 ml each) were prepared containing 80 mM glycine-NaOH, pH 9.5, 8 mM $MgCl_2$, and 2.5 mM $\beta$-mercaptoethanol. To one (reaction mixture a) were added single-stranded DNA chains with $H^3$-AMP at their 3'-termini (60 μg DNA, 800 cpm), 17 units of exonuclease I,[15] and 5 units of phosphatase.[23] To the other two were added heat-denatured DNA-adenylate (19 μg DNA, 250 cpm) and either exonulcease I and phosphatase (b) or phosphatase alone (c). The reaction mixtures were incubated at 37° for 30 min, then chromatographed for 18 hr in the 1-propanol-ammonia-water system. The paper was dried, the nucleotides were identified, and the radioactivity was determined as described in *Methods*. The values shown correspond to the total number of counts, corrected for background recorded for each strip.

were in phosphodiester linkage at the 3' end of the DNA, it should be released as free AMP.[15]

The isolated DNA-adenylate intermediate and a control DNA preparation bearing AMP at its 3' terminus were denatured, treated with exonuclease I and alkaline phosphatase, and then chromatographed. In the case of the DNA with AMP at its 3' terminus, the only product formed was adenosine (Fig. 3); AMP must therefore have been released by the exonuclease I treatment. On the other hand, treatment of the isolated intermediate with exonuclease I and alkaline phosphatase yielded products which formed a rather broad radioactive peak on the chromatogram at the position expected of a mixture of trinucleotides (Fig. 3). Digestion of synthetic poly dT-adenylate (see below) with these enzymes produced a similar peak of radioactivity in the trinucleotide region of the chromatogram. These data indicate that the AMP is linked at the 5' end of the isolated DNA-adenylate. Moreover, the finding that the product of exonuclease I digestion migrated to the position occupied by a trinucleoside triphosphate even after phosphatase treatment is consistent with the presence of an internal pyrophosphate group linking the AMP to the 5'-phosphoryl terminus of the DNA.

### Activity of synthetic poly dT-adenylate as a substrate for the joining enzyme

To determine directly whether the polynucleotide-joining reaction involves formation of a pyrophosphate linkage between AMP and the 5'-phosphoryl terminus of the polynucleotide chain, the presumptive intermediate was synthesized and tested as a substrate for the joining enzyme.

Poly dT-adenylate was prepared in which the 5'-terminal phosphate of the poly dT was labeled with $P^{32}$ and the adenylate labeled with $H^3$. The double label permitted simultaneous measurement of the release of AMP from the polynucleotide and the incorporation of the terminal phosphate of poly dT into phosphodiester linkage.

When poly dT-adenylate was incubated with joining enzyme in the absence of DPN, the $H^3$ was released as an acid-soluble product, identified chromatographically as AMP, and a nearly equivalent amount of the $P^{32}$ was converted to a form which was insensitive to alkaline phosphatase after heating in 1 N HCl at 100° for 15 minutes (Table 3). Upon degradation of the product to 3'-mononucleotides by the combined action of micrococcal nuclease and spleen phosphodiesterase, all of the $P^{32}$ was found to be associated with 3'-dTMP, a result which is consistent with its incorporation

TABLE 3
*Reactivity of synthetic poly dT-adenylate in the polynucleotide-joining reaction*

| Enzyme | Poly dA | $P^{32}$ in phosphodiester linkage ($\mu\mu$moles) | $H^3$-AMP released ($\mu\mu$moles) |
|---|---|---|---|
| − | + | 5 | 7 |
| + | + | 38 | 46 |
| + | − | 2 | 3 |

The reaction mixtures contained, in a final volume of 0.1 ml, 10 mM Tris-HCl, pH 8.1, 3mM $MgCl_2$, 1 mM EDTA, 10% glycerol, 10 µg bovine plasma albumin, 0.42 µM (in dT termini) $P^{32}$-poly dT-$H^3$ adenylate (200 cpm of $P^{32}$ and 13 cpm of $H^3$ per µµmole of termini), 48 µM (in dAMP residues) poly dA, and 6 units of joining enzyme as indicated. After 1 hr at 30°, 0.015-ml aliquots were removed from each reaction and heated with 0.05 ml 1 N HCl at 100° for 15 min. 2 M Tris-HCl, pH 8.1 (0.1 ml), and 0.12 unit of phosphatase[23] were then added and the reaction mixtures incubated at 37° for 30 min. The fraction of $P^{32}$ adsorbable to Norit after treatment with phosphatase was then measured as described previously.[1] To the remainder of the incubation mixtures were added 0.01 ml 0.95 mM poly dA, 0.1 ml 0.1 M pyrophosphate, 0.1 ml calf thymus DNA (2.5 mg/ml), and 0.5 ml 3.5% perchloric acid—0.35% uranyl acetate. After 15 min at 0°, the mixtures were centrifuged and the radioactivity of the supernatant fluid was determined.

into a phosphodiester bond. Thus, there is a stoichiometric correspondence between cleavage of the pyrophosphate bond linking poly dT and AMP on the one hand and the phosphodiester bond formation on the other. Both the release of AMP and the incorporation of the $P^{32}$ into phosphodiester linkage required that poly dA be present.

The reaction mechanism for the polynucleotide-joining enzyme as proposed in Figure 1 predicts that adenylylation of the enzyme (to form E-AMP) would render it inactive in the reaction with poly dT-adenylate. Consistent with this prediction was the observation that preincubation of the enzyme with 0.3 mM DPN before the addition of poly dT-adenylate resulted in 95 per cent inhibition of phosphodiester bond formation.

### Discussion

The data presented here show that an intermediate in which AMP is bound in pyrophosphate linkage to the 5′-phosphoryl termini of DNA is formed in the polynucleotide-joining reaction. The steady-state concentration of this intermediate is extremely low and it may be accumulated in detectable amounts only under rather restricted conditions (i.e., brief incubation at 0° in the presence of high concentrations of enzyme and single-strand breaks in DNA). This is presumably the result of the extreme rapidity of the final step in the reaction sequence (reaction (3), Fig. 1), in which attack of the activated DNA phosphate by the 3′-hydroxyl group of the neighboring DNA chain occurs.

The mechanism of formation of phosphodiester bonds by the DNA polymerase and polynucleotide-joining enzyme are basically similar. Thus, chain growth from the 3′-hydroxyl end of DNA by the polymerase involves an attack by the 3′-hydroxyl group of the DNA on the activated α-phosphate of the incoming deoxynucleoside triphosphate.[19] Analogously, synthesis of a phosphodiester bond in the joining reaction occurs by attack of the 3′-hydroxyl group of one DNA chain on the activated 5′-phosphoryl group of another (apposing) chain.

A closer analogy to the joining reaction is to be found in phosphodiester bond formation in the phospholipids. For example, phosphoryl choline is first linked to CMP *via* a pyrophosphate bond to form CDP-choline. An attack by the α-hydroxyl group of the diglyceride on the activated choline phosphate follows, resulting in the synthesis of the phosphodiester bond of phosphatidyl choline and the liberation of CMP.[20]

In the first step of the T4-ligase-catalyzed reaction, E-AMP is formed by reaction of the enzyme with ATP.[21,22] It would seem likely that once E-AMP has been generated, the phage-induced ligase reaction would proceed through a similar DNA-adenylate intermediate.

### Summary

DNA with AMP in pyrophosphate linkage at the 5′-phosphoryl termini was identified as a component of the polynucleotide-joining reaction. The isolated DNA-adenylate was active as a substrate for the joining enzyme in the absence of DPN. Synthetic poly dT-adenylate was also active in the joining reaction in the absence of DPN, provided that poly dA were added. The liberation of AMP and phosphodiester bond synthesis was stoichiometric. These findings are consistent with a mechanism for the joining reaction in which (1) E-AMP is formed by a reaction of enzyme with DPN, (2) AMP is transferred to the 5′-phosphoryl terminus of a DNA chain to form a new pyrophosphate bond, and (3) the activated 5′-phosphate group is attacked by the 3′-hydroxyl group of the apposing DNA chain, displacing the AMP and producing a phosphodiester bond linking the two chains.

**REFERENCES**

1. Olivera, B. M., and I. R. Lehman, these Proceedings **57**:1426 (1967).
2. *Ibid.*, p.1700.
3. Zimmerman, S. B., J. W. Little, C. K. Oshinsky, and M. Gellert, these Proceedings **57**:1841 (1967).
4. Gefter, M., A. Becker, and J. Hurwitz, these Proceedings **58**:240 (1967).
5. Little, J. W., S. B. Zimmerman, C. K. Oshinsky, and M. Gellert, these Proceedings **58**:2004 (1967).
6. Olivera, B. M., Z. W. Hall, Y. Anraku, J. R. Chien, and I. R. Lehman, in Cold Spring Harbor Symposia on Quantitative Biology **33** (1968), in press.
7. Glynn, I. M., and J. B. Chappel, Biochem. J. **90**:147 (1964).
8. Kornberg, A., J. Biol. Chem. **182**:779 (1950).
9. Khorana, H. G., A. F. Turner, and J. P. Vizsolyi, J. Am. Chem. Soc. **83**:686 (1961).
10. Kaiser, A. D., and D. S. Hogness, J. Mol. Biol. **2**:392 (1960).
11. Weiss, B., and C. C. Richardson, these Proceedings **57**:1021 (1967).
12. Richardson, C. C., and A. Kornberg, J. Biol. Chem. **239**:242 (1964).
13. Moffatt, J. G., and H. G. Khorana, J. Am. Chem. Soc. **83**:649 (1961).
14. Yomeda, M., and F. J. Bollum, J. Biol. Chem. **240**:3385 (1965).
15. Lehman, I. R., and A. L. Nussbaum, J. Biol. Chem. **239**:2628 (1964).
16. Anraku, Y., and I. R. Lehman, to be published.
17. Olivera, B. M., and I. R. Lehman, J. Mol. Biol., in press.
18. Hanes, C. S., and F. A. Isherwood, Nature **164**:1107 (1949).
19. Kornberg, A., Science **131**:1503 (1960).
20. Kennedy, E. P., Federation Proc. **20**:934 (1961).
21. Weiss, B., and C. C. Richardson, J. Biol. Chem. **242**:4270 (1967).
22. Becker, A., G. Lyn, M. Gefter, and J. Hurwitz, these Proceedings **58**:1996 (1967).
23. Malamy, M. H., and B. L. Horecker, Biochemistry **3**:1893 (1964).

# 13 ENZYMATIC SYNTHESIS OF DNA, XXIV. SYNTHESIS OF INFECTIOUS PHAGE φX174 DNA

*Mehran Goulian**
*Arthur Kornberg*
*Robert L. Sinsheimer*
*Department of Biochemistry*
*Stanford University School of Medicine*
*Palo Alto, California, and*
*Division of Biology*
*California Institute of Technology*
*Pasadena, California*

Past attempts at *in vitro* replication of transforming factor present in DNA have given negative or inconclusive results.[1-3] Rigid proof was lacking that template material had been excluded from the synthetic product. Even if a rigorous demonstration of net synthesis of transforming factor for a given genetic marker were forthcoming, it would still prove only that some relatively short sequence of nucleotides, sufficient for replacement of the mutant locus, had been synthesized. If enzymatic synthesis of infectious bacteriophage DNA were achieved, it would be made clear at once that relatively few, if any, mistakes had been made in replicating a DNA sequence of several thousand nucleotides.

*Escherichia coli* DNA polymerase can replicate single-stranded circular DNA from phage M13 or φX174[4] and in conjunction with a polynucleotide-joining enzyme produces a fully covalent duplex circle.[5] Analyses of this product by equilibrium and velocity sedimentation and by electron microscopy have shown it to be indistinguishable, except for supercoiling, from replicative forms (RF)[6] of the viral DNA.[5] By substitution of bromouracil for thymine in the complementary strand ((−) circle), it should be possible on the basis of density difference to isolate this strand from the duplex circle and determine whether it has the infectivity known to reside in (−) circles.[7,8]

This report will describe: (1) the isolation of infective, synthetic (−) circles from the partially synthetic replicative form, (2) the ability of the isolated (−) circles to serve as templates for the production of infective, completely synthetic duplex circles, and (3) the isolation of infective, synthetic (+) circles from the latter.

Thus, DNA polymerase carries out the relatively error-free synthesis of the φX174 genome from the four deoxyribonucleoside triphosphates on direction from phage DNA templates.

## Results[9]

### Isolation of synthetic (−) circle and test of infectivity

A duplex circle was synthesized by replicating $H^3$-φX174 DNA with DNA polymerase in the presence of a polynucleotide-joining enzyme. Details for the production and isolation of this partially synthetic RF, containing $\overline{BU}$ and $P^{32}$ in the (−) circle, were described in an earlier report.[5] Separation of the synthetic (−) circle from the duplex form followed the plan outlined in Figure 1. The duplex circles were exposed to pancreatic DNase to an extent sufficient to produce a single scission in one of the strands in about half of the molecules. The resulting mixture of intact and nicked molecules was denatured by heating. The mixture, which now contained circular and linear $H^3$-T (+) strands, and $P^{32}$-$\overline{BU}$ (−) strands, in addition to intact RF, was fractionated by equilibrium density-gradient sedimentation in CsCl (Fig. 2).[10] Three peaks of radioactivity were evident, corresponding, in order of decreasing density, to single-stranded DNA containing $\overline{BU}$, a

---

From Proceedings of the National Academy of Sciences, U.S.A. **58**:2321-2328, 1967. Reprinted with permission.

We gratefully acknowledge the expert assistance of Mrs. Gloria Davis of the Division of Biology at the California Institute of Technology in performing the spheroplast assays for infectivity.

This research was supported by grants from the National Institutes of Health and the National Science Foundation.

*USPHS special fellow. Present address: Department of Medicine, University of Chicago.

Synthesis of DNA in vitro   113

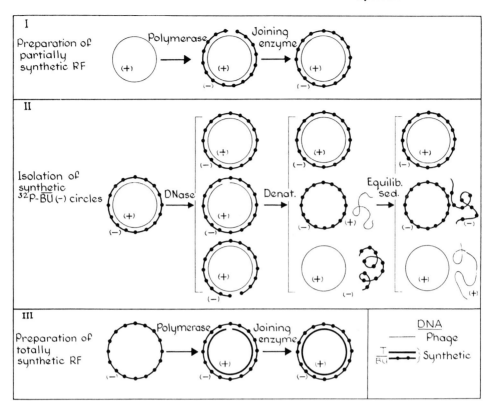

**Fig. 1.** Schematic representation of the preparation of synthetic (—) circles and RF. For details see text and Figs. 2 and 6.

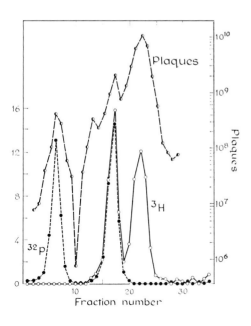

**Fig. 2.** Equilibrium density-gradient sedimentation analysis of partially synthetic RF after limited DNase action and denaturation. Partially synthetic RF, with $P^{32}$ and $\overline{BU}$ in the synthetic (—) strand,[5] was incubated for 20 min at 20° at a concentration of 0.1 mM, in 0.2 ml of 10% glycerol-10 mM Tris HCl (pH 7.6)-2 mM $MgCl_2$-0.25 mμg/ml pancreatic DNase. (The DNase (Worthington 1 × recrystallized), 5 mg/ml in 0.01 N HCl, was stored at $0°^{11}$ and diluted immediately before use in 10 mM Tris acetate (pH 5.5)-5 mM $MgCl_2$-0.2 M KCl-50% glycerol.) The reaction was stopped by addition of EDTA to 8 mM. The mixture was heated at 90° for 2 min and adjusted to a volume of 9.8 ml with 0.01 M Tris HCl (pH 7.6); EDTA was added to 1 mM, as well as 1 mg of bovine plasma albumin and 9.961 gm of CsCl. Centrifugation of this mixture ($\rho$ = 1.750) was carried out in the Spinco no. 50 angle rotor at 45,000 rpm at 25° for 50 hr. Aliquots from each fraction were assayed for radioactivity on filter paper disks,[5] and for infectivity by the spheroplast assay of Guthrie and Sinsheimer.[12] Inhibition by CsCl in the spheroplast assay was avoided by dilution.

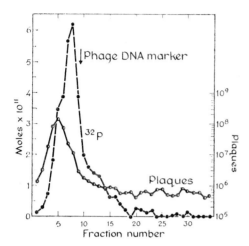

**Fig. 3.** Velocity sedimentation of $P^{32}$-$\overline{BU}$, (−) synthetic DNA derived from partially synthetic RF. The $P^{32}$-$\overline{BU}$ peak fractions were pooled, dialyzed against 2 mM Tris HCl (pH 7.6)-0.2 mM EDTA and then concentrated fivefold to a volume of 0.1 ml by rotary evaporation under reduced pressure. An aliquot of 20 μl was centrifuged in a 5-20% sucrose gradient in 5 mM NaCl-5 mM Tris HCl (pH 7.6)-1 mM EDTA, at 60,000 rpm and 10° for 360 min. $H^3$ was not detectable (<0.3 μμmole/fraction). The position of φX174 DNA was obtained from a separate tube containing this DNA as marker.

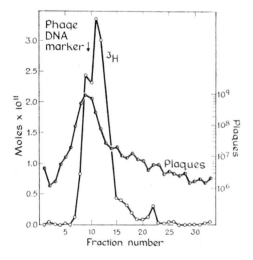

**Fig. 4.** Velocity sedimentation of $H^3$-T, (+) phage DNA derived from partially synthetic RF. The $H^3$-T peak (Fig. 2) was treated as described in Fig. 3 for $P^{32}$-$\overline{BU}$ except that half as much $H^3$-T was placed on the sucrose gradient.

duplex hybrid of $\overline{BU}$ and T, and single strands containing T, with mean densities of 1.809, 1.747, and 1.722 gm/ml, respectively. These values may be compared to the values of 1.732 and 1.725 previously determined[13] for the native hybrid and T (+) single strands prepared *in vivo* or to the calculated values[14] of 1.815 and 1.753 for the $\overline{BU}$ (−) strands, and the hybrid, respectively. In addition to the three peaks, there was an area on the heavy side of the hybrid zone, which in other experimental trials appeared as a more distinct shoulder and is attributable to some duplex circles that had failed to renature after the heat treatment (Fig. 2).

Inasmuch as the (−) circle is infectious in the spheroplast assay,[7,8] it was possible to test the enzymatically synthesized material directly for biologic activity. Four peaks of infectivity were found (Fig. 2). One corresponded to the position of heavy, $P^{32}$-$\overline{BU}$, synthetic (−) single strands and another to that of light, $H^3$-T (+) single strands. Specific infectivity values for the single-stranded regions could not be determined from these data because there was an unknown quantity of linear strands. The $P^{32}$-$\overline{BU}$ and $H^3$-T peaks were therefore each subjected to velocity sedimentation in a neutral, low-salt sucrose gradient to give a partial separation of the circles from linear forms. As seen in Figure 3 for the $P^{32}$-$\overline{BU}$ (−) strands, and in Figure 4 for the $H^3$-T (+) strands, the infective material was found, in each case, in the leading shoulder of the peak which contains the circles because of their more rapid sedimentation. Because of their content of $\overline{BU}$, the (−) circles had a distinctly higher sedimentation rate than their T (+) complements (compare sedimentation values relative to the DNA marker in Figs. 3 and 4). The specific infectivities estimated for the synthetic (−) circles and template (+) circles were 0.074 and 0.80, respectively (Table 1).

The other two peaks of infectivity in Figure 2 corresponded to the position of denatured and native forms of duplex hybrid molecules. Their respective specific infectivities were 0.066 and 0.012 (Table 1).

*Proof that the infectivity of the $P^{32}$-$\overline{BU}$ peak resides in the enzymatically synthesized DNA*

(1) A peak of infectivity coincides with the $P^{32}$-$\overline{BU}$ peak[17] in the density gradient (Fig. 2)

**TABLE 1**
*Infectivities of natural and synthetic φX174 DNA*

| | Plaques ($ml^{-1} \times 10^{-8}$) | DNA ($\mu\mu mole\ ml^{-1}$) | Specific infectivity (plaques/particle) | Relative infectivity | Ref. |
|---|---|---|---|---|---|
| (+) Circle, natural | 37 | 64 | 0.80* | 1.0 | Fig. 4 |
| (−) Circle, natural | | | | ~0.2 | 8 |
| "      ", synthetic | 6.2 | 150 | 0.074* | 0.09 | Fig. 3 |
| RF (native), natural | | | | 0.05 | 15 |
| "      ", part. synthetic | 6,000 | 200,000 | 0.058 | 0.07 | † |
| "      ", part. synthetic | 61 | 9,100 | 0.012 | 0.01 | Fig. 2 |
| RF (denat.), natural | | | | 1.0 | 16 |
| "      ", part. synthetic | 9.5 | 260 | 0.066 | 0.06 | Fig. 2 |
| "      ", fully synthetic | 24 | 120 | 0.36 | 0.3 | Fig. 6 |

*Specific infectivity* was calculated on the basis of $1.1 \times 10^8$ particles/$\mu\mu$mole of nucleotide residues for single-stranded molecules and half that value for the duplexes. *Relative infectivity* of the natural (+) circle was arbitrarily taken as 1.0 and the other figures adjusted, with inclusion of a correction, for variations between different assays, from phage DNA standards that were run in each assay. Technical difficulties resulting from the low concentrations of DNA have thus far prevented reliable estimates of the specific infectivities of native, fully synthetic RF and synthetic (+) circles.
*Includes a correction for estimated contamination with linear forms.
†Sample assayed prior to exposure to DNase as in Fig. 2.

and is separated from neighboring peaks. (2) Phage (+) circles are absent from the single-stranded, $P^{32}$-$\overline{BU}$ peak as judged by the absence of detectable $H^3$-labeled material. In view of the sensitivity of the radioactivity measurements, the *upper* limit for the amount of template material in the synthetic peak is 8 $\mu\mu$moles/ml; this concentration is one-tenth of that necessary to account for the infectivity of the peak. (3) In velocity sedimentation in sucrose gradients, the peak of infectivity corresponds to the position of intact $P^{32}$-$\overline{BU}$ (−) circles, and sediments more rapidly because of the presence of $\overline{BU}$ than the analogous peak of intact $H^3$-T (+) circles. (4) The photoinactivation of $P^{32}$-$\overline{BU}$ (−) DNA as compared with $H^3$-T (+) DNA (Fig. 5) demonstrates the more rapid inactivation of most of the infectious particles in the CsCl gradient peak corresponding to $P^{32}$-$\overline{BU}$ (−) strands and is consistent with the known greater photosensitivity of $\overline{BU}$-containing DNA.[18] The presence of approximately 5 per cent of the infectious material displaying an inactivation rate similar to that of T DNA[19] (Fig. 5) indicates the extent of contamination by T (+) circles. Inasmuch as the (+) circles of Figure 2 have about ten times the specific infectivity of these (−) circles, the residual content of phage DNA in the $\overline{BU}$ fraction is estimated to be closer to 0.5 per cent than 5 per cent.

*Replication of synthetic (−) circle, isolation of fully synthetic replicative forms, and a test of infectivity*

The synthetic, $P^{32}$-$\overline{BU}$ (−) circles, separated from phage (+) circles, could now be used as templates for the production of fully synthetic RF (Fig. 1) which proved to be infective. Incubation conditions for synthesis of the RF were as previously employed,[5] except that $H^3$-dCTP was the labeled substrate, dTTP replaced d$\overline{BU}$TP, and the pooled $P^{32}$-$\overline{BU}$ (−) peak from the CsCl gradient (Fig. 2) was the template. Evidence that a duplex circle was synthesized was obtained by velocity sedimentation analysis in an alkaline sucrose gradient (Fig. 6). The hybrid peak contained $H^3$ and $P^{32}$ in approximately equimolar amounts and had the S value expected of a covalent duplex circle in alkali. The infectivity coincided exactly with the radioactivity, and the specific infectivity values were within the range expected for the denatured form of natural RF (Table 1). Additional evidence for the covalent duplex structure was obtained by density-gradient centrifugation in the presence of ethidium bromide (Fig. 7). This analysis was preceded by an initial density-gradient centrifugation with ethidium bromide in which a peak of higher buoyant density was identified as corresponding to the duplex covalent zone by alkaline

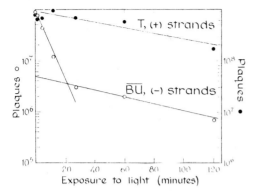

**Fig. 5.** Photoinactivation of synthetic $P^{32}$-$\overline{BU}$ DNA and $H^3$-T, phage DNA. The $P^{32}$-$\overline{BU}$, (−) and $H^3$-T, (+) peaks (Fig. 2), dialyzed and concentrated as described in Fig. 3, were each diluted into 10 mM Tris HCl (pH 7.6)-1 mM EDTA. The diluted DNA's were exposed in identical fashion to a 15-watt daylight fluorescent tube at a distance of 3 cm; 0.05-ml aliquots were placed in the dark at the indicated times and subsequently assayed for infectivity. The ratio of the initial slope for $\overline{BU}$-DNA to that for T-DNA is 12.6.

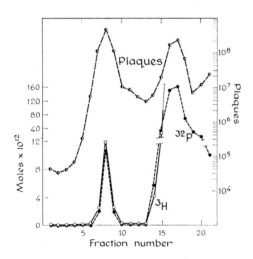

**Fig. 6.** Alkaline sucrose-gradient sedimentation of fully synthetic RF. $P^{32}$-$\overline{BU}$, (−) strands (40 μl of peak sample from Fig. 2, dialyzed and concentrated as described in Fig. 3) were replicated in a volume of 0.1 ml as described previously;[5] the labeled nucleotide was $H^3$-dCTP (Schwarz BioResearch, 1000 cpm/μμmole), and dTTP rather than d$\overline{BU}$TP was used. After 180 min, the mixture was made 20 mM in EDTA, 0.1 $M$ in NaOH, and centrifuged in a sucrose gradient in 0.2 $M$ NaOH–0.8 $M$ NaCl-1 mM EDTA, at 60,000 rpm and 1° for 100 min. The fractions were neutralized with 1 $M$ Tris citrate (pH 5) before being assayed for radioactivity and infectivity.

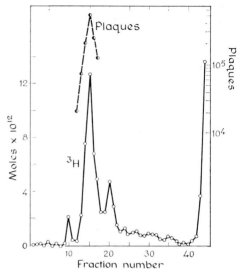

**Fig. 7.** Density-gradient sedimentation of synthetic RF in the presence of ethidium bromide. The synthetic RF, prepared as described in Fig. 6, was purified by a preliminary density-gradient centrifugation in CsCl-ethidium bromide, as described previously.[5] The covalent duplex zone, identified by alkaline sucrose-gradient sedimentation of aliquots from the fractions, was collected and refractionated in the same type of CsCl-ethidium bromide gradient with results shown above. Fractions were diluted 200-fold for the spheroplast assay but were not otherwise treated to remove CsCl or ethidium bromide. $P^{32}$ in the $\overline{BU}$, (−) template was not measurable due to radioactive decay and low recoveries.

sucrose gradient analysis of each fraction (legend to Fig. 7).

## Isolation of a synthetic (+) circle from the fully synthetic replicative form

A procedure similar to the one employed to separate synthetic (−) circles from partially synthetic RF forms was used (Fig. 1). A limited digestion by pancreatic DNase, followed by alkaline denaturation, achieved the release of $H^3$-T (+) circles from the fully synthetic RF containing $H^3$-T (+) and $P^{32}$-$\overline{BU}$ (−) circles. The mixture was fractionated directly in an alkaline sucrose gradient (Fig. 8). The synthetic $H^3$-T (+) circles, complementary to the synthetic $P^{32}$-$\overline{BU}$ (−) circles and now corresponding in structure to the original (+) phage DNA template, were evident as a $H^3$-labeled shoulder, with corresponding infectivity, partially separated from the slower sedimenting linear

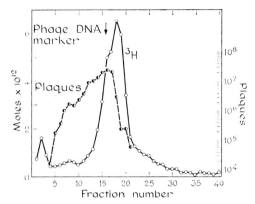

**Fig. 8.** Identification of synthetic, (+) circles by alkaline sucrose-gradient sedimentation of synthetic RF exposed to limited DNase action. The rapidly sedimenting fractions of synthetic RF in the alkaline sucrose gradient (Fig. 6) were pooled, dialyzed against 10 mM Tris HCl (pH 7.6)-1 mM EDTA and incubated for 20 min at 20° (final volume of 2 ml) at a concentration of 50 μμmoles/ml in 10 mM Tris HCl (pH 7.6)-5 mM MgCl$_2$-0.1 mg/ml bovine plasma albumin-1.2 mμg/ml pancreatic DNase. The mixture was then made 15 mM in EDTA, reduced in volume to 0.15 ml by rotary evaporation under reduced pressure, brought to pH 12 with NaOH, and centrifuged in a sucrose gradient in 0.2 M NaOH-0.8 M NaCl-1 mM EDTA, at 60,000 rpm and 10° for 240 min. The bottom of the tube was punctured with a hollow needle and the contents were displaced by saturated CsCl solution (containing Blue Dextran from Pharmacia) using a peristaltic pump. The sucrose-gradient fractions were collected from the top of the tube via a fine polyethylene tube in a stopper at the top. The fractions were neutralized (as in Fig. 6) prior to assays. The fractions were numbered in the reverse order of their collection, in order that the direction of sedimentation conform to the illustration of velocity sedimentations in the other figures. P$^{32}$ in the $\overline{BU}$, template strand was not measurable due to radioactive decay and the small amounts of DNA employed.

forms. Trailing from this infective (+) strand peak obscured the position of the more rapidly sedimenting and less infective $\overline{BU}$ (−) template circles (Fig. 8; also note legend for method of collecting the fractions); all hybrid molecules which had remained were pelleted under the conditions used.

## Discussion

Physical studies on the partially synthetic RF prepared by enzymatic replication of phage DNA showed its structure to be like that of the RF-form I isolated from infected cells.[5] The only distinction was the relative absence of supercoiling in the partially synthetic molecule and this can be attributed, at least in part, to the difference between the *in vitro* and *in vivo* conditions of salt and temperature[20] at the time of strand closure to form the circular duplex. The test of infectivity is a more rigorous and meaningful measure of the accuracy of replication and ring closure of phage DNA by polymerase and joining enzyme. Numerous φX174 mutants are known[21] in which the change of a single nucleotide results in loss of infectivity under the assay conditions employed. The fact that isolated synthetic circles and fully synthetic RF forms made with these circles as templates had specific infectivity values in the range measured for natural forms of viral DNA (Table 1) attests to the precision of the enzymatic operation.

It should now be possible to apply the techniques used in this work to the synthesis of the duplex circular genomes of other viruses, such as phage λ and animal viruses, and DNA molecules of comparable structure from cellular organelles. Such synthetic efforts will permit the insertion of base and nucleoside analogues in a manner and variety not attainable with *in vivo* systems. In addition, base changes generated by replication of the DNA with defective polymerases can now easily be studied in combination with standard genetic tools. It is of interest that DNA of approximately normal specific infectivity has been synthesized here without the use of any methylated nucleotide. This result may be related to the lack of host modification or restriction in the *E. coli* C-K12 pair and might not be applicable to other viral DNA's.

Since the conversion of phage DNA to RF-form I is accomplished *in vivo* by host enzymes, and since the DNA polymerase and polynucleotide-joining enzyme are so effective in converting phage DNA to RF-form I *in vitro*, it appears likely that these enzymes are used by infected *E. coli* cells to carry out this conversion *in vivo*. Although the predominant pathway of phage replication appears to involve the open Rf-form II,[22] the two forms are in fact interconvertible *in vivo*. Questions of the roles of these enzymes and the replicative forms in the production of (+) circles for progeny phage require further study.

The fact that *E. coli* DNA polymerase can synthesize biologically active DNA does not establish its function in the replication of the bacterial chromosome. However, the effectiveness of the combined action of the polymerase and the polynucleotide-joining enzyme in forming infective DNA may have considerable significance for chromosomal replication. In an earlier paper,[4] a mechanism was suggested whereby polymerase, with a then hypothetical polynucleotide-joining enzyme, might function in the simultaneous replication of both strands of helical DNA. The subsequent discovery of this joining enzyme, the requirement for it in phage T4 DNA synthesis,[23] its persistence in the most purified *E. coli* and phage T4 DNA polymerase preparations,[5] as well as the current demonstration of its conjoint action with polymerase, all strengthen the suggestion of this replication mechanism.[4]

## Summary

A partially synthetic, closed replicative form (RF) of $\phi$X174 DNA, consisting of phage DNA as the (+) circle and a bromouracil-containing complement synthesized by DNA polymerase as the (−) circle, was used as the source of synthetic (−) circles. The latter were separated from template strands by limited DNase action on the RF followed by denaturation and density-gradient equilibrium sedimentation. The isolated (−) circles were infectious and had the buoyant density, sedimentation velocity, and radiation sensitivity expected for DNA containing bromouracil. These (−) circles served as templates for a second round of replication which produced a fully synthetic RF with the specific infectivity of natural RF. Infective synthetic (+) circles, corresponding to the original phage DNA, were isolated from the synthetic RF after DNase treatment, as in the previous isolation of synthetic (−) circles. These results imply a relatively error-free synthesis of the $\phi$X174 genome by DNA polymerase.

*Note added in proof:* A study by Okazaki, R., T. Okazaki, K. Sakabe, and K. Sugimoto (*Jap. J. Med. Sci. Biol.* **20**:255 (1967)) of DNA replication in *E. coli* supports a mechanism of discontinuous 5′ → 3′ chain growth on the 5′ template strand (see *Discussion*).

## REFERENCES AND NOTES

1. Litman, R. M., and W. Szybalski, Biochem. Biophys. Res. Commun. **10**:473 (1963).
2. Richardson, C. C., C. L. Schildkraut, H. V. Aposhian, A. Kornberg, W. Bodmer, and J. Lederberg, in Informational Macromolecules, ed. H. J. Vogel, V. Bryson, and J. O. Lampen (New York: Academic Press, 1963), p. 13.
3. Richardson, C. C., R. B. Inman, and A. Kornberg, J. Mol. Biol. **9**:46 (1964).
4. Mitra, S., P. Reichard, R. B. Inman, L. L. Bertsch, and A. Kornberg, J. Mol. Biol. **24**:429 (1967).
5. Goulian, M., and A. Kornberg, these Proceedings **58**:1723 (1967).
6. Abbreviations used are: RF for replicative form; T for thymine; $\overline{BU}$ for bromouracil; dCTP, d$\overline{BU}$TP, and dTTP for the deoxyribonucleoside triphosphates of cytosine, $\overline{BU}$, and T, respectively; (+) circle for phage DNA; (−) circle for complementary copy of (+) circle.
7. Rust, P., and R. L. Sinsheimer, J. Mol. Biol. **23**:545 (1967).
8. Siegel, J. E. D., and M. Hayashi, J. Mol. Biol. **27**:443 (1967).
9. Experimental procedures were as described previously[5] or as detailed in the figure legends.
10. In this figure, and in all succeeding ones, the ordinate values represent the total moles of nucleotide or plaques per fraction. The fractions, except where indicated otherwise, are numbered in the order of their collection from the bottom of the tube.
11. Elson, E. L., thesis, Stanford University, Stanford (1966).
12. Guthrie, G. D., and R. L. Sinsheimer, Biochim. Biophys. Acta **72**:290 (1963).
13. Denhardt, D. T., and R. L. Sinsheimer, J. Mol. Biol. **12**:647 (1965).
14. The $\rho$ values for $\overline{BU}$-containing DNA were calculated from the base composition (Sinsheimer, R. L., J. Mol. Biol. **1**:43 (1959)), and the figure of 0.2 gm/ml determined by Baldwin and Shooter (J. Mol. Biol. **7**:511 (1963)) for the difference in $\rho$ between dAT and dA$\overline{BU}$.
15. Sinsheimer, R. L., M. Lawrence and C. Nagler, J. Mol. Biol. **14**:348 (1965).
16. Burton, A., and R. L. Sinsheimer, J. Mol. Biol. **14**:327 (1965).
17. The possibility of a facilitative effect of the synthetic DNA upon the infectivity of a small contaminant of natural DNA was tested by mixing synthetic DNA molecules ($P^{32}$-$\overline{BU}$; Fig. 2; amber mutant) and natural DNA ($\gamma$h, temperature-sensitive mutant).[13] The plaque count for the amber mutant was 398 for the $\overline{BU}$ DNA alone and 459 for the mixture, whereas the corresponding figures at similar dilutions for the temperature-sensitive mutant, alone and mixed, were 88 and 126, thus indicating the lack of interaction.

18. Denhardt, D. T., and R. L. Sinsheimer, J. Mol. Biol. **12**:674 (1965).
19. This relatively large amount is unexplained and surprising in view of the low level to which infectivity dips between the $P^{32}$-$\overline{BU}$ peak and the denatured hybrid duplex region of the CsCl gradient (Fig. 2).
20. Wang, J. C., D. Baumgarten, and B. M. Olivera, in preparation.
21. Sinsheimer, R. L., C. Hutchison, and B. H. Lindqvist, in The molecular biology of viruses, ed. J. S. Colter (New York: Academic Press, in press.)
22. Lindqvist, B. H., and R. L. Sinsheimer, in preparation.
23. Fareed, G. C., and C. C. Richardson, these Proceedings **58**:665 (1967).

# QUESTIONS FOR CHAPTER 3

## Paper 10

1. a. In metabolic studies *in vivo* the incorporation of radioactive isotopes is often reported as "percent uptake." Is this a significant figure for reporting the results of *in vitro* labeling experiments? Explain your answer.
   b. How should *in vitro* incorporation data be correctly reported?
2. a. What is the purpose of carrying out the deoxynucleotide kinase assay?
   b. From the given reaction mixture, can you write down the reaction(s) taking place during this assay?
   c. From the data given in the paper, what is the specificity of the semen phosphatase?
   d. What is the "hexokinase assay" mentioned in the paper?
   e. Can you think of any other assays to determine the ATP selectivity in a mixture of deoxynucleoside triphosphates?
3. In the isolation and processing of the $^{32}$P-DNA, what is the purpose or significance of the following:
   a. The drying of the cells
   b. The digestion with 1N NaOH
   c. The addition of exactly 0.23 ml of 5N perchloric acid
   d. Keeping the acid solution cold
   e. The repeated precipitations with acid
   f. The addition of thymus DNA as a carrier in determining the extent of digestion by DNAase.
   g. The use of both DNAase and snake venom phosphodiesterase for digesting the $^{32}$P-DNA.
4. If the search for the DNA polymerase had been carried out in 1967 instead of 1957:
   a. Name the assays that would not have been necessary to carry out.
   b. How would the deoxynucleoside triphosphates have been separated?
   c. Would the precipitation of the triphosphates as the barium salts have been necessary?
   d. Would the use of Sephadex have been advantageous, and where?
5. Table III in paper 10 gives the complete flow sheet for the purification of the polymerase. From the data given, calculate the following:
   a. Total protein present at step IV in the purification
   b. Percent recovery of enzyme
   c. The specific activity of the enzyme if a mistake of 10% had been made in the protein determinations
6. How do you explain the increase in activity between steps I and II in Table III of paper 10?
7. Is the loss of total activity due to mechanical losses or to inactivation of the enzyme? Specify for each step.

## Paper 11

1. Calculate the following from the data given in Table I:
   a. The specific activity (in units per milligram protein) of the enzyme used
   b. The actual number of counts obtained in the synthesis experiment
   c. The actual number of counts in the degradation of poly d(A-T)
   d. The mμmoles of PP$_i$ released in the presence of 2mM PP$_i$, instead of 1mM as given in the table
2. The paper states that a phosphonate derivative of a nucleoside triphosphate does not inhibit pyrophosphate exchange. What would be its expected action on synthesis?
3. a. In Fig. 4, what is the maximal activity of the DNA polymerase employed?
   b. What would Fig. 4 look like if hydrolysis and pyrophosphorolysis were noncompetitive?
4. It is repeatedly stated that no enzyme-nucleotide intermediate is involved in PP$_i$ exchange.
   a. What is the evidence for that statement?
   b. Does this mean that the enzyme does not bind any nucleotide at all?
5. How is the distinction made between hydrolysis from the 5' and the 3' ends of the DNA primer?

## Paper 12

1. Describe two possible assays for the polynucleotide joining enzyme.
2. Write down the reactions catalyzed by the following enzymes used in the paper:
   a. DPN pyrophosphorylase
   b. Alkaline phosphatase
   c. Polynucleotide kinase
   d. Deoxynucleotidyl transferase
3. Phenol extraction was used in this paper to separate the DNA from the proteins.
   a. Is the DNA in the phenol or in the water?
   b. Should the pH during the extraction be acid, neutral, or basic, and why?
   c. Is the addition of EDTA before the extraction critical?
   d. Can the proteins be recovered after a phenol extraction?
   e. When a phenol extraction is performed on a

crude cellular extract, there is often a precipitate at the interphase between the phenol and the buffer. What does it contain?
4. Why are the following procedures used in this paper?
   a. CsCl centrifugation
   b. G-25 Sephadex chromatography
   c. Pancreatic DNAase treatment
   d. High-voltage electrophoresis
5. Could it be possible that the DNA-adenylate was a side product and not an intermediate? Give all the evidence.

## Paper 13

1. In previous papers we saw that the DNA polymerase needs a 3'OH group to start its synthetic action. How does it get started when copying the single-stranded, circular $\phi$X174 DNA template?
2. In the production of the (−)-strand:
   a. Why was Br-deoxyuracil used?
   b. Would F- or I-deoxyuracil, or dUTP serve just as well?
   c. Would it be better to use a natural density label, such as $^{15}N$, $^{13}C$, or $^{3}H$.
3. a. Why was it necessary to treat the duplex with DNAase in order to separate the synthetic (−)-strand from the (+)-strand?
   b. How was the DNAase action restricted to the (+) strand?
   c. How many peaks are obtained when the denatured duplex is centrifuged on CsCl? Identify them and give the order, starting from the heaviest.
   d. If any strand were nicked more than once, would the pieces form a separate boundary?
   e. The paper states certain densities at which radioactive bands are observed. How are these densities determined?
4. a. Would you expect the (−)-strands to be infective? How would their infectivity compare with that of the (+)-strands?
   b. Would the $\overline{BU}$-labeled strands be infective?
   c. Would the nicked strands be capable of generating the whole viral DNA?
   d. Would a duplex be more or less infective than the (+)-strands?
   e. Do (−)-strands generate whole virus?
5. In velocity sedimentation in sucrose gradients:
   a. Are the closed single-stranded circles separated from nicked ones, and why?
   b. What is the increase in molecular weight of the synthetic (−)-circles containing $\overline{BU}$?
   c. Would the RF sediment slower or faster in alkali?
   d. Why is alkali used to demonstrate a covalently closed duplex circle?
6. What is the purpose of using ethidium bromide in conjunction with CsCl?
7. a. Is the *in vitro* synthesis of viral DNA identical to its synthesis *in vivo*?
   b. Since the synthetic RF differed from natural RF in the degree of supercoiling, does this mean that coiling is not important for the synthesis of viral strands?
   c. Does the fact that infective viral DNA can be synthesized by DNA polymerase and polynucleotide ligase prove that these enzymes have the same function *in vivo*?

# ANSWERS FOR CHAPTER 3

## Paper 10

1. a. The term "percent uptake" dates from *in vivo* isotopic studies, where the percentage of material retained by the animal indicated the facility with which a given compound entered the metabolic cycle. The term is also significant in working with whole cells, for the amount of a given compound that is passively taken up is usually only a fraction of that taken up when it is actively metabolized. In *in vitro* studies, however, percent uptake is not a significant figure. Enzymes must be saturated with substrate, which means that an excess of substrate must be present. The percentage taken up into a given macromolecule could therefore vary tremendously, depending on the amount of excess substrate present, without actually changing the total amount of compound incorporated. The largest percent uptake would actually be obtained when not enough substrate is present to saturate the enzyme. Occasionally, one still encounters the practice of using small amounts of labeled substrates for the purpose of increasing the "efficiency" of the incorporation, although the total amount incorporated would no doubt be decreased. A related practice, which also has its roots in metabolic studies with intact cells, is to use a given number of counts in the incubation mixture. Since the specific radioactivity of commercial compounds has been increasing over the years, this means that less and less actual substrate is added. Again, in *in vivo* studies, the relative uptake would hardly be affected, or might even be increased, but the isolated enzyme does not "see" counts, only molecules, and it would therefore be operating under suboptimal conditions.
   b. The correct way of reporting the incorporation of a radioactive substrate is the same as that of any other enzyme activity—namely, as some fraction of moles of substrate changed, that is, incorporated, per milligram enzyme protein.
2. a. The deoxynucleotide kinase assay was used to follow the purification of the enzyme deoxynucleotide kinase from *E. coli*, and to ascertain the activity of the final preparation used for the conversion of the labeled deoxynucleotides isolated from $^{32}$P-DNA to the deoxynucleoside triphosphates, the substrates for the polymerase.
   b. The following reactions take place during the deoxynucleotide kinase assay:

   $$ATP + dXMP \rightarrow dXDP + ADP$$
   $$ATP + dXDP \rightarrow dXTP + ADP$$
   $$Acetyl\text{-}P + ADP \xrightarrow{acetokinase} ATP + Acetate$$

   c. The semen phosphatase must be specific for monophosphates.
   d. In the hexokinase assay ATP is assayed by the change in adsorption at 340 m$\mu$ when TPN is transformed into TPNH$_2$, according to the two associated reactions:

   $$Glucose + ATP \xrightarrow{hexokinase} Glucose\text{-}6\text{-}phosphate + ADP$$
   $$Glucose\text{-}6\text{-}phosphate + TPN \xrightarrow{Zwischenferment} 6\text{-}Phosphogluconic\ acid + TPNH_2$$

   e. Theoretically, any reaction that requires ATP can be the basis of an assay. In practice, however, there are three more requirements. One is that it must be absolutely specific for ATP, and there are quite a number of reactions in which ATP is the natural substrate, but which are able to utilize other triphosphates at a slower rate. The second requirement is that the product of the reaction must be readily and specifically determinable. Finally, the third requirement is that the ATP-requiring reaction proceed at very low levels of ATP, for in a contamination assay of the type required here, it is necessary to determine small amounts. Needless to say, the reaction measured must be proportional to the amount of ATP.

   One of the most specific and sensitive assays for ATP is the so-called luciferase assay, luciferase being the enzyme responsible for causing fireflies to light up in a reaction between luciferin and ATP:

   $$Luciferase + H_2\text{-}luciferin + ATP \xrightarrow{Mg^{++}}$$
   $$Luciferase \cdot H_2\text{-}luciferyl\text{-}adenylate + PP_i$$
   $$Luciferase\text{-}H_2\text{-}luciferyl\text{-}AMP + O_2 \longrightarrow$$
   $$Luciferase\text{-}luciferyl\text{-}AMP + H_2O + \textit{Light}$$

3. a. The main reasons for drying the cells are to allow the volumes to be kept small and to accurately control the molarity of the reagents being added. For example, if wet cells were to be macerated with 1N NaOH, the amount of concentrated NaOH to be added would depend on a proper guess of the amount of dilution afforded by the cells. Another reason might be that the drying would denature some of the proteins and prevent them from

going into solution. The drying, however, is probably not an essential step in the procedure.

b. The digestion with alkali hydrolyzes the RNAs to 2'(3')-mononucleotides but leaves the DNA and the proteins intact.

c. According to the protocol, the DNA is in 2.0 ml of 0.12N NaOH; the addition of 0.23 ml of 5N perchloric acid gives a final concentration of 0.4N perchloric acid, which is sufficient to neutralize the NaOH and afford the proper acidity to precipitate the DNA. A smaller amount might not afford enough acidity to precipitate the DNA, while a larger amount might cause some danger of hydrolysis; the exact amount, however, is hardly critical.

d. It is very important to keep the acid solution at 0° C, for otherwise the DNA might be slightly hydrolyzed by the acid.

e. The repeated precipitations with acid are done to free the DNA of coprecipitated ribonucleotides.

f. The precipitation of DNA with acid obeys the law of mass action, so that a successively larger relative amount remains in solution as the total amount of DNA is decreased by the DNAase action. When carrier is added, the same amount of DNA remains in solution, but this is now a very small proportion of the total DNA, and since the labeled DNA is diluted by a large amount of unlabeled DNA, its precipitation is essentially quantitative. For example, let us assume for the sake of argument that the solubility of DNA in acid is 0.01 mg/ml. If 1.0 mg DNA is present at the beginning of the assay, the precipitation is 99% complete, that is, essentially quantitative. If, however, in the course of the assay the DNA is reduced to 0.1 mg, this will precipitate only to 90%. Now, if 1.0 mg carrier DNA is added to the assay all along, the precipitation will be 99.5% complete all along, as far as total DNA is concerned; and the same is true for the labeled one. At the beginning, since half of the total DNA is labeled, only 0.005 mg labeled DNA is in solution; by the time the labeled DNA has been reduced to 0.1 mg, the ratio of total to labeled DNA is 20, and so there is only 0.0005 mg labeled DNA in solution, that is, also 0.5%.

g. The DNAase is an endonuclease that works better on large molecules and depolymerizes them into short polydeoxynucleotides, which are an ideal substrate for the phosphodiesterase, which works from the ends. Neither enzyme alone gives 100% hydrolysis to monodeoxynucleotides.

4. a. If the search for the DNA polymerase had been carried out in 1967, it would not have been necessary to synthesize the labeled deoxynucleoside triphosphates, for these would have been available commercially. Hence, the deoxynucleotide kinase and the hexokinase assays would have been unnecessary. If the commercial substrates should be tested for purity, this would have been done by paper or thin-layer chromatography.

b. Modified celluloses, such as diethylaminoethylcellulose (DEAE), are superior to the "hard" ion exchangers used previously. Triphosphates are best separated on DEAE cellulose in the carbonate form, using a volatile salt such as triethylammonium bicarbonate for elution.

c. The isolated triphosphates are no longer precipitated as the Ba salt because the use of volatile solvents for the elution makes direct lyophilization of the fractions possible.

d. It is possible that Sephadex would have been helpful in the isolation of the polymerase, but there is no obvious step where it should be used.

5. a. Table III in paper 10 gives the essential information for the construction of the complete purification scheme in Table 1, below, which one would construct in the laboratory.

**TABLE 1**
*Purification of polymerase*

| Step | Total volume of fraction (ml) | Protein concentration (mg/ml) | Total Protein (mg or [g]) | Activity Units (/ml) | Activity Total (units) | Specific (units/mg protein) |
|---|---|---|---|---|---|---|
| I | 8,400 | 20.0 | [168] | 2.0 | 16,800 | 0.1 |
| II | 1,500 | 3.0 | 4,500 | 13.0 | 19,500 | 4.3 |
| III | 1,500 | 1.80 | 2,700 | 12.1 | 18,100 | 6.7 |
| IV | 800 | 0.78 | 624 | 15.4 | 12,300 | 19.8 |
| V | 90 | 4.90 | 440 | 110 | 9,900 | 22.4 |
| VI | 9 | 8.40 | 75.5 | 670 | 6,030 | 80.0 |
| VII | 30 | 0.60 | 18 | 120 | 3,600 | 200.0 |

b. From these data, recovery calculations such as those given in Table 2, below, can be made.
c. The 10% error could be ±. Therefore, the protein concentration would have been 0.54 mg or 0.66, and the specific activity, thus, could range from 182 to 222.
6. The increase in activity between steps I and II is probably due to a loss of DNAases in whose presence the activity appears lower because they digest some of the newly synthesized DNA as soon as it is made. The removal of any other inhibitor would have a similar effect.
7. By comparing the "expected" purification of Table 2, above (obtained by dividing the total protein in one step by that of the next and multiplying by the overall recovery of protein for that step), with the "observed" purification (obtained from the ratio of the specific activities in the two steps), it can be seen that there is no "apparent" loss of activity from step to step. However, in order to know whether the incomplete recovery of activity in each step is due to incomplete fractionation or to inactivation of the enzyme, it is necessary to assay the discarded fractions as well. This presumably was done by the authors, but the data are not given in the paper.

## Paper 11

1. a. One unit of enzyme activity is defined as every 10 nmoles that become acid-insoluble in the standard assay time of 30 min. Since 0.05 μg of enzyme was used, the specific activity is $1000 \times 4.24 / 0.05 \times 10 = 8,480$ units/mg.
   b. For synthesis, $\alpha$-$^{32}$P-dTTP with a specific activity of 9000 cpm/nmole was used; hence there were 38,160 cpm.
   c. For the degradation studies, $^3$H-d(A-T) with 1000 cpm/nmole was used; hence there were 1,660 cpm.
   d. It is stated in the text that 2 mM $PP_i$ inhibited by about 20% the values obtained with 1 mM $PP_i$. Hence, $4.38 \times 0.8 = 3.50$ nmoles are released under these conditions.
2. The phosphonate derivative of a triphosphate looks very much like the real thing, for the methylene groups joining the P have close to the same dimensions as the bridge oxygens. Furthermore, the base pairing ability of the nucleotide should not be affected by the methylene. It would seem likely, therefore, that the phosphonate would bind in the triphosphate site of the enzyme, thereby inhibiting the binding of a productive triphosphate and synthesis of a phosphodiester bond.
3. a. In Fig. 4 is represented a Lineweaver-Burk plot, where the highest activity ($V_{max}$) corresponds to the intercept on the $1/v$ axis, that is, $1/1.5 = 0.67$ nmoles.
   b. In a noncompetitive situation the lines would not have a common $V_{max}$, but instead, they would converge on the same spot on the substrate axis and have a common $K_m$.
4. a. If there were a nucleotide-enzyme complex involved in the $PP_i$ exchange reaction, then any analogue that can be recognized by the enzyme, but that cannot be incorporated, should inhibit the exchange. It is found, however, that $\alpha,\beta$-dTTP-methylene diphosphonate does not affect the exchange. Furthermore, the two analogues 2′,3′-dideoxy-TTP and 2′,3′-dideoxy 2′,3′-dehydro-TTP inhibit the exchange below the level present in the absence of any triphosphates. This means that they must be protecting the 3′OH ends of the primer against reaction with $PP_i$. The only way they could do that is by being incorporated into the primer and, since they cannot receive another nucleotide, thereby inhibiting all further reactions. If, however, they are already incorporated into the chain, they must have left the triphosphate site on the enzyme before they exerted their inhibitory effect.

**TABLE 2**
*Recovery calculations*

| | Percent recovery | | | | Purification | | | |
| | Protein | | Activity | | Expected | | Observed | |
| Step | Overall | Step | Overall | Step | Overall | Step | Overall | Step |
|---|---|---|---|---|---|---|---|---|
| I | 100 | | 100 | | 0 | | 0 | |
| II | 2.7 | 2.7 | 116 | | 37 | | 43 | |
| III | 1.6 | 60 | 108 | 93 | 62.6 | 1.55 | 67 | 1.72 |
| IV | 0.38 | 23.8 | 73 | 68 | 260.0 | 2.85 | 198 | 2.96 |
| V | 0.26 | 70.5 | 59 | 80 | 390.0 | 1.13 | 224 | 1.13 |
| VI | 0.045 | 17.2 | 36 | 61 | 2220.0 | 3.52 | 800 | 3.55 |
| VII | 0.0107 | 23.8 | 21.5 | 60 | 9350.0 | 2.5 | 2000 | 2.5 |

b. The noninvolvement of a nucleotide-enzyme complex in the exchange reaction does not mean that the enzyme does not bind any triphosphate, for it does, indeed, do this. It only means that the synthetic and exchange reactions take place at different sites.
5. Since only the 3'OH group of the primer can accept further nucleotides, it is possible to label the 3' end specifically with $\alpha$-$^{32}$P-dXTP. Hydrolysis of that end would then be signaled by the appearance of $^{32}$P in the mononucleotide fraction. Hydrolysis from the 5' end, on the other hand, can be monitored by using a $^{3}$H-labeled primer.

## Paper 12

1. Polynucleotide ligase catalyzes a number of partial reactions, each of which can form the basis of an assay. Another approach is to measure the restoration of biological activity of DNA previously inactivated by "nicking." The most widely used assays, however, measure the polynucleotide joining activity on substrates especially prepared for assay purposes.
    a. An assay devised by Richardson and colleague (Weiss, B., and Richardson, C. C.; Proc. Nat. Acad. Sci. U.S.A. **57**:1021, 1967) uses DNA nicked by pancreatic DNAase in which the 5' terminal phosphates were labeled by means of polynucleotide kinase and $\gamma$-$^{32}$P-ATP; it measures the acquisition of resistance of these labeled phosphates to E. coli alkaline phosphatase as they pass into phosphodiester bonds. Operationally, alkaline phosphatase is added at the end of the polynucleotide ligase incubation and allowed to act for a given period of time; the amount of phosphate liberated is determined by the amount of acid-soluble radioactivity.
    b. An assay devised by Kornberg and colleagues (Cozzarelli, N. R., Melechen, N. E., Jovin, T. M., and Kornberg, A.: Biochem. Biophys. Res. Commun. **28**:578, 1967) uses a synthetic dI-dC duplex, in which the dC strands are interrupted and comprise both unlabeled and $^{3}$H-labeled pieces. The unlabeled strands are covalently linked to a cellulose matrix. When this duplex is denatured by means of alkali and filtered, the $^{3}$H-dC deoxypolynucleotides will be in the filtrate, unless they have been joined to the cellulose-bound strands by the ligase. The assay thus consists simply in filtering the incubation mixture at the end of the reaction and determining the radioactivity in the cellulose.
2. a. DPN pyrophosphorylase catalyzes the condensation of NMN (nicotinamide mononucleotide) with ATP to yield DPN (diphosphopyridine nucleotide) or NAD (nicotinamide adenine dinucleotide):

$$NMN + ATP \rightleftharpoons DPN + PP_i$$

b. Alkaline phosphatase of E. coli specifically removes monophosphates from a large variety of substances.

$$R\text{-}P \rightarrow R + P_i$$

c. Polynucleotide phosphorylase transfers $P_i$ from a nucleoside triphosphate (ATP, GTP, UTP, CTP) specifically to a 5'-hydroxyl group of polynucleotides:

$$ATP + OH\text{-}Xp(Xp)_n \rightarrow ADP + P\text{-}O\text{-}Xp(Xp)_n$$

d. Deoxynucleotidyl transferase is a mammalian enzyme that transfers deoxynucleotidyl residues to the 3'OH ends of DNA, or synthesizes deoxyribopolynucleotides when presented with a suitable initiator. It differs from DNA polymerase in that it does not require a template:

$$pTpTpT + ndATP \rightarrow pTpTpTpA(pA)_n + nPP_i$$

3. The separation of polynucleotides from proteins by phenol extraction depends on the fact that highly charged molecules remain in the aqueous phase. Therefore:
a. The DNA is in the aqueous phase.
b. The idea is to maximize the charges on the nucleic acid and to minimize the charges on the protein. Therefore, the extraction would never be carried out at acid pH. Usually, a neutral pH would be most suitable, but if the proteins are highly basic ones, an alkaline pH would be best.
c. In this particular case the phenol is being added to free the DNA of the DNAase used to "nick" the DNA. It is therefore not just an admixture of nucleic acid and protein, but an enzyme-substrate complex that has to be broken. Since $Mg^{++}$ is required for the formation of the complex, it is especially important to remove it by means of EDTA. Thus, the addition of the chelator is critical.

In general, metal chelators increase the yield of DNA from phenol extractions, probably because $Mg^{++}$ is part of the structure of nucleoproteins, and because divalent ions, $Mg^{++}$ in particular, are very effective counter ions for the charges on the nucleic acids.
d. Proteins are denatured by the phenol and cannot be recovered after phenol extraction. This method of separating nucleic acids from proteins is therefore not suitable when the

protein as well as the nucleic acid has to be recovered.
e. The precipitate at the interphase between phenol and the aqueous phase contains denatured protein, insoluble nucleoprotein, and a host of undefined materials. It seems to be particularly prone to forming when the nucleic acid is complexed with lipoproteins, as with cellular membranes. It is always advisable to reextract the interphase separately if a complete yield of nucleic acid is desired.

4. a. CsCl density gradient centrifugation is used to separate native from denatured λ DNA, and to identify the product as pertaining to native λ DNA.
b. G-25 Sephadex was used to separate labeled free DPN from DNA-adenylate.
c. Limited treatment with pancreatic DNAase is used to produce single-strand "nicks" in the DNA.
d. According to the title of Fig. 3, high-voltage electrophoresis was used to identify the products of digestion of the DNA-adenylate, but according to the legend this was accomplished by paper chromatography.

5. The DNA-adenylate was an intermediate in the reaction and not a by-product. The fact that it did appear only when rather high enzyme concentrations were used could conceivably argue for the latter, but together with all the other evidence, it, too, argues in favor of the intermediate. The DNA-adenylate fulfills all necessary and sufficient conditions of an intermediate: it does not accumulate, but exists in a steady, low concentration, and the isolated, or in this case synthetic, intermediate serves as a substrate for the end reaction. A further argument particular to this case is that the enzyme-adenylate complex is inactive with the DNA-adenylate complex.

## Paper 13

1. As shown in the quoted references, the DNA polymerase cannot start copying the circular ɸX DNA unless it is provided with a deoxypolynucleotide starter. Almost any polynucleotide, including homopolynucleotides of short length can serve as a starter. This is probably so because almost any short polynucleotide can find a homologous sequence in the virus with which it can pair. Furthermore, even if the fit is not completely perfect, it can be digested away by the polymerase, since, as we have seen, it has both 3'- and 5'-hydrolase activity. The fact that almost any polynucleotide can serve as a starter also shows that there is no special initiating sequence on the viral DNA, and that copying can start anywhere.

2. a. Br-deoxyuracil is used as a density label because the bromine approximates the methyl group of thymine closely enough to be acceptable to most enzyme systems; yet at the same time it is much denser than the methyl group.
b. Fluoro-deoxyuracil is sometimes used as a density label, for the F, although smaller than the methyl group, is acceptable to the DNA polymerase, as are, in fact, all uridine derivatives and uridine itself. However, F and H do not differ in density from the methyl group as much as Br, and thus do not afford as clean a separation from the natural species as Br. I is too big to be physiological. F-deoxyuridine cannot be used for in vivo density labeling experiments, however, for it is a specific metabolic poison for some enzyme systems. As for using in vitro labeled F-dU-containing DNA, it has been found that the F-derivative is excessively mutagenic. For all these reasons, Br-deoxyuridine is the density label of choice.
c. $^3$H cannot be used as a density label because of its radioactivity ($^3$H-labeled radioactive compounds have such a small proportion of the H in $^3$H that their density is not affected). $^{15}$N and $^{13}$C are very expensive, and are not available as organic compounds. They have been used for in vivo density labeling, where the bacteria could be grown on simple ammonia and carbonate salts. But even for in vivo labeling they have been superseded by the much more convenient Br-deoxyuridine.

3. a. One of the characteristics of double-stranded, circular DNA is that it renatures instantly when cooled after melting, which means that the strands do not separate. This can easily be demonstrated with a model of two twisted strings, which also shows that all that is necessary to separate the two strands is a single scission in one of the strands, the result then being one single-stranded circle and one linear strand.
b. There is no way of restricting DNAase action to one of the strands. The method depends on recovering some portion of intact (+)-strands, not all of them. Although only enough enzyme is added to nick half the strands, the enzyme will work randomly, nicking some of the strands several times, and some duplexes not at all.
c. If radioactivity is determined as is stated in the paper, there are four boundaries. One, containing $^{32}$P, is formed by the $\overline{\text{BU}}$-containing single strands. A second one, containing both $^{32}$P and $^3$H, obviously contains unreacted duplex. A third boundary on the heavy side of

the duplex peak, mentioned in the text, is not apparent from the radioactivity peak in Fig. 2, but it is apparent from the infectivity assay. It is said to contain imperfectly renatured duplex. A fourth boundary contains only $^3$H, and is formed by single (+)-strands. The circular and linear single strands are not separated.

d. If any strand were nicked more than once, the pieces would not form a separate boundary because the average density of a molecular species is essentially independent of its molecular weight. Separate boundaries could be formed only if the pieces were vastly different in base composition.

e. The buoyant density of a DNA is obtained by determining the Cs$^+$ concentration at the peak from the refractive index, and looking up the correspondent density from tables in the *Handbook of Chemistry and Physics*.

4. a. Since the (−)-strands can serve as templates for viral strands *in vitro*, there is no *a priori* reason why they should not be able to do the same *in vivo*. The only catch is that the (−)-strands lack the attachment mechanism of the whole virus, and are therefore not infective for whole *E. coli* cells. They are, however, infective for spheroplasts, which permit naked DNA to enter the cell. Again, *a priori*, it should be just as easy to generate replicative duplexes from (+)- as from (−)-strands. However, since in nature a cell would never see a free (−)-strand, it seems fair to assume that it would be less infective than the viral strand. Table I does, in fact, state that the (−)-circle has a relative infectivity of 0.2 as compared to 1 by the (+)-circle.

b. The BU-labeled (−)-circle would be expected to be infective, since it is capable of generating (+)-strands. It would be less infective than the natural strand, however, for although the Br-dU is much less mutagenic than the F derivative, it is nevertheless mutagenic. Furthermore, $\overline{BU}$-containing strands are sensitive to light, as shown in Fig. 5. According to Table I, the relative infectivity is a low 0.09.

c. Although polynucleotide ligase cannot join single strands, it would be expected that a few nicked circles would nevertheless be repaired, possibly after being replicated, and would therefore generate viral DNA. The infectivity would, however, be expected to be quite low, since the nicked strands are susceptible to exonucleases as well as endonucleases.

d. The replicative intermediate might be expected to be more infective than single circles because it is much more resistant to nuclease action. In fact, however, Table I of paper 13 shows that the natural RF has a relative infectivity of only 0.05. A likely reason for the low infectivity might be that double-stranded DNA penetrates the cells less readily than do single strands.

e. All the DNAs generate whole virus in the spheroplasts.

5. a. In velocity sedimentation, as opposed to equilibrium sedimentation, closed circles sediment faster than open strands, even though they have the same molecular weight. The more compact the shape of a molecule, the faster it sediments, as can easily be demonstrated with a pot of water and any number of objects.

b. Mature $\phi$X174 DNA contains 24.6% A, which means that the (−)-strand contains 24.6% T, or $\overline{BU}$, as the case may be. The molecular weight of $\phi$X is $1.7 \times 10^6$, the total number of nucleotides is 5,300, and the difference in molecular weight between T and $\overline{BU}$ is 64.9. The difference in molecular weight is therefore $(0.246)5,300 \times 64.9 = 84,617$.

c. In alkali the two circles of the RF would partially separate, making the structure less compact; therefore, it would sediment slower than in a neutral sucrose density gradient.

d. When a covalently closed duplex circle is subjected to alkali, it denatures, but the two circles cannot separate. It therefore yields a single boundary. If one of the circles is nicked, on the other hand, alkali will produce a closed circle and a linear molecule, which will separate into two boundaries in an alkaline sucrose density gradient.

6. When certain dyes, such as ethidium bromide, bind to DNA they cause a certain unwinding of the duplex structure. Any such unwinding in a circular duplex causes the number of superhelical turns to change, so that there is no change in the total number of turns. This produces considerable mechanical stress on the circular duplex, causing the molecule to be more ordered. As a result, fewer dye molecules can be bound to a closed duplex circle than to a nicked one or a linear molecule; since the density is inversely proportional to the number of dye molecules bound, a closed duplex circle has a greater density in ethidium bromide than the other forms. The method thus serves to separate nicked and closed circles, which is not possible in CsCl alone.

7. a. Although the end result is the same, the mode

of synthesis of the viral DNA *in vitro* does not correspond to the *in vivo* process. Although the mode of replicating the virus in the cell has not been completely elucidated, it is fairly certain that viral strand (+)-strands are not made by dissociating the previous viral strand and laying down a new complementary strand to the (−)-strand, as was done in the test tube. A number of different replicative intermediates have been isolated, the most interesting of which is a duplex circle with a single-stranded tail corresponding to the length of about one viral DNA. This suggests that the viral strands are produced on a double-stranded RF, by building onto the 3'OH end of the (−)-strand to form a long single strand which would then be cut to size and closed into a circle during the maturation process. Such a mechanism would solve the problem of how a few RFs can produce a large number of viral strands, and it would also solve the problem of initiating the synthesis of a new strand.

b. The difference in supercoiling between the natural and the synthetic RFs does not provide any information on the importance of this feature for the synthesis of viral DNA *in vivo*, because the synthetic RF was not used to synthesize (+)-strands. The low infectivity of the synthetic RF might conceivably reflect the fact that it cannot be directly utilized to synthesize viral strands *in vivo*, although other explanations for the low infectivity are possible (see 7a).

c. The fact that viral DNA can be synthesized with DNA polymerase and polynucleotide ligase *in vitro* does not prove that the same enzymes are responsible for the job *in vivo*, but it does suggest that they are. Since the final step in viral DNA synthesis, the generation of (+)-strands from the RF, was not carried out *in vitro*, this step might conceivably require a different enzyme.

# CHAPTER 4

# THE STRUCTURES OF RNAs

Ribonucleic acid differs chemically from DNA in that it has an OH group at the 2' position of the sugar, and in that it has uracil instead of thymine as a major base constituent. The presence of adjacent hydroxyls renders RNA more labile than DNA, especially to alkali, by which it is degraded with great facility.

RNAs are not nearly as large as DNAs and exist in the form of a large variety of well-defined molecules, more of which are being discovered all the time. The division of RNA into messenger, ribosomal, and transfer RNA is no longer valid. Not all rRNA is of high molecular weight, and one, the so-called 5S RNA, resembles tRNA in size. Then, in the nucleus of mammalian cells, for example, there are many different RNAs that do not belong to these three categories. There are also small RNAs of distinct composition found on chromosomes. Furthermore there are a large number of different viral RNAs.

RNAs are almost universally prepared by the phenol extraction method, with such variations as the use of hot phenol or extraction in the presence of a detergent, such as sodium dodecyl sulfate (SDS).

The fractionation of different RNAs is usually accomplished in steps: first the major molecular weight classes are separated from each other by chromatography on methylated bovine serum albumin–Kieselguhr (MAK) or Sephadex G-75, G-100, or G-200, by sedimentation in a density gradient of sucrose or other suitable materials, or by preparative electrophoresis on polyacrylamide gels, and then each fraction, especially the 4S-5S one, is further submitted to more discriminating techniques. Countercurrent distribution in a variety of solvent systems, chromatography on benzoylated DEAE columns, and reverse phase chromatography on diatomaceous earth with Freon have made it possible to isolate many individual tRNAs in pure form.

Most RNAs, in addition to the four usual constituents, possess a large number of minor bases. There are all possible methyl and dimethyl derivatives of the four common bases, and many kinds of adenine derivatives with different acids, including amino acids, combined with the 6-$NH_2$ group; then there are pseudouridine, dihydrouridine and dihydrocytosine, thiouridines, and several others. In some kinds of RNA 2'-methylribose is common.

The presence of the minor bases has greatly facilitated the determination of nucleotide sequences, especially in transfer RNAs, of which eleven have been completely sequenced. The presence of minor bases is not, however, a prerequisite for sequence determinations, for the complete primary structures of two 5S RNAs, which do not contain any minor bases at all, have been elucidated. The use of highly radioactive RNA, labeled either *in vivo* with $^{32}P$ or *in vitro* with polynucleotide kinase, has made it possible to do sequence determinations on a small amount of RNA in a relatively short time. Structure determinations on the larger RNAs, however, are still in the future, although partial sequences of 16S and 23S ribosomal RNAs and of many viral RNAs have been determined.

When dealing with the larger RNAs, especially messenger RNA and the large nuclear RNAs of unknown function, it is often sufficient for the time being to establish identity, or the degree of relatedness, by hybridization with DNA. Competition hybridization, in which the unknown, highly labeled RNA is hybridized against homologous DNA in the presence of different amounts of unlabeled standards, is capable of considerable sophistication in determining the differences and similarities between two RNAs, or even two families of RNAs.

RNAs have a less extreme shape than DNA, if only because they are much smaller, and it is therefore easier to determine their molecular weights from hydrodynamic constants. However, now that careful sedimentation studies have been done on a number of RNAs, they can be used as standards, and it has become customary to determine the molecular weight by molecular sieving, either by chromatography on Sephadex columns or by electrophoresis on polyacrylamide gel. In both of these cases, the log of the molecular weight is inversely proportional to the distance traveled in the gel.

The ribonucleic acid chains are folded into a large variety of structures: they may be single-stranded as in messenger and some viral RNAs; the strand may fold back upon itself to form a double helix, as is largely the case with ribosomal RNA; there may be a series of hairpin loops and the strand may be folded in a complicated pattern as in the "cloverleaf" structures of transfer RNAs; or there may be a bonafide double-stranded helix, as in reovirus RNA. The latter is quite similar to the structure of double-stranded DNA, except that the bases are somewhat more inclined with respect to the axis of the helix, and there are eleven bases instead of ten in one complete turn.

RNAs resemble DNA in many of their physicochemical properties. Thus, RNAs melt when heated, but instead of a single $T_m$, a plot of the hyperchromicity versus temperature often shows a series of steps, each with its own $T_m$. The steps do not necessarily show heterogeneity of the nucleic acid but are likely to indicate the sequential melting of different parts of the molecule. RNAs do not show as large a hyperchromic effect as DNA, because their helicity content is less than that of double-stranded DNA. On the other hand, even a random coil or small oligonucleotide shows considerable hypochromicity compared to a mixture of nucleotides. Another aspect in which the melting curves of RNAs differ from those of most DNAs is that they are reversible. This is due to the fact that RNAs have but a single strand folded and doubled back over itself, so that no collision is necessary for renaturation to occur. Salts have the same effect on RNA as on DNA, that is, an increase in ionic strength or the presence of divalent ions increases the $T_m$, and an increase in pH or the presence of certain organic solvents lowers it.

An important parameter in the identification of any RNA is its content of double-stranded regions, or helicity, which is usually expressed as the percentage of bases engaged in hydrogen bonding. The $T_m$ is one index of helicity, but there are a large number of additional chemical and physical methods capable of giving that information. For example, formaldehyde will react only with the amino group of free bases; therefore, the reactivity toward formaldehyde (the amount reacted can be simply determined spectrophotometrically) is a good indication of helicity. Enzymes can also be used for determining the amount of single- versus double-stranded regions. All RNAases, in the presence of $Mg^{++}$, digest only single-stranded chains, and phosphodiesterases and polynucleotide phosphorylase are specific for single-stranded ends.

As we have already seen in the case of DNA, salts have the property of rendering polynucleotides more compact, by virtue of their counteracting the electrostatic repulsions between the various parts of the molecule. The degree of change of the hydrodynamic constants in the presence of salt is, therefore, also an index of the amount of secondary structure of which the RNA is capable. Furthermore, density is also dependent on secondary structure, single strands being far denser than double ones.

In addition to the extinction coefficient, there are a number of other optical properties which are sensitive to double-strandedness. For example, the optical rotation is very dependent on the orientation of the bases, and there are also changes in the optical rotatory dispersion when new helices are formed.

Some RNAs, especially tRNAs, are unique among polynucleotides in possessing a definite tertiary structure. (It has been said by Crick that "transfer RNA is an RNA that makes like a protein.") This was first discovered by Fresco and colleagues, who found that some RNAs could lose activity under certain circumstances when no hydrolysis of the chain could be detected, and, moreover, that they could regain activity after being annealed at some critical temperature in the presence of $Mg^{++}$. Tertiary structure in RNAs is undoubtedly due to the folding of the helices. Among the optical methods, circular dichroism is the best index of tertiary structure. A molecule is said to have positive dichroism when the electric vector of the absorbed light and the long axis of the molecule coincide; negative dichroism results when the direction of the light and the long axis of the molecule are perpendicular to each other. In nucleic acids, the light is absorbed in the plane of the bases. As would be expected, DNA has strong negative dichroism, since the planes of the bases are perpendicular to the axis of the fiber. RNAs, on the other hand, when oriented in an electric field, have positive dichroism. One way this can come about is to have a surplus of short helices oriented perpendicularly to the long axis of the molecule, so that the planes of the bases are parallel to the direction of the light.

Recently, several laboratories were able to produce tRNA crystals suitable for x-ray crystallography, from which detailed conformational information can eventually be expected. None of the studies under way is as yet complete, but it is already certain that the arms of the tRNA "cloverleaf" are not open, but folded in such a way as to yield a rodlike molecule.

# 14 A SIMPLIFIED PROCEDURE FOR THE SIMULTANEOUS ISOLATION OF 4S AND 5S RNA

H. I. Robins
I. D. Raacke
Department of Biology
Boston University
Boston, Massachusetts

In recent years it has become apparent that most of the commonly used methods used for the preparation of tRNA, such as phenol extraction of whole cells (Monier, Stephenson & Zamecnik, 1960), centrifugation of total phenol-extracted RNA on a sucrose density gradient (e.g., Gilbert, 1963), or differential precipitation with salt (e.g., Smith, 1960), yield preparations contaminated with other species of low molecular weight RNA (Rosset & Monier, 1963; Reynier, Aubert & Monier, 1967). In order to prevent contamination of tRNA with 5S ribosomal RNA it is necessary to completely free the extract of ribosomes and ribosomal subunits, but even so, the low molecular weight RNA is still heterogeneous (Schleich & Goldstein, 1966; Reynier et al., 1967). On the other hand, it is not possible to obtain 5S RNA free of 4S by any of the above procedures, because even highly purified ribosomes still contain some tRNA.

Of the methods available for separating the various classes of low molecular weight RNA from each other, such as chromatography on methylated albumin-Kieselguhr (MAK) columns (e.g., Brown & Littna, 1966), electrophoresis on polyacrylamide gels (e.g., Gould, 1966), and chromatography on long columns of Sephadex G-100 or G-200 (e.g., Schleich & Goldstein, 1966; Reynier et al., 1967), only the latter is suitable as a preparative procedure.

It is common practice to deproteinize the RNA prior to the application of separation techniques. Despite the great usefulness of phenol extraction, however, its use on a large scale is cumbersome and time-consuming. In the case of crude extracts, containing a large excess of protein, multiple extractions are usually required, and emulsion problems or loss of RNA in the precipitate forming at the interphase are not uncommon. We were further struck by the usual apparent absence of peptidyl-RNA even in samples prepared from actively growing cells, which should contain an appreciable amount of this form. The partition characteristics of peptidyl-RNAs in the phenol-water system have not been studied, and it is possible that RNAs with larger peptides attached to them are extracted by phenol. Bresler et al. (1966) isolated peptidyl-RNA by direct chromatography on G-200 Sephadex of ribosomes in sodium dodecyl sulfate (SDS), but their columns were too short to resolve different classes of low molecular weight RNA.

We sought to develop a method for the separation of different classes of RNA which would not require prior deproteinization of the RNA, and which would allow the recovery of all the RNA in an extract, including that covalently bound to nascent protein. We found that direct chromatography of ribosomes or of crude extracts on G-75 Sephadex, in the presence of small amounts of SDS, was fast and efficient and yielded separate peaks of 5S and 4S RNA. These peaks are contaminated by some low molecular weight proteins and lipids, but these contaminants can easily be removed by a single phenol extraction and/or precipitation with ethanol.

### Experimental

An arg$^-$, met$^-$, RC$^{rel}$ mutant (687) of E. coli K$_{12}$ was used for the preparation of crude extracts. (The mutant was kindly supplied by Dr. B. D. Davis' laboratory.) The bacteria were grown in minimal medium with various amounts of the two essential amino acids. SDS (99% pure) was obtained from Pfaltz and Bauer, New York. The use of pure SDS is essential, as the usual commercial product (lauryl

---

From Biochemical and Biophysical Research Communications 33:240-246, 1968. Reprinted with permission.

Part of this material was originally submitted to the Boston University Graduate School by one of us (H. I. R.) in partial fulfillment of the requirements for the degree of Master of Arts.

We are grateful to the American Cancer Society, Inc. for financial support.

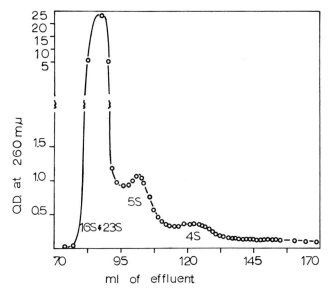

**Fig. 1.** Chromatography of ribosomes on G-75 Sephadex in the presence of SDS. Purified ribosomes corresponding to 11.2 mg of RNA were chromatographed on a column measuring 60 × 2.5 cm in a buffer containing 0.01M Tris, 0.1M NaCl and 0.25% SDS, at pH 7.3. Recovery of UV-absorbing material in the three peaks was 92%. The balance of the optical density appeared in the region of small molecules.

sulfate, 93% pure) leads to degradation of RNA (see also Maaløe & Kjeldgaard, 1966; Crestfield, Smith & Allen, 1955).

The bacteria were disrupted in a French Press at high pressure (15,000-17,000 psi) so as to disrupt polysomes mechanically (McQuillen, Roberts & Britten, 1959). The extract was then centrifuged to yield an S-30 fraction (Matthaei & Nirenberg, 1961). Ribosomes were prepared from frozen *E. coli* B according to the method of Kurland (1966).

Both ribosomes and crude extracts were adjusted to a given SDS concentration just prior to chromatography on Sephadex columns which had been preswollen in SDS for several days and prewashed with a given buffer. The concentration of SDS is not critical, but 0.25% usually was sufficient to clear turbid extracts, although with some ribosome suspensions the sample had to be made 0.5% in SDS in order to achieve clearing. Concentrations as low as 0.5% SDS are sufficient to cause disruption of ribosomes. All runs were made at room temperature to prevent precipitation of the SDS.

## Results

Preliminary runs on G-100 of crude extracts in the presence of SDS had shown that the RNA pattern obtained was exactly like that obtained with phenol-isolated total RNA (see Reynier *et al.*, 1967). It was also found that most of the protein in the extract is in the excluded portion. However, the flow rate of columns of the required length (180 cm) is extremely slow, several days being required for complete elution.

G-75 Sephadex, which has not yet been used in the literature for the isolation of 5S RNA, probably because of fear of exclusion of this RNA by the gel, was tested with purified ribosomes, which yielded the pattern shown in Fig. 1. From the number and distribution of the peaks it is apparent that 5S RNA is not excluded by G-75; in addition it was found to give better separation between the 5S and 4S regions than G-100, and allowed the use of a shorter column (60-100 cm versus 180 cm). G-75 also has a better flow rate. Therefore, G-75 was used in all subsequent experiments.

It is well-known that *E. coli* 5S RNA is devoid of methylated bases (Brownlee, Sanger & Barrell, 1967), so that this is a suitable property for checking on the identity and purity of the putative 5S RNA peak.

A crude extract prepared from bacteria methylated with $^{14}CH_3$-methionine in the absence of protein synthesis was chromatographed on G-75 in the presence of SDS. The results are shown in Fig. 2. It is seen that contrary to the results obtained with preiso-

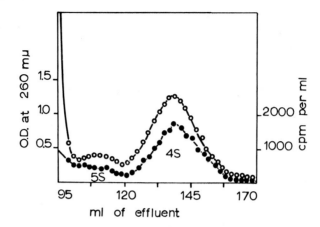

**Fig. 2.** Chromatography on G-75 Sephadex of a SDS-treated crude extract of *E. coli* labeled with $^{14}CH_3$-methionine in the absence of protein synthesis. Bacteria were grown in minimal medium in the presence of methionine (5 µg/ml) and arginine (25 µg/ml) to an optical density at 650 mµ of 0.4. They were then washed in minimal medium and resuspended. After 10 min at 37°, chloramphenicol was added to a concentration of 100 µg/ml and the incubation was continued for one additional hour. At this point $C^{14}$-Met was added (0.01 mc, or 0.014 mg/liter of culture). After all the labeled methionine had been taken up, the cells were harvested and washed in minimal medium, then resuspended in complete medium and grown to an optical density of 0.7. A crude extract was prepared and an aliquot containing 7.35 mg of RNA was made 0.1% in SDS and chromatographed on a column measuring 100 × 1.8 cm in 0.1M $NH_4Ac$, pH 5.2, and 0.1% SDS (Galibert, Larsen, Lelong & Boiron, 1966). The recovery of O.D. in the macromolecular components was 92%. Another 9% was recovered in smaller material. Open circles represent optical density; closed circles represent radioactivity.

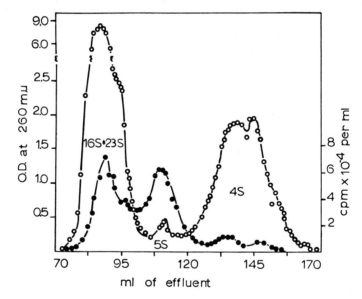

**Fig. 3.** Chromatography on G-75 of an SDS-treated crude extract of *E. coli* pulse-labeled with $^{14}CH_3$-methionine in complete medium. Bacteria were grown in minimal medium containing 2.5 mg/liter of Met and 100 mg/liter of arginine. Growth leveled off at an O.D. of 0.45, at which point 0.175 mg $C^{14}$-Met, containing $1.7 \times 10^7$ cpm, were added to the 1 liter culture. When all the labeled methionine had been taken up, 5 mg $C^{12}$-Met were added and growth allowed to continue to an O.D. of 0.6.

Crude extract corresponding to 8 mg of RNA was chromatographed under the conditions described in Fig. 1. The recovery was 97%. Open circles represent optical density; closed circles represent radioactivity.

lated RNA, there are some counts in the 5S region. The fractions were therefore pooled, and the RNA precipitated with 67% ethanol at $-20°$. In the case of the 5S peak no counts would be recovered in the precipitate, whereas the 4S region, as expected, contained $C^{14}$-methyl groups. Whatever counts there were in the 5S region can therefore be attributed to contamination by methylated lipid—and would thus not be seen if, for example, the discs had been dried by ethanol-ether.

In order to test for the degree of protein contamination in the 5S region, bacteria were labeled with a limited amount of $C^{14}$-methionine in the presence of arginine, i.e., under conditions where radioactivity from the methionine would be expected to be used for methylation as well as for protein synthesis. When such an extract was chromatographed on G-75, the results shown in Fig. 3 were obtained. It is seen that now the 5S peak is substantially labeled. Each peak was pooled, and submitted to a single extraction with 90% phenol. In the case of the 5S peak, 100% of the counts were extracted into the phenol, whereas in the case of the 4S, 74,000 cpm per mg RNA remained in the aqueous phase.

Preliminary experiments with extracts containing peptidyl-RNA labeled both *in vivo* and *in vitro* have shown that complete recovery of label is possible.

In summary, then, our results show that deproteinization is unnecessary for the chromatographic separation of different RNA fractions on Sephadex columns in the presence of SDS. The method can be applied either preparatively, or analytically for labeled RNAs and peptidyl-RNAs. Besides representing a very substantial saving in time and effort, the method has the advantage of allowing essentially complete recovery of both counts and UV absorbing material even from turbid or partially precipitated extracts. Furthermore, the SDS inhibits ribonuclease action (Crestfield *et al.*, 1955).

All RNA fractions were contaminated with protein and with lipid, but these contaminants could be removed preparatively by a single phenol extraction of the pooled fractions, after chromatography, and analytically by proper washing procedures of the discs used for the determination of radioactivity, according to Mans and Novelli (1961).

## REFERENCES

1. Bresler, S., Grajevskaja, R., Kirilov, S., Śaminski, E. & Shutov, F., Biochim. Biophys. Acta 123:534 (1966).
2. Brown, D. D. & Littna, E., J. Mol. Biol. 20:95 (1966).
3. Brownlee, G. G., Sanger, F. & Barrell, B. G., Nature 215:735 (1967).
4. Crestfield, A. M., Smith, K. C. & Allen, F. W., J. Biol. Chem. 216:185 (1955).
5. Galibert, F., Larsen, C. J., Lelong, J. C. & Boiron, M., Bull. Soc. Chim. Biol. 48:21 (1966).
6. Gilbert, W., J. Mol. Biol. 6:389 (1963).
7. Gould, H., Biochemistry 5:1103 (1966).
8. Kurland, C. G., J. Mol. Biol. 18:90 (1966).
9. Maaløe, O. & Kjeldgaard, N. O., Control of Macromolecular Synthesis, New York: W. A. Benjamin, Inc. (1966) p. 253.
10. Mans, R. J. & Novelli, G. D., Arch. Biochem. Biophys. 94:48 (1961).
11. Matthaei, H. & Nirenberg, M. W., Proc. Natl. Acad. Sci., U.S. 47:1580 (1961).
12. McQuillen, K., Roberts, R. B. & Britten, R. J., Proc. Natl. Acad. Sci., U.S. 45:1437 (1959).
13. Monier, R., Stephenson, M. L. & Zamecnik, P. C., Biochim. Biophys. Acta 43:1 (1960).
14. Reynier, M., Aubert, M. & Monier, R., Bull. Soc. Chim. Biol. 49:1205 (1967).
15. Rosset, R. & Monier, R., Biochim. Biophys. Acta 68:653 (1963).
16. Schleich, T. & Goldstein, J., J. Mol. Biol. 15:136 (1966).
17. Smith, K. C., Biochim. Biophys. Acta 40:360 (1960).

# 15 RIBONUCLEIC ACID FROM ESCHERICHIA COLI; III.
## THE INFLUENCE OF IONIC STRENGTH AND TEMPERATURE ON HYDRODYNAMIC AND OPTICAL PROPERTIES

R. A. Cox
U. Z. Littauer
*The Weizmann Institute of Science*
*Rehovoth (Israel)*

*Summary.* Values of viscosity, sedimentation coefficient, absorbancy at 258 m$\mu$, and optical rotatory dispersion of *E. coli* ribosomal RNA were measured at different sodium chloride concentrations at 25°. It was found that increasing the sodium chloride concentration led to a decrease in viscosity and absorbancy, and to an increase in sedimentation coefficient and optical rotatory dispersion. Over the range of $5 \cdot 10^{-4}$ to $1 \cdot 10^{-2}$ $M$ NaCl the ionic-strength dependence of the viscosity of RNA was found to be greater than could be attributed to the normal polyelectrolyte nature of the molecule, indicating an anomalous contraction of the RNA molecules. The changes in absorbancy and optical rotatory dispersion were also found to take place over this same range of sodium chloride concentrations. It is possible to infer from these results that as organised secondary structure is formed the hydrodynamic volume is diminished. This is consistent with the stabilisation of ordered regions by intramolecular bonds.

In support of this interpretation it was found that as the temperature was increased the dependence of the viscosity upon sodium chloride concentration approached normal polyelectrolyte behaviour. An increase in absorbancy at 258 m$\mu$ was also noticed. Furthermore when other electrolytes such as magnesium chloride and tetramethylammonium chloride were added to *E. coli* RNA solutions it was found that changes in absorbancy at 258 m$\mu$ and anomalous changes in viscosity were concomitant.

## Introduction

Studies of the optical properties of RNA and other polynucleotides have shown that RNA may have a partly helical structure depending upon temperature[1,2] and ionic strength[3-6]. Doty *et al.*[1,2] and Fresco *et al.*[7] have proposed a model for RNA in solutions of moderate ionic strength which attributes secondary structure to interactions between base residues located on different segments of the same molecule. The formation of intramolecular bonds would be expected to have a marked effect upon the hydrodynamic volume of the RNA coil. It has been shown that in solutions of moderate electrolyte concentration RNA behaves as a random coil that is impermeable to solvent (*cf.* ref. 6). Furthermore it has been found that RNA is extended in salt-free solutions[8-14], and that the addition of electrolyte decreases the viscosity to a greater extent than could be attributed to the polyelectrolyte nature of the molecule, indicating an anomalous contraction of the RNA molecules[3,15,16]. In the present studies a correlation between this anomalous contraction of high-molecular-weight RNA isolated from *E. coli* and its helical structure has been sought.

The results presented below are consistent with the studies of organised secondary structure in TMV-RNA[17-19,5,20,21], calf-liver microsomal RNA[1,2,22,23] and in yeast RNA[24].

## Materials and methods
### Isolation of RNA

*E. coli* RNA was prepared by a mild method in which the cell wall was first degraded with lysozyme and RNA was then extracted from the "protoplasts" with a phenol-water mixture and precipitated by ethanol[10]. The dissolved precipitate revealed three boundaries in the ultracentrifuge, having sedimentation constants of 4.1, 16.5 and 23.7 (determined by ultraviolet-absorption optics). The two faster-moving components were separated from the slower one by precipitation with ammonium sulphate at 4°. The

---

From Biochimica et Biophysica Acta 61:197-208, 1962. Reprinted with permission.

The authors wish to thank Mr. D. Givon for his technical assistance.

This research was supported in part by a United States Public Health Service research grant RG-5217.

Abbreviation: TMV, tobacco mosaic virus.

RNA (300 mg) was dissolved in 25 ml of Tris buffer (0.01 $M$, pH 7.4), and small amounts of insoluble matter were removed by centrifugation for 20 min at 10,000 × g. To the clear solution $(NH_4)_2SO_4$ (0.365 g/ml) was added, after 30 min at 4° the solution was centrifuged for 10 min at 10,000 × g. The supernatant contained the lighter soluble-RNA component (4.1 S). The precipitate which contained the heavier ribosomal-RNA components (16.5 S and 23.7 S) was dissolved in 25 ml Tris buffer (0.01 $M$, pH 7.4) and clarified by centrifugation for 20 min at 10,000 × g. Both fractions were dialyzed 36 h against four changes of NaCl($10^{-3}$ $M$) and then lyophilized.

Good separation of these components by $(NH_4)_2SO_4$ fractionation was thus achieved by using high RNA concentrations (10-15 mg/ml). It was indicated previously that in more dilute solutions better separation occurred in the presence of phenol[10].

Measurements have been made principally on the fraction containing the two faster-moving components. In one instance the 23.7-S component was absent and a homogeneous sample was obtained of $s_{20}$ = 16.5 S. The molecular weights of the components computed from the values of sedimentation velocity and viscosity using Mandelkern-Flory equation[25], are $5.6 \cdot 10^5$ and $1.1 \cdot 10^6$, using values of 0.57 and $2.25 \cdot 10^6$ for partial specific volume and $\beta$, respectively[26].

The viscosity of the 23.7-S component was estimated from the value found for the mixture of both components, assuming this to be an average value and knowing the viscosity of the slower-sedimenting fraction. The high-molecular-weight RNA isolated by this procedure may originate from the ribosomes since the sedimentation constants of the two components correspond to the values found for RNA isolated from the 30-S and 50-S ribosomal particles[26,27].

## Analytical methods

All methods were performed as previously described[10].

## Ultraviolet spectroscopy

Absorption measurements were made with a Beckman model-DU spectrophotometer, equipped with Beckman thermospacers, and using stoppered cells. An aqueous stock solution of RNA (1 mg/ml) was prepared and 0.1-ml aliquots were diluted with 2.9 ml of the desired salt solution; three independent dilutions were made in each case. When the dependence of absorbancy on temperature was studied each set of triplicate dilutions was exposed to the desired temperature for not more than 5 min before the absorbancy was measured and was then cooled quickly to 1.0° and another measurement was made. The observed readings were corrected for changes in concentration as a result of the thermal expansion of water.

## Optical rotation

Optical rotations and optical rotatory dispersions were measured with a Rudolph high-precision ultraviolet polarimeter, model 80, equipped with a Rudolph photoelectric-polarimeter attachment and an oscillating polarizer prism. A xenon compact arc lamp or mercury lamp provided with the instrument was used. The RNA concentration was about 3 mg/ml.

## Sedimentation in the ultracentrifuge

A Spinco Model-E ultracentrifuge fitted with ultraviolet-absorption optics was used.

## Viscosity measurements

Measurements were performed with Ostwald-Fenske viscosimeters requiring 10 ml or 1.0 ml and having a water-flow time of about 240 or 260 sec, respectively, at 25°. (Cannon Instrument Co., State College, Pa.)

## Results

### The effect of ionic strength upon hydrodynamic properties

The hydrodynamic properties of high-molecular-weight *E. coli* RNA were studied at 25° at various concentrations of added sodium chloride (Fig. 1). The viscosity decreased and sedimentation velocity increased upon increasing the electrolyte concentration above $5 \cdot 10^{-4}$ $M$, as reported previously[3,16]. The major changes were observed over the range $5 \cdot 10^{-4}$ to $1 \cdot 10^{-2}$ $M$. The dependence of viscosity on the concentration of added sodium chloride was characterized by three distinct regions (Fig. 5): first, CD, ($H_2O$ to $5 \cdot 10^{-4}$ $M$) the addition of sodium chloride had little effect on the viscosity; secondly, an anomalous region, BC, ($5 \cdot 10^{-4}$ to $1 \cdot 10^{-2}$ $M$) where the viscosity decreased inversely with sodium chloride concentration; and finally, AB, (at higher concentrations of sodium chloride) the viscosity was found to be inversely proportional to the square root of the salt concentrations.

The observed increase in the sedimentation coefficients of the two components of high-molecular-weight RNA (characterized by values of $s_{20}$ of 16.5 and 23.7 at 0.2 $M$ NaCl) coincided with the decrease in viscosity. The amount of the two components did not depend upon the electrolyte concentration; the two

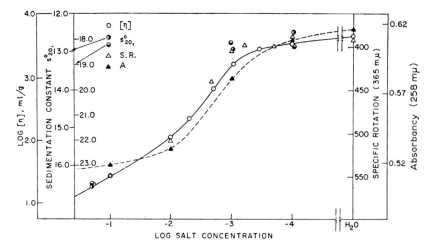

**Fig. 1.** The dependence of hydrodynamic and optical properties of RNA on the concentration of added electrolyte: ○──○, intrinsic viscosity, ordinate is on the extreme left; ◓──◓, sedimentation constant of RNA at 20°, highest-molecular-weight fraction; ◐──◐, sedimentation constant of RNA at 20°, lower-molecular-weight fraction, the ordinate is second from the left, the appropriate scales are shown in the figure; △──△, specific rotation [α] of RNA at 25° measured at λ = 365 mμ; the ordinate is on the right-hand side of the figure; ▲──▲, absorbancy of RNA at 25° measured at 258 mμ; the ordinate is on the extreme right.

sedimenting boundaries were always sharply defined.

Values of the viscosity of the 16.5-S component obtained at various salt concentrations are given in Table I; they were about 0.7 lower in high salt than the preparation containing both components (16.5 and 23.7 S). The characteristic dependence of viscosity on salt concentration found for *E. coli* ribosomal RNA has also been observed for rat-liver microsomal RNA[11, 16, 28]. The viscosity of calf-liver microsomal RNA[22, 23] and of yeast RNA[29, 30] has also been found to depend on electrolyte concentration. The sedimentation velocity of TMV-RNA has been found to depend upon both the concentration and nature of the electrolyte[5, 6].

### Optical properties

The specific rotation at wavelengths 350-700 mμ of *E. coli* RNA was measured at different concentrations of sodium chloride. The product of the square of the wavelength and specific rotation was a linear function of the specific rotation at all concentrations of added electrolyte, showing[31] that the Drude equation applies. The dispersion constant, $\lambda_c$, decreased with increasing salt concentrations. The rotation constant, specific rotation and dispersion constant varied with the concentration of salt,

TABLE I

*The dependence of viscosity of the "16.5-S" component upon sodium chloride concentration*

| Sodium chloride concentration (M) | ηsp/c (ml/g) |
|---|---|
| 5·10⁻⁴ | 1700 |
| 1·10⁻³ | 1080 |
| 2·10⁻³ | 600 |
| 3·10⁻³ | 370 |
| 5·10⁻³ | 215 |
| 1.2·10⁻² | 83 |
| 1.2·10⁻¹ | 26 |

TABLE II

*The dependence of the specific rotation ([α]) and the dispersion constant ($\lambda_c$) on the concentration of sodium chloride at 25°*

| Sodium chloride concentration (M) | [α] | | | | | | $\lambda_c$ |
|---|---|---|---|---|---|---|---|
| | Wavelength (mμ) | | | | | | |
| | 365 | 405 | 435 | 546 | 579 | 589 | |
| 0.0 | 418 | 271 | 216 | 111 | 93 | 86 | 272 |
| 1·10⁻⁴ | 422 | 278 | 219 | 103 | 94 | 87 | 269 |
| 1·10⁻³ | 436 | 288 | 231 | 126 | 109 | 102 | 262 |
| 1·10⁻² | 515 | 350 | 279 | 151 | 129 | 122 | 254 |
| 1·10⁻¹ | 565 | 387 | 316 | 174 | 149 | 142 | 245 |
| 1.0 | 578 | 412 | 325 | 187 | 166 | 159 | 221 |

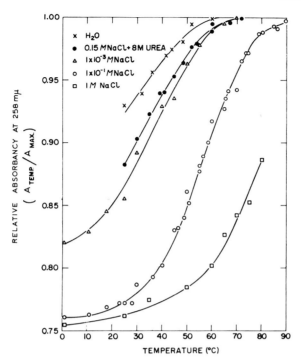

**Fig. 2.** The dependence of the absorbancy of RNA at 258 mμ on temperature at various salt concentrations. All measurements were made at about pH 7.0.

there being a relatively small change over the range 0 to $5 \cdot 10^{-4}$ M, a rapid increase between $5 \cdot 10^{-4}$ and $1 \cdot 10^{-2}$, and a slow increase on further addition of salt (Table II). The specific rotation found in the absence of added salt ($[\alpha]_D^{25} = 86°$) may be due to the already appreciable ionic strength of the sodium ribonucleate at the concentrations used.

The hypochromic effect exhibited by *E. coli* RNA solution[10] was studied. The absorbancy at 258 mμ decreased when the concentration of added sodium chloride was increased from $5 \cdot 10^{-4}$ to $1 \cdot 10^{-2}$ M (Fig. 1). The molar-extinction values with respect to phosphorus $\epsilon(P)$, within the range 220-310 mμ decreased by about the same proportion as $\epsilon(P)_{258}$ values, provided the pH was maintained at neutrality. A close correlation between changes in absorbancy, sedimentation coefficient, optical rotation and viscosity with ionic strength is apparent from Fig. 1.

The absorbancy at 258 mμ of neutral RNA solution rose significantly upon increasing the temperature (Fig. 2). The $\epsilon(P)_{258}$ increased from about 7300 in 0.1 M sodium chloride at 25° to about 9600 at 90°, which is still less than the value of 11,600 found on alkaline hydrolysis. The changes in the absorbancy were found to be reversible and the original values were obtained upon cooling the RNA solutions to 1.0°. The absorbancy temperature curves were displaced to higher temperatures with increasing salt concentrations. It has tentatively been assumed that the change in absorbancy found upon heating solutions of RNA arises from the unfolding of helical regions[1,2]. The limiting value, $\epsilon(P)_{258}$ of about 9600, should on this basis characterize RNA in the completely unfolded form. The further increase in $\epsilon(P)_{258}$ found upon alkaline hydrolysis is attributed to interactions between base residues of the type reported for oligonucleotides[32].

At low sodium chloride concentrations ($10^{-3}$ M) it was found that the absorbancy of samples which were cooled to 1.0°, then warmed to 26°, showed lower absorption than those which were not previously cooled to 1.0° but kept at 26°. Thus in an experiment the absorbancy was found to be 0.579 and 0.607, respectively.

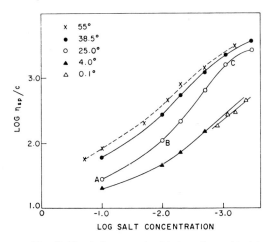

**Fig. 3.** The influence of added sodium chloride on the viscosity of RNA at various temperatures.

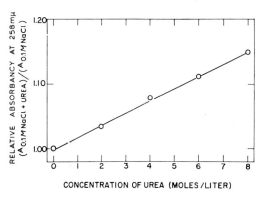

**Fig. 4.** The influence of added urea on the absorbancy of RNA measured at 258 m$\mu$ at 25°.

when the solution was made 8 $M$ with respect to urea. However, in $1 \cdot 10^{-1}$ $M$ NaCl solutions the viscosity number ($\eta_{sp}/c$) was about 67 ml/g compared with 34 ml/g when urea was absent. By contrast 4 $M$ urea had little effect at either salt concentration[10]. It was noted that the viscosity dropped slowly with time when high concentrations of urea were present, indicating that under these conditions the molecule was more susceptible to degradation.

The absorbancy at 258 m$\mu$ of RNA at 25° in 0.1 $M$ sodium chloride increased linearly with increasing urea concentrations and a rise of 14% was observed in 8 $M$ solutions (Fig. 4). The absorption at all wavelengths was increased by about the same ratio. A further 14% increase in the absorbancy at 258 m$\mu$ was found when the temperature was raised from 25 to 70° (Fig. 2).

The increase in absorbancy and viscosity upon the addition of urea to *E. coli* RNA solution, in the presence of salt, was also observed for TMV-RNA (see refs. 20 and 21). However, in the case of yeast RNA[29,30] the absorbancy increased upon the addition of urea whereas the viscosity decreased. The authors considered their sample of yeast RNA to be in the form of aggregates which were degraded into smaller fragments upon the addition of urea. However, the procedure used for the isolation of yeast RNA could have led to degradation of the RNA during its isolation.

### The effect of different electrolytes upon viscosity and absorbancy of RNA

The viscosity and absorbancy at 258 m$\mu$ of RNA solutions was found to depend on the nature as well as the concentration of added electrolyte (Figs. 5 and 6). The curve RCD, Fig. 5, denotes the form of the dependence of viscosity upon electrolyte concentrations noticed for synthetic polyelectrolytes. The curves found for RNA differed from RCD over the range corresponding to BC, $B_1 C_1$, etc. where the viscosity decreased most rapidly upon increasing the concentration of electrolyte. The viscosity of RNA did not depend simply upon the ionic strength ($I$) of the solution. For example, the addition of only $1 \cdot 10^{-4}$ $M$ MgCl$_2$ ($I = 3 \cdot 10^{-4}$) curve $A_2 B_2 C_2 D$ led to a striking decrease in viscosity. Values of viscosity in solutions of sodium chloride (curve ABCD) and potassium phosphate, pH 7.0

### The dependence of viscosity upon temperature

The dependence of viscosity upon sodium chloride concentration was studied at 0.1, 4.0, 25, 38.5 and 55° (Fig. 3). The viscosity of RNA solutions increased with temperature. The changes in the viscosity were found to be reversible and the orginal values were obtained upon cooling the RNA solution to 1.0°. The abrupt fall in viscosity brought about by increasing the electrolyte concentration from $5 \cdot 10^{-4}$ to $1 \cdot 10^{-2}$ $M$ (region BC, Fig. 3) diminished as the temperature was raised above 25°.

### The effects of urea

The viscosity of *E. coli* RNA at 25° in $1 \cdot 10^{-3}$ $M$ sodium chloride was not changed

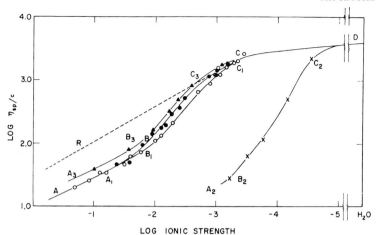

**Fig. 5.** The influence of added electrolyte on the viscosity of high-molecular-weight RNA at 25°. ○——○, sodium chloride at neutral pH (curve ABCD); ●——●, potassium phosphate buffer, pH 7 (curve $A_1B_1C_1D$); ×—×, magnesium chloride at neutral pH (curve $A_2B_2C_2D$) (see ref. 10); ▲——▲, tetramethylammonium chloride at neutral pH (curve $A_3B_3C_3D$). The curve RCD is that expected for simple-polyelectrolyte behaviour assuming $m = 0.5$ in Eqn. I (ref. 15).

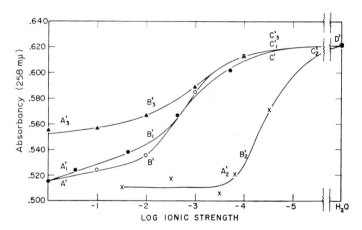

**Fig. 6.** The influence of added electrolyte on the absorbancy of RNA at 258 mμ at 25°; ○——○, sodium chloride (curve A'B'C'D'); ●——●, potassium phosphate buffer, pH 7.4 (curve $A'_1B'_1C'_1D'$); ×—×, magnesium chloride (curve $A'_2B'_2C'_2D'$); ▲——▲, tetramethylammonium chloride at neutral pH (curve $A'_3B'_3C'_3D'$).

(curve $A_1B_1C_1$), were of the same order of magnitude. (A value of $\eta_{sp}/c = 110$ in $5 \cdot 10^{-4}$ $M$ potassium phosphate reported earlier[10] is now believed to be spurious probably as a result of contamination of the buffer by heavy-metal ions.) The values of viscosity of RNA in tetramethylammonium chloride solutions were in general higher than for the same concentration of sodium chloride.

Values of $\epsilon(P)_{258}$ were found to change over the range of electrolyte concentration $B'C'$, $B'_1C'_1$, etc. (Fig. 6) corresponding to BC, $B_1C_1$, etc. (Fig. 5). The limiting value of $\epsilon(P)_{258}$ was found to be 7300 in both $1 \cdot 10^{-4}$ $M$ magnesium chloride and 1 $M$ sodium chloride. A higher value, $\epsilon(P)_{258} = 7800$, was found in 1 $M$ tetramethylammonium chloride solutions. Values of $\epsilon(P)_{258}$ were also found to decrease upon the addition of Tris to $\epsilon(P)_{258} = 7800$ in 1 $M$ solutions.

### Discussion

The present studies contribute to the accumulating evidence that RNA is a single-stranded coil capable of forming, once electrostatic repulsions have been diminished, a variety of partly organised configurations that are stabilised by intramolecular bonds. It was found that increasing the sodium chloride concentrations led to a decrease in viscosity and to an increase in sedimentation coefficients. The complementary changes in viscosity and sedimentation constants show that the molecular weight is not affected by ionic strength. Thus the changes in hydrodynamic properties of *E. coli* ribosomal RNA with ionic strength at 25° reveal the changing shape of the polymer coil.

Furthermore the decrease in the viscosity of RNA solution with increasing sodium chloride concentration is several times greater than that found for simple polyelectrolytes. In the latter case the changes in viscosity are, to a close approximation, inversely proportional to the square root of the electrolyte concentration[15]. Referring to Figs. 1 and 5, a slope of 0.5 would be expected for normal polyelectrolyte behaviour, and the changes found for the RNA over the range $5 \cdot 10^{-4}$ to $1 \cdot 10^{-2}$ $M$ salt are in excess of this. These results indicate that the addition of salt to RNA solution leads to an anomalous contraction of the molecules to form a highly compact configuration. The formation of helical regions stabilised by hydrogen bonds in RNA, is believed to be accompanied by a considerable increase of optical rotation and a similar decrease in ultraviolet absorption[1,6]. Fig. 1 reveals that these optical properties change most rapidly over the range $5 \cdot 10^{-4}$ to $1 \cdot 10^{-2}$ $M$ sodium chloride, which is the same range of electrolyte concentration causing the anomalous changes in viscosity (*i.e.* contraction of the molecules). These data, therefore, suggest that the anomalous changes in viscosity found upon the addition of electrolyte are due to the formation of an ordered structure probably involving helical regions.

The changes observed upon increasing the salt concentration above $1 \cdot 10^{-2}$ $M$ are apparently largely due to the polyelectrolyte nature of RNA. This behaviour is consistent with an intermittent distribution of helical regions, the flexibility required for folding being derived from non-helical segments. Furthermore, factors that increase $\epsilon(P)_{258}$ (*i.e.* decrease helical content) also increase the hydrodynamic volume and result in a closer approach to normal polyelectrolyte behaviour. Thus the abrupt fall in viscosity brought about by increasing the electrolyte concentration from $5 \cdot 10^{-4}$ to $1 \cdot 10^{-2}$ $M$ (region BC, Fig. 3) diminished as the temperature was raised above 25°. In addition it was noticed that increasing urea concentration led to an increase in both the absorbancy and viscosity of RNA.

The observed reduction in hydrodynamic volume with increasing ionic strength is probably caused by two factors: first the polyelectrolyte nature of RNA and secondly the formation of secondary structure. An attempt was therefore made to estimate the changes in the hydrodynamic volume of RNA due to its secondary structure at various salt concentrations. This was carried out by measuring the viscosity and absorbancy at various salt concentrations and taking into account the changes in these properties owing to the normal polyelectrolyte behaviour of these molecules.

The viscosity of RNA solution in $5 \cdot 10^{-4}$ $M$ NaCl was taken as a basis for calculation. On increasing the concentration of added sodium chloride the reduction of the viscosity due to the polyelectrolyte nature of RNA was calculated by using the following equation[15]:

$$[\eta]_{\text{calc.}} = [\eta_0] (c_{\text{salt}}/c^0_{5 \cdot 10^{-4}})^{-0.5} \quad (1)$$

where $[\eta]_{\text{calc.}}$ is the limiting viscosity number in a given salt concentration ($c_{\text{salt}}$), and $[\eta_0]$ is the limiting viscosity number in $5 \cdot 10^{-4}$ $M$ NaCl ($c^0_{5 \cdot 10^{-4}}$). As indicated above, values of $[\eta]_{\text{calc.}}$ at different NaCl concentrations are due to the polyelectrolyte nature of RNA; these values are higher than the observed values $[\eta]_{\text{obs.}}$. We therefore assumed that the deviation from the calculated viscosity values is due to the formation of secondary structure. At a given salt concentration the ratio $[\eta]_{\text{obs.}}/[\eta]_{\text{calc.}} = f$ provides a measure of the anomalous contraction of the RNA coil due to secondary structure, and has been related to $\epsilon(P)_{258}$ in Fig. 7. From the study of the dependence of the absorbancy of RNA on temperature it was found that a limiting value of $\epsilon(P)_{258} = 9600$ was obtained at 90°. This is the value given by an RNA molecule which is devoid of secondary structure[1]. The lower value of $\epsilon(P)_{258} = 8500$ found at 25° in $5 \cdot 10^{-4}$ $M$ NaCl indicates that appreciable secondary structure may be present under these condi-

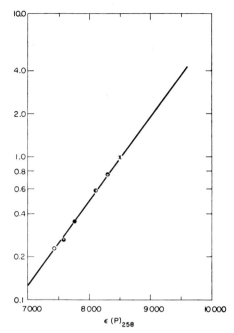

**Fig. 7.** The dependence of the ratio $f = [\eta]_{obs.}/[\eta]_{calc.}$ on the $\epsilon(P)_{258}$ values at various sodium chloride concentrations: ×—×, $5 \cdot 10^{-4}$ M; ◐—◐, $1 \cdot 10^{-3}$ M; ◑—◑, $2 \cdot 10^{-3}$ M; ●—●, $5 \cdot 10^{-3}$ M; ◒—◒, $1 \cdot 10^{-2}$ M; ○—○, $1 \cdot 10^{-1}$ M. For the method of calculating $f$ see text.

tions. Added evidence for the presence of secondary structure is the significant optical rotatory disperion found in the presence of $5 \cdot 10^{-4}$ M NaCl. Therefore the values of $f$ given in Fig. 7 are relative; the absolute values are significantly lower and could be found if the magnitude of the anomalous contraction in $5 \cdot 10^{-4}$ M NaCl were known. However, extrapolation to $\epsilon(P)_{258} = 9600$ (Fig. 7) provides a provisional estimate of about 4. Thus in order to obtain the corrected value of $f$ ($f_{corr.}$) at a given salt concentration one has to divide the $f$ values given in Fig. 7 by 4. Based on this estimate $f_{corr.}$ in 0.1 M NaCl is about 0.0625, suggesting that in the complete absence of secondary structure the viscosity would be about 500 ml/g (a 16-fold increase). Viscosities of this order have been reported for simple electrolytes such as polyphosphates of comparable molecular weight[33].

Experimental confirmation of this estimate of $f_{corr.}$ has not been obtained in these studies although a ten-fold increase in the viscosity of $1 \cdot 10^{-3}$ M NaCl solutions of RNA has been noticed upon increasing the temperature from 4 to 55°. The corresponding $\epsilon(P)_{258}$ values increased from about 7900 to about 9500. Whatever the precise value might be it is evident that in 0.1 M NaCl $f_{corr.}$ is greater than about 10 and that the volume of the RNA coil is considerably diminished by its secondary structure.

The correlation noted between changes in $\epsilon(P)_{258}$ and anomalous changes in viscosity upon changing the sodium chloride concentration was also found for magnesium chloride, potassium phosphate, tetramethylammonium chloride and Tris. Relative low concentrations of $Mg^{2+}$ were found highly effective in stabilizing the secondary structure of RNA. It required much higher concentration of $Na^+$ than $Mg^{2+}$ to achieve a comparable helix stability. Monovalent cations having a large ionic radius such as tetramethylammonium and Tris were found to be less effective in inducing secondary structure than sodium ions. The large organic ions seem to be less efficient than sodium ions in neutralizing the charge of the polyanion. On the other hand divalent ions probably interact very strongly with the RNA chains and thereby reduce the effective charge[34]. Thus the hydrodynamic volume and optical properties of RNA depend both on the nature and concentration of added electrolyte.

## REFERENCES

1. P. Doty, H. Boedtker, J. R. Fresco, R. Haselkorn and M. Litt, Proc. Natl. Acad. Sci., U.S. **45**:482 (1959).
2. P. Doty, H. Boedtker, J. R. Fresco, B. D. Hall and R. Haselkorrn, Ann. N. Y. Acad. Sci. **81**:693 (1959).
3. R. A. Cox and U. Z. Littauer, Nature **184**:818 (1959).
4. R. A. Cox and U. Z. Littauer, J. Mol. Biol. **2**:166 (1960).
5. R. Haschemeyer, B. Singer and H. R. Fraenkel-Conrat, Proc. Natl. Acad. Sci., U.S. **45**:313 (1959).
6. H. Boedtker, J. Mol. Biol. **2**:171 (1960).
7. J. R. Fresco, B. M. Alberts and P. Doty, Nature **188**:98 (1960).
8. H. Eisenberg and U. Z. Littauer, Bull. Research Council of Israel **7A**:115 (1958).
9. U. Z. Littauer and H. Eisenberg, Bull. Research Council of Israel **7A**:114 (1958). Abstracts of the 4th International Congress of Biochemistry, Vienna, Pergamon Press Ltd., 1958, p. 31.

10. U. Z. Littauer and H. Eisenberg, Biochim. Biophys. Acta **32**:320 (1959).
11. R. Laskov, E. Margoliash, U. Z. Littauer and H. Eisenberg, Biochim. Biophys. Acta **33**:247 (1959).
12. U. Z. Littauer, D. Danon and Y. Marikovsky, Biochim. Biophys. Acta **42**:435 (1960).
13. D. Danon, Y. Marikovsky and U. Z. Littauer, J. Biophys. Biochem. Cytol. **9**:253 (1961).
14. Y. Kawade, Ann. Rep. Inst. Virus Res., Kyoto Univ. **2**:219 (1959).
15. R. A. Cox, J. Polymer. Sci. **47**:441 (1960).
16. U. Z. Littauer and R. J. C. Harris, Protein Biosynthesis, Academic Press, Inc., New York, 1960, p. 143.
17. A. Gierer, Nature **179**:1297 (1957).
18. A. Gierer, Z. Naturforsch. **136**:477 (1958).
19. H. Boedtker, Biochim. Biophys. Acta **32**:519 (1959).
20. A. S. Spirin, L. P. Gavrilova, S. E. Bresler and M. I. Messevitsky, Biokhimiya **24**:938 (1959).
21. A. S. Spirin, L. P. Gavrilova and A. N. Belozerski, Doklady Akad. Nauk. S.S.S.R. **125**:568 (1959).
22. B. D. Hall and P. Doty, Microsomal Particles and Protein Synthesis, Washington Acad. Sci., 1958, p. 27.
23. B. D. Hall and P. Doty, J. Mol. Biol. **1**:111 (1959).
24. I. Watanabe and Y. Kawade, Abstracts of the IV International Congress of Biochemistry, Vienna, Pergamon Press Ltd., 1958, p. 35.
25. P. J. Flory, Principles of Polymer Chemistry, Cornell University Press, Ithaca, 1953, Chapter XIV.
26. C. G. Kurland, J. Mol. Biol. **2**:183 (1960).
27. A. Tissières, J. D. Watson, D. Schlessinger and B. R. Hollingworth, J. Mol. Biol. **1**:221 (1959).
28. U. Z. Littauer, in preparation.
29. J. P. Hummel and G. Kalnitsky, J. Biol. Chem. **234**:1517 (1959).
30. G. Kalnitsky, J. P. Hummel, H. Resnick, J. R. Carter, L. B. Barnett and C. Dierks, Ann. N. Y. Acad. Sci. **81**:542 (1959).
31. J. T. Yang and P. Doty, J. Am. Chem. Soc. **79**:761 (1957).
32. A. M. Michelson, J. Chem. Soc., (1959) 1371.
33. G. Saini and L. Trossarelli, J. Polymer. Sci. **23**:563 (1957).
34. A. Katchalsky, Endeavour **12**:90 (1953).

# 16 MOLECULAR WEIGHT ESTIMATION AND SEPARATION OF RIBONUCLEIC ACID BY ELECTROPHORESIS IN AGAROSE-ACRYLAMIDE COMPOSITE GELS

*Andrew C. Peacock*
*C. Wesley Dingman*
*Chemistry Branch*
*National Cancer Institute*
*National Institutes of Health*
*Bethesda, Maryland*

*Abstract.* An electrophoretic method has been developed for the analysis of ribonucleic acids (RNAs) ranging in size from $10^4$ to $10^8$ daltons. The method depends on the use of acrylamide gels strengthened with agarose for analysis of the larger RNAs. The resolving power of the method permitted individual characterization of RNAs in mixtures containing multiple species of RNA, without prior purification of each species; RNA molecules which differed in molecular weight by only a few per cent could be clearly distinguished, and the molecular weight of each estimated. This unusual application of electrophoretic methods for the determination of molecular weight is based on the observation that, for RNAs smaller molecules migrate more rapidly than larger ones. The mobility and the logarithm of the molecular weight are inversely related and this relationship is approximately linear. The molecular weights estimated by this technique, although numerically dependent on values assigned to known RNA standards, are highly reproducible in gels of various composition, and are at present the best means of identification of species resolved by gel electrophoresis. By this means, liver 18S RNA is identified as a doublet of RNAs of 0.66 and 0.62 $\times 10^6$ daltons and the analogous 16S of *Escherichia coli* as a doublet of 0.58 and 0.54 $\times 10^6$ daltons, while liver 5S RNA has a molecular weight of 38,000 daltons.

Several recent publications have described the usefulness of acrylamide gels in studying the electrophoretic behavior of ribonucleic acids of mol wt $10^4$-$10^6$ daltons (Loening, 1967; Loening and Ingle, 1967; Mills *et al.*, 1967; Bishop *et al.*, 1967a,b; Peacock and Dingman, 1967). The resolution of RNA species in this molecular weight range appears much superior to that afforded by other techniques, such as centrifugation in density gradients or chromatography on methylated albumin kieselguhr.

We have been able, by incorporation of agarose in the gel mixtures, to prepare acrylamide gels of very low concentration, which still retain sufficient strength to permit easy handling. These new porous gels provide a medium suitable for the high-resolution analysis of RNAs up to $10^8$ daltons, and are thus useful for the study of nuclear RNA (nRNA) (Dingman and Peacock, 1968). The addition of agarose also improves the handling characteristics of the 3.5% gels which we previously described as being useful for the study of cRNA (Peacock and Dingman, 1967).

Uriel (1966) and Uriel and Berges (1966) have described the preparation of composite agarose-acrylamide gels for protein electrophoresis, but they could not prepare gels with less than 3.0% acrylamide. Agar (Bachvaroff and Tongur, 1966; Tsanev, 1965) and agarose (McIndoe and Munro, 1967) have been used for RNA electrophoresis in 2-5% gels. We have used just enough agarose to provide mechanical support for an acrylamide gel otherwise too weak to retain its shape. The size separation depends principally on the acrylamide component of the gel.

A very important property of these composite gels is that the mobility of each species is inversely related to its molecular weight. Thus it is possible to obtain on a single electrophoretogram (1) qualitative information as to type and distribution of RNA species present, (2) quantitative data on isotope incorporation,

---

From Biochemistry 7:668-674, 1968. Copyright (1968) by the American Chemical Society. Reprinted with permission.

We thank Miss Sylvia L. Bunting for the excellent technical support she has provided in the performance of the electrophoretic analyses.

and (3) an estimate of the molecular weight of each species.

## Experimental section
### Materials

The electrophoretic cell of Raymond (1962) was purchased from E-C Apparatus Co., Philadelphia, Pa. Acrylamide (catalog no. X-5521, for electrophoresis), Bis,[1] and DMAPN were obtained from Eastman Kodak, Distillation Product Industries, Rochester, N. Y. Tris was purchased from Calbiochem, Los Angeles, Calif. Agarose used in most of the experiments was "Seakem," distributed by Bausch and Lomb. It provided clear solutions at 1.0% concentrations, which gelled at about 32-33°. Agarose from Behringwerke was also satisfactory, but two lots of agarose from Mann Research Laboratories were not satisfactory for the present purpose.

Liver cRNA and a similar preparation from *Escherichia coli* were prepared as described earlier (Peacock and Dingman, 1967). The 30S, 18S, 5S, and 4S components were identified on the gels by electrophoretic analysis of fractions separated by centrifugation, and has been fully described previously (Peacock and Dingman, 1967).

## Methods
*Preliminary experiments.* A warm solution of agarose which also contains all the reagents required for an acrylamide gel (acrylamide cross-linking reagent, accelerator, and catalyst) may be handled in either of two ways. (1) It may be kept above 35° to prevent the agarose from gelling until acrylamide polymerizes, and then subsequently cooled (method of Uriel and Berges, 1966), or (2) the solution may be cooled to 20° to permit the agarose to gel first with the acrylamide gelling last. We tried both of these approaches. At acrylamide concentrations above 3.0%, where a definite acrylamide gel formed, it made no difference which method was used. At lower acrylamide concentrations, however, where the polymerized acrylamide product was still fluid, the prior gelation of the agarose was important.

*Preparation of composite agarose-acrylamide gels.* Four solutions were used in the preparation of the gels: (1) 20% acrylamide monomer (19 g of acrylamide and 1 g of Bis in 100 ml of water; (2) DMAPN, 6.4% in water; (3) ammonium persulfate, 1.6% in water; and (4) buffer consisting of Tris (108 g), disodium EDTA (9.3 g), and boric acid (55 g), in 1 l (pH 8.3).

Gel solution (106 ml) was prepared. Water was added to 0.8 g of agarose in an amount which varied according to the amount of the acrylamide solution used. For example, for 2.0% acrylamide content, 0.8 g of agarose was placed in an erlenmeyer flask with 113 ml of water at room temperature. The mixture of agarose and water was stirred vigorously by magnetic stirring, connected to a condenser, and refluxed at 100° for 15 min. The agarose solution was cooled to 40° with running tap water set at approximately 30° (somewhat warm, to avoid local undercooling and premature gelation). Buffer (16 ml), DMAPN (10 ml), and acrylamide (16 ml) were mixed and warmed to 35°. The agarose and acrylamide solutions were mixed, the temperature was adjusted to 35°, and 5 ml of 1.6% ammonium persulfate was added. To ensure the gelation of the agarose prior to the acrylamide, we have decreased the concentration of persulfate compared to our earlier report (Peacock and Dingman, 1967). The completed gel solution was mixed well and poured rapidly into an electrophoretic cell previously equilibrated at 20°. The slot former (precooled in ice water) was inserted in place and a chilled glass rod was placed in the front part of the cell to improve cooling of this area. The agarose gelled rapidly, and after 1 hr, the acrylamide had polymerized and the excess gel was removed from the cell.

*Electrophoresis.* The cell was placed vertically, buffer (diluted one-tenth) was added to the top and bottom buffer reservoirs, and the slot former was removed. The temperature of the circulating water was then reduced gradually over the next 0.75 hr to approximately 5°. During this time, 200 v was applied to the cell, constituting the prerun. Samples were applied at the end of this time, and the electrophoresis was commenced at 200 v. The time of the run was varied depending upon the circumstances, but in general was 1.5-2 hr for the 2% gels and up to 4 hr for the 10% gels. When the run was over, the gels were stained in 0.2% methylene blue as described before (Peacock and Dingman, 1967). Destaining was most conveniently done by pouring off the excess stain and covering the gel overnight with distilled water. During the day, the gel was rinsed by continuous flow of fresh water at approximately 25°. The rate at which gels of various concentrations destain was variable and frequent inspection was required to determine when destaining had been satisfactorily completed.

*Determination of mobilities.* Mobilities were calculated by first measuring the migration distance of the species in question. The voltage used in calculating the voltage gradient was the voltage indicated across the electrode (200 v usually), and the distance used was the 17-cm length of the gels (assuming the voltage gradient in the buffer was negligible). The temperature of the coolant was 5° entering the cell, and 7° leaving the cell. Mobilities are thus those observed at approximately 6° and are expressed as $cm^2 v^{-1} sec^{-1}$.

---

[1] Abbreviations used: Bis, $N,N'$-methylenebisacrylamide; DMAPN, dimethylaminopropionitrile.

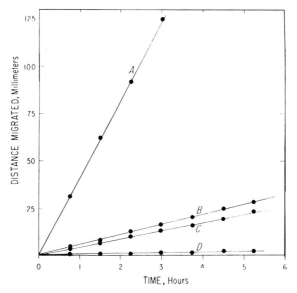

**Fig. 1.** Migration distance as a function of time in 3.5% acrylamide gel. A similar result was found in a composite gel with 3.0% acrylamide and 0.5% agarose. The RNA species (from liver cytoplasm) are: A, 4 S; B, 18 S, faster component; C, 18 S, slower component; and D, 30 S.

## Results

### Physical properties of composite agarose-acrylamide gels

Acrylamide gels of all concentrations containing 0.5% agarose are firmer and tougher than gels not containing agarose, and swell in water not nearly as much as do simple acrylamide gels. Most important, composite gels formed from 1.5 to 2.5% acrylamide were as easy to handle as the more concentrated gels, although simple acrylamide gels are fluid at these concentrations.

Gels containing 2% acrylamide and 0.5% agarose were easily cut with a multiple slicer described by Dingman and Peacock (1968). In the range 3-5% acrylamide with 0.5% agarose, the gels were more easily sectioned by removing individual sections sequentially with a microtome. All sections retained their shape and were easily transferred from the cutting device to vials for liquid scintillation counting.

### Electrophoretic properties of RNA in composite acrylamide-agarose gels

The migration of RNA molecules in these gels was linear with time even when this mobility was slow (Figure 1). The log of the (assumed) molecular weight of the RNA was inversely related to its mobility in these composite gels (Figure 2). A similar relation has been observed by Bishop et al. (1967a) in simple acrylamide gels. We have studied this relationship in composite gels having different concentrations of acrylamide (Figure 2). Results typical of those observed in composite gels containing 2.0 and 3.0% acrylamide are shown in Figure 3. The lines in Figure 2 are all described by the relationship $\log M = \log M_0 + m\mu$, where $M$ is the molecular weight of the species under study, $M_0$ is the intercept molecular weight obtained by extrapolation of the linear portion of the curve to the ordinate, $\mu$ is the observed mobility, and $m$ is the slope. Both the intercept molecular weight ($M_0$), and the slope ($m$) are a function of the concentration of acrylamide (Figure 4). It is apparent that $M_0$ rises sharply as the concentration of acrylamide drops. The highest value of $M_0$ in the present experiments is determined by the properties of the agarose support. The intercept molecular weight ($M_0$) is a reproducible characteristic of each gel. It is not, however, the molecular weight of the largest RNA that will enter the gel; RNA species of molecular weight substantially above this limit entered the gels and had measurable migration velocities. The

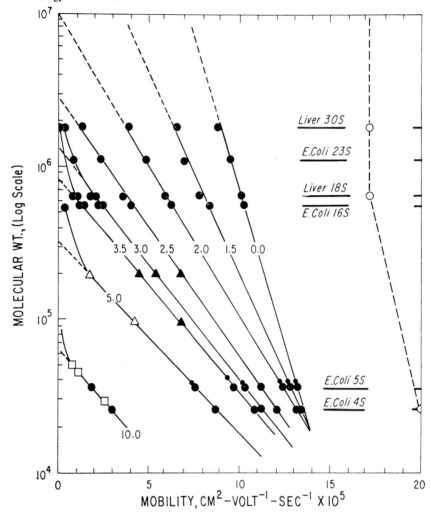

**Fig. 2.** Relations between the molecular weight of RNA species in 0.5% agarose gels containing various acrylamide concentrations and their mobilities. The molecular weights used for the *E. coli* RNAs were cited by Bishop *et al.* (1967a); values for liver 30S and 18S RNA were calculated by the formula of Kurland (1960). The acrylamide concentrations are shown on each line. The dotted line represents the mobilities in free solution (Olivera *et al.*, 1964). The significance of △, ▲, □, and small circles is described in the text.

relationship of the slope ($m$) to gel concentration (Figure 4) mirrors that of $M_0$, becoming rapidly more negative as the acrylamide concentration decreases.

Numerous other gels of varying composition were studied and the slope and intercept values found characteristic for these materials are shown in Table I. The amount of Bis used as a cross-linking agent made a modest difference in $M_0$, the more cross-linked gels being somewhat less porous. The resolution of the gels with no cross-linking was not as good as with the standard amount of Bis. Simple agarose gels also sieved; $M_0$ increased with decreasing the agarose concentration. The slope ($m$) is less for a simple agarose gel of a given $M_0$ than it is for a composite gel of the same $M_0$. Resolution on agarose gels was notably poorer than on composite gels having a similar $M_0$.

*Estimation of molecular weight.* The molecular weight of RNA molecules of unknown size, even if they occur as mixtures, may be estimated by use of an appropriate graph in which the log of the molecular weight of two or

The structures of RNAs 149

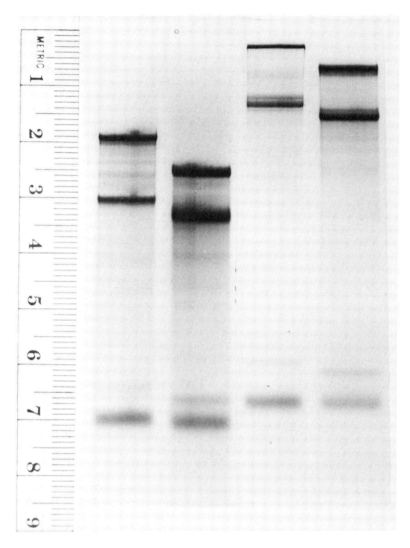

**Fig. 3.** Photograph of a group of four different electrophoretograms. The origin is at 0.0 mm on the ruler. From left to right: slot 1, liver RNA in 2.0% acrylamide plus 0.5% agarose; slot 2, *E. coli* RNA in a similar gel; slot 3, liver RNA in 3.0% acrylamide plus 0.5% agarose; and slot 4, *E. coli* RNA in a similar gel. Run 90 min at 200 v; temperature approximately 2°. Major RNA classes are identified as follows. Slot 1 (liver): 19 mm, 30 S; 31 mm, 18 S; 64 mm, 5 S; and 70 mm, 4 S. Slot 2 (*E. coli*): 26 mm, 23 S; 34 mm, 16 S; 67 mm, 5 S; and 70 mm, 4 S. Slot 3 (liver): 3 mm, 30 S; 13, 14 mm, 18S doublet; 60 mm, 5 S; and 67 mm, 4 S. Slot 4 (*E. coli*): 7 mm, 23 S; 15, 16 mm, 16S doublet; 62 mm, 5 S; and 67 mm, 4 S.

three reference RNAs is plotted against the observed mobility (as in Figure 2). An illustration of the application of this method is shown in Figure 2. The open triangles are observed mobilities of two minor RNA species from liver cytoplasm (see Peacock and Dingman, 1967), plotted, where data were available, on the 3.5, 3.0, and 2.5% lines formed by the reference RNAs (*E. coli* 23, 16, 5, and 4 S). The molecular weights of these two unknown species of RNA were then estimated from the ordinate as 2 and 0.94 $\times$ 10$^5$ daltons. These same RNA species could now be used as standards on the 5.0% acrylamide gel (closed triangles, Figure 2). A similar process was used to obtain the molecular weights of nRNA

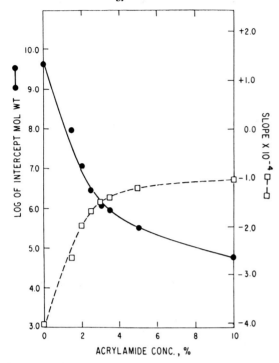

**Fig. 4.** The intercept molecular weight ($M_0$) and slope ($m$) of the line relating log $M$ and mobility as a function of acrylamide concentration in gels containing 0.5% agarose.

### TABLE I
*Characteristics of various gel formulations for RNA electrophoresis*

| Gel composition (%) | | | | |
|---|---|---|---|---|
| agarose | Total acrylamide[a] | Bis | $M_0$ | $m$ ($\times 10^{-4}$) |
| 1.0 | | | $1.5 \times 10^8$ | −3.4 |
| 2.0 | | | $6.5 \times 10^6$ | −4.9 |
| 1.0 | 1.0 | 0.05 | $5.0 \times 10^6$ | −2.0 |
| 0.5 | 3.5 | 0.0 | $2.2 \times 10^6$ | −2.0 |
| 0.5 | 3.5 | 0.05 | $1.3 \times 10^6$ | −1.5 |
| 0.5 | 3.5 | 0.01 | $1.1 \times 10^6$ | −1.5 |
| 0.5 | 3.5 | 0.175 | $0.9 \times 10^6$ | −1.4 |
| | 3.5 | 0.175 | $1.2 \times 10^6$ | −2.3 |

[a]Includes weight of acrylamide and bisacrylamide.

species (Dingman and Peacock, 1968), shown as open squares on the 10% gel line. The small circles in Figure 2, just above points for *E. coli* 5S RNA, are data points for liver 5S RNA. They fall, with small variation, on a line corresponding to a molecular weight of 38,000 daltons, somewhat larger than *E. coli* 5S RNA.

Figure 3 and Peacock and Dingman (1967) show that there are two components in liver 18S RNA and in *E. coli* 16S RNA. We estimate from their mobilities in 3.0 and 3.5% gels (Figure 2) that the molecular weight of the two components of liver 18S are 0.66 and 0.62 $\times$ $10^6$ daltons. The molecular weight of the two components of the *E. coli* 16S are 0.58 and 0.54 $\times$ $10^6$ daltons.

### Discussion

The physical properties of dilute acrylamide gels when supported by 0.5% agarose are surprising in view of the characteristics of each of these gels separately. Thus, the agarose at 0.5% is just able to maintain itself without flaking and crumbling and the acrylamide possesses so little structure that it is very nearly fluid. As has been found with other reinforced materials such as fiberglass and plastic, these composite gels possess physical properties unexpected from the properties of each constituent separately. Agarose and acrylamide differ in the chemical structure of the basic polymeric unit. Agarose, which is polyhydroxylic, is more hydrophilic than acrylamide and it may be that an appropriate balance between these two characteristics is an important feature in producing gels suitable for electrophoresis of RNA.

We do not know the mechanism by which these gels produce the size discrimination which results in the electrophoretic separation. One possibility is that there are tunnels throughout the gel through which RNA molecules pass with a tumbling motion proportional to their chain length. That is, the longer strands of RNA require a tunnel with a somewhat larger cross-section. The frictional resistance of the solution in the tunnel presumably acts as the viscous force limiting the velocity. It appears from the data in Figure 2 and Table I that agarose has a substantial effect on this viscosity, for the velocities are reduced appreciably. The acrylamide, on the other hand, has some effect on the width of the tunnel openings because increasing the acrylamide concentration results in gels which are progressively less permeable to large molecules (Figure 2). In addition, Figure 2 shows that the effect of acrylamide concentration is not as great on the smaller RNA molecules as it is on the larger ones.

The relation between the log of the molecular weight and the mobility is clearly an empirical one for which we have no satisfactory explanation. Fortunately, this relationship changes in a consistent way from one gel concentration to another, with simple identifiable changes in the parameters $M_0$ and $m$ (Figures 2 and 4). Calibration of these gels may be accomplished by determining the mobilities of RNAs of known molecular weight (1) by the presence in the sample of certain suitable calibrating standards, such as, liver cytoplasmic RNA which contains 30S, 18S, and 5S RNA, or (2) by the preparation of additional samples containing such standards run in another slot on the same gel. The molecular weight of an unknown RNA species, determined from a plot of the above data, as shown in Figure 2, is a more reproducible property than its relative mobility since the latter varies with gel concentration. For this reason, the estimated molecular weight is a better identification than the relative mobility. The precision with which the true molecular weights of various RNAs fit into this relationship is not known at present. In addition to errors in the measurement of mobilities on the gel, there are uncertainties in reported values of the molecular weight of several species.

Figure 2 shows that RNA molecules larger than the intercept molecular weight limit enter the gel with definite mobilities. One interpretation of this result is that the first few millimeters of the gel possess a structure more porous than the rest of the gel. This interpretation is unlikely in view of Figure 1, which shows that the migration distance is proportional to the time, even for large species which spend a substantial portion of the time in this first part of the gel. However, the use of the first portion of the gel for the determination of molecular weights is not recommended.

It appears from Figure 2 that the maximum resolution between two species of RNA with similar molecular weights occurs when the mean molecular weight is approximately one-half of the intercept molecular weight ($M_0$). Thus, to determine a suitable circumstance for the resolution of the *E. coli* 16S doublet whose mean molecular weight is approximately $5.4 \times 10^5$, we should choose a gel whose $M_0$ is approximately $10^6$, that is, one containing between 3 and 3.5% acrylamide. To discriminate between liver 5S and *E. coli* 5S in which the mean molecular weight is of the order of $3.5 \times 10^4$, we should choose a 10% gel in which the $M_0$ is approximately $7 \times 10^4$. On the other hand, nRNAs of the order of $4.0-5.0 \times 10^6$ will be best resolved on 2% gels, in which the $M_0$ is $10^7$. The estimation of molecular weights of two species of 18S RNA found in liver and the two species of 16S RNA found in *E. coli* affords one type of distinction that may be made between species of RNA occurring in the same tissue, whereas the 10% difference observed between *E. coli* 5S and rat liver 5S RNA indicates how species of RNA from different sources may be compared. The assignment of absolute values of molecular weight to various RNAs is, of course, limited by the standards employed and thus the method can only be used as a secondary method for the determination of molecular weights. Nonetheless, for many of the less common species of RNA which we have described (Peacock and Dingman, 1967), there is as yet no other readily applied method for the determination of the molecular weight.

Separation of RNAs by electrophoresis in gels is different in several important ways from separation by ultracentrifugation. (1) The properties of an RNA molecule which determine its mobility in an electrical field are different from the properties which determine its sedimentation velocity in a gravitational field. (2) In centrifugation, the large molecules move faster, while in the electrophoretic analysis, the small molecules move faster. (3) The electrophoretic method has much higher resolution, and each fraction is very much less contaminated with other RNAs than is the case for centrifugation. These differences suggest that new information concerning metabolic patterns may emerge from electrophoretic studies. Such studies should complement, and perhaps, by virtue of the higher resolution, even supplant earlier results based on less complete analysis.

**REFERENCES**

1. Bachvaroff, R., and Tongur, V. (1966), Nature **211**:248.
2. Bishop, D. H. L., Claybrook, J. R., Pace, N. R., and Spiegelman, S. (1967b), Proc. Nat. Acad. Sci. U. S. **57**:1474.
3. Bishop, D. H. L., Claybrook, J. R., and Spiegelman, S. (1967a), J. Mol. Biol. **26**:373.

4. Dingman, C. W., and Peacock, A. C. (1968), Biochemistry 7:659.
5. Kurland, C. G. (1960), J. Mol. Biol. 2:283.
6. Loening, U. E. (1967), Biochem. J. 102:251.
7. Loening, U. E., and Ingle, J. (1967), Nature 215:363.
8. McIndoe, W., and Munro, H. N. (1967), Biochim. Biophys. Acta 134:458.
9. Mills, D. R., Peterson, R. L., and Spiegelman, S. (1967), Proc. Nat. Acad. Sci. U. S. 58:217.
10. Olivera, B. M., Baine, P., and Davidson, N. (1964), Biopolymers 2:245.
11. Peacock, A. C., and Dingman, C. W. (1967), Biochemistry 6:1818.
12. Raymond, S. (1962), Clin. Chem. 8:455.
13. Tsanev, R. (1965), Biochim. Biophys. Acta 103:374.
14. Uriel, J. (1966), Bull. Soc. Chim. Biol. 48:969.
15. Uriel, J., and Berges, J. (1966), Compt. Rend. 262:164.

# 17 A QUANTITATIVE ASSAY FOR DNA-RNA HYBRIDS WITH DNA IMMOBILIZED ON A MEMBRANE

*David Gillespie*
*S. Spiegelman*
*Department of Microbiology*
*University of Illinois*
*Urbana, Illinois*

---

An improved method for the formation of DNA-RNA hybrids is described. The procedure involves immobilizing denatured DNA on nitrocellulose membrane filters, hybridizing complementary RNA to the membrane-fixed DNA, and eliminating RNA "noise". Denatured DNA and hybridized RNA remain on the filter throughout the procedure. Unpaired RNA is removed by washing, and RNA complexed over short regions is eliminated by RNase treatment.

Many samples can be easily handled, permitting kinetic and saturation studies. Large amounts of DNA can be loaded on a filter without interfering with the efficiency of the annealing reaction. Moreover, the "noise" can be depressed to a level (0.003% of the input RNA) permitting the identification of small regions of DNA complementary to a given RNA species. The results are quantitatively more certain than annealing in liquid, since the competing DNA renaturation reaction is suppressed.

## 1. Introduction

The usefulness of hybridizations between DNA and radioactive RNA as a test for complementarity has stimulated the search for procedures which combine reliability and convenience. Initially, Hall & Spiegelman (1961) used annealing in solution and equilibrium density-gradient centrifugation (Meselson, Stahl & Vinograd, 1957) in preparative rotors for hybrid detection.

To exploit this technique for the identification of small (less than 0.1%) complementary segments, ribonuclease treatment was introduced (Yankofsky & Spiegelman, 1962a) to eliminate low levels of adventitious contamination. Here advantage is taken of the resistance to ribonuclease of RNA complexed to DNA. This permitted the detection and measurement of the proportion of DNA complementary to the two ribosomal RNA components (Yankofsky & Spiegelman, 1962b, 1963) and the S-RNA molecules (Giacomoni & Spiegelman, 1962; Goodman & Rich, 1962).

All of the above studies used density-gradient separation of hybrids; this required lengthy centrifugations. The full realization of the potentialities of the hybridization technique clearly required the design of a procedure less costly in time and expensive equipment. The first success came with the introduction by Bautz & Hall (1962) of nitrocellulose columns to which glucosylated DNA was attached. This method was quickly generalized by Bolton & McCarthy (1962), who recognized that mechanical immobilization *per se* would suffice; and from this concept they developed agar columns containing denatured DNA trapped in the solidified gel. Britten (1963) fixed DNA by ultraviolet irradiation to synthetic polymers which were then employed as columns for RNA hybridization. Most recently, the ultimate in convenience and capacity for handling many samples was provided by the discovery (Nygaard & Hall, 1964) that nitrocellulose filters strongly adsorbed single-stranded DNA along with any hybridized RNA.

To appreciate the purpose of the present investigation, it is important to recognize that hybridization involves both hybrid formation and hybrid detection. Except for the two column procedures, all the methods described anneal the DNA and RNA in solution and then assay the amount of hybrid complex. Hybridi-

---

From Journal of Molecular Biology **12**:829-842, 1965. Reprinted with permission.

This investigation was supported by Public Health Service Research Grant No. CA-01094 from the National Cancer Institute and the National Science Foundation. One of us (D. G.) is a United States Public Health Predoctoral Trainee in Microbial and Molecular Genetics, 5-T1-GM-319-05.

zation in solution has one obvious disadvantage, stemming from the fact that RNA-DNA formation must compete with the re-formation of the DNA-DNA complexes. The existence of these competitive interactions can introduce serious errors, particularly in experiments determining saturation plateaux.

In certain instances the complications of annealing in solution can be surmounted by incubating at temperatures well below the melting temperature ($T_m$) of the DNA, a strategy used by Yankofsky & Spiegelman (1963, and unpublished observations). However, low temperature is not a universally available solution. RNA molecules possessing an extensive secondary structure will not hybridize until their own melting temperature is approached, a situation encountered with S-RNA (Giacomoni & Spiegelman, 1962).

In principle, immobilization of the DNA during the hybridization provides a logical method of avoiding these unwanted interactions. The obvious answer was to fix the denatured DNA irreversibly to nitrocellulose membrane filters, and to carry out the hybridization on the filters. This should serve to eliminate, or greatly reduce, DNA-DNA interactions while retaining the almost indispensable convenience of the filter method. To these advantages can be added RNase treatment to eliminate low levels of contamination with unpaired RNA, which can become crucial in many types of investigations.

After exploring several more involved procedures, it was discovered that irreversible fixation of DNA to nitrocellulose membranes was readily achieved by thorough drying at moderate temperatures. The result is a simple and conveniently flexible method of hybrid assay possessing a vanishingly small noise level combined with high accuracy. It is the primary purpose of the present paper to provide the details and describe the use of the resulting procedure.

## 2. Materials and methods
### (a) Bacterial strains

Thymine and uracil auxotrophs of *Bacillus megaterium*, strain KM, were isolated and kindly supplied by Dr J. T. Wachsman. An auxotrophic derivative of *Escherichia coli* 15T$^-$ requiring histidine and uracil as well as thymine was used to prepare labeled *E. coli* RNA and DNA.

### (b) Media and buffers

A basal medium (Mangalo & Wachsman, 1962) supplemented with 20 $\mu$g/ml. of uridine or 5 to 10 $\mu$g/ml. of thymidine was used. The phosphate concentration was dropped from $2.4 \times 10^{-2}$ M to $1 \times 10^{-4}$ M for $^{32}$P incorporation.

SSC buffer contains 0.15 M-NaCl and 0.015 M-sodium citrate (pH 7.0). 1/100 SSC, 2 × SSC and 6 × SSC are buffers containing one one-hundredth, two times and six times these concentrations, respectively.

### (c) Enzymes

The enzymes used in this study were purchased from the following companies: 5 times crystallized RNase, Sigma Chemical; electrophoretically pure DNase, Worthington Biochemical; lysozyme and bovine serum albumin, Armour Pharmaceutical; pronase, Calbiochem. The RNase was heated to 90°C for 10 min to remove DNase activity.

### (d) Incorporation of $^{32}$P and isolation of RNA

The uracil-requiring strain was grown overnight to late log-phase in basal medium supplemented with uridine. The cells were washed once in basal medium lacking phosphate and resuspended in a medium containing $1 \times 10^{-4}$ M-PO$_4$ at an O.D.$_{660}$ of about 0.01. When the cells had reached an O.D.$_{660}$ of 0.05, 50 $\mu$C/ml. pyrophosphate-free, neutralized [$^{32}$P]orthophosphate was added to the culture and the cells were allowed to grow for 2 doublings. The culture was then made 0.024 M in [$^{31}$P]orthophosphate and growth was allowed to occur for one more generation. The cells were harvested, concentrated, converted to protoplasts and the RNA extracted as described by Yankofsky & Spiegelman (1962a,b). The RNA prepared by this procedure had an initial specific activity of about $2 \times 10^6$ cts/min/$\mu$g.

### (e) RNA purification

The RNA was freed from DNA and other contaminating phosphorus compounds in the following way. To the RNA solution in 0.05 M-PO$_4$ (pH 6.8) was added 20 $\mu$g/ml. of electrophoretically purified DNase (Worthington Biochemicals) and the mixture was incubated at room temperature for 10 min. The RNA was then loaded on a column of methylated albumin coated on kieselguhr (MAK†) that had been previously washed with 25 to 50 vol. of 0.05 M-phosphate buffer. After loading, the column was again washed with a similar volume of 0.05 M-phosphate buffer, and then the RNA was eluted from the column with NaCl gradients ranging from 0.1 to 1.5 M in the same buffer. The DNase and chromatography

---

†Abbreviations used: MAK, methylated albumin kieselguhr; TCA, trichloroacetic acid.

procedures were repeated until the counts alkali-stable material were less than 0.05%.

## (f) $^3H$ labeling and DNA isolation

Thymidine-requiring cells (B. megaterium) were grown to late log-phase in basal medium supplemented with 20 µg/ml. of thymidine. A portion of this culture was inoculated into basal medium containing tritiated thymidine (1 µC/7 µg/ml.) at an initial O.D.$_{660}$ of 0.01. The cells were allowed to grow until the O.D.$_{660}$ reached 0.6, whereupon the cells were harvested, concentrated and converted to protoplasts; the DNA was extracted by Marmur's procedure (1961). An additional step included was treatment with pronase to remove RNase. The pronase used was self-digested for 2 hr at 37°C, after which it was added at a level of 50 µg/ml. to the DNA and incubated for 2 hr at 37°C. The solution was then made 0.5% in Dupanol and deproteinized with chloroform until no detectable RNase activity remained. The latter was monitored by 20-hr assays at 37°C with radioactive RNA as the substrate. Ethanol precipitations were used throughout. The final DNA precipitate was resuspended in 1/100 SSC and extensively dialyzed against this buffer.

## (g) Denaturation of DNA

Native DNA preparations were denatured by alkali by bringing the pH of the solution to 13, and after standing for some 10 min, the solution was neutralized. Denaturation was monitored by following the increase in O.D.$_{260}$. Sedimentation velocity analysis also confirmed the single-strandedness of the DNA. Denatured DNA was kept in 1/100 SSC at 100 µg/ml.

## (h) Hybridization with immobilized DNA

The procedure, in outline, consists of: binding the DNA to nitrocellulose membrane filters (type B-6, coarse, of Schleicher & Schuell); hybridizing RNA to the fixed DNA; and removing RNA "noise", i.e., unpaired RNA and RNA complexed over short regions. The details of each step are presented below.

*Step I. Immobilization of denatured DNA on nitrocellulose membrane filters.* Denatured DNA solutions were diluted to 5 ml. with 2 × SSC and passed through a membrane filter (presoaked in 2 × SSC for 1 min and washed with 10 ml. of the same buffer), then washed with 100 ml. of 2 × SSC. The DNA filters were subsequently dried at room temperature for at least 4 hr and at 80°C for an additional 2 hr in a vacuum oven. In some cases, the procedure was carried out with 6 × SSC. Prior drying at low temperature was instituted to avoid renaturation of the DNA in the early stages. Filtration of the DNA should be carried out at moderate speeds, particularly with smaller DNA fragments.

*Step II. Hybridization.* Hybrids were formed by immersing the DNA filters in scintillation vials containing 5 ml. of [$^{32}$P]RNA in either 2 × SSC or 6 × SSC as specified. Annealing was generally carried out at 66°C without shaking, after which the vials were chilled in an ice bath. The volume can be cut down to 0.5 ml., if found convenient, by using small tubes and rolling up the filters.

*Step III. Elimination of RNA "noise."* The filters were removed from the hybridization fluid and *each side* was washed with 50 ml. of 2 × SSC by suction filtration. RNA not completely complexed is destroyed by immersing the filters for 1 hr at room temperature in 5 ml. of 2 × SSC containing 20 µg/ml. of heated pancreatic RNase. After RNase treatment, the vials were again chilled and the filters were rewashed on each side as described above. Finally, the filters were dried and counted in a Packard Tri-Carb scintillation counter.

It should be remarked that the RNase treatment must be carried out in 2 × SSC, whereas the other steps may be carried out at the higher (6 × SSC) buffer concentration.

## (i) Hybridization with DNA in solution

Hybrids were formed in liquid medium and, after the incubation, the tubes were chilled and the contents collected on membrane filters, followed by washing with 100 ml. buffer (2 × SSC or 6 × SSC). The filters were then placed, without drying, in 5 ml. of a 2 × SSC solution containing 20 µg/ml. of boiled pancreatic RNase. This is then followed by washing with 50 ml. of 2 × SSC on each side. It should be noted that no DNA is lost from a filter during the enzyme treatment in 2 × SSC or the subsequent washing.

## 3. Results

Knowledge of the proportions of DNA and RNA in the complex is essential for quantitative interpretation. In all the experiments to be described, $^3$H-labeled DNA and $^{32}$P-labeled RNA are employed. This permits us to monitor with certainty the amounts of the two reactants on the membranes at every stage of the procedures used.

## (a) Retention of DNA by membrane filters

It was first necessary to check how well the fixed DNA withstands the various steps required for the hybridization and "noise" elimination with RNase. Table 1 demonstrates that single-stranded DNA is retained by the membrane filters under conditions of hybridization and no detectable DNA is lost during any

156  Molecular biology of DNA and RNA

TABLE 1
Retention of DNA by nitrocellulose membrane filters

| Sample no. | Treatment of the DNA filters | Percentage DNA remaining on the filter | |
|---|---|---|---|
| | | 6 × SSC | 2 × SSC |
| 1 | Incubate at 66°C without shaking | 97, 98 | 101, 97 |
| 2 | Incubate at 66°C with shaking | 100 | 101 |
| 3 | Incubate at 66°C without shaking; wash, RNase, wash | 99 | 99 |

12 μg of [$^3$H] DNA (500 cts/min/μg) were loaded on membrane filters and washed with 100 ml. of the indicated buffer. The DNA filters were dried at room temperature overnight, then at 80°C for 2 hr. They were then incubated in the same buffer for 24 hr at 66°C. The filters were then either rinsed in fresh buffer or carried through the washing and RNase treatments as described in the Materials and Methods section. Each value is an average of three trials with a 2σ of 5%.

TABLE 2
Elimination of RNA noise

| Sample no. | Salt | Radioactive material remaining (cts/min) | Percentage input radioactive material remaining (cts/min) |
|---|---|---|---|
| 1 | 2 × SSC ------ | 13,838 | 1.8 |
| 2 | 2 × SSC wash | 290 | 0.059 |
| 3 | 2 × SSC wash; RNase; wash | 25 | 0.0033 |
| 4 | 6 × SSC wash; RNase; wash | 108 | 0.0076 |
| 5 | 6 × SSC wash; RNase; wash (50 μg DNA; 0°C; 6 hr) | 99 | 0.0066 |
| 6 | TMS† ------ | 81,038 | 10.0 |
| 7 | TMS† wash | 23,528 | 2.9 |

Unloaded or DNA-containing filters were immersed in buffer containing 2 μg RNA (400,000 cts/min/μg) for 24 hr at 53°C, except for samples 4 and 5. These latter samples were incubated in buffer containing 10 μg RNA (100,000 cts/min/μg) for 6 hr at 66°C. All filters were removed from the RNA solutions and some were washed and/or treated with RNase as outlined in Materials and Methods.
†TMS = 0.3 M-NaCl; 0.005 M-MgCl$_2$; 0.001 M-tris (pH 7.3).

of the subsequent steps. The fact is, as measured by the $^3$H (counts per minute) retained, no detectable DNA is lost during the RNase treatment in 2 × SSC even without prior drying of the filter. Finally, it can be seen (sample 2, Table 2) that shaking during hybridization does not lead to significant removal of DNA from the membrane filters.

Because of its possible convenience, an examination was made of loading the DNA on the filters, not by filtration, but by spotting a known volume of a DNA solution in 2 × SSC on a dry filter, allowing it to spread, and then to dry. Even with this precaution, the spotting method fails to give a completely irreversible fixation. Some DNA is lost with time (about 0.8%/hr at 66°C) during hybridization at elevated temperatures. Further, the rate at which DNA is lost is greatly increased (about 3%/hr at 66°C) when the system is shaken during hybridization. Finally, the amount of DNA which can be loaded on a filter is much more limited than with the filtration procedure.

*(b) Elimination of unpaired RNA*

An obvious advantage conferred by fixing the DNA on the filter is the ease with which contaminating RNA is removed. As may be seen from Table 2, simply lifting the membrane out of the hybridizing mixture leaves 98% of the non-complexed RNA behind. The washing procedure brings the contamination down to 0.06% of the input and the RNase treatment reduces this further to 0.003%. Fixed DNA on the filter has no effect on the capacity to remove non-hybridized RNA. It should be

TABLE 3
*Some parameters affecting RNA noise level*

| Sample no. | µg DNA | µg RNA | Temp. (°C) | Time (hr) | RNA (cts/min) on filter |
|---|---|---|---|---|---|
| 1 | 50 | 2 | 4 | 6 | 37 |
| 2 | 50 | 10 | 4 | 6 | 62 |
| 3 | 50 | 50 | 4 | 6 | 94 |
| 4 | 5 | 10 | 4 | 8 | 53 |
| 5 | 0 | 10 | 66 | 5 | 36 |
| 6 | 0 | 10 | 66 | 10 | 71 |

The standard procedure using 6 × SSC was employed to prepare DNA filters. The filters were immersed in 6 × SSC for the indicated times, then removed, washed, treated with RNase and rewashed as usual. DNA = 5,000 cts/min/µg; RNA = 100,000 cts/min/µg.

TABLE 4
*RNA noise level of liquid hybridization techniques*

| Sample no. | Method for RNA noise elimination | Percentage of input of RNA remaining |
|---|---|---|
| 1 | No RNase | 0.0718 |
| 2 | RNase before filtration | 0.2005 |
| 3 | RNase after filtration | 0.0034 |

Reaction mixtures containing 50 µg DNA and 50 µg RNA (600,000 cts/min/µg) were made up at room temperature in 2 × SSC and collected by filtration on nitrocellulose membrane filters. RNase (20 µg/ml.) treatment was carried out at room temperature for 1 hr. All filters were washed with 100 ml. of 2 × SSC after collection of the reaction mixtures. Sample 3 was additionally washed with 100 ml. of 2 × SSC after RNase treatment (Materials and Methods, section (h)).

TABLE 5
*Effect of proteins on the RNA noise level*

| Sample | Protein addition | Percentage of input RNA remaining |
|---|---|---|
| 1 | None | 0.108 |
| 2 | None | 0.071 |
| 3 | Lysozyme | 74.9 |
| 4 | Methylated albumin | 37.9 |
| 5 | Bovine serum albumin | 0.061 |
| 6 | Pronase | 0.029 |
| 7 | DNase | 0.043 |

Proteins (20 µg/ml.) were added to 5 ml. of 2 × SSC solutions containing 50 µg each of RNA and DNA and the mixtures were incubated at room temperature for 1 hr. The reaction mixtures were passed through membrane filters and washed as described for samples 1 and 2 of Table 4 (RNA = 120,000 cts/min/µg).

noted from samples 6 and 7 of Table 2 that the presence of magnesium is to be avoided if low contamination levels are desired.

Table 3 compares the amount of radioactive material (in counts) which survives the purification procedure at different levels of RNA and DNA inputs held for various times and temperatures. The actual counts finally observed for the filters are recorded in the last column. They are all less than 100 counts per minute with input counts ranging from $2 \times 10^5$ to $5 \times 10^6$ cts/min. Increasing the input of RNA by 25-fold resulted only in a 2.5-fold increase in residual counts. The presence of DNA at non-complexing temperature does not increase the contamination. The procedure is clearly effective in eliminating non-complexed RNA.

### (c) Modification of the Nygaard-Hall technique to lower noise level

We wanted to compare hybridizations carried out with DNA in solution with others in which the DNA is immobilized. The Nygaard-Hall (1963) method omits RNase and, depending on the amount of RNA, has a noise level ranging between about 0.1 and 1% of the input. This amount of contamination with unpaired RNA was not important, since their experiments involved comparatively massive amounts of hybrid formation. It was, however, too high for our purpose, and we consequently introduced (Materials and Methods section) the RNase step *after* collecting the hybrids formed in liquid on the membrane. Table 4 shows that this modification lowers the noise level from 0.08 to 0.003%.

Another useful fact is recorded in Table 4. Note that RNase treatment *before* collection of the complex on the filters *increases* the "noise" from 0.08 to 0.2%. The RNase, being a basic protein, tends to remain on the membrane when the treated complexes are filtered, and it adsorbs small fragments of RNA which would otherwise be washed away.

Table 5 dramatically illustrates the difficulty introduced by the presence of any basic protein at the filtration step. Note that with either lysozyme or methylated albumin present, between 37 and 75% of all the input RNA appears as "noise" on the filter. The non-basic proteins do not have this unfortunate effect.

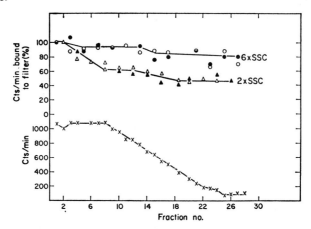

**Fig. 1.** Retention of DNA as a function of size. 0.5 ml. native DNA (725 μg/ml. in 1/100 SSC) was sheared by passing it 10 times through a Tomac 1 ml. disposable syringe. The sheared DNA was mixed with 0.5 ml. unsheared DNA and heated to 95°C for 5 min. The denatured DNA was layered over a 2.5 to 15% sucrose gradient in 0.1 M-NaCl–5 × 10⁻³ M-MgCl₂–0.01 M-tris (pH 7.3) and spun at 25,000 rev./min for 8 hr at 19°C in the SW25 head of the Spinco model L centrifuge. Samples collected and assayed for TCA-precipitable radioactive material are indicated by the lower curve (—X—X—). (—●—●—, —▲—▲—) Percentage retention on filtration; (—○—○—, —△—△—) percentage of the DNA remaining after the hybridization and purification steps. The results included in Fig. 1 are corrected for losses of radioactive material (in counts) during TCA precipitation; about 20% more radioactive material is retained on membrane filters when 6 × SSC is used than when TCA is used. Control experiments revealed that this was not a case of the TCA absorbing radioactive disintegration substances, but a case of differential DNA retention.

### (d) Retention of DNA by membrane filters as a function of DNA size

In most of our experiments, the DNA used for hybridization had an $S_{20,w}$ value of about 35 s. There are situations where it is desirable, or necessary, to use smaller DNA fragments. It was, therefore, of interest to see whether size influenced the capacity of the membranes to retain DNA. Accordingly, sheared and unsheared DNA were combined and fragments of different size classes were separated on a sucrose gradient. Each fraction collected was then filtered onto membrane filters and the amount of DNA retained on the filters compared with the total amount of DNA in each sample.

The reversibility of the fixation was tested by carrying each loaded membrane through all the stages of the hybridization tests. The results of these experiments are summarized in Fig. 1. The lowest curve (cts/min) gives the size distribution of the ³H-labeled DNA used. A comparison was made of retention before (not incubated) and after (incubated) the hybridization steps employing two buffer concentrations. Three facts are evident: (1) more DNA is retained by the filters when 6 × SSC is used than when 2 × SCC is used; (2) larger pieces are retained somewhat more readily than small ones; and (3) once fixed, no DNA, big or small, is lost during the hybridization and washing steps.

### TABLE 6
*Availability of DNA for hybridization*

| Hybridization technique | Average complexed (cts/min) | Percentage DNA hybridized ($\bar{x} \pm 2\sigma$) |
|---|---|---|
| DNA in solution | 11,711 | 0.314 ± 0.027 |
| Membrane-fixed DNA | 1702 | 0.315 ± 0.008 |
| CsCl |  | 0.303 |

The annealing in liquid was carried out in 6 × SSC, using 40 μg [³H] DNA (5000 cts/min/μg) and 40 μg [³²P] RNA (100,000 cts/min/μg) in 1.5 ml. Hybridization was performed at 43°C for 12 hr; subsequent steps follow the procedure outlined in Materials and Methods.

The membrane-fixed DNA technique employed a filter containing 5 μg DNA (5000 cts/min/μg) in a 6 × SSC solution containing 10 μg RNA (100,000 cts/min/μg). Hybridization was carried out at 66°C for 8 hr. Purification of the hybrid is described in Materials and Methods. Each number is the average of three experiments ± 2σ.

## (e) Availability of the DNA for hybridization

The next question we wanted to resolve was whether the sequences of the immobilized DNA were as freely available for hybridization as in solution. Since the saturation plateau for ribosomal RNA had already been carefully examined, it was adopted as a useful test system. Comparisons were made between annealing with DNA in solution and with membrane-fixed DNA. To minimize interference with DNA-DNA interactions, the liquid hybridizations were carried out at 43°C as had been done previously (Yankofsky & Spiegelman, 1963). In all cases, the ratio of RNA to DNA employed was sufficient to attain the saturation plateau, and the times required were determined from preliminary kinetic experiments. Table 6 summarizes the results obtained with dissolved and immobilized DNA, each number representing the average of three determinations. The saturation value obtained previously (Yankofsky & Spiegelman, 1963) by the CsCl density-gradient method with the same biological materials is also recorded. The average plateau values of all four are in agreement, indicating that immobilization of the DNA does not interfere with its capacity to hybridize.

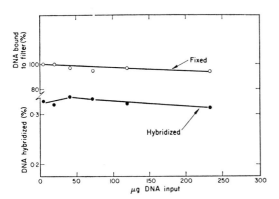

**Fig. 2.** Filter capacity and availability for hybridization at different levels of DNA. DNA filters were prepared in 2 × SSC. Hybridizations were carried out in 0.8 M-NaCl and 0.05 M-potassium phosphate buffer (pH 6.8) at 66°C for 10 hr. All filters were washed, treated with RNase and rewashed as usual. The specific activity of the DNA varied from 411 cts/min/µg (5 µg DNA on the filter) to 354 cts/min/µg (234 µg DNA on the filter); that of the RNA was 280,433 cts/min/µg.

## (f) Capacity of membranes to retain DNA and availability of the DNA for hybridization at low and high DNA inputs

The capacity of the membranes to retain DNA at different levels of input was examined along with the capacity of the fixed DNA to pair with ribosomal RNA. The results obtained with DNA inputs up to 250 µg per filter are summarized in Fig. 2. It is clear that, within the range tested, the capacity of the membranes for irreversible adsorption of DNA has not been exceeded. Further, as evidenced by the comparative constancy of the plateau values, there is no numerically significant decline in the availability of the DNA for hybridization as increasing amounts are fixed per filter.

## (g) A comparison of kinetics and plateau stability between liquid and immobilized DNA hybridization

Figures 3 through 6 give some typical outcomes of hybridizations carried out at various input ratios and amounts of RNA and DNA. We may first focus attention on the

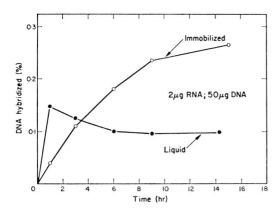

**Fig. 3.** Hybridizations with DNA in solution and immobilized.
*Immobilized-DNA technique.* DNA filters containing 50 µg DNA were prepared using 6 × SSC and immersed in 5 ml. of a 6 × SSC solution containing 2 µg RNA. Hybridization was carried out at 66°C without shaking. Purification of the hybrids is described in Materials and Methods.
*Liquid technique.* Reaction mixtures containing 50 µg DNA and 2 µg RNA in 1.5 ml. were made up in 6 × SSC. The mixtures were held at 66°C, after which the hybrids were purified as described in Results section (c). DNA = 5000 cts/min/µg; RNA = 100,000 cts/min/µg.

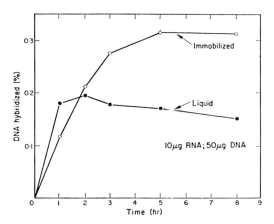

**Fig. 4.** Hybridizations with DNA in solution and immobilized. The procedures are the same as those described in the legend for Fig. 3, except that 50 μg DNA was hybridized with 10 μg RNA.

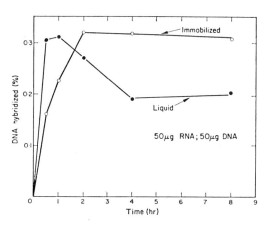

**Fig. 5.** Hybridizations with DNA in solution and immobilized. The procedures are the same as those described in the legend for Fig. 3, except that 50 μg DNA was hybridized with 50 μg RNA.

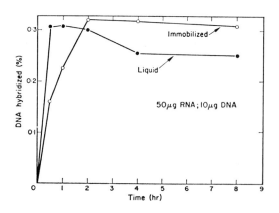

**Fig. 6.** Hybridizations with DNA in solution and immobilized. The procedures are the same as those described in the legend for Fig. 3, except that 10 μg DNA was hybridized with 50 μg RNA.

results obtained with immobilized DNA. Comparing Figs. 3 and 4 reveals, not unexpectedly, that the rate of approach to the plateau is influenced by the concentration of RNA. At 10 μg of RNA, the hybridization reaches the expected plateau in five hours, a value not attained in 15 hours at 2 μg. On the other hand (Figs. 5 and 6), the kinetics of complex formation is very slightly influenced by increasing the amount of DNA fixed to the membrane.

Comparison of these findings with the results obtained with DNA in solution dramatically illustrates the advantages of using immobilized DNA. We see (Fig. 3) that at the lower input of RNA the plateau is never reached in the liquid hybridizations. At a level of 50 μg RNA (Fig. 6), the saturation plateau is almost attained within a few hours; but this is then followed by a decomposition of formed hybrid, a phenomenon reported by Nygaard & Hall (1964). In general, the loss of hybrids formed in liquid increases in severity with the concentration of DNA present. This phenomenon is effectively suppressed by immobilization of the DNA. At high levels of DNA per filter (50 μg and greater) and low ionic strength (2 × SSC) some loss of hybrid is detected in the immobilized hybridizations. However, it is much less than that observed in annealing to DNA in solution and is negligible unless the hybridization is prolonged excessively.

## 4. Discussion
### (a) Some general precautions in hybridization experiments

Unless care is exercised, basic proteins can occur as contaminants of DNA and RNA preparations. We have already pointed out that their presence can generate considerable difficulty by virtue of their ability to adsorb RNA, and the efficiency with which the resulting complex is retained by the filter. Nucleic acid preparations purified on MAK columns can contain enough methylated albumin to cause this sort of trouble. It can be avoided by exhaustive washing of the columns before use.

It may be useful to others if we explicitly record here some of the other complications which can be encountered in carrying out hybridizations and the precautions we have adopted as a consequence. In most instances they center on the components of the system, and the details are best discussed in terms of the individual reactants and the steps involved.

*DNA preparations.* The DNA used must be completely available for hybridization if saturation plateaux are to be interpreted with confidence. It follows that the denaturation of the DNA employed must be shown to be complete.

A difficulty commonly encountered with DNA preparations is contamination with ribonuclease, a protein well known for its ability to bind to DNA and for its resistance to inactivation. The usual preparative procedures rarely provide DNA completely free of ribonuclease. Passage through an MAK column is extremely helpful since RNase remains behind. A useful alternative or supplement is to incubate the DNA with pronase to destroy residual RNase activity. The pronase can then be removed by deproteinization with phenol. It is important here first to subject the pronase to "self-digestion" to remove any nucleases it might contain. In any event, no DNA preparations should be employed which have not been assayed for ribonuclease activity for the time periods and under the conditions employed for hybridization.

*RNA preparations.* Assay for and removal of nuclease activity is, of course, equally important in the case of RNA. A frequently encountered difficulty is contamination with DNA fragments resulting from incomplete removal or digestion. This is true for RNA labeled with either $^{32}$P or uridine, since many cells can convert the latter to thymine. It is a particularly serious complication with material containing glucosylated DNA (e.g., cells infected with T2, T4, or T6 bacteriophages) because these types of DNA are much more resistant to enzymic digestion. The DNase treatment must be continued until the residue of radioactive material stable in alkali is at an acceptable level. Even small amounts of contaminating labeled DNA can make the interpretation of many kinds of hybridization experiments impossible.

*Enzymes.* The RNase and DNase used must be assayed to be certain that each is free of significant contamination by the other. DNase can be readily freed of ribonuclease activity by column chromatography on DEAE (Polatnick & Bachrach, 1961). Deoxyribonuclease activity in RNase is readily destroyed by heating a solution adjusted to pH 5.0 to 90°C for 10 minutes. Assays for contaminating enzymes must be adequately sensitive and of a duration comparable to that employed in the relevant hybridization step. The use of radioactive polynucleotide as substrate makes it possible to achieve any desired level of sensitivity in the assay for contaminating activity.

*Enzymic elimination of unpaired RNA.* One of the most common pitfalls stems from the use of commercial RNase without checking for DNase contamination. No resistant plateaux of hybrid are observed if DNase activity is detectable. The other is the failure to treat the hybrid under conditions specified for stability. Although stability may vary, in our experience temperatures above 37°C and salt concentrations below 0.25 M should be avoided. We obtain reproducibly an absolutely resistant hybrid fraction when enzyme treatments are carried out with 10 μg of RNase per ml. at 30°C in 0.3 M-NaCl. Wherever possible, internal controls of free RNA appropriately labeled with a radioactive isotope should be included, in order to be certain that the ribonuclease is functioning properly.

## (b) Methods of hybrid detection and assay

Of the available methods of hybrid detection, two permit the actual isolation of the complex and allow its further characterizations. These involve the use of equilibrium density-gradients of CsCl or $Cs_2SO_4$, and the more recently developed chromatographic separations on MAK columns (Hayashi, Hayashi & Spiegelman, 1965). The latter method has the advantage in both capacity and convenience. Further, it has been shown in the same study that the column fractionates hybrids on the basis of the ratio of RNA to DNA in the complex.

The agar columns (Bolton & McCarthy, 1962) and nitrocellulose columns (Bautz &

Hall, 1962) are particularly useful for preparative fractionation of particular classes of complementary RNA. The membrane filter is clearly the method of choice when large numbers of assays for complementarity are required.

## (c) Annealing with dissolved versus immobilized DNA

The experiments reported here clearly establish that hybridizations with immobilized DNA are superior, conferring certainty and accuracy by avoiding DNA-DNA interactions. Hybridizations at elevated temperatures ($> 50°C$) in liquid which are not monitored by a kinetic analysis are likely to be in error. If they are extended, they are almost certain to be low estimates.

The introduction by Nygaard & Hall (1963) of membrane filters for hybrid detection greatly expanded the number of samples it was feasible to analyze. The use of membrane-fixed DNA both for annealing and detecting the hybrid product generates still other advantages of convenience.

The technique is inherently flexible and can be adapted to almost any volume. For example, one need not use the entire filter, since after fixing a specified amount of DNA, circular subsections of known area can be punched out and employed. Further, more than one filter can be placed in the same hybridization mixture, and if desired, individual ones removed at suitable intervals and analyzed. Alternatively a loaded filter can be left in for a fixed time, removed and replaced by a new one to allow further hybridization, permitting a simple search for heterogeneity in hybridizability. In addition, with the DNA immobilized, a wide temperature range is available uncomplicated by DNA renaturation. One can thus study the hybridization of molecules possessing high degrees of secondary structure and which require higher temperatures for annealing analysis.

The RNA hybridized to the fixed DNA can be recovered for further analysis by elution at elevated temperatures and low ionic strength. Finally, the combined use of RNase and the membrane-immobilized DNA makes it possible to push the "noise" level down easily to almost any desired level and it does not require extraordinary effort to achieve levels corresponding to 0.003% of the input RNA.

It may be noted in conclusion that this method has been successfully employed (Ritossa & Spiegelman, 1965) to establish that the "nucleolar organizer" segment of the X-chromosome of *Drosophila melanogaster* contains the DNA complements of ribosomal RNA. The experiments required that 25% differences in saturation plateaux be reliably detected at levels of 0.3% of the DNA input. The method was found adequate.

## REFERENCES

1. Bautz, E. K. F. & Hall, B. D. (1962). Proc. Nat. Acad. Sci., Wash. **48**:400.
2. Bolton, E. T. & McCarthy, B. J. (1962). Proc. Nat. Acad. Sci., Wash. **48**:1390.
3. Britten, R. J. (1963). Science **142**:963.
4. Giacomoni, D. & Spiegelman, S. (1962). Science **138**:1328.
5. Goodman, H. M. & Rich, A. (1962). Proc. Nat. Acad. Sci., Wash. **48**:2101.
6. Hall, B. D. & Spiegelman, S. (1961). Proc. Nat. Acad. Sci., Wash. **47**:137.
7. Hayashi, M., Hayashi, M. N. & Spiegelman, S. (1965). Biophys. J. **5**:231.
8. Mangalo, R. & Wachsman, J. T. (1962). J. Bact. **83**:27.
9. Marmur, J. (1961). J. Mol. Biol. **3**:208.
10. Meselson, M., Stahl, F. W. & Vinograd, J. (1957). Proc. Nat. Acad. Sci., Wash. **43**:581.
11. Nygaard, A. P. & Hall, B. D. (1963). Biochem. Biophys. Res. Comm. **12**:98.
12. Nygaard, A. P. & Hall, B. D. (1964). J. Mol. Biol. **9**:125.
13. Polatnick, J. & Bachrach, H. L. (1961). Analyt. Biochem. **2**:161.
14. Ritossa, F. M. & Spiegelman, S. (1965). Proc. Nat. Acad. Sci., Wash. **53**:737.
15. Yankofsky, S. A. & Spiegelman, S. (1962*a*). Proc. Nat. Acad. Sci., Wash. **48**:106.
16. Yankofsky, S. A. & Spiegelman, S. (1962*b*). Proc. Nat. Acad. Sci., Wash. **48**:146.
17. Yankofsky, S. A. & Spiegelman, S. (1963). Proc. Nat. Acad. Sci., Wash. **49**:538.

# 18 NUCLEOTIDE SEQUENCE OF KB CELL 5S RNA

*Bernard G. Forget**
*Sherman M. Weissman***
*Metabolism Branch*
*National Cancer Institute*
*Bethesda, Maryland*

*Abstract.* The nucleotide sequence of 5S RNA derived from KB carcinoma cell ribosomes has been determined. The molecule has a length of either 120 or 121 nucleotides with uridine at its 3'-terminus and guanylic acid at its 5'-terminus. If, in addition to Watson-Crick base-pairing, one accepts occasional base-pairing of guanylic acid to uridylic acid, long sequences of complementary nucleotides can be identified within the molecule. Two regions of the molecule contain sequences complementary to four or five bases in the pentanucleotide sequence guanylic acid, ribothymidylic acid, pseudouridylic acid, cytidylic acid, guanylic acid, which is common to most transfer RNA molecules. This is the first time the sequence of an animal-cell RNA has been determined.

In many cell species, the ribosomes contain a low-molecular-weight ribonucleic acid (5S RNA) whose function is unknown (1, 2). The introduction by Sanger et al. of an improved procedure for the separation and identification of $P^{32}$-labeled oligonucleotides by two-dimensional paper electrophoresis (3) has facilitated the study of the primary structure of 5S RNA and other low-molecular-weight homogeneous RNA species. Enzymatic digests of 5S RNA yield only a small number of oligonucleotides that are specific for the cell species of its origin and are present in quantities, which suggests that 5S RNA has a single nucleotide sequence, or a limited number of closely related sequences (4-6).

We used this oligonucleotide fractionation procedure (i) to isolate and establish the sequence of all the oligonucleotides present in complete pancreatic and Tl ribonuclease digests of 5S RNA derived from KB cells (human epidermoid carcinoma line), and (ii) to identify the products of its partial digestion by ribonuclease Tl after their isolation and purification by column chromatography. As a result we were able to establish the entire nucleotide sequence of 5S RNA from KB cells.

Unlike previous studies (7-10) on the sequence of tRNA's (11), our work did not require large amounts of RNA, since the final identification of the sequences depended only on the radioactivity of the RNA and not on its optical density. Extensive procedures to purify the RNA were not needed since 5S RNA is homogeneous when separated from other RNA by methylated-albumin-kieselguhr column chromatography. The 5S fraction contains less than 7 percent contaminating tRNA as judged by its content of pseudouridine, and its internal homogeneity is indicated by the fact that enzymatic digests of KB cell 5S RNA isolated by this method yield only a small number of specific oligonucleotides (5) which are present in molar amounts consistent with a single molecule, or a small number of very similar molecules. Three trinucleotides, however, are present in the pancreatic ribonuclease digest in traces (5), from 0.1 to 0.15 mole. These could be derived from another RNA contaminating the 5S RNA (4), or could result from oversplitting of large oligonucleotides of the 5S RNA itself.

The $P^{32}$-labeled KB cell 5S RNA was obtained as described (5). The RNA was completely digested by pancreatic and Tl ribonuclease, the oligonucleotides were isolated, and their sequences were determined by techniques similar to those of Sanger et al. (3, 4).

Figure 1 shows the radioautographs given by the two-dimensional paper-electrophoretic fractionation of pancreatic and Tl ribonuclease digests of KB cell 5S RNA. Because of the

---

From Science **158**:1695-1699, 1967. Copyright © 1968 by the American Association for the Advancement of Science. Reprinted with permission.

*Present address: Massachusetts General Hospital, Boston 02114.

**Present address: Department of Internal Medicine, Yale University School of Medicine, New Haven, Connecticut 06510.

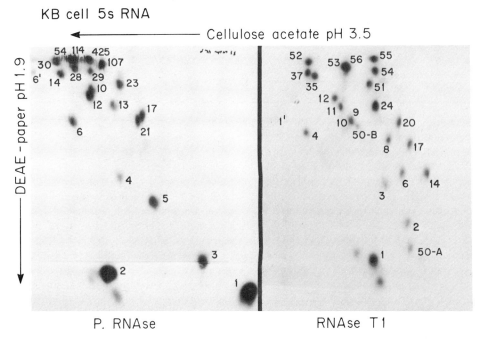

**Fig. 1.** Radioautographs given by two-dimensional paper-electrophoretic fractionation of pancreatic ribonuclease and ribonuclease T1 digests of $P^{32}$-labeled KB cell 5S RNA. The sequence of the oligonucleotides is given in Table 1.

overlapping of oligonucleotides Nos. 9 and 10, and Nos. 53 and 56 in the fractionation of the T1 ribonuclease digest, simultaneous digestion of 5S RNA with this enzyme and alkaline phosphomonoesterase (PME) was carried out. Fractionation resulted in separation of the previously overlapping oligonucleotides (Fig. 2) and permitted their individual analysis.

Table 1 lists the code numbers and the corresponding sequence of the oligonucleotides shown in Figs. 1 and 2. The following procedures were used to establish their sequence (3, 4). The spots were eluted, their base composition was determined by alkaline hydrolysis, and, for the short oligonucleotides, their sequences were assigned according to the results of enzymatic digestion by pancreatic or T1 ribonuclease, micrococcal nuclease, or snake-venom phosphodiesterase in the case of dephosphorylated oligonucleotides. To establish the sequences of the longer oligonucleotides it was necessary to digest them partially with spleen or snake-venom phosphodiesterase and analyze the products by alkaline hydrolysis or by digestion with pancreatic ribonuclease.

The longest oligonucleotide (No. 55) has 13 nucleotides; its sequence is based on a large number of separate analyses by alkaline hydrolysis and pancreatic ribonuclease digestion of the intact oligonucleotide and of the many digestion products resulting from partial degradation by spleen and snake-venom phosphodiesterase.

The molar yield of each oligonucleotide was calculated by measuring its radioactivity in a low-background gas-flow counter and dividing the result by the number of component bases multiplied by the amount of the radioactivity corresponding to 1 mole of phosphate in the particular fractionation. This last value was calculated by dividing the total recovered radioactivity by 120. The experimental molar yield obtained by this method for each oligonucleotide was compared with the corresponding theoretical molar yield (Table 1) based on the sequence which was ultimately determined for KB cell 5S RNA (Fig. 3). With the following exceptions, the experimental and theoretical molar yields are in close agreement.

The 5'-terminal oligonucleotides in both

## KB Cell 5s RNA - PME + T₁ RNAse

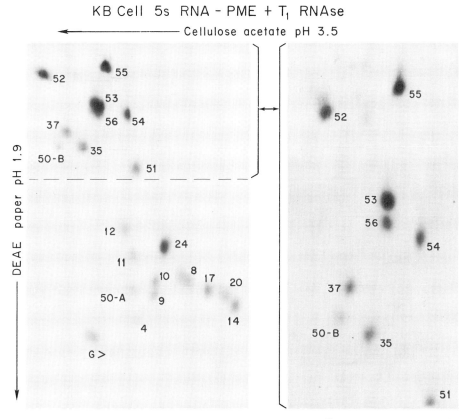

**Fig. 2.** Radioautographs given by two-dimensional fractionation of P³²-labeled KB cell 5S RNA digested simultaneously with T1 ribonuclease and alkaline phosphomonoesterase (PME). In the fractionation result shown on the right, the second-dimension electrophoresis was continued for 12 hours instead of the usual 3½ hours. G > represents cyclic guanosine 2′,3′-phosphate. The oligonucleotides are numbered as in the T1 ribonuclease digest of Fig. 1, and their sequence is that given in Table 1, without the terminal 2′-phosphate.

enzymatic digests, spots 1′ and 6′, were recovered in only half-molar amounts. However, when an alkaline hydrolyzate of KB cell 5S RNA, labeled with C¹⁴-nucleoside precursors, was chromatographed on a Dowex-1-formate column (1), the only 3′,5′-diphosphate identified coincided with the pGp marker and was recovered in a yield of 0.9 mole. Another unusual feature is that in fractionations of both the pancreatic and Tl ribonuclease digests two spots are present, corresponding to the 5′-end groups. On alkaline hydrolysis both yield only pGp in the Tl ribonuclease digest, and only Up and pGp in the pancreatic ribonuclease digest. Whether this duplication is due to an artifact of the experimental procedures, or to an undetermined structural feature of the molecule, at this point is not yet known.

In the Tl ribonuclease digest there are two other oligonucleotides each present in nearly half-molar amounts: spots 50-A and 50-B. Presumably they represent alternate forms of the 3′-end groups of the molecule since they both contain a free 3′-hydroxyl group and have nearly identical sequences, differing by only one base.

Many of the larger oligonucleotides, in the pancreatic ribonuclease digest more than in the Tl ribonuclease digest, are present in yields lower than those expected: from 0.6 to 0.8 mole. This was true not only in analyses of whole 5S RNA digests but also in analyses of purified partial-digestion fragments of KB cell 5S RNA. This suggests that the low recovery of these oligonucleotides is due to the methods of RNA digestion and fractionation rather than to

166  Molecular biology of DNA and RNA

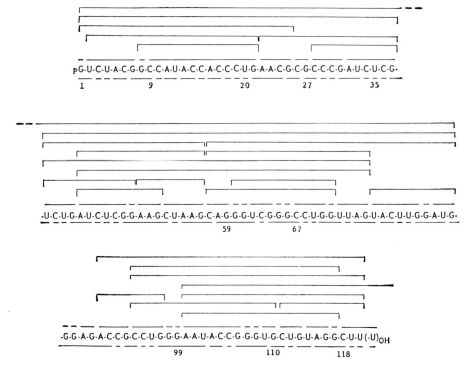

**Fig. 3.** Sequence of KB cell 5S RNA. The lines that overscore the sequence indicate the ribonuclease T1 digestion products, and the lines that underscore the sequence indicate the products of pancreatic ribonuclease digestion. The brackets indicate the composition in oligonucleotides of various T1 ribonuclease partial-digestion fragments, determined by their complete digestion by pancreatic or T1 ribonuclease.

heterogeneity of the nucleotide sequence (4).

The molar yields of the mononucleotides Cp and Up in the pancreatic ribonuclease digests were also lower than expected. This low yield may be due to the conditions of digestion which release variable quantities of 2′,3′-cyclic nucleoside monophosphates. A more reliable estimate of the total content of Cp and Up in the RNA can be obtained by determining the sum of these nucleotides in all the T1 ribonuclease digestion products (4).

Aside from the alternate 3′-end groups, quantitative analysis of the molar yields of oligonucleotides from both pancreatic and T1 ribonuclease digests shows no other clear evidence of heterogeneity in the sequence of KB cell 5S RNA.

Comparison of Table 1 to the list of oligonucleotides obtained from digests of 5S RNA from *Escherichia coli* (4) reveals that there are definite differences in the two cell species. The 3′-terminal nucleoside is uridine in both KB cell 5S RNA and *E. coli* 5S RNA; but the 5′-terminal nucleotide in the *E. coli* 5S RNA is uridylic acid, while in the KB cell 5S RNA it is guanylic acid. The sequences of most of the larger oligonucleotides also differ in the two cell species.

To arrange all the oligonucleotides obtained from digests of KB cell 5S RNA into one linear sequence, it was necessary to study a large number of partial-digestion fragments of the RNA. These were obtained by techniques similar to those used by Zachau *et al.* (9). The KB cell 5S RNA was digested with T1 ribonuclease at 0°C for 1 hour, in a solution of 0.2M tris-HCl, pH 7.5, containing no magnesium, and in the presence of a large excess of nonradioactive *E. coli* tRNA. The digests were extracted with phenol and applied to columns (200 by 0.6 cm) of microgranular DEAE-cellulose, which were eluted by linear gradients of NaCl in 7M urea and 0.02M tris-HCl, pH 7.5. Each individual peak of radioactivity obtained

## TABLE 1
*Sequence and molar yield of the oligonucleotides derived from pancreatic and T1 ribonuclease digestion of KB cell 5S RNA*

The code numbers (spot No.) of the oligonucleotides correspond to those of Figs. 1 and 2. The sequence is given from the 5'-end (left) to the 3'-end (right) of the oligonucleotide. Each molar experimental yield (exp.) was determined as described in the text, and is listed opposite the molar theoretical yield (theor.) derived from the sequence ultimately determined for KB cell 5S RNA. The hyphens indicate phosphate groups.

| Spot no. | Sequence | Moles (no.) Exp. | Moles (no.) Theor. | Spot no. | Sequence | Moles (no.) Exp. | Moles (no.) Theor. |
|---|---|---|---|---|---|---|---|
| Pancreatic ribonuclease digestion products | | | | T1 Ribonuclease digestion products | | | |
| 1 | C- | 13.2 | 17 | 1 | G- | 13.8 | 15 |
| 2 | U- | 10.1 | 13-14 | 2 | C-G- | 1.27 | 1 |
| 3 | A-C- | 4.8 | 5 | 3 | A-G- | 1.27 | 1 |
| 4 | A-U- | 1.3 | 1 | 4 | U-G- | 1.08 | 1 |
| 5 | G-C- | 4.2 | 4 | 6 | C-A-G- | 0.89 | 1 |
| 6 | G-U- | 2.2 | 2 | 8 | A-A-G- | 0.95 | 1 |
| 10 | A-G-U- | 1.1 | 1 | 9 | U-C-G- | 0.99 | 1 |
| 12 | G-A-U- | 1.9 | 2 | 10 | C-U-G- | 0.99 | 1 |
| 13 | G-G-C- | 0.94 | 1 | 11 | U-A-G- | 1.0 | 1 |
| 14 | G-G-U- | 1.15 | 1 | 12 | A-U-G- | 1.0 | 1 |
| 17 | A-A-G-C- | 1.03 | 1 | 14 | C-C-C-G- | 1.02 | 1 |
| 21 | G-A-A-C- | 0.95 | 1 | 17 | A-C-C-G- | 0.87 | 1 |
| 23 | A-G-G-C- | 0.73 | 1 | 20 | A-A-C-G- | 0.85 | 1 |
| 28 | G-G-A-U- | 1.07 | 1 | 24 | C-C-U-G- | 2.12 | 2 |
| 29 | G-G-G-C- | 0.76 | 1 | 35 | U-C-U-G- | 1.19 | 1 |
| 30 | G-G-G-U- | 1.01 | 1 | 37 | U-U-A-G- | 0.89 | 1 |
| 54 | A-G-G-G-U- | 0.59 | 1 | 51 | C-U-A-A-G- | 0.86 | 1 |
| 107 | G-G-A-A-G-C- | 0.97 | 1 | 52 | U-A-C-U-U-G- | 0.99 | 1 |
| 114 | G-G-G-A-A-U- | 0.8 | 1 | 53 | A-U-C-U-C-G- | 1.84 | 2 |
| 425 | G-G-G-A-G-A-C- | 0.57 | 1 | 56 | U-C-U-A-C-G- | 0.85 | 1 |
| 6' | pG-U- | 0.5 | 1 | 54 | A-A-U-A-C-C-G- | 0.8 | 1 |
| | | | | 55 | C-C-A-U-A-C-C- -A-C-C-U-G- | 0.58 | 1 |
| | | | | 1' | pG- | 0.5 | 1 |
| | | | | 50-A | C-U-U$_{OH}$ | 0.65 | 0.5 |
| | | | | 50-B | C-U-U-U$_{OH}$ | 0.47 | 0.5 |

from this first column was chromatographed again on either (i) a column (200 by 0.45 cm) of DEAE-Sephadex that was eluted by a linear gradient of NaCl in 7$M$ urea, $pH$ 3.5, or (ii) on a heated (55°C) column (200 by 0.6 cm) of DEAE-cellulose that was eluted by linear gradients of NaCl in 7$M$ urea and 0.02$M$ tris-HCl, $pH$ 7.5. Some of the earlier peaks from the first column were also fractionated by two-dimensional paper electrophoresis in the same fashion as the complete RNA digests. These partial-digestion fragments were then analyzed, and in most instances they proved to be homogeneous segments of the molecule, containing 1 mole, or multiples thereof, of specific oligonucleotides. The oligonucleotide composition of each partial-digestion fragment was determined by digesting portions of it completely with pancreatic or T1 ribonuclease. The complete digests were then fractionated by a one- or two-dimensional paper-electrophoresis procedure and compared to a similar fractionation of a whole 5S RNA digest to identify the oligonucleotides present. The sequences of the oligonucleotides were not determined individually in each partial-digestion fragment. Each oligonucleotide was given the previously determined sequence of the oligonucleotide occurring in the corresponding position of a similar fractionation of a whole 5S RNA digest. In doubtful cases, however, the identification of the oligonucleotide was confirmed by alkaline hydrolysis or by appropriate enzymatic digestion.

Over 35 different partial-digestion fragments were analyzed. All the oligonucleotides can be grouped into three large partial-digestion fragments which correspond to three general regions of the molecule (Fig. 3). Together the three fragments contain all of the oligonucleotides listed in Table 1 except for No. 425 of the pancreatic ribonuclease digest. Smaller partial-digestion fragments permitted the arrangement of the oligonucleotides into a single possible logical linear sequence within each of the three large fragments or regions of the molecule. In most cases the order of the oligonucleotides in the smaller fragments could be established by complete digestion of portions of the fragment by both pancreatic and Tl ribonuclease. The results of both digestions are usually compatible with only one logical linear sequence. Longer fragments, overlapping these short fragments of known sequence, lengthen the sequence by progressive addition of oligonucleotides to the short fragments and order all the oligonucleotides into only one possible sequence within each of the three regions.

In two regions (residues 57 through 75 and residues 100 through 117) there was increased resistance to splitting by Tl ribonuclease, with the result that no short partial-digestion fragments were obtained after the primary partial digestion. This allowed alternate arrangements of the oligonucleotides in these regions. To obtain the necessary short fragments, we subjected the fragments from the resistant regions to partial digestion for a second time with Tl ribonuclease at $0°C$ for 45 minutes, under the same conditions as in the primary partial digestion. The digest was then extracted with phenol; the extract was placed on a column of DEAE-Sephadex and eluted with a linear NaCl gradient in $7M$ urea, $p$H 3.5. This procedure yielded shorter partial-digestion fragments which established the sequence of the doubtful areas.

The results of both pancreatic and Tl ribonuclease digestion of the individual bracketed fragments, and the various overlaps between the different brackets, establish the complete sequence of the molecule—a single logical linear sequence—with the following exception. The fragments obtained allow two alternate positions for the dinucleotide CpGp placed in positions 26 and 27. By the data shown, it could also be located in positions 36 and 37, at the far right-hand end of the large fragment encompassing residues 1 to 37. This alternate position was disproved by the following experiment. The large fragment of the region was treated with $0.1N$ HCl to open the cyclic terminal $2',3'$-phosphate bonds, neutralized, and then treated with PME, which splits off the terminal phosphate. The digest was extracted with phenol and then completely digested with Tl ribonuclease and fractionated by two-dimensional paper electrophoresis. The dinucleotide CpGp was not dephosphorylated, and therefore it must be placed internally within the fragment. That 75 to 80 percent of oligonucleotide No. 53 was dephosphorylated is confirmation of its position at the end of the fragment.

Once the sequence of each of the three large regions was established, it was possible to arrange them in relation to one another. One contained the $5'$-end group, another had the $3'$-end group, and the third had no end group. One very large partial-digestion fragment contained all of the $5'$-end and central regions, and provided an overlap confirming the arrangement of these two large regions of the molecule. There is no fragment which overlaps the central and $3'$-end regions. The point which joins these two regions of the molecule must be very susceptible to the action of Tl ribonuclease. The missing overlap, however, is given by the oligonucleotide GpGpGpApGpApCp (No. 425) derived from complete digestion of the RNA by pancreatic ribonuclease. It is the only oligonucleotide from Table 1 not already accounted for in the partial-digestion fragments obtained.

Since it is the only pancreatic ribonuclease product ending in ... GpApCp, it must overlap with the oligonucleotide ApCpCpGp which is the only Tl ribonuclease digestion product beginning in ApCp ... and is present at the far left of the $3'$-end region of the RNA. In summary, the results shown in Fig. 3 indicate the complete linear sequence of KB cell 5S RNA.

The presence of regions in the molecule that show marked and reproducible variability in susceptibility to digestion by Tl ribonuclease, as noted above, suggests that the molecule has a specific secondary structure. If, in addition to classic Watson-Crick base-pairing of nucleotides, one proposes occasional Gp-to-Up base-pairing, as assumed in the cloverleaf model for tRNA (7, 8), it is possible to construct different models of the possible secondary structure of KB cell 5S RNA. The molecule contains two sets of complementary sequences, nine nucleotides in length. Residues 1 through 9 (Fig. 3) can form base pairs with residues 118 through 110, thus joining the 5'-end to the 3'-end of the molecule as in the cloverleaf model for tRNA (7, 8); and residues 27 through 35 can form base pairs with residues 67 through 59. There are a number of additional complementary sequences containing from 3 to 6 nucleotides. They can be arranged in different patterns. Selection of a final model of the secondary structure of KB cell 5S RNA will depend on the detailed comparison of its sequence to that of 5S RNA in other cell species, such as that of *E. coli* 5S RNA *(12)*. In *E. coli* 5S RNA there is also the possibility for base pairing of the 5'-end of the RNA with its 3'-end for a length of 10 nucleotides.

In KB cell 5S RNA there are two sequences, one of seven nucleotides and another of six nucleotides, which occur twice in the molecule: the sequence GpApUpCpUpCpGp (residues 31 to 37 and 41 to 47) and the sequence GpCpCpUpGpGp (residues 66 to 71 and 93 to 98). In *E. coli* 5S RNA there are sequences of ten and eight nucleotides which also occur twice *(12)*. However, in *E. coli* 5S RNA these sequences are repeated in different halves of the molecule, while in 5S RNA from KB cell they are repeated in the same half of the sequence.

In most models of the secondary structure of KB cell 5S RNA, the sequences GpApApUp (residues 99 to 102) and UpGpApApCp (residues 20 to 24) are not involved in base pairing. These sequences are complementary in an "antiparallel" fashion to four or five bases in the pentanucleotide sequence GpTp$\psi$pCpCp which is common to most tRNA's and is tentatively located in the loop of the limb of the molecule nearest its 3'-end *(7, 13)*. This association may be fortuitous or it may be related to the function of 5S RNA in the ribosome. It is of interest that *E. coli* 5S RNA contains the sequences GpApApCp (residues 107 to 110) and CpGpApApCp (residues 43 to 47) which are also complementary to four or five bases in the pentanucleotide sequence GpTp$\psi$pCpGp of the tRNA *(12)*.

## REFERENCES AND NOTES

1. R. Rosset, R. Monier, J. Julien, *Bull. Soc. Chim. Biol.* **46**:87 (1964).
2. J. Marcot-Queiroz, J. Julien, R. Rosset, R. Monier, *ibid.* **47**:183 (1965); D. G. Comb, N. Sarkar, J. DeVallet, C. J. Pinzino, *J. Mol. Biol.* **12**:509 (1965); F. Galibert, C. J. Larsen, J. C. Lelong, M. Boiron, *Nature* **207**:1039 (1965).
3. F. Sanger, G. G. Brownlee, B. G. Barrell, *J. Mol. Biol.* **13**:373 (1965).
4. G. G. Brownlee and F. Sanger, *ibid.* **23**:337 (1967).
5. B. G. Forget and S. M. Weissman, *Nature* **213**:878 (1967).
6. J. Hindley, personal communication.
7. R. W. Holley, J. Apgar, G. A. Everett, J. T. Madison, M. Marquisee, S. H. Merrill, J. R. Penswick, A. Zamir, *Science* **147**:1462 (1965).
8. J. T. Madison, G. A. Everett, H. Kung, *ibid.* **153**:531 (1966).
9. H. G. Zachau, D. Dutting, H. Feldmann, *Z. Physiol. Chem.* **347**:212 (1966).
10. U. L. RajBhandary, S. H. Chang, A. Stuart, R. D. Faulkner, R. M. Hoskinson, H. G. Khorana, *Proc. Nat. Acad. Sci. U. S.* **57**:751 (1967).
11. Abbreviations used: tRNA, transfer ribonucleic acid; PME, alkaline phosphomonoesterase; DEAE, diethylaminoethyl; tris-HCl, tris (hydroxymethyl) aminomethane hydrochloride; T, ribothymidine; $\psi$, pseudouridine; A, adenosine; C, cytidine; G, guanosine; U, uridine; the subscript OH is used to indicate the presence of a 3'-hydroxyl group; p, on the left of A, C, and the like, indicates a 5'-phosphate; and on the right, 1 3'-phosphate. In Fig. 3 and Table 1, hyphens are used instead of p.
12. G. G. Brownlee, F. Sanger, B. G. Barrell, *Nature* **215**:735 (1967).
13. A. Zamir, R. W. Holley, M. Marquisee, *J. Biol. Chem.* **240**:1267 (1965).
14. Supported in part by grant ACS-E456 of the American Cancer Society. Miss Carol Hybner and Mrs. Iris Morton contributed technical assistance.

# QUESTIONS FOR CHAPTER 4

## Paper 14

1.  a. It is mentioned in the paper that a common method for preparing tRNA is extraction of whole cells by phenol. Why is only low molecular weight RNA extracted by this procedure?
    b. When "total phenol-extracted RNA" is centrifuged on a sucrose density gradient, how many peaks are obtained?
    c. Suppose that the solubility of tRNA in 1M NaCl is 200 µg/ml; suppose, further, that 1 gm of *E. coli* was extracted by 10 ml of buffer and the RNA prepared by phenol extraction. Would it be possible to prepare tRNA by the salt fractionation method in 90% yield? How concentrated would the RNA have to be in order to obtain a 95% yield?
2.  It is stated in the paper that some methods are and others are not suitable as a preparative procedure. Determine the capacity of the following:
    a. An MAK column
    b. A column of polyacrylamide gel for electrophoresis
    c. A 5 × 150 cm column of G-100 Sephadex
    d. A 1.5 × 45 cm column of DEAE cellulose
3.  The ribosomes used for the experiment reproduced in Fig. 1 were extensively purified and treated with puromycin.
    a. What is the purpose of the puromycin treatment?
    b. What is the molar ratio of 16S + 23S:5S:4S in Fig. 1?
    c. What is your explanation for the presence of tRNA on these highly purified ribosomes?
4.  a. What is the "disc method" for the determination of radioactivity used in this paper?
    b. What other method is commonly used for the determination of radioactivity?
    c. What are the advantages of the "disc method"?
    d. Is there any occasion when the disc method cannot be used?
5.  In using the disc method for the determination of radioactivity, how are the following determined:
    a. Protein in the presence of labeled nucleic acid
    b. DNA in the presence of labeled RNA and protein
    c. RNA in the presence of labeled protein and lipid
    d. Lipid in the presence of labeled RNA

## Paper 15

1.  a. In fractionations with ammonium sulfate, fractions are usually characterized as precipitating at so many % SAS (saturated ammonium sulfate). At what % SAS does the high molecular weight RNA precipitate?
    b. 16S + 23S RNA precipitate at 1M NaCl. What is the minimum volume of water the lyophilized RNA has to be dissolved in to be soluble?
    c. Why does the concentration of RNA have to be 10 to 15 mg/ml in order to achieve good fractionation with SAS?
    d. The paper states that fractionation with SAS is more effective in the presence of phenol. Why?
2.  a. What is the Mandelkern-Flory equation?
    b. What is the difference between $s$ and $S$?
    c. What is the partial specific volume?
    d. What is the molecular weight of an RNA having an $S$ value of 10?
3.  a. Explain why the sedimentation constant increases with salt concentration.
    b. Why does the $s$ value plateau around $10^{-3}$M salt?
    c. Explain why the viscosity increases with salt concentration.
    d. Why does the specific rotation change with salt concentration?
    e. Which way does the concentration of the RNA solution influence the $s$ value?
    f. The sedimentation was carried out at 25°, yet the $s$ is reported at 20°. Which value is higher?
4.  a. What does the dispersion constant, $\lambda c$, measure?
    b. What is a hypochromic effect?
    c. Is $\epsilon$ sensitive to pH?
5.  a. How do you explain the increase of the viscosity of RNA with temperature?
    b. Would tRNA, with a molecular weight of 25,000, show a greater or smaller change than the ribosomal RNA used in Fig. 2?
    c. Why do different salt concentrations affect the change in viscosity in different ways?
    d. Is the change in viscosity greater or smaller at higher temperatures?
    e. Why does urea affect the viscosity at 0.1M Na$^+$, but has no effect at 0.001 M Na$^+$?
    f. What would be the effect of urea on the sedimentation if the RNA were present in the form of aggregates?
6.  a. Calculate the ionic strengths of $10^{-4}$M MgCl$_2$, 1M NaCl, and 5 × $10^{-4}$M KH$_2$PO$_4$.
    b. Could you distinguish between different RNAs from the shape of the absorbancy curves as a function of ionic strength?

c. Can you distinguish between different RNAs by the shape of the curve of absorbancies as a function of temperature, at a given ionic strength?
d. Is the effect of $Mg^{++}$ on various properties of RNA proportional to its ionic strength? Give reasons for your answer.
e. Since $Mg^{++}$ has such a large effect on the properties of RNA, can you be sure from the method of preparation that the RNA used for these experiments is free of $Mg^{++}$? Can you suggest a method which would assure freedom from $Mg^{++}$?

## Paper 16

1. a. How is electrophoretic mobility described?
   b. Give two sets of measurable variables that can be used in the calculation of mobility.
   c. Are the authors justified in assuming that the potential gradient between the electrodes and the ends of the gel is zero?
   d. Can electrophoresis be carried out in free solution?
   e. Can molecules be made to have zero mobility?
   f. Are molecules with zero mobility necessarily immobile in an electric field in solution? Explain your answer.
   g. What determines the mobility of a molecule in solution?
   h. How does molecular weight affect mobility?
2. a. Can different RNA molecules be separated by free electrophoresis?
   b. What causes the separation according to molecular weight in acrylamide gel electrophoresis?
   c. Can the mobilities obtained in acrylamide gel be used for the determination of isoelectric points?
   d. What is the necessary prerequisite for molecules to have a mobility proportional to molecular weight in acrylamide gel?
3. a. Are the results shown in Fig. 1 a necessary prerequisite for the results in Fig. 2?
   b. Is the prerun mentioned in the experimental section important for obtaining results like those shown in Fig. 1? Why or why not?
   c. Is the prerun important in some other way for obtaining good results with the electrophoresis of RNA?
   d. What properties of the system are influenced by the agarose concentration?
4. a. Does a change in $M_0$ and $m$ affect the utility of the system for determining the molecular weight of RNAs?
   b. Which gives the better separation of different RNA species, a gel with high or one with low $M_0$?
   c. Would the determination of mobility be more accurate in a gel with high $m$ or in one with low $m$?
   d. It is said that of two gels with the same $M_0$, a simple agarose gel shows much poorer resolution than a composite gel. What is the relative magnitude of the $m$ of the two gels?
5. a. What is the minimum number of known RNAs needed to standardize the gels?
   b. Is it sufficient to standardize the gels once and for all, or must some reference RNA be included in every run? In case one wanted to do the latter, what is the minimum number of standards that have to be used per run?
   c. What is the theoretical significance of the observed fact that the mobility is inversely proportional to the log of the molecular weight?
   d. When analyzing an unknown RNA on a given gel, it is possible that one or more components do not penetrate the gel, and will thus barely move out of the starting slot. How can one determine if a band near the start has migrated a distance proportional to its molecular weight, or whether it has not penetrated the gel?
   e. *E. coli* 5S RNA possesses 120 nucleotides. Is the molecular weight given in this paper correct?
   f. Does the fact that the *E. coli* 16S and the liver 18S RNA appear as doublets necessarily mean that the two bands have a different molecular weight?
   g. Would electrophoresis on acrylamide gels be a suitable method for determining the molecular weights of different proteins?

## Paper 17

1. How does the membrane method differ in principle from the following:
   a. The method of Hall and Spiegelman
   b. The column methods of Bautz and Hall, Bolton and McCarthy, and Britten
   c. The nitrocellulose filter method of Nygaard and Hall
2. a. What are the practical advantages of the present membrane method compared with the above methods?
   b. Can you think of a further improvement of the present method?
3. a. What does heating at 90° do to the RNAase activity?
   b. Mononucleotides and small oligonucleotides come straight through an MAK column. Why is the RNA still contaminated with alkali-stable $^{32}P$ counts after one passage through the column?

172   Molecular biology of DNA and RNA

   c. Would the different RNA fractions be separated on the MAK column run as described, or would they come out in a single peak?
   d. Is it necessary or even preferable to treat the crude nucleic acid preparation with DNAase, in order to free the RNA from DNA?
   e. Why does alkali-denatured DNA remain single-stranded after it is neutralized?
4. a. Why is SSC buffer used for the hybridization?
   b. Why are the filters presoaked in SSC used in binding the single-stranded DNA?
   c. Why should filtration of the DNA solution proceed slowly?
   d. Why was the hybridization carried out at 66° C?
   e. Under what conditions can the temperature of hybridization be reduced?
   f. Why is the salt concentration critical for the RNAase digestion?
   g. Since the RNAase digestion is carried out at room temperature, is there any danger that some double-stranded RNA might escape digestion?
   h. Are DNA-RNA hybrids absolutely resistant to RNAase?
   i. Is the use of $^3$H-DNA critical, or would $^{14}$C-DNA do just as well?
   j. Why is the presence of $Mg^{++}$ deleterious?
5. a. Why does hybridization in solution proceed so very much more rapidly than with immobilized DNA?
   b. Could the double-reciprocal plot for determining the maximum amount of RNA hybridized be used for the data pertaining to hybridization in solution in Figs. 3 to 6?
   c. Compare the maximum amount of RNA hybridizable with 50 µg of DNA by the direct method used in Figs. 3 to 5.
   d. To what do you attribute the loss of hybrid observed with liquid hybridization?
6. In checking a commercial DNAase preparation for contamination by RNAase, $^{32}$P-RNA was used as a substrate. When it was freshly prepared, it had a specific activity of $2 \times 10^6$ cpm/µg. Two weeks later, an incubation of 30 min with 1 mg DNAase released 500 cpm. Assume that under the conditions of assay pure RNAase has a specific activity of 20 µmoles/min/mg protein. How much RNAase is there in 1 mg of DNAase?
7. a. In detecting the nucleic acid hybrid, why is $Cs_2SO_4$ used instead of CsCl?
   b. Is equilibrium centrifugation in $Cs_2SO_4$ a reliable method for determining the amount of hybrid?
   c. When a nucleic acid hybrid is chromatographed on an MAK column, where would it emerge relative to the DNA, and why?

Paper 18
1. As is stated in the paper, the function of 5S RNA is unknown. Why, then, have the structures of two 5S RNAs, that of E. coli and that of KB cells, been determined, when the structures of many tRNAs remain unknown?
2. a. What are the end groups of oligonucleotides produced by RNAase A and $T_1$ respectively?
   b. Which of the following can be produced by either enzyme, AMP, 3'AMP, GMP, 3'GMP, (2'→3')GMP, 2'UMP, 3'UMP, (2'→3')UMP, 3'CMP, pAp, pCp, pGp, pUp, adenosine, uridine, guanosine, or cytidine?
3. In order to obtain a nucleotide sequence, the total amount of RNA, from which a given collection of oligonucleotide was produced, has to be known. Which of the following is more useful:
   a. The total amount of RNA used for the nuclease digestion
   b. The total number of counts placed on the paper for separation of the nucleotides
   c. The sum of the counts in all the oligonucleotides on the paper
4. a. Is it necessary to know the specific activity of the starting RNA or of the $^{32}$P in the individual oligonucleotides?
   b. If so, how are the necessary specific activities determined?
5. a. Why is it necessary to obtain and analyze the products of both partial digestion and complete digestion of an RNA by a given nuclease?
   b. Why was ribonuclease $T_1$ chosen for obtaining a partial digest, rather than pancreatic RNAase?
   c. What is the longest oligonucleotide whose sequence can be obtained by alkaline hydrolysis? Explain.
   d. Oligonucleotides were dephosphorylated before digestion with snake venom phosphodiesterase. Do natural RNAs also have to be dephosphorylated before digestion with this enzyme?
   e. Does digestion with spleen phosphodiesterase also require dephosphorylation?
   f. Which of the phosphodiesterases yields results more useful for sequence determination? Explain.
6. Suppose an RNAase A digest yielded four spots on a two-dimensional chromatogram. The spots were cut out and counted, yielding results as follows: spot 1, 3800 cpm; 2, 2200; 3, 900; and 4, 2100. The spots were digested with alkali and each was separated on a separate chromatogram. Spot 1 yielded Up, Gp, and what appeared to be pAp, containing 800, 900, and 2000 cpm, respectively. Spot 2 yielded only Up and contained 2200 cpm. Spot 3 yielded on Cp and

contained 800 cpm, and spot 4 yielded on Gp and Ap and contained 1000 cpm each. The intact RNA had the following base composition: 33.3% A, 22.3% G, 11.1% C, and 33.3% U. The specific activity of the phosphate that day was 1000 cpm per µµmole.

a. Write down all the structural information obtainable from these data. Are they sufficient to obtain the whole sequence?
b. Why have no nucleosides, signifying the 3'OH end group, been obtained from the alkaline digests?
c. Another aliquot of the above $^{32}$P-RNA was digested with $T_1$ and submitted to two-dimensional paper chromatography; three spots were obtained, 11, 12 and 13, and they contained 2500 cpm, 4000 cpm, and 800 cpm, respectively. The spots were eluted and digested with alkali. Spot 11 yielded pAp and Gp; spot 12 yielded Up containing 2300 cpm, Cp with 750 cpm, and Gp, also with 750 cpm. Spot 13 yielded only Ap with 780 cpm. What would be the most useful next step for obtaining the complete sequence of nucleotides?
d. Write down the structural information obtained from the $T_1$ digest and combine it with the information obtained from the RNAase A digest.

# ANSWERS FOR CHAPTER 4

Paper 14

1. a. The preparation of tRNA (now known to be contaminated by other low molecular weight RNAs) by phenol extraction of intact cells depends on the fact that the glycopeptide saccule of the cell wall serves as a sieve that allows low molecular weight RNA to pass, but retains DNA and the high molecular weight ribosomal RNAs.

   b. "Total phenol-extracted RNA" refers to the RNA obtained when a crude cellular extract is extracted with phenol. When this RNA is centrifuged on a sucrose density gradient (SDG), only three peaks are obtained: 23S, 16S, and a peak barely separated from the meniscus, which contains both tRNA and 5S RNA. Messenger RNA does not appear as a separate peak but is present throughout the gradient. If more than these three peaks appear on an SDG, it usually means that the RNA is degraded.

   c. When a material is precipitated from solution, it always retains some finite solubility. This means that the efficiency of the precipitation decreases as the total concentration of material is decreased and as the volume is increased. *E. coli* has a total RNA content of 6%, 15% of which is tRNA; assuming complete extraction, 1 gm will yield a maximum of 9 mg of tRNA in 10 ml of buffer. For the phenol extraction, an equal volume of 90% phenol is added to the extract, and the RNA remains in the aqueous phase; let us assume that there is no change in volume (although, strictly speaking, there is an increase, due to solution of some phenol in the water). Of the 900 µg/ml, 200 µg will remain in solution; therefore, the yield in the precipitate is only 78%.

   In order to obtain a 95% yield, the 200 µg must constitute only 5% of the total concentration per milliliter; that is, there must be 4 mg/ml.

2. a. The capacity of a regular-sized methylated albumin–Kieselguhr (MAK) column is of the order of 2 to 3 mg total RNA, which would yield only 300 to 450 µg of tRNA.

   b. The maximum amount of RNA to be put on each polyacrylamide electrophoresis tube is 25 µg. The preparative setup takes 1 mg.

   c. A G-100 Sephadex column of the stated dimensions, which is about the maximum feasible size, can tolerate up to 50 mg of RNA.

   d. A regular DEAE cellulose column (about $1.0 \times 45$ cm) can be run with 10 mg of crude tRNA.

3. a. Ribosomes are incubated with puromycin under conditions of protein synthesis in order to discharge the growing peptide chains in the form of a peptidyl-puromycin.

   b. A crude calculation of the relative amounts of the different RNAs in the ribosomes can be made from Fig. 1. For this, the O.D. of each point is multiplied by the volume eluted between it and the previous point, and the total O.D.s per fraction are then added up. The peak of high molecular weight RNA trails toward the light side; this factor has to be taken into account in calculating the O.D. pertaining to this peak. As for the 5S peak, the O.D. due to the trailing 16S RNA must be subtracted, as well as the O.D. due to partial overlapping of the 4S peak in this particular chromatograph. (Complete separation between the two peaks is possible by this method, as can be seen from Figs. 2 and 3.) The total O.D.s are then divided by 23 or 24 to obtain the amount of RNA (a 0.1% solution of ribosomal RNA has an O.D. at 260 mµ of 23, and tRNA, of 24), and the amount of RNA is divided by the molecular weights to obtain the molar quantities.

   The total O.D.s for the three peaks are 196, 5.5, and 7.2, respectively, and the corresponding amounts of RNA are 8.5 mg, 0.24 mg, and 0.3 mg. The molecular weight of 23S + 16S is $1.8 \times 10^6$, that of 5S, 40,000, and that of 4S, 25,000 on the average. The relative molar amounts are thus 4.7:6.0:12 mµmoles, which is close enough to 1:1:2.

   c. When the nascent peptide chains are displaced from the ribosomes with puromycin, the tRNA to which the peptide was bound remains on the ribosomes. It does not seem possible to displace this tRNA as long as the ribosome remains in the 70S form. In addition, according to some recent experiments it is necessary also to have an unacylated tRNA in order to form 70S ribosomes from 50S and 30S. The reason for this is not well understood. The presence of exactly 2 tRNAs per ribosome, as found in Fig. 1, is probably accidental, since not all ribosomes would have been bearing nascent peptide chains, and others might have been bearing more than two tRNAs.

4. a. The "disc" method for determining radioactivity was first introduced by Bollum (J. Biol. Chem. **234**:2733, 1959), who used it for nucleic acids, and it was adapted to proteins by Mans and Novelli (Arch. Biochem. Biophys. **94**:48, 1961). It can be used quite generally. It consists of applying a small aliquot of the incubation mixture to a 3 MM Whatman filter-paper disc of 2.3 cm in diameter, and precipitating the proteins and nucleic acids in the fibers of the paper by dropping the disc into a beaker of TCA or other precipitant. The material thus stays on the disc, and can be subjected to a number of different washing procedures. After washing, the disc is dropped into scintillation fluid and counted in a liquid scintillation counter.

b. The more traditional method for determining radioactivity in proteins or RNA consists of precipitating the material in a test tube, filtering it through glass-fiber discs or Millipore membranes, and washing it by suction filtration.

c. The "disc" method has manifold advantages over the traditional method. Because the material is precipitated in the fibers of the paper, there is no recovery problem, and very small amounts of material can be precipitated without a carrier. Since in a regular incubation mixture 10 to 50 $\mu$l is sufficient, it is possible to take several samples from an incubation mixture, so that the zero time and several time samples all can be taken from the same tube. In the traditional method, it is necessary to have a separate incubation mixture for every time point. The washing of the discs can be done in bulk, so that literally hundreds of discs can be processed together. Since precipitation in the fibers is essentially quantitative, it is unnecessary to determine the amount of material in the precipitate; one simply assumes complete recovery. The saving in time is phenomenal. Last, but not least, it is not necessary to correct for self-absorption with $^{32}P$ and $^{14}C$, efficiencies of 60% being achieved for the latter; with $^3H$, the efficiency is less than when the sample is dissolved in the scintillation fluid. In all cases, however, the counts are strictly additive if several discs are added to the same vial; therefore, although 100 $\mu$l is the absolute limit of volume of incubation mixture that can be placed on a disc, it is possible to increase the capacity by using several discs. If the method is used simply for determining radioactivities of, say, a column effluent, rather than for an assay, several 100 $\mu$l applications can be made to the same disc, if it is dried between applications. A further advantage is that with most materials the radioactivity does not leach out from the discs, so that the scintillation fluids can be used over and over.

d. Given the advantages of the "disc" method, there are few occasions when filtration through Millipore filters would be used. One such occasion would be in binding assays, in which the Millipore is used not to retain a precipitate but to specifically adsorb some materials. For example, Millipore filters retain ribosomes and anything bound to them.

Another reason for using the filtration method would be the unavailability of a liquid scintillation counter. By using larger amounts of material on the discs, it might, however, be possible to glue the discs to planchettes and count them in a gas-flow counter, although the counting efficiency would probably be quite low. Even if no saving in materials would be accomplished this way, one would still retain the advantages of bulk processing.

5. a. Protein in the presence of labeled nucleic acid is determined by keeping the discs in TCA at 90° for half an hour to hydrolyze the nucleic acid, and then removing the labeled nucleotides by washing with cold TCA.

b. In order to determine DNA in the presence of labeled RNA and protein, it is necessary to incubate the sample in dilute alkali at 37° for an hour, prior to placing an aliquot on paper and precipitating it with cold TCA. The TCA does not interfere with the counting, so the discs can be dried and counted to determine the DNA + protein. The discs are then treated with hot TCA, washed with cold TCA, and counted to determine the protein, which is then subtracted from the first counts to obtain the DNA by difference.

c. RNA in the presence of labeled protein and lipid is determined by precipitating the sample in cold TCA, and washing it with hot methanol:chloroform, followed by alcohol:ether and ether. This removes the labeled lipid, and a count will yield the labeled RNA + protein. The discs are then treated with hot TCA, washed with cold TCA, and counted to yield the protein. The amount of RNA is determined by difference between the two counts.

d. Labeled lipid, provided it is precipitable with TCA, is determined by first determining the counts on a cold TCA precipitate, and then treating the discs with organic solvents as above, counting again, and determining the lipid by difference.

## Paper 15

1. a. The solubility of ammonium sulfate at 0° C is 70.6 gm/100 ml; therefore, 0.365 gm/ml, which precipitates the high molecular weight ribosomal RNA, corresponds to 51.7% SAS.
   b. The RNA was lyophilized out of 25 ml of $10^{-3}$M NaCl, and therefore contained 25 μmoles of NaCl. The total concentration of Na$^+$ could be larger, however, because there must be some counter ions for the negative phosphate charges of the RNA. There were 300 mg of RNA, of which about 80% would have been high molecular weight RNA, that is, 240 mg; this would contain 240/0.320, or 750 μmoles of phosphates. The exact number of counter ions would be dependent on the pH of the solution, which is not given; however, since the RNA was dialyzed against several changes of NaCl, one might assume that all the phosphates are saturated with Na$^+$, and that, in addition, there are the 25 μmoles of NaCl. Therefore, as long as the lyophilized RNA is dissolved in a volume larger than 775 μl, the concentration of NaCl would be less than 1M, and the RNA would be soluble.
   c. It was explained in answer 1c, paper 14, that in any precipitation there remains a finite amount of material in solution. This must be such in 51.7% SAS that it would constitute a significant proportion of the total RNA if it were less than 10 mg.
   d. Phenol is soluble in water to about 10%. Its presence would reduce the solubility of any strongly ionized substance, and for the reasons stated above, a decrease in solubility of the RNA would increase the efficiency of any fractionation by precipitation.

2. a. The Mandelkern-Flory equation is

$$M_W = \left[ \frac{S^0_{20,w} (\eta)^{1/3} \eta_0 N \, 10^{-13}}{\beta (1-\bar{v}\rho)} \right]^{3/2}$$

   where $M_W$ is the molecular weight, $S^0_{20,w}$ is the sedimentation coefficient in svedbergs at zero solute concentration in pure water at 20°, $\eta$ is the intrinsic viscosity at zero shear of the solution in dl/g, $\eta_0$ is the solvent viscosity, N is Avogadro's number (6.02 × $10^{23}$), $\beta$ is a shape factor defining the relation between molecular weight and sedimentation-viscosity, $\bar{v}$ is the partial specific volume, and $\rho$ is the density of the solvent.

   b. $s$ is the sedimentation coefficient, and is equal to

$$\frac{1}{\omega^2 x} \cdot \frac{dx}{dt}$$

   where $\omega$ is the angular velocity, x is the distance of the sedimenting species from the meniscus, and dx/dt is the rate of sedimentation. It has the units cm/sec/dyne/gm or sec.
   S is a svedberg unit, and equals $s \times 10^{13}$ second.
   c. Partial specific volume, $\bar{v}$, equals

$$\frac{\bar{V}}{M}$$

   where $\bar{V}$ is the increase in volume when one mole of solute is added to a large volume of solution.
   d. It is not possible to calculate the molecular weight from the S value alone. One needs, additionally, to know the $\bar{v}$ and the viscosity or the diffusion constant. Moreover, 10S does not say whether the S was obtained from $s^0_{20,w}$, for only this corrected value can be used for molecular weight determinations. The fact that S is not directly proportional to molecular weight can be best seen from the fact that 4S, 5S, 16S, and 23S RNAs have molecular weights of $2.5 \times 10^4$, $4.0 \times 10^4$, $6 \times 10^5$, and $1.2 \times 10^6$, respectively. Assuming, however, that 10S is equal to $10 \, s^0_{20,w} \times 10^{-13}$, an approximate molecular weight can be calculated from the expression:

$$M = K S^2$$

   For the four RNAs mentioned above, K averages $1.87 \times 10^3$, so that the molecule sedimenting with 10S has a molecular weight of $1.87 \times 10^5$.

3. a. When the sedimentation coefficient changes without an accompanying change in molecular weight, it means that the molecule changed its shape. An increase in $s$ value means that the molecule has become more compact, that is, it offers less resistance to sedimenting. The sedimentation coefficient depends not only on the sedimenting macromolecule but also on the medium through which sedimentation takes place. Normally, as the concentration of molecules in the medium increases, $s$ also increases, for the sedimenting molecule collides with molecules in the medium and takes more time to sediment; furthermore, the viscosity is also greater than that of water, and it thus offers greater resistance to sedimentation. These effects, however, are fairly minor, and would not account for

changes of the mangitude recorded in Fig. 1. The major effect of salt, therefore, must be that it changes the shape of the RNA so as to render it more compact. This means that the polynucleotide chain is folded more tightly. One way in which salt could accomplish this would be by lowering the electrostatic repulsions between different parts of the chain.

b. The relatively small differences in s value between $10^{-4}$ and $10^{-3}$M, compared to the large differences between similar changes in concentration with larger amounts of salt, means that these concentrations are not high enough to cause a change in shape. (The higher s values compared to $s^{20}$ are probably due to the reasons duscussed in 3a, above.) The concentration of RNA in the sedimentation experiments is 33 μg/ml, or about 0.1 μmole of phosphate per ml. Since $10^{-2}$M NaCl is needed to elicit a significant change in shape, it means that the counter ion concentration has to be at least 100 times greater than the phosphate ion (of $10^{-4}$ here) to produce a significant reduction in electrostatic repulsion. If this interpretation is correct, the position of the plateau should vary with the concentration of RNA.

c. Other things being equal, the more compact a molecular structure, the lower is the viscosity. The latter is therefore lowered by salt, because salt causes the RNA to fold up more tightly.

d. As mentioned in the discussion of the paper, the tightening up of the RNA structure means the formation of more and more stable double-stranded, helical regions. Since the bases in helices are highly oriented, it is easy to see why the specific rotation increases.

e. As the concentration of a sedimenting species increases, the frequency of collisions between the molecules will also increase. Each collision means a deviation from a straight path, and the molecule will take longer to sediment a given distance. Hence, the sedimentation constant will decrease with increasing concentration of solute. For randomly coiled RNA,

$$s^0 = s(1 + 1.70[\eta]\, c),$$

where $[\eta]$ is the intrinsic viscosity, that is,

$$\frac{(\eta/\eta_0) - 1}{c}$$

and c is the concentration in gm/dl.

f. The viscosity of the medium is lower at 25° than at 20°, and therefore the s value is higher at the higher temperature.

4. a. Optical rotatory dispersion is the dependence of rotation on wavelength. At any wavelength, λ, the specific rotation is given by

$$[\alpha]_\lambda = \frac{k}{\lambda^2 - \lambda_c^2}$$

where $\lambda_c$ signifies the so-called dispersion constant and k is also a constant.

b. A hypochromic effect is the decrease in absorbancy of a nucleic acid at a given wavelength, due to a change in structure.

c. The molar absorbancy, $\epsilon$, is very sensitive to pH, for the spectra of the individual bases comprising the nucleic acid vary drastically at different pH values. Even though the changes of some bases are somewhat compensated for by changes in the opposite direction of other bases, still the overall changes are considerable.

5. a. We have already seen that with increasing temperatures, the double helices "melt out." Similarly, increasing temperatures cause the double-stranded portions of RNA to loosen up and eventually to "melt out," yielding as a limiting situation a single polyribonucleate strand as a random coil. The RNA helices are much shorter than those in DNA and contain fewer hydrogen bonds, because G-U pairs, united by a single hydrogen bond and completely unpaired bases, are included in the double-stranded regions. Therefore, any increase in temperature is likely to produce an increase in volume of the RNA molecule, which, in turn, results in an increase in viscosity.

b. From answer 5a, above, it can be seen that the maximum possible increase in viscosity depends on the number of helices that can be melted out before the random coil stage is reached. At equal molar concentrations, then, a larger molecule would provide a greater absolute change than a smaller one, provided that the relative amount of helicity, that is, the percentage of bases in hydrogen bonded form, is the same for both. The rate of change in the viscosity depends on the composition of the different helices, and may or may not be the same for small and large molecules. Thus, it can be expected that tRNA would show a smaller absolute change in viscosity than the mixture of rRNAs shown in Fig. 2, not only because of the difference in molecular weight, but also because rRNA has a higher helix content than tRNA. Since the midpoint of the curve probably occurs at about the same temperature, the rate of change for tRNA must occur less steeply than for rRNA.

c. We have seen that salt renders the RNA

structure more compact by allowing the folded chains to approach closer to each other. The higher the salt concentration, therefore, the lower is the starting viscosity of RNA. On the other hand, the viscosity of fully unfolded, random coils should not be affected by temperature, as is also apparent from Fig. 2. In the presence of salt, therefore, the absolute change in viscosity in going from 0° to 90° must be much greater.

d. It can be seen from Fig. 3 that temperature has no great influence on the rate of viscosity change, although the total change is slightly greater at the higher temperature.

e. We have seen that 0.001M NaCl is too low a concentration to produce any contraction of the molecular volume. Since 8M urea, which breaks H bonds, does not produce any change in viscosity, the molecule in 0.001M Na$^+$ must be so spread out that the breaking of the helices causes no further increase in volume. In 0.1M NaCl, on the other hand, the helices are folded together; their breaking by urea nullifies the effect of the salt and causes the molecule to expand from a very compact form to the fully extended random coil, thus producing a large concomitant increase in viscosity.

f. If the RNA were present in the form of aggregates, liberation of the subunits by urea would cause a rather large decrease in viscosity; on the other hand, as we have seen, the effect of urea on the individual molecules is to increase the viscosity. The net effect, then, might be an increase, a decrease, or no change at all, depending on the magnitude of the individual effects.

6. a. The ionic strength, $\mu$, is defined as half the sum of the terms obtained by multiplying the concentration of each ion by the square of its valence. Therefore,

$\mu_{MgCl_2, 0.1mM}$ = ½(0.0001$^2$ + 0.0001 + 0.0001) = **0.0001;**

$\mu_{NaCl, 1M}$ = **1.0;**

and $\mu_{KH_2PO_4, 0.0005M}$ = ½(0.0005 + 0.0005 + 0.0005 + 0.0005$^3$ + 4(0.0005)$^2$) = **0.00075.**

b. In comparing Fig. 2 with Fig. 5, one sees immediately that the shapes of the curves are quite similar. Thus, viscosity and absorbancy measure in a similar way the changes in shape which the salt produces in the RNA. Different RNAs possessing a different number of helices of varying composition must be affected in different ways by the salt, changes which are reflected quantitatively as changes in absorbancy.

c. The melting out of helices by an increase in temperature affects the shape of the RNA in the same way as does the withdrawal of salt. Just as with DNA, each RNA has a characteristic "melting point," that is, the point at which half the helices are broken up, depending on the total number of hydrogen bonds, which in turn depends on the base composition of the helices.

d. From Figs. 5 and 6 it is apparent that the Mg$^{++}$ concentration necessary to effect a certain change in shape is at least two orders of magnitude smaller than that of monovalent salt needed to accomplish a similar change. Furthermore 10$^{-4}$M Mg$^{++}$ lowers the $\epsilon_{258}$ to the same value as 1M NaCl. Since the changes effected by the monovalent salts are very close to those expected of any polyelectrolyte exposed to the given ionic strengths, it follows that the effect of Mg$^{++}$ is much larger than what would be expected on the basis of ionic strength. One explanation for this large effect is that the divalent Mg$^{++}$ can aid in the folding by combining with one phosphate on each side, thus holding two chains together.

e. The exhaustive dialysis against dilute NaCl preceding the lyophilization of the RNA was done for the purpose of replacing all extraneous counter ions by Na$^+$. Since, however, the Mg$^{++}$ binds much more strongly than the Na$^+$, one cannot be sure that all Mg$^{++}$ ions were replaced, especially those in a protected position. One could be sure to remove all Mg$^{++}$ by treating the RNA with EDTA prior to the dialysis, or by chromatographing it on a chelating resin.

## Paper 16

1. a. The electrophoretic mobility of a substance is equal to the distance it can migrate in a unit field strength, that is, 1 volt/cm per unit time (sec).

$$\mu = \frac{d}{F\,t}$$

In practice, the mobility is thus determined by dividing the distance migrated per second by the field strength, the latter being equal to the potential (in volts) across the electrodes divided by the distance between them. The dimensions of $\mu$ are thus cm$^2$/volt/sec.

b. The field strength can be expressed as V/cm; however, by taking advantage of Ohm's law, one can also use the current and the conductivity per unit cross section, and

$$F = \frac{i\,R}{A\,k}$$

where i is the current, R the specific resistance in ohms, A the cross section, and k the conductivity constant of the conductivity cell.

In zone electrophoresis, however, it is extremely difficult to determine the effective cross section of the electrophoretic tube, that is, that occupied by buffer and not by the gel particles. Furthermore, it is often found that the current and the conductivity vary throughout a run, whereas the voltage remains relatively constant, and even if it should vary, it is easy enough to determine the voltage many times in the course of a run. However, it is not possible to continuously monitor the conductivity, for it would be necessary to extract the actual buffer in the electrophoretic tube and measure its conductivity.

c. The actual potential drop across the buffer between the electrodes and the ends of the electrophoretic tube depends on the construction of the apparatus and the nature of the buffer. In some, where the electrode chamber is separated from the bulk of the buffer by a series of baffles, the drop may amount to several volts per centimeter. In zone electrophoresis experiments it is not practical, however, to measure the actual distance between the electrodes, and in any case, it would not be valid to divide the voltage across the electrodes by the total distance, because the conductivity of the buffer in the chamber is very different from that of the electrophoretic tube. In a case like the present, where the same apparatus and the same buffers are used throughout, perfectly acceptable relative mobilities can be obtained by ignoring the actual distance between the electrodes. If different buffers were to be used, the mobilities would probably be slightly different. If absolute mobilities were desired, one should take the voltage across the tube only by applying the leads of a voltmeter directly to the two ends of the tube.

d. Electrophoresis can be carried out in free solution, provided the boundaries formed by the migration of the different ionic species can be stabilized. In the classical Tiselius apparatus a special U tube is used. It is first filled about half full with buffer, and the solution is then carefully layered on in one arm. As the different molecules migrate into the buffer, they form discrete fronts that are visualized by means of schlieren optics. More recently, electrophoresis has been carried out in vertical columns, where the migrating zones are stabilized by a sucrose gradient.

e. Molecules suspended in a buffer with a pH equal to their isoelectric point have zero mobility in an electric field.

f. Even though molecules may have zero mobility, they might still move due to the phenomenon of electroosmotic flow. This is the net transport of fluid from one electrode to the other. All dissolved molecules are passively carried by the fluid.

g. The main determinant of electrophoretic mobility is the charge/mass ratio.

h. Molecular weight affects mobility inasmuch as it changes the charge/mass ratio. In an oligonucleotide series, for example, the highest mobility is shown by a mononucleotide that has two charges per subunit mass. In a dinucleotide, there are three charges per two subunits, and so on. Therefore, as the molecular weight increases, the charge/mass and thus the mobility decrease. It is easy to see that after the addition of relatively few subunits, the change in the charge/mass becomes very small, and the addition of further subunits would affect the mobility very little if at all.

2. a. Because of what was said in answer 1h, above, there is essentially no difference in the charge/mass ratio of different RNA molecules. As can also be seen from Fig. 2, the mobilities of different RNAs in free solution are very similar.

b. The separation of RNAs of different molecular weight is thus not due to the electric field, but it is, rather, a property of the acrylamide gel. The electric field is used mainly as a driving force, similar to the action of gravity on a chromatographic column with Sephadex, for example.

c. Since the mobilities of RNAs on the gel are not due to their intrinsic mobility in an electric field, the separations obtained are largely independent of pH. Isoelectric points can therefore not be obtained.

d. In order to obtain a sieving effect strictly proportional to molecular weight, it is necessary for all the molecules to be affected uniformly by the driving force, in this case the electric field. Therefore all the molecules to be separated must have identical charge/mass ratios.

3. a. Linearity of mobilities with time is absolutely necessary for obtaining separations proportional to molecular weight, as shown in Fig. 2.

b. The purpose of the prerun is to free the gel of all components not found in the electrophoresis buffer, especially persulfate. The presence of persulfate would alter the conductivity of the buffer; as the current is turned on, the

persulfate ions would start migrating, so that some parts of the gel would be free of persulfate, while others are not. This would cause local changes in conductivity, and the RNAs would change their mobilities when they pass from a zone containing persulfate to one not containing this ion.
c. The presence of persulfate is deleterious in another way as well: it often causes precipitation of material in the gel.
d. It is stated in the paper that the extrapolated molecular weight ($M_0$) varied with different amounts of agarose.

4. a. A change in $M_0$ and $m$ (the slope) does not affect the utility of the system for analyzing RNAs within rather large limits. If $M_0$ and $m$ increase too much, however, the resolution of the RNAs will be adversely affected.
b. A gel with a low $M_0$ will give the better resolution, as can be seen from Fig. 2.
c. It can be seen from Fig. 2 that all RNAs of the same molecular weight are on the same horizontal, regardless of $m$. The only thing is that as $m$ becomes extreme, the resolution of the different bands might not be good enough to determine their position accurately. This, however, does not seem to have been a problem in the work described in this paper.
d. For two gels with the same $M_0$, the resolution varies inversely with the slope: the higher the slope, the poorer the resolution, as can be seen from Fig. 2. The agarose gel must therefore have a higher $m$ than the composite gel. (The opposite statement is made in the paper, but the difference is one of semantics only: what the authors mean is that the slope of the agarose gel is less because it is a negative slope, and $-3$ is less than $-1$. I find this somewhat misleading, since the slope for 0% acrylamide in Fig. 2 is obviously steeper than that for 5% acrylamide, for example.)

5. a. An absolute minimum of two RNAs are needed to standardize a gel, but as can be seen from Fig. 2, a more accurate slope is obtained when more standards are run.
b. The paper states that the value of $M_0$ is a reproducible quantity, but there are small variations in the slope. The authors therefore recommend running a standard containing several species in a different slot on the same gel. If necessary, however, the slope could be determined from $M_0$ and one other reference RNA added to the sample.
c. There is no theoretical explanation for the relationship between mobility and molecular weight. It does indicate, however, that the electric field acts as an unbiased driving force,

for in molecular sieve columns, the effluent volume is also proportional to the log of the molecular weight.
d. If the migration of a molecular species is proportional to its molecular weight, then it will fall on the line obtained from the migration of the standards. Fig. 2 shows several examples where the migration was not proportional to molecular weight.
e. The average molecular weight of a nucleotide is 321.5, which yields a molecular weight of 39,000, and not 35,000, as shown in Fig. 2.
f. The appearance of doublets in the 16S and 18S bands does not necessarily mean that the two species differ in molecular weight. Macromolecules change their mobilities as they change shape, for in this process some charges may become more or less available. It is therefore just as likely to assume that the doublets reflect two different conformations in these RNAs.
g. We have already seen that the gel electrophoresis works so well for the determination of the molecular weights of RNAs, because they all have a similar charge/mass ratio. Natural proteins, on the other hand, all have different charge/mass, and so would not migrate uniformly in an electric field.

Paper 17

1. a. Hall and Spiegelman (Proc. Nat. Acad. Sci. U.S.A. **47**:137, 1961) used hybridization in solution, and detection of the hybrid by equilibrium centrifugation in a CsCl gradient.
b. Bautz and Hall (Proc. Nat. Acad. Sci. U.S.A. **48**:400, 1962) prepared a column in which the DNA was bound chemically to phosphocellulose, and the hybridization occurred by slowly passing the labeled RNA through the column at a medium temperature. The hybridized RNA was recovered by eluting the column at a temperature above the $T_m$.

Bolton and McCarthy (Proc. Nat. Acad. Sci. U.S.A. **48**:1390, 1962) trapped DNA in agar, passed it through a mesh and poured it into a column. The RNA was then introduced and allowed to hybridize in a high salt concentration (2 × SSC). The nonhybridized RNA was washed out in 2 × SSC, and the DNA-specific RNA was eluted at a very low salt concentration (0.01 SSC).

Britten (Science **142**:963, 1963) fixed the DNA onto synthetic polymers by UV irradiation and also made a column, after which he proceeded as Bolton and McCarthy.
c. The nitrocellulose filter method of Nygaard and Hall (J. Mol. Biol. **9**:125, 1964) was the direct forerunner of Gillespie and Spiegel-

man. The difference was that the hybridization proceeded in solution at a critical temperature and salt concentration. The hybrid was recovered on nitrocellulose filters, because single-stranded DNA and the hybrid were bound to them.

2. a. The greatest practical advantages are the enormous saving in time, making it possible to obtain literally hundreds of points in the time it previously took to run and analyze a single column, and the saving in material, both of DNA and RNA. Furthermore, despite its speed and convenience, the method is more accurate and more reproducible than previous ones, and avoids the problem of DNA-DNA interactions. It can also be adapted to almost any scale. While other methods also permit the recovery of the hybridized RNA, the present one permits one to do this with maximum convenience, by simply eluting the filters at low salt and high temperature.

  b. The method is nearly perfect. It might be possible to avoid the incubation at 66°, and thereby obviate the need for an accurate water bath or incubator, by carrying out the hybridization in the presence of an organic solvent such as formamide or dimethylsulfoxide, as has been done in the case of DNA-DNA hybrid formation (see Chapter 1, paper 4). In determining the maximum amount hybridized, it is not necessary to carry out the points until an actual plateau is achieved. The maximum can be determined from a double reciprocal plot, as the point where the straight line crosses the ordinate, in a manner akin to the determination of $V_{max}$ by a Lineweaver-Burk plot of enzyme kinetics (see Chapter 1, paper 4, questions and answers 4a to 4e). This not only saves material but at the same time gives a kind of $K_m$ that might be useful as a relative measure of the tightness of the hybrid. The only time the membrane method cannot be used is when it is necessary to isolate the hybrid.

3. a. Pancreatic RNAase is a very stable protein and can be boiled for several minutes without appreciable loss of activity. Heating it to 90°, therefore, destroys contaminating enzymatic activities, especially DNAase, without affecting the RNAase.

  b. Small oligodeoxynucleotides bind quite tightly to complementary sequences in the RNA and are eluted with the latter. Repeated passage through the column eliminates the contamination because the RNA is dialyzed and lyophilized, or precipitated with alcohol, before a rerun, and only a part of the contaminating oligonucleotides have a chance to reform a hybrid. Instead of repeated passage through the column, it might be more efficient to repeatedly precipitate the RNA with alcohol from a high salt solution before fractionating it on the column.

  c. It is not possible to ascertain the amount of fractionation obtained on the column because the volume of the gradient is not given, and one therefore cannot tell how steep the gradient was. The shallower the gradient, the greater the fractionation. The fractionation also depends on the size of the column, also not given, but one can probably assume that it is the same as that in the original reference (Mandell, J. D., and Hershey, A. D.: Anal. Biochem. **1**:66, 1960).

  d. Macromolecular, but sheared DNA is very cleanly separated from the various RNA fractions on an MAK column. Because of the adsorption of smaller oligodeoxynucleotides to RNA, many authors therefore find it preferable not to digest the DNA with DNAase.

  e. The alkaline solution containing single-stranded DNA is neutralized rapidly, so that the complementary DNA strands do not have time to find each other and hybridize.

4. a. SSC buffer is used almost universally in physicochemical studies of DNA, because the citrate is a good chelator of deleterious divalent ions, and at the same time it is a buffer. The NaCl provides the necessary ionic strength to conserve the helical structure of DNA.

  b. Dry filters of any kind often bind a certain amount of material nonspecifically and irreversibly. Any double-stranded DNA present in the DNA solution would thus also be bound, whereas presoaked filters would bind only single-stranded DNA. Dry filters sometimes also trap air, so that the whole surface of the filter is not available for binding DNA.

  c. Adsorption, even irreversible adsorption, is a time-dependent process. If filtration occurs too fast, the DNA will not be hung up on the filter as well as it should be.

  d. Because the DNA is already in a single-stranded form, the high temperature is not needed to prevent DNA-DNA hybridization, as in other methods; however, the RNA is partly double-stranded and needs to be made single-stranded. The temperature of hybridization should be such that perfect hybrids will not melt but that imperfect hybrids have a chance to come apart and reform with a more complementary partner.

  e. Reducing the salt concentration or adding

organic solvents to the mixture will lower the $T_m$ and therefore permit the hybridization to be carried out at a lower temperature.

f. The salt concentration controls the tightness of the helices; at a concentration below $2 \times$ SSC, even perfect hybrids may be somewhat digested, whereas at $6 \times$ SSC probably even imperfect hybrids will be resistant.

g. There should be no double-stranded RNA present, for any RNA that can form double strands at room temperature must be RNA that could not hybridize, and that would have been washed away.

h. DNA-RNA hybrids are resistant to RNAase only under carefully tested conditions. Decreasing the salt concentrations, and increasing the pH and the enzyme concentration, would make it possible for the enzyme to gain access to the DNA-RNA hybrid through a fringed end, which can be digested away, leaving another end, which can come apart and be digested, and so on.

i. $^3$H-DNA is used in conjunction with the $^{32}$P-RNA, because the counts from these two isotopes do not overlap and can thus be counted separately, without having to be corrected. $^{14}$C and $^{32}$P can also be determined, but the $^{14}$C counts have to be corrected for overlapping $^{32}$P. This, besides being a lot of work, does introduce a small element of uncertainty into the $^{14}$C counts.

j. $Mg^{++}$, even in low concentrations, causes a large increase in the $T_m$ of nucleic acids. In the presence of this ion, the RNA would therefore not be single-stranded at the ordinary temperature of hybridization, and furthermore, it would cause a substantial amount of nonspecific hybridizations.

5. a. Hybridization in solution proceeds much more rapidly than on the filters, because in solution the DNA molecules are available to all the RNA molecules at the same time, whereas with the filters, only the RNA molecules in the immediate vicinity of the filter are available for hybridization.

b. The data in Figs. 3 to 6 are time curves and therefore not suitable for double reciprocal plots, for which data recording the hybridization of increasing amounts of RNA are needed. Furthermore, a curve of the shape of those representing liquid hybridization in Figs. 3 to 6 could not be used in a double-reciprocal plot, because only the points before the plateau is reached can be taken.

c. In order to make a double reciprocal plot, the reciprocal of the maximum amount of DNA hybridized at each concentration of RNA (2, 10, and 50 μg) is plotted against the reciprocal of the RNA concentration. The figure below shows that a perfect straight line is obtained, which indicates that this is a true maximum and not a spuriously high amount of hybrid that is later corrected. The maximum amount of hybrid formed according to Fig. 5 is 0.32%, whereas the figure obtained from the reciprocal plot below is 0.325%. In this case, therefore, the two methods give comparable results; on the other hand, the true maximum would still be obtained by the reciprocal plot if the hybridization had not been carried out at 50 μg of RNA, whereas in the case of the direct method, saturation would not have been reached.

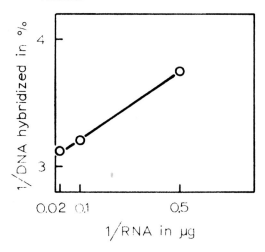

d. Hybridization is carried out at a temperature in which helices are continuously made, broken, and remade. In time, there is an increasingly greater chance that two complementary DNA strands meet, and since the DNA-DNA helices are more stable than the DNA-RNA ones, the DNA double strands accumulate, leading to a progressive loss of hybrid. The fact that high DNA concentration favors the loss of hybrid corroborates this interpretation.

6. $^{32}$P is not a stable isotope but has a half-life of 14 days. At the time of the assay, therefore, 1 μmole of nucleotide (MW 335) contained $335 \times 10^6$ cpm; the 500 cpm liberated in 30 min therefore corresponded to 1.53 μμmoles, or 0.05 μμmoles/min. Since 1 mg of RNAase liberates about 20 μmoles per min, the 0.05 μμmoles correspond to 2.5 mμg. This corresponds to a contamination of only 0.025%, and illustrates the sensitivity of the assay.

7. a. Cs$_2$SO$_4$ is more dense than CsCl and is therefore preferable for use with RNA, which is considerably denser than DNA. In a CsCl gradient designed for DNA, RNA pellets, and a hybrid is very close to the bottom. Since the gradients are usually analyzed by punching a hole through the bottom, all the fractions have to move through the pelleted RNA and may easily be contaminated. Cs$_2$SO$_4$ or Cs$_2$SO$_4$-CsCl mixed gradients can be constructed in which the RNA does not pellet.
   b. Equilibrium centrifugation through Cs gradients often leads to artifacts because of a tendency of the RNA to aggregate and to precipitate in these gradients. This can be prevented by adding formaldehyde to the Cs$_2$SO$_4$, which reacts with all available amino groups and prevents further hydrogen bond-related interactions. Formaldehyde does not react with amino groups already hydrogen-bonded, and will therefore not decompose any RNA-DNA hybrid present.
   c. When a DNA-RNA hybrid is chromatographed through an MAK column, it should emerge just after the DNA, provided both are of comparable molecular weight. Chromatography on the MAK column is dependent on the number of hydrogen bonds the material can form with the column. Its elution is therefore dependent on both the base composition and the molecular weight. The smaller the material, the earlier it emerges, and for comparable molecular weight, and hydrogen bonding, the higher the GC content, the later it will elute. Single-stranded molecules, such as mRNA, emerge last of all. Since the DNA-DNA interaction is stronger than that of a DNA-RNA hybrid, the latter can break more hydrogen bonds that can be reformed with the column and that will therefore elute later than the DNA.

Paper 18
1. 5S RNA is very easily obtained in a homogeneous state, whereas obtaining a homogeneous species of tRNA is extremely difficult. Moreover, 5S RNA is the only known RNA completely devoid of methylated or any other minor bases. It therefore presented a special challenge to see whether it was possible to obtain the complete sequence of an RNA devoid of special markers.
2. a. RNAase A is specific for pyrimidines, and all oligonucleotides obtained by RNAase digestion will therefore have a pyrimidine, U or C, at the 3'phosphate end. RNAase T$_1$ specifically splits after guanine nucleotides, and all oligonucleotides obtained by this enzyme will therefore terminate in a 3'GMP.
   b. When two or more pyrimidine or guanine nucleotides are clustered, free 3'UMP and 3'CMP are obtained from RNAase A digests, and 3'GMP from RNAase T$_1$ digests. No 5'P nucleotides or 3'AMP is obtained. Since a 2'→3' cyclic nucleotide is an intermediate in all digests by a true RNAase, cyclic UMP and GMP can be found. Nucleoside diphosphates of the pXp type signify the 5'P end group, but only pUp, pCp, and pGp can be obtained by the two enzymes under consideration. The liberation of a nucleoside identifies it as the 3'OH end group; any nucleoside can be obtained, provided the adjacent nucleotide can be split by one of the two enzymes.
3. a to c. To obtain the total amount of RNA, it is of little use to determine the amount of RNA used for digestion, for it is virtually impossible and unnecessarily laborious to transfer the digest quantitatively onto the paper. Moreover, there is always some irreversible adsorption of material to the origin; and, in the case of $^{32}$P-labeled material, there will be considerable decay of radioactivity from the time of digestion to the completion of the two-dimensional chromatogram or column chromatography. It is therefore more practical and more desirable to use the total amount recovered, obtained by adding all the counts found in the individual spots.
4. a. Because of the decay of $^{32}$P there is not much point in determining the specific material in the starting material; rather, the specific activity of the oligonucleotides in the spots on the paper should be determined.
   b. The amount of material is far too small to yield any optical density, and therefore it is impossible to determine the specific activity of a nucleotide on that basis. To determine the specific activity, then, several standards containing a known amount of phosphate are carried along and counted at the beginning, the middle, and the end of the experimental series. The specific activity is then calculated by averaging all the values.
5. a. A complete digest gives a number of individual nucleotides, but even when all the sequences in these are known, there is no information about how to line them up. For that, it is necessary to obtain large oligonucleotides that encompass the sequences obtained in the smaller ones. Only when a sequence encompassing the 3'OH end of one oligonucleotide and the 5'P end of the next exists, can the two be set side by side.

b. The sensitivity of G residues to $T_1$ digestion can be radically altered by carrying out the digestion at 0° or under some other suboptimal conditions, thus allowing for the production of very large oligonucleotides. It is more difficult to obtain partial digests with RNAase A, and even at best there results a mixture of complete and partial digestion products, which makes the separation of the various oligonucleotides very much more difficult.

c. A trinucleotide is the longest oligonucleotide whose structure can be obtained from the products of an alkaline digestion: it will yield one diphosphate, and two nucleotides as 2′,3′mononucleotides, but from the specificity of the nuclease that produced the oligonucleotide, the nature of the nucleotide at the 3′P end can be ascertained. If the trinucleotide is of the form pXpYpZ, X will be obtained as pXp, Y as Yp, and Z as a nucleoside.

d. The oligonucleotide products of RNAase action have to be dephosphorylated before digestion with snake venom phosphodiesterase because this enzyme is an exonuclease that needs a free 3′OH group to start. Natural RNAs, not the products of nuclease action, have a free 3′OH and can be digested with snake venom diesterase directly.

e. Spleen phosphodiesterase is also an exonuclease, but it requires a free 5′OH group and yields 3′nucleotides. All oligonucleotides in an RNAase digest, except the one containing the 5′P end of the parent RNA, possess a free 5′OH group and can thus be digested without dephosphorylation.

f. The main utility of both phosphodiesterases is that under suitable conditions they can be made to work slowly enough so that the sequential release of nucleotides can be followed. Otherwise, the spleen phosphodiesterase is more useful because it yields the 3′OH end as a free nucleoside, whereas the snake venom enzyme yields all nucleotides as 5′nucleotides.

6. Since spot 1 yielded pAp, it must contain the 5′phosphate end; since it was produced by RNAase A, Up must be at the 3′P end. From the counts it can be seen that it contains only one A, U, and G, so that the structure is pApGpUp. Since spot 2 contained only Up, the counts indicate that there must be two moles of Up. Spot 3 contains a single Cp residue. Spot 4 does not contain a pyrimidine and must therefore contain the 3′OH end.

a. So far, we have the following structures: pApGpUp, 2Up, 1Cp, and (GpAp)X; the polynucleotide thus contains 9 nucleotides. The recovered bases account for 22.3% A, 22.3% G, 11.1% C, and 33.3% U. These values all coincide with the base composition determined on the intact material, except for A, which indicates that X must be A. In order to obtain free Up, it must come from the sequence UpUp or CpU; similarly, Cp must come from UpC, since there is only one C. From the information obtained so far, three possible sequences can be written as follows:

pApGpUpCpUpUpGpApA,
pApGpUpUpCpUpGpApA, or
pApGpUpUpUpCpGpApA

b. Nucleosides cannot be recovered from $^{32}$P-labeled RNA digests because they are unlabeled.

c. The most useful next step would probably be to dephosphorylate spot 12 with alkaline phosphatase and attempt a stepwise chemical analysis with periodate, which splits the *cis*-glycol of the terminal nucleotide to yield a dialdehyde. This undergoes β-elimination in the presence of a primary amine, and eliminates the base of the first nucleotide, leaving a 3′-phospho-oligonucleotide shorter by one nucleotide. The latter can be recovered, again treated with alkaline phosphatase, and again reacted with periodate. The reaction yields are not 100%, so that the detection of the eliminated base becomes more difficult at every step; however, the method is good for at least three nucleotides, which would be sufficient to reveal the sequence of the oligonucleotide in spot 12. As a matter of fact, as soon as a C would be eliminated, the rest could only be U.

d. The $T_1$ digest yields pApG, Up(Up,Up,Cp)Gp, and ApA. It adds no information to the structure.

CHAPTER 5

# GENE TRANSCRIPTION IN VIVO

The transcription of DNA into RNA is an essential step in the expression of the genetic material. Sometimes, as with the genes for tRNAs and the ribosomal RNAs, the transcription product itself, or its derivative, is the end product of a gene; in the great majority of cases, however, the transcribed RNA functions as a "messenger" of the gene to the ribosomes, where the RNA is translated into proteins. As far as we know, DNA itself never serves as a messenger *in vivo*, although under certain conditions single-stranded DNA can be translated into peptide chains in cell-free systems.

The "messenger" hypothesis was first proposed by Jacob and Monod (J. Mol. Biol. 3:318) in 1961 on the basis of genetic data on the *lac* operon. Essentially, the hypothesis stated that each functional unit of genes (previously named an operon by Benzer) produced a short-lived messenger that was capable of instructing the ribosomes to synthesize the proteins coded for by the operon. (Previously, it had been believed for several years that the ribosomal RNA itself held the information for making proteins, so that new ribosomes had to be synthesized before an induced enzyme could be formed. For lack of something better, this hypothesis persisted even though it was obvious that the synthesis of ribosomes, especially in nucleated cells, was far too slow to account for the kinetics of enzyme induction.) Jacob and Monod further hypothesized that the transcription of an operon was regulated by a region on the chromosome called the operator. This had a specific affinity for a regulatory molecule called a repressor, which was specified by a regulatory gene, i, usually located outside the operon it controls. The repressor was thought to be RNA but is now known to be a protein. Two repressors, *lac* and $\lambda$, have been isolated.

In a system like the *lac* operon, said to be inducible, the operon cannot be transcribed when the repressor is combined with the operator, but it can be "opened" by a substance called an inducer, which specifically combines with the repressor, thereby lessening its affinity for the operator. In an alternative mode of regulation, called repression, the free repressor, or aporepressor, is inactive, but it becomes active when combined with a specific small molecule, called the corepressor. Both, inducer and corepressor, are effectors. In general, it can be said that repressors are passive unless combined with an effector. The majority of operons are under negative control, that is, the product of the regulatory gene serves to prevent the expression of the genes. There are, however, examples of positive control (the arabinose genes and genes on phage $\lambda$) where the product of the regulatory gene is needed for the expression of the operon.

From the beginning it was speculated that the "messenger" was RNA, and the demonstration of the actual existence of mRNA followed closely after the publication of the messenger hypothesis. Brenner, Jacob, and Meselson (paper 19) took advantage of the fact that infection with $T_2$ phage stops all host RNA synthesis and starts synthesis of a different RNA, presumably phage-specific. In order to lay to rest the idea that ribosomal RNA itself was the messenger, it was necessary to show that the new, phage-specific RNA can be found on ribosomes synthesized prior to the infection. This was done by transferring density-labeled bacteria to normal medium at the time of infection. It could thus be shown unequivocally that new RNA was associated with heavy ribosomes, and that no new ribosomes at all were made in the infected cells. The existence in normal *E. coli* of rapidly labeled RNA capable of reversible association with ribosomes was demonstrated at about the same time by Gros, Hiatt, Gilbert, Kurland, Risebrough, and Watson (Nature **190**:581, 1961).

Basically, all RNA synthesized in normal cells is copied from DNA templates by the enzyme RNA polymerase. These RNA molecules, however, suffer some later modifications (see Chapter 6), and there can therefore be some incorporation of labeled precursors which is independent of DNA. There is a region on the chromosome between the operator and the start of the first gene in the operon called the promotor, without which there can be no expression of the operon. It is therefore likely that the promotor is the binding site for the RNA polymerase that transcribes the entire operon to yield a polycistronic mRNA. This can then combine with many ribosomes in the course of translation to form polysomes. The beginnings and ends of the proteins corresponding to each cistron are signified by appropriate codons (AUG, UUG, and GUG for beginnings, and UAA, UAG, and UGA for chain termination). Natural mRNAs also seem to contain some nontranslatable regions, the significance of which is not known.

Messenger RNAs are very fragile, due to their great length. The prevalence of mRNAs in the 8S-14S range in the early reports is due to degradation of larger molecules.

Only one of the two strands of DNA is usually transcribed into RNA, as can be proven by hybridization in those cases in which the two DNA strands can be separated by density centrifugation or chromatography on MAK columns. In the case of the phage λ, however, both strands have transcribable regions. The same can be surmised to be the case with some bacteria in which the direction of expression of some operons is contrary to that of the majority. In any case, however, it can be stated as a general rule that if a stretch of DNA is transcribed, its complement will not be transcribed. Therefore, although the two strands of DNA are genetically equivalent, only one of the strands "makes sense" phenotypically, the other being nonsense. The lack of complementary transcription also assures the single-strandedness of mRNA, which is essential for translation.

Since, then, RNA is transcribed from one strand only, there is no reason why mRNA should have the same base composition as the DNA from which it was transcribed; early findings to that effect are therefore purely coincidental. Another early expectation—that mRNA was necessarily short-lived—also turned out not to be generally true. Although some mRNAs do, in fact, decay rapidly, others are quite stable. This is particularly true in higher cells, and one of the best examples can be found in the reticulocyte, which continues to synthesize hemoglobin long after it has lost its DNA. The stability of some mRNAs also means that not all mRNAs are "rapidly labeled," another criterion frequently used for identifying mRNA. (Moreover, there are rapidly labeled RNAs in nuclei, and these RNAs are not mRNAs.) The only valid test for mRNA, then, is its ability to serve as a messenger for a well-defined protein. (The ability of an RNA to stimulate amino acid incorporation *in vitro* is not an acceptable criterion, for any single-stranded RNA has this ability.)

Nascent transfer RNA does not contain any minor bases, nor does it have the pCpCpA end. The latter can be reversibly added or removed by an enzyme in the cytoplasm, and numerous other enzymes specifically modify certain bases (see Chapter 6). Bacterial genomes possess about 40 genes for tRNA, distributed all over the chromosome.

The small ribosomal RNA (5S) is transcribed in a slightly longer form, and since it does not contain any minor bases, maturation involves only the loss of a few nucleotides from the 5' end. The large ribosomal RNAs (16S and 23S for bacteria, 18S and 25S for plants, 18S and 27S for reptiles, and 18S and 28S for mammals) are transcribed together in the form of a large precursor, which is later methylated and then split a number of times to give rise to the mature rRNAs, and to other pieces that are discarded. The best-defined sequence is that of mammals, which have a 45S ($4.4 \times 10^6$ daltons) precursor. This is then split through 41S and 36S intermediates to 20S and 32S particles, and a discarded piece of $1.3 \times 10^6$ daltons. The 20S particle is further transformed to the RNA of the small subunit—18S, or $0.7 \times 10^6$ daltons. The 32S ($2.2 \times 10^6$ daltons) particle is further split to yield a discarded piece of $0.45 \times 10^6$ daltons and the RNA of the large subunit, 28S, or $1.75 \times 10^6$ daltons. Thus, almost half of the original precursor (a total of $2 \times 10^6$ daltons) is lost in the maturation process.

The genes for ribosomal RNA occupy anywhere from 0.1% to more than 1% of the genome. They are thus highly redundant, numbering from about 10 in *E. coli* to many thousands in man. The genes for the small and the large subunit are strictly alternating, as required by the fact that they are transcribed together. The genes for 5S RNA are also highly redundant but are not adjacent to the other ribosomal genes.

The amount of tRNA per genome stays constant throughout the life cycle of bacteria, but the amount of rRNAs varies widely, depending on the nutrients present. The concentration of ribosomes is generally proportional to the rate of protein synthesis, but the exact nature of the RNA regulator is not known. Ever since Pardee enunciated the famed "RNA for protein synthesis" hypothesis in the mid-1950s evidence for the dependence of RNA synthesis on protein synthesis has been sought. Although such a dependence could never be convincingly demonstrated, and it is now known not to exist, it was believed that aminoacyl-tRNAs were the regulators. The evidence for this was that in the wild type of bacteria, called *stringent,* RNA synthesis was dependent on the presence of amino acids. The *str* locus locus could be mutated to a *relaxed* form, in which RNA synthesis proceeded independently of the presence of amino acids. It has now been shown, however, that amino acids do not control the rate of RNA synthesis directly, but that they act by controlling the synthesis of UTP. This still might not be the whole story, for it seems that the nucleotide pools are not depleted rapidly enough to account for the abrupt cessation of RNA synthesis upon the withdrawal of amino acids.

There are a number of inhibitors of DNA-dependent RNA synthesis, but by far the inhibitor most used is an antibiotic called actinomycin D.

Although in normal cells all RNA is transcribed from DNA, this is not so if a cell has been infected with an RNA virus. The RNA genome, acting as its own messenger, binds to ribosomes and specifies the synthesis of an RNA polymerase (commonly called RNA synthetase to distinguish it from the DNA-dependent RNA polymerase), which transforms the single-stranded viral RNA into a double-stranded structure, on which progeny viral RNAs are synthesized. In animal cells, it has recently been demonstrated that the viral RNA was able to direct the synthesis of a complementary DNA, thus contradicting the unidirectional flow of information from DNA to RNA specified by the "central dogma."

# 19 AN UNSTABLE INTERMEDIATE CARRYING INFORMATION FROM GENES TO RIBOSOMES FOR PROTEIN SYNTHESIS

S. Brenner
*Medical Research Council Unit for Molecular Biology*
*Cavendish Laboratory*
*University of Cambridge*
*Cambridge, England*

F. Jacob
*Institut Pasteur*
*Paris, France*

M. Meselson
*Gates and Crellin Laboratories of Chemistry*
*California Institute of Technology*
*Pasadena, California*

---

A large amount of evidence suggests that genetic information for protein structure is encoded in deoxyribonucleic acid (DNA) while the actual assembling of amino-acids into proteins occurs in cytoplasmic ribonucleoprotein particles called ribosomes. The fact that proteins are not synthesized directly on genes demands the existence of an intermediate information carrier. This intermediate template is generally assumed to be a stable ribonucleic acid (RNA) and more specifically the RNA of the ribosomes. According to the present view, each gene controls the synthesis of one kind of specialized ribosome, which in turn directs the synthesis of the corresponding protein—a scheme which could be epitomized as the one gene—one ribosome—one protein hypothesis. In the past few years, however, this model has encountered some difficulties: (1) The remarkable homogeneity in size[1] and nucleotide composition[2] of the ribosomal RNA reflects neither the range of size of polypeptide chains nor the variation in the nucleotide composition observed in the DNA of different bacterial species[2,3]. (2) The capacity of bacteria to synthesize a given protein does not seem to survive beyond the integrity of the corresponding gene[4]. (3) Regulation of protein synthesis in bacteria seems to operate at the level of the synthesis of the information intermediate by the gene rather than at the level of the synthesis of the protein[5].

These results are scarcely compatible with the existence of stable RNA intermediates acting as templates for protein synthesis. The paradox, however, can be resolved by the hypothesis, put forward by Jacob and Monod[5], that the ribosomal RNA is not the intermediate carrier of information from gene to protein, but rather that ribosomes are non-specialized structures which receive genetic information from the gene in the form of an unstable intermediate or 'messenger'. We present here the results of experiments on phage-infected bacteria which give direct support to this hypothesis.

When growing bacteria are infected with a virulent bacteriophage such as $T2$, synthesis of DNA stops immediately, to resume 7 min. later[6], while protein synthesis continues at a constant rate[7]. After infection many bacterial enzymes are no longer produced[8]; in all likelihood, the new protein is genetically determined by the phage. A large number of new enzymatic activities appears in the infected cell during the first few minutes following infection[9], and from the tenth minute onwards some 60 per cent of the protein synthesized can be accounted for by the proteins of the phage coat[7]. Surprisingly enough, protein synthesis after infection is not accompanied, as in growing cells, by a net synthesis of RNA[10]. Using isotopic labelling, however, Volkin and Astrachan[11] were able to demonstrate high turnover in a minor RNA fraction after phage infection. Most remarkable is the fact that this

---

From Nature **190**:576-581, 1961. Reprinted with permission.

This work was initiated while two of us (S. B. and F. J.) were guest investigators in the Division of Biology, California Institute of Technology, Pasadena, during June 1960. We would like to thank Profs. G. W. Beadle and M. Delbrück for their kind hospitality and financial support.

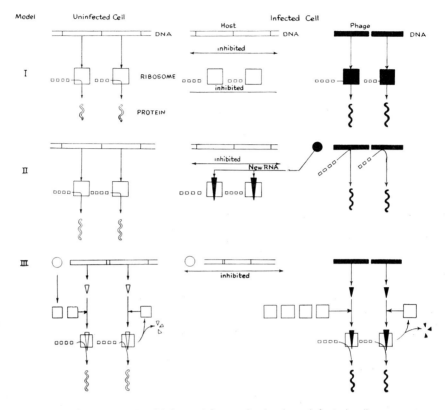

**Fig. 1.** Three models of information transfer in phage-infected cells.

RNA fraction has an apparent nucleotide composition which corresponds to that of the DNA of the phage and is markedly different from that of the host RNA[11]. Recently, it has been shown that the bulk of this RNA is associated with the ribosomes of the infected cell[12].

Phage-infected bacteria therefore provide a situation in which the synthesis of protein is suddenly switched from bacterial to phage control and proceeds without the concomitant synthesis of stable RNA. *A priori*, three types of hypothesis may be considered to account for the known facts of phage protein synthesis (Fig. 1). Model I is the classical model. After infection the bacterial machinery is switched off, and new ribosomes are then synthesized by the phage genes. The *ad hoc* hypothesis has to be added that these ribosomes are unstable, to account for the turnover of RNA after phage infection. This is, in fact, the model favoured by Nomura *et al.*[12]. Model II assumes that in the particular case of phage the proteins are assembled directly on the DNA; the new RNA is a special molecule which enters old ribosomes and destroys their capacity for protein synthesis. At the same time, synthesis of ribosomes is switched off. Model III implies that a special type of RNA molecule, or 'messenger RNA', exists which brings genetic information from genes to non-specialized ribosomes and that the consequences of phage infection are two-fold: (*a*) to switch off the synthesis of new ribosomes; (*b*) to substitute phage messenger RNA for bacterial messenger RNA. This substitution can occur quickly only if messenger RNA is unstable; the RNA made after phage infection does turn over and appears, therefore, as a good candidate for the messenger.

It is possible to distinguish experimentally between these three models in the following way: Bacteria are grown in heavy isotopes so that all cell constituents are uniformly labelled 'heavy'. They are infected with phage and

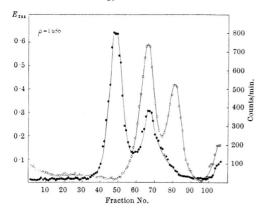

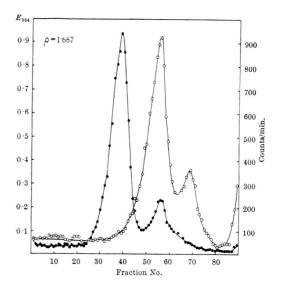

**Fig. 2.** Distribution of heavy and light ribosomes in a density gradient. *E. coli* B, grown in 5 ml. of a medium containing $^{15}$N (99 per cent) and $^{13}$C (60 per cent) algal hydrolysate and $^{32}$PO$_4$, were mixed with a fifty-fold excess of cells grown in nutrient broth, the ribosomes extracted by alumina grinding in the presence of 0.01 $M$ Mg$^{++}$ and purified by centrifugation. 1 mgm. of ribosomes was centrifuged in 3 ml. of caesium chloride buffered to pH 7.2 with 0.1 $M$ tris and containing 0.03 $M$ magnesium acetate for 35 hr. at 37,000 r.p.m. in the *SW*39 rotor of the Spinco model *L* ultracentrifuge. After the run, a hole was pierced in the bottom of the tube and drops sequentially collected. Ultra-violet absorption at 254 m$\mu$ detects the excess of light ribosomes (O), $^{32}$P counts detect the heavy ribosomes (●).

**Fig. 3.** Distribution of randomized heavy and light ribosomes in a density gradient. The mixture of $^{15}$N$^{13}$C$^{32}$P and $^{14}$N$^{12}$C$^{31}$P ribosomes was dialysed first for 18 hr. against 0.0005 $M$ magnesium acetate in 0.01 $M$ phosphate buffer pH 7.0, and then for 24 hr. against two changes of 0.01 $M$ magnesium acetate in 0.001 $M$ tris buffer pH 7.4. 1 mgm. of ribosomes was centrifuged for 38 hr. at 37,000 r.p.m. in caesium chloride containing 0.03 $M$ magnesium acetate. The drops were assayed for ultra violet absorption (O) and $^{32}$P content (●).

transferred immediately to a medium containing light isotopes so that all constituents synthesized after infection are 'light'. The distribution of new RNA and new protein, labelled with radioactive isotopes, is then followed by density gradient centrifugation[13] of purified ribosomes.

Density gradient centrifugation was carried out in a preparative centrifuge, and the ribosomes were stabilized by including magnesium acetate (0.01–0.06 $M$) in the caesium chloride solution. Ribosomes show two bands, a heavier $A$ band and a lighter $B$ band, the relative proportions of which, for a given preparation, depend on the magnesium concentration used. The lower the magnesium concentration, the smaller the proportion of $B$ band ribosomes and the larger the proportion of $A$ band ribosomes.

In order to show that there is no aggregation of ribosomes during preparation and density gradient centrifugation, an experiment was carried out on ribosomes extracted from a mixture of $^{15}$N$^{13}$C and $^{14}$N$^{12}$C bacteria. The results are shown in Fig. 2, from which it can be seen that ribosomes of different isotopic compositions band independently and that there are no intermediate classes. The same preparation was then dialysed against low magnesium to dissociate the ribosomes into their 50 $S$ and 30 $S$ components and then against high magnesium to re-associate the sub-units[14]. This should have resulted in distributing heavy 30 $S$ and 50 $S$ sub-units into mixed 70 $S$ and 100 $S$ ribosomes. Surprisingly enough, density gradient centrifugation of this preparation (Fig. 3) yields the same bands as found in the original ribosomes except for a decrease in the proportion of the $B$ bands. This means that both bands contain units which do not undergo reversible association and dissociation and that the mixed 70 $S$ ribosomes prepared by dialysis separate into their components in the density gradient. Other experiments to be reported elsewhere suggest that the

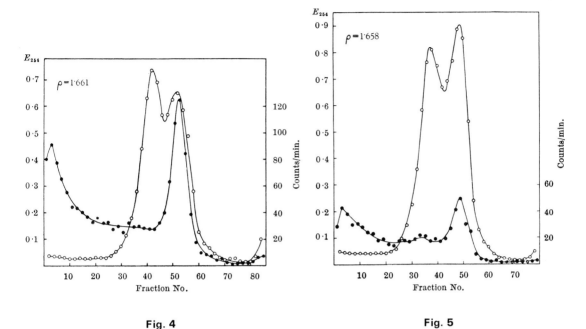

**Figs. 4** and **5**. Distribution and turnover of RNA formed after phage infection. A 600 ml. culture of *E. coli* B6 (mutant requiring arginine and uracil) was infected with *T4D* (multiplicity 30) and fed $^{14}$C-uracil (10 mc./m*M*) from third to fifth min. after infection. One half of the culture was removed and ribosomes prepared (Fig. 4). The other half received a two hundred-fold excess of $^{12}$C-uridine for a further 16 min. and ribosomes prepared (Fig. 5). In both experiments approximately 3 mgm. of purified ribosomes were centrifuged for 42 hr. at 37,000 r.p.m. in caesium chloride containing 0.05 *M* magnesium acetate. Alternate drops were collected in *tris*-magnesium buffer for ultra-violet absorption (O) and on to 0.5 ml. of frozen 5 per cent trichloroacetic acid. These tubes were thawed, 1 mgm. of serum albumin added, and the precipitates separated and washed by filtration on membrane filters for assay of radioactivity (●).

*A* band is composed of free 50 *S* and 30 *S* ribosomes and that the *B* band contains undissociated 70 *S* particles.

The bulk of the RNA synthesized after infection is found in the ribosome fraction, provided that the extraction is carried out in 0.01 *M* magnesium ions[12]. We have confirmed this finding and have studied the distribution of the new RNA among the ribosomal units found in the density gradient. Fig. 4 shows that this RNA, labelled with $^{14}$C-uracil, bands in the same position as *B* band ribosomes. There is no peak corresponding to the *A* band. In addition, there is radioactivity at the bottom of the cell. This is free RNA, as its density is greater than 1.8, and, moreover, it must have a reasonably high molecular weight to have sedimented in the gradient. Lowering of the magnesium concentration in the gradient, or dialysing the particles against low magnesium, produces a decrease of the *B* band and an increase of the *A* band. At the same time, the radioactive RNA leaves the *B* band to appear at the bottom of the gradient. This shows that the uracil has labelled a species of RNA distinct from that of the bulk of *B* band ribosomes, since the specific activity of the RNA at the bottom of the cell is much higher than that of the *B* band. Fig. 5 shows that this RNA turns over during phage growth. There is a decrease by a factor of four in the specific activity of the *B* band after 16 min. of growth in $^{12}$C-uridine. Similar results have been obtained using $^{32}$PO$_4$ as a label.

These results do not distinguish between a messenger fraction and a small proportion of new ribosomes which are fragile in caesium chloride and which are also metabolically unstable. In order to make the distinction, the experiment was carried out with an isotope transfer, in the following manner: Cells grown in a small volume of $^{15}$N$^{13}$C medium were infected with *T*4, transferred to $^{14}$N$^{12}$C medi-

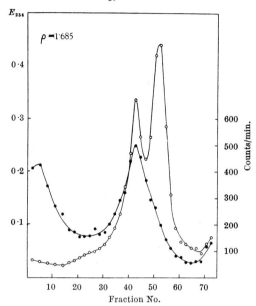

**Fig. 6.** RNA after isotope transfer in phage-infected cells. *E. coli* B grown in 10 ml. of $^{15}$N (99 per cent) and $^{13}$C (60 per cent) algal hydrolysate, medium were starved in buffer, infected with T4 and growth initiated by addition of glucose and dephosphorylated broth ($^{14}$N$^{12}$C). $^{32}$PO$_4$ was fed from the second to seventh min. after infection. The culture was mixed with a fifty-fold excess of *E. coli* B grown in nutrient broth, infected in buffer and then grown for 7 min. in dephosphorylated broth medium. 1 mgm. of purified ribosomes was centrifuged for 36 hr. at 37,000 r.p.m. in caesium chloride containing 0.03 $M$ magnesium acetate. Ultra-violet absorption (O) detects the $^{14}$N$^{12}$C$^{31}$P carrier; radioactivity (●) detects the new RNA in the heavy cells transferred to light medium.

um and fed $^{32}$PO$_4$ from the second to the seventh minute. They were mixed with a fifty-fold excess of cells grown and infected in $^{14}$N$^{12}$C$^{31}$P medium. Fig. 6 shows that the RNA formed after infection in the heavy cells has a density greater than that of the *B* band of the carrier. Its peak corresponds exactly with the density of the *B* band of $^{15}$N$^{13}$C ribosomes (Fig. 2) although it is skewed to lighter density, and its response to changing the magnesium concentration was that of a *B* band. There is no radioactive peak corresponding to the *B* band of the carrier: this means that no wholly new ribosomes are synthesized after phage infection. As already shown, the new RNA does not represent random labelling of *B* band ribosomes; therefore it constitutes a fraction which is added to pre-existing ribosomes the bulk of the material of which has been assimilated before infection. This result conclusively eliminates model I.

To distinguish between models II and III an experiment was carried out to see whether pre-existing ribosomes participate in protein synthesis after phage infection. Cells were grown in $^{15}$N medium, infected with phage, transferred to $^{14}$N medium and fed $^{35}$SO$_4$ for the first 2 min. of phage growth. Fig. 7 shows that only the *B* band of preexisting ribosomes becomes labelled with $^{35}$S and there is no peak corresponding to a $^{14}$N *B* band. All this label can be removed by growth in nonradioactive sulphate and methionine (Fig. 8). In this experiment, the incorporation of $^{35}$S into the total extract was measured and the amount in the *B* band found to correspond to 10 sec. of protein synthesis. This is probably an overestimate since it is unlikely that pool equilibration was attained instantaneously. This value corresponds quite closely with the ribosome passage time of 5-7 sec. for the nascent protein in uninfected cells[15]. In addition, electrophoresis of chymotrypsin digests of *B* band ribosomes shows that the radioactivity is already contained in a variety of peptides. It would therefore appear that most, if not all, protein synthesis in the infected cell occurs in ribosomes. The experiment also shows that pre-existing ribosomes are used for synthesis and that no new ribosomes containing stable sulphur-35 are synthesized. This result effectively eliminates model II.

We may summarize our findings as follows: (1) After phage infection no new ribosomes can be detected. (2) A new RNA with a relatively rapid turnover is synthesized after phage infection. This RNA, which has a base composition corresponding to that of the phage DNA, is added to pre-existing ribosomes, from which it can be detached in a caesium chloride gradient by lowering the magnesium concentration. (3) Most, and perhaps all, protein synthesis in the infected cell occurs in pre-existing ribosomes.

These conclusions are compatible only with model III (Fig. 1), which implies that protein synthesis occurs by a similar mechanism in uninfected cells. This, indeed, appears to be the case: exposure of uninfected cells to a 10-sec. pulse of $^{32}$PO$_4$ results in labelling of the RNA

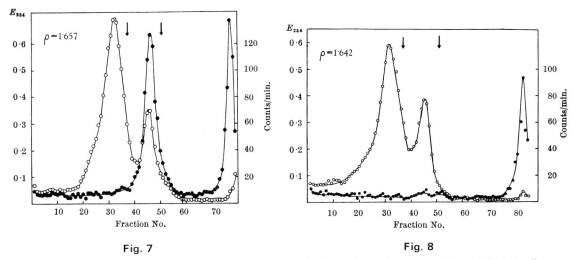

**Figs. 7** and **8**. Distribution and turnover of newly synthesized protein in ribosomes of phage-infected cells after transfer. *E. coli* B grown in 600 ml. of a salt glucose medium containing $^{15}NH_4Cl$ (99 per cent) were starved in buffer infected with T4 and transferred to $^{14}NH_4Cl$ medium. $^{35}SO_4$ was fed for the first 2 min. of infection and one half of the culture removed (Fig. 7). The other half received an excess of $^{32}SO_4$ and $^{32}S$-methionine and growth was continued for a further 8 min. (Fig. 8). 1 mgm. of purified ribosomes was centrifuged for 39 hr. at 37,000 r.p.m. in caesium chloride containing 0.05 M magnesium acetate. Drops were assayed for ultra-violet absorption (O) and for radioactivity (●). The arrows mark the expected positions for the peaks of $^{14}N$ A and B bands. The radioactivity at the top of the gradient is contaminating protein.

in the B band and not in the A band of ribosomes, and this RNA can be detached from the ribosomes by lowering the concentration of magnesium ions. Similarly the nascent protein can be labelled by a short pulse of $^{35}SO_4$; it is located in the B band and most of the label is removed by growth in non-radioactive sulphate. In contrast to what was observed in infected cells, residual stable radioactivity is found in both bands, reflecting the synthesis of new ribosomes.

In order to act as an intermediate carrier of information from genes to ribosomes, the messenger has to fulfil certain prerequisites of size, turnover and nucleotide composition. In the accompanying article, Gros *et al.*[16] have analysed the distribution of pulse-labelled RNA in sucrose gradients. They have shown that in uninfected cells there is an RNA fraction which has a rapid turnover and which can become attached reversibly to ribosomes depending on the magnesium concentration. The T2 phage-specific RNA shows the same behaviour and both are physically similar, with sedimentation constants of 14-16 S. We have carried out similar experiments[17] independently and our results confirm their findings. These suggest that, although the messenger RNA is a minor fraction of the total RNA (not more than 4 per cent), it is not uniformly distributed over all ribosomes, and may be large enough to code for long polypeptide chains. When ribosomes, from phage-infected cells labelled with $^{32}PO_4$ for five min., are separated by centrifugation in a sucrose density gradient[18] containing 0.01 M $Mg^{++}$, most of the messenger is found in 70 and 100 S ribosomes, contrary to previous reports[12]. When the magnesium concentration is lowered, the radioactivity is found in three peaks of roughly equal amount: (1) corresponding to a small residual number of 70 S ribosomes; (2) corresponding to 30 S ribosomes; (3) a peak of very high specific activity at 12 S. Separation of the RNA extracted from such ribosomes with detergent shows all the counts to be located in a peak at 12 S, skewed towards the heavier side. These results suggest that the messenger is heterogeneous in size and may have a minimum molecular weight of about ¼ to ½ million. Similar results have been obtained in uninfected cells[17].

The undissociable 70 S ribosomes are enriched for messenger RNA over the total ribosomes of this type, and it has been shown that they are also enriched for the nascent protein[17]. These ribosomes have been called "active 70 S" ribosomes, by Tissières et al.[19], and they appear to be the only ribosomes which preserve the ability to synthesize protein *in vitro*. This leads one to suspect that there is a series of successive events involved in protein synthesis, and that at any time we investigate a temporal cross-section of the process.

The exact determination of the rate of turnover of the messenger RNA should give information about the process of protein synthesis. This might be stoichiometric, in the sense that each messenger molecule functions only once in information transfer before it is destroyed. Its rate of turnover should then be the same as that of the nascent protein; but experiments to test this idea have been limited by difficulties in pool equilibration with nucleotide precursors.

It is a prediction of the hypothesis that the messenger RNA should be a simple copy of the gene, and its nucleotide composition should therefore correspond to that of the DNA. This appears to be the case in phage-infected cells[11,20], and recently Ycas and Vincent[21] have found a rapidly labelled RNA fraction with this property in yeast cells. If this turns out to be universally true, interesting implications for coding mechanisms will be raised.

One last point deserves emphasis. Although the details of the mechanism of information transfer by messenger are not clear, the experiments with phage-infected cells show unequivocally that information for protein synthesis cannot be encoded in the chemical sequence of the ribosomal RNA. Ribosomes are non-specialized structures which synthesize, at a given time, the protein dictated by the messenger they happen to contain. The function of the ribosomal RNA in this process is unknown and there are also no restrictions on its origin in the cell: it may be synthesized by nuclear genes or by enzymes or it may be endowed with self-replicating ability.

**REFERENCES**

1. Hall, B. D., and Doty, P., J. Mol. Biol. **1**:111 (1959). Littaucr, U. Z., and Eisenberg, H., Biochim. Biophys. Acta **32**:320 (1959). Kurland, C. G., J. Mol. Biol. **2**:83 (1960).
2. Belozersky, A. N., Intern. Symp. Origin of Life. 194 (Publishing House of the Academy of Sciences of the U.S.S.R., 1957).
3. Chargaff, E., The Nucleic Acids **1**:307 (Academic Press, New York, 1955). Lee, K. Y., Wahl, R., and Barbu, E., Ann. Inst. Pasteur **91**:212 (1956).
4. Riley, M., Pardee, A. B., Jacob, F., and Monod, J., J. Mol. Biol. **2**:216 (1960).
5. Jacob, F., and Monod, J., J. Mol. Biol. (in press).
6. Cohen, S. S., J. Biol. Chem. **174**:218 (1948). Hershey, A. D., Dixon, J., and Chase, M., J. Gen. Physiol. **36**:777 (1953). Vidaver, G. A., and Kozloff, L. M., J. Biol. Chem. **225**:335 (1957).
7. Koch, G., and Hershey, A. D., J. Mol. Biol. **1**:260 (1959).
8. Monod, J., and Wollman, E., Ann. Inst. Pasteur **73**:937 (1947). Cohen, S. S., Bact. Rev. **13**:1 (1949). Pardee, A. B., and Williams, I., Ann. Inst. Pasteur **84**:147 (1953).
9. Kornber, A., Zimmerman, S. B., Kornberg, S. R., and Josse, J., Proc. U.S. Nat. Acad. Sci. **45**:772 (1959). Flaks, J. G., Lichtenstein, J., and Cohen, S. S., J. Biol. Chem. **234**:1507 (1959).
10. Cohen, S. S., J. Biol. Chem. **174**:281 (1948). Manson, L. A., J. Bacteriol. **66**:703 (1953).
11. Volkin, E., and Astrachan, L., Virology **2**:149 (1956). Astrachan, L., and Volkin, E., Biochim. Biophys. Acta **29**:544 (1958).
12. Nomura, M., Hall, B. D., and Spiegelman, S., J. Mol. Biol. **2**:306 (1960).
13. Meselson, M., Stahl, F. W., and Vinograd, J., Proc. U.S. Nat. Acad. Sci. **43**:581 (1957).
14. Tissières, A., Watson, J. D., Schlessinger, D., and Hollingworth, B. R., J. Mol. Biol. **1**:221 (1959).
15. McQuillen, K., Roberts, R. B., and Britten, R. J., Proc. U.S. Nat. Acad. Sci. **45**:1437 (1959).
16. Gros, F., Hiatt, H., Gilbert, W., Kurland, C. G., Risebrough, R. W., and Watson, J. D., Nature **190**:581 (1961).
17. Brenner, S., and Eckhart, W. (unpublished results).
18. Britten, R. J., and Roberts, R. B., Science **131**:32 (1960).
19. Tissières, A., Schlessinger, D., and Gros, F., Proc. U.S. Nat. Acad. Sci. **46**:1450 (1960).
20. Volkin, E., Astrachan, L., and Countryman, J. L., Virology **6**:545 (1958).
21. Ycas, M., and Vincent, W. S., Proc. U.S. Nat. Acad. Sci. **46**:804 (1960).

# 20 EVIDENCE FOR A NONRANDOM READING OF THE GENOME

T. Kano-Sueoka*
S. Spiegelman
Department of Microbiology
University of Illinois
Urbana, Illinois

Recently accumulated data suggest that the transcription of an individual cistron into a ribopolynucleotide and its subsequent translation into a polypeptide are ordered and oriented processes. Thus, the ingenious genetic experiments of Crick et al.[1] and Champe and Benzer[2] are readily interpretable if the reading of the base sequence starts from a fixed point and continues until the end of the cistron is indicated. Further, the elegant analysis of haemoglobin synthesis by Dintzis[3] implies that the growth of a polypeptide chain is the result of a sequential addition of residues starting at the $NH_2$-terminal amino acid and finishing at the free carboxyl end.

Since every cistron is presumed to have its own beginning, each can, in principle, be read independently of the others. The experiments cited do not, therefore, provide an answer to the following question: *Is the transcription of the genome random or nonrandom?* By "random" we mean that the probability of transcription for any particular cistron is invariant with time and independent of its location in the genome.

The fact that specific proteins appear sequentially in phage-infected bacteria[4-7] does not answer the question since one can have an ordered use of randomly produced genetic messages.

An approach toward the solution of the problem may be seen by recognizing that if the reading is random, then any sample of RNA messages is equivalent to any other. If it is not random, a sample taken at one time should be distinguishable from that taken at another. To decide between these alternatives, three experimental conditions are required. One is a situation synchronized with respect to the onset in the transcription of a known genome. A second is the suppression of possible selective degradation of particular messages due to use in protein synthesis. The third is a method of distinguishing the polyribonucleotide transcriptions of one set of cistrons from another. The first and second were readily provided by the *E. coli*-T2 system and the use of chloramphenicol. The third requisite was attained by the combined use of simultaneous column chromatography and double labeling. It is the purpose of the present paper to describe the relevant experiments. The data obtained support the conclusion that transcription of the genome is not a random process.

## Methods and materials

*Materials.* Worthington DNAase (2 × crystallized) was freed of contaminating RNAase chromatographically.[8] Crystalline lysozyme from Armour and Company and herring sperm DNA from California Corporation for Biochemical Research were used. Chloramphenicol was a gift of Parke Davis and Company. For preparation of columns, Fraction V from bovine plasma (Armour and Company) and "Hyflo Supercel" (Johns-Manville Products Corporation) were employed. Tritiated uridine (1,000 µc/74 γ) and $C^{14}$-uridine (1 µc/270 γ) were obtained from Schwarz Laboratories.

*Cells and viruses. Escherichia coli* BB and bacteriophage T2 were used. Conditions of growth and medium were as described previously.[9] Virus adsorption was carried out without aeration in 10 times the ultimate cell concentration at 25°C for 2 min. The suspension was then diluted into 9 volumes of prewarmed (37°C) medium and aerated. Zero time was taken as the beginning of aeration. Multiplicity of infection was 10-15, and, in all the experiments reported, the per cent of noninfected cells was assayed at less than 0.1%. When chloramphenicol was employed, it was added to a final concentration of 50 γ/ml. In labeling experiments, $H^3$-uridine at 1 or 0.05 µc/γ/ml and $C^{14}$-uridine at 0.004 µc/γ/ml were present during the pulse.

---

From Proceedings of the National Academy of Sciences U.S.A. **48**:1942-1949, 1962. Reprinted with permission.

This investigation was aided by grants from the U.S. Public Health Service, the National Science Foundation, and the Office of Naval Research.

*Predoctoral trainee in molecular genetics (USPHS 2G-319).

*Isolation of RNA.* Cells were lysed in the presence of lysozyme and DNAase according to the protocol detailed by Hayashi and Spiegelman.[9] RNA was isolated by the phenol procedure[10] and further purified as described earlier.[9]

*Characterization of RNA.* Size distribution was examined in linear sucrose gradients.[11] Methylated-albumin-kieselguhr columns were prepared according to Mandell and Hershey.[12] Linear NaCl concentration gradients were used and the elution patterns of variously labeled RNA preparations compared. Columns, 10 cm in height and 2 cm in width, were loaded with 0.7–1.5 mg RNA. Recovery of RNA was between 90 and 100%. Gradients were monitored by refractive index measurements on selected eluent fractions.

*Optical density and radioactivity measurements.* After the UV absorption at 260 m$\mu$ of each fraction was read, aliquots were removed for precipitation with trichloracetic acid (TCA) with the addition of 100 $\mu$g of DNA as a carrier. Washing was done with 10% TCA on millipore filters.[13] Double counting of $C^{14}$ and $H^3$ was accomplished using the two channels[14] of the Packard Tri-Carb Scintillation Counter. All data are reported as per cent of total cpm of the two isotopes with respect to each other, which makes for easier visual comparison of the profiles.

## Components of the experiment

*1. Nature of system.* The *E. coli*-T2 system possesses obvious advantages for an investigation into the details of the reading mechanism. Transcription from the viral genome is readily initiated simultaneously in a population of cells. Further, earlier studies[13,15,16] have shown that the synthesis of all types of host RNA is virtually completely suppressed. Thus, if one introduces a labeled precursor subsequent to infection, the labeled RNA is exclusively T2-specific. Finally, the comparison is simplified since the number of genetic messages being examined is likely to be much smaller than would obtain for genetically more complicated organisms.

*2. Chromatographic fractionation of RNA.* To distinguish one type of RNA molecule from another, recourse was had to the methylated-albumin-kieselguhr (MAK) columns introduced by Lerman[17] and subsequently developed and used by Mandell and Hershey[12] to separate DNA from RNA and size fractionation of DNA. Others[18,19] have recently employed these columns to detect and isolate RNA molecules differing in size. The discovery by Sueoka and Cheng[20] that these columns can recognize base composition of DNA encouraged us to investigate whether a similar recognition occurs with RNA. A detailed description of the results obtained with T2-complementary RNA will be given elsewhere.[21] For present purposes, two features of these columns may be noted. They do fractionate RNA according to base composition. The higher the per cent of AU, the greater is the molarity of NaCl required for elution. They also separate according to size, the smaller fragments eluting at lower salt concentrations. Despite the fact that these two processes are occurring simultaneously, base compositional differences were readily detectable in different parts of the chromatographic profile of total T2-RNA. However, an increase in the resolution with respect to base composition was obtained by prior fractionation with respect to size in linear surcrose gradients.

*3. Simultaneous chromatography with two isotopic labels.* To insure a ready and sensitive detection of chromatographic differences, use was made of double labeling and simultaneous chromatography. Thus, consider two RNA preparations, one identified by $H^3$-label and the other by $C^{14}$. If a mixture of the two is loaded on a column, the elution profiles of $H^3$ and $C^{14}$ should be identical if the two preparations are the same and should differ if one contains one or more components absent from the other. To insure the comparability of RNA labeled *in vivo* for short periods of time, it is obviously necessary that the two labels be on the same precursor. One avoids thereby complications of varied pool sizes and different paths of entry. In the present experiments, $C^{14}$ and $H^3$-uridine were used.

## Outcome of the experiment

The plan of the experiment is obvious from the above discussion. Two aliquots of a culture are infected with T2. Each is then allowed to incorporate one or the other of the labeled uridines for equal intervals at various times after the infection. The total RNA is then isolated by the procedures described under *Methods* and mixed in proportions determined by the specific activities of the two isotopes. The properties of the counting system made it convenient to keep the ratio of $H^3$ to $C^{14}$ cpm at about 1.5. The mixture is then eluted from the column and the profiles of the two isotopes compared. In all experiments reported here, the period of incorporation was restricted to two min.

*1. The effect of protein synthesis.* As a preliminary to comparing different periods, it was of interest to see whether selective breakdown due to use during the period of labeling was going to introduce a complication in the comparisons. Figure 1 describes an experiment examining this question. Both cultures were labeled during the same two-min period. In one, however, chloramphenicol (CM) was added one min prior to introducing the radioactive uridine to prevent protein synthesis during incorporation. As in all the figures, the dotted curves

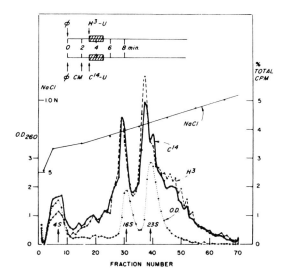

**Fig. 1.** MAK column chromatography of T2-RNA. As in all the other figures, the protocol and timing of each experiment is diagrammed. Two cultures were infected with T2 (φ) at zero time. At 3 min, H³-uridine (H³-U) was introduced into one and C¹⁴-uridine (C¹⁴-U) into the other. To the latter was also added chloramphenicol (CM) at 2 min. The period of incorporation is indicated by hatched rectangle. At 5 min, the incorporation was terminated and the RNA isolated from each and purified. A mixture of the two samples was chromatographed. The O.D. profile identifies pre-existent cellular stable components.

identify the pre-existent stable cellular RNA components (4S, 16S, 23S). Comparison of the $H^3$ and $C^{14}$ profiles reveals that they are essentially the same. As will be seen from subsequent experiments, the slight discernible differences are comparatively insignificant. It would appear that the effect of protein synthesis for a two-min period is not readily detectable in terms of the observable RNA components. If the comparison is extended for longer intervals (20-30 min), striking differences can be observed[21] in the presence and absence of protein formation, but these do not concern us in the present investigation.

**2. Comparison of RNA synthesized in different periods.** Figure 2 compares the RNA synthesized in the same and in different periods. In every case of this experiment, chloramphenicol was added one min prior to addition of the radioactive uridine. It will be seen from the control experiment of Figure 2B that the correspondence of the $H^3$ and $C^{14}$

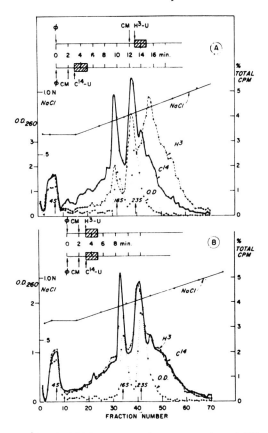

**Fig. 2.** MAK column chromatography of T2-RNA. **A,** Mixture of two RNA samples; H³-labeled coming from the culture pulsed between 13-15 min after infection and the C¹⁴-labeled RNA derived from culture exposed to C¹⁴-uridine between 3-5 min. **B,** Control mixture of two RNA samples independently labeled during the same period. All symbols and other details as in Fig. 1.

profiles is excellent. The agreement provides convincing evidence of the combined reproducibility of the various steps and operations required for such experiments. Figure 2A shows clearly that one can readily distinguish the RNA synthesized between 3 and 5 min from that which is formed between 13 and 15 min. This means that the mechanism which produced these two samples contains a nonrandom element.

**3. Chromatographic comparison of RNA of restricted size ranges.** The RNA used in the experiments of Figure 2 were total samples. As pointed out, MAK columns fractionate according to both size and base composition. It was of

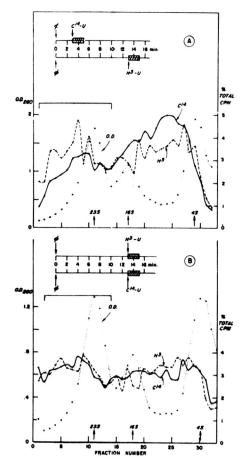
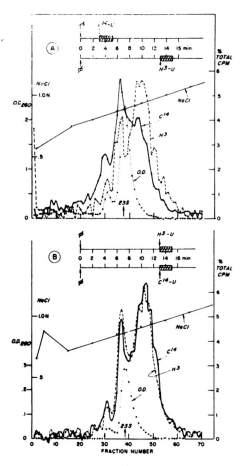

**Fig. 3.** Linear sucrose gradient analysis of mixed RNA samples. Linear gradients were 3-20 per cent sucrose buffered with Tris 0.01 $M$ at pH 7.3 and containing $5 \times 10^{-3}$ $M$ Mg$^{++}$. Centrifugation was at 25,000 rpm for 6.5 hr at 10°C in a Spinco SW-25 swinging bucket rotor. **A,** RNA samples from 3-5 min (C$^{14}$-labeled) and 13-15 min (H$^3$-labeled) after infection. **B,** Control mixture from same period (13-15 min) labeled independently. Brackets over the profiles delineate regions pooled and subjected to chromatographic analysis in Figure 4. All other symbols as in Figure 1.

**Fig. 4.** MAK column chromatography of a selected size region. **A,** Sample originates from pooled fractions enclosed within bracket of linear sucrose gradient analysis of Figure 3,A. **B,** Sample originates from pooled fractions enclosed within bracket of linear sucrose gradient analysis of control experiment of Figure 3,B.

obvious interest to inquire whether the difference observed in Figure 2B could be explained solely by the synthesis in the 13-15 min period of a greater proportion of large-size RNA molecules. A comparison of size distribution was made in linear sucrose gradients as described in Figure 3. In this case, the control period being exhibited is the 13-15 min interval. Comparison of Figures 3A and 3B shows that the size distributions of the RNA synthesized in the two periods are detectably different. There is evidently a somewhat higher proportion of larger RNA molecules synthesized in the later (H$^3$-labeled) than in the earlier (C$^{14}$-labeled) interval. However, our previous experience would suggest that the magnitude of the difference in size distribution is not adequate to explain the striking discordancy observed in the chromatographic profiles of Figure 2A.

A more direct check on this question can be made by performing the chromatographic comparison on RNA previously fractionated for size by simultaneous centrifugation in a sucrose gradient. To minimize the size effect, the segment selected *should, on a percentage basis, contain within it* approximately the same size distribution for both labeled RNA samples. The percentage for each point is calculated on the basis of the *total number of relevant counts included in the region chosen.* The chromatograms are then compared on the same percentage basis. A region of this type is indicated by the bracket of Figure 3A. It contains relatively minor differences in size distribution of the two labeled RNA preparations. The fractions included within the bracket were pooled and chromatographed. For purposes of comparison, the bracketed fractions in the control experiment of Figure 3B were treated in the same way. The results are described in Figure 4. Again, we see in Figure 4B correspondence of $H^3$ and $C^{14}$ profiles in the control preparation. On the other hand, Figure 4A exhibits the previously observed clear distinction between RNA synthesized in the two periods subjected to labeling. This difference should have been almost eliminated if it had been principally due to size. We may conclude that RNA molecules differing either in base sequence or composition are being synthesized in various periods of the infection. Direct evidence for base compositional differences will be reported in a subsequent communication.

### Discussion

The experiments described here have been confined to the periods 3-5 and 13-15 min after infection. However, other periods have been examined with equivalent results. The data obtained show that, using the methods devised, it is possible to differentiate the set of RNA molecules synthesized in one period of infection from those made in another. Since this distinction can be observed when the RNA is synthesized in the presence of chloramphenicol, the difference cannot be attributed to the degradation of particular RNA components as a result of protein synthesis. The data are consistent with an ordered transcription of the genome. The only element of uncertainty introduced is due to the unlikely possibility that there exists a mechanism of *selective* destruction of message RNA completely unconnected with protein synthesis.

It is of interest to note that Green and Hall[22] have recently examined competitive interaction during hybridization between RNA preparations derived from different portions of the latent period after T2 infection. Their data are consistent with the conclusion that such preparations are detectably different.

The methods developed during the present investigation possess great flexibility and can be applied to a wide variety of problems. The resolving power can be increased considerably beyond the level found adequate for the present investigation. Clearly, much finer time slices of the transcription process in the coli-T2 system can be examined with the hope of identifying individual messages for particular proteins. The use of the two labels and simultaneous chromatography permits the identification and isolation of any detectable persistent differences in the chromatograms. Any given region can be magnified by suitable adjustment of the NaCl gradient. These devices allow one to analyze the factors which permit or hinder progress of the reading process. They also make possible informative comparisons between normal and mutant strains possessing deletions at certain loci. The results of experiments along these and similar lines will be reported subsequently.

### Summary

The present paper poses the question of whether the reading of the entire genome is a random or ordered process. The T2-*E. coli* system was used as a convenient source of a controlled commencement in the transcription of a known genome. The use of two isotopic labels to identify RNA synthesized in various periods, coupled with simultaneous chromatography on one column, permitted an unambiguous decision between difference and similarity of any two samples. They were always easily distinguishable if they were synthesized at different times and identical if they came from the same period. This is consistent with an ordered transcription of the genome.

The methods developed for the present investigation possess a degree of flexibility and resolving power which makes them useful for the experimental analysis of a number of problems.

## REFERENCES

1. Crick, F. H. C., L. Barnett, S. Brenner, and R. J. Watts-Tobin, Nature **192**:1227 (1962).
2. Champe, S. P., and S. Benzer, J. Mol. Biol. **4**:288 (1962).
3. Dintzis, H. M., these Proceedings **47**:247 (1961).
4. Luria, S. E., Science **136**:685 (1962).
5. Kornberg, A., J. B. Zimmerman, S. R. Kornberg, and J. Josse, these Proceedings **45**:772 (1959).
6. Dirksen, M. L., J. S. Wiberg, J. F. Koerner, and J. M. Buchanan, these Proceedings **46**:1425 (1960).
7. Wiberg, J. F., M. L. Dirksen, R. H. Epstein, S. E. Luria, and J. M. Buchanan, these Proceedings **48**:293 (1962).
8. Polatnick, J., and H. L. Bachrach, Analyt. Biochem. **2**:161 (1961).
9. Hayashi, M., and S. Spiegelman, these Proceedings **47**:1564 (1961).
10. Gierer, A., and G. Schramm, Nature **177**:702 (1956).
11. Britten, R. J., and R. B. Roberts, Science **131**:32 (1960).
12. Mandell, J. D., and A. D. Hershey, Analyt. Biochem. **1**:66 (1960).
13. Hall, B. D., and S. Spiegelman, these Proceedings **47**:137 (1961).
14. Okita, G. T., J. J. Kabara, F. Richardson, and G. V. LeRoy, Nucleonics **15**:111 (1957).
15. Astrachan, L., and T. M. Fischer, Federation Proc. **20**:359 (1961).
16. Nomura, M., B. D. Hall, and S. Spiegelman, J. Mol. Biol. **2**:306 (1960).
17. Lerman, L. S., Biochim. et Biophys. Acta **18**:132 (1955).
18. Philipson, L., J. Gen. Physiol. **44**:899 (1961).
19. Otaka, E., H. Mitsui, and S. Osawa, these Proceedings **48**:425 (1962).
20. Sueoka, N., and Ts'ai-Ying Cheng, J. Mol. Biol. **4**:161 (1962).
21. Kano-Sueoka, T., and S. Spiegelman (in manuscript).
22. Green, M., and B. D. Hall (personal communication).

# 21 GENE-SPECIFIC MESSENGER RNA: ISOLATION BY THE DELETION METHOD

*Ekkehard K. F. Bautz*
*Eugene Reilly*
Institute of Microbiology
Rutgers University
New Brunswick, New Jersey

*Abstract.* Messenger RNA molecules, homologous to a small portion of the genome of bacteriophage T4, have been isolated. RNA fragments specific to the rII A and the rII B cistrons have been separated by hybridization with DNA isolated from appropriate deletion mutants. An RNA species homologous in nucleotide sequence to a defined part of the rII A cistron has been identified.

The separation of a gene-specific messenger RNA (mRNA) species from a multitude of structurally similar, but informationally different, messengers can be accomplished in principle by either of two procedures. In some instances, a limited number of genes can become integrated into the genome of a transducing phage, and as such be separated from the bacterial genome; this DNA can be made to form complexes selectively with its homologous RNA (1). In those cases, where an episomal element is not available for the selection of a particular segment of DNA, a deletion mutant lacking the gene to be studied can be used to remove, by hybridization, all but the desired mRNA species.

The method of complex formation between mRNA and episomal DNA is straightforward once the selection of the proper DNA piece has been achieved. This selection is, however, so far limited to only three experimental systems (1). Since there are many more genes known for which deletion mutants are available, the method of removal by hybridization would provide a wider range of application. The experimental difficulty encountered here is that the first step of hybridization, namely, elimination of all undesired mRNA species, has to be quantitative.

A previous attempt (2) has revealed the feasibility of purifying RNA, specific for the rII region of bacteriophage T4, by this approach. In this report we describe a general procedure for the isolation and the detection of gene-specific mRNA. The procedure applies the observations that nitrocellulose binds single-stranded DNA (3) and that the bound DNA is still capable of forming complexes with homologous RNA (4).

DNA was extracted by shaking concentrated bacteriophage stocks three times with phenol saturated with 0.01$M$ tris ($pH$ 7.2) and then by precipitating the DNA from the aqueous layer with two volumes of ethanol. The DNA was collected on a glass rod and stored as dried fibers. Before use a portion of the DNA was dissolved in low-salt buffer (5) and denatured by heat at a concentration of 0.3 mg/ml; the denatured DNA was cooled rapidly, adjusted to high-salt buffer, and pipetted into a suspension of nitrocellulose powder in high-salt buffer; this mixture was then rapidly stirred for 5 minutes. The suspension of nitrocellulose (type RS, Hercules Powder Co.) was prepared by grinding it in high-salt buffer in a mortar, passing the slurry through a 40-mesh stainless-steel sieve, decanting to remove fine particles, and washing with high-salt buffer for 2 hours at room temperature. A slurry of 10 mg of DNA of bacteriophage T4r⁺ (wild type) on nitrocellulose was poured into a jacket column of 15 mm diameter to yield a column height of 12 cm. A small amount of plain nitrocellulose was packed on top of it, and on this layer was placed a 12-cm column of nitrocellulose containing 10 mg of DNA of the mutant phage r1272 [a deletion spanning the entire rII region (6)]. The three layers were visibly divided by small plugs of glass wool inserted between them. With these precautions, cross contamination between the two types of DNA was avoided.

---

From Science 151:328-330, 1966. Copyright 1966 by the American Association for the Advancement of Science. Reprinted with permission.

Supported by grants PHS-GM 10395 and NSF-GB 1882. We thank Drs. S. Benzer and S. Champe for the phage stocks.

## TABLE 1
*Fractionation of T4r$^+$ RNA-H$^3$ by passage through DNA nitrocellulose*

RNA, pulse labeled with uracil-H$^3$ (specific activity 2.7 c/mole) for 3 minutes after infection of E. coli B with T4r$^+$ or with mutant T4r1272, was passed through a double column as indicated. Portions of RNA eluted from the bottom column with low-salt buffer were incubated in high-salt buffer with nitrocellulose filters containing 100 μg of T4r$^+$ or T4r1272 DNA for 9 hours at 65°C, washed, and counted (4).

| Infecting phage | DNA in column | | RNA input (10$^5$ count/min) | Hybrid recovered from bottom column (10$^4$ count/min) | Total hybrid (%) | Radioactivity on filters (count/min) | | Fraction of radioactivity bound to r$^+$ DNA exclusively r$^+$−r1272 |
|---|---|---|---|---|---|---|---|---|
| | Top | Bottom | | | | T4r$^+$ | T4r1272 | r$^+$ |
| T4r$^+$ | T4r1272 | T4r$^+$ | 7.1 | 2.614 | 4.8 | 607 | 358 | 0.41 |
| T4r$^+$ | T4r$^+$ | T4r1272 | 3.5 | 0.292 | 1.3 | 131 | 141 | 0 |
| T4r1272 | T4r1272 | T4r$^+$ | 9.12 | 2.518 | 3.0 | 759 | 800 | 0 |

## TABLE 2
*Fractionation of H$^3$-labeled T4r$^+$ RNA on DNA from deletions of different sizes and map locations*

The RNA, labeled with uridine-H$^3$ (specific activity 8 c/mole) from 6 to 8 minutes after infection of E. coli B with T4r$^+$, was fractionated first on a column containing T4r1272 DNA (top) and T4r$^+$ DNA (bottom). A sample of the RNA eluted from the part of column I containing T4r$^+$DNA was applied to column II. A portion of the RNA eluted from the top portion of this column, containing T4r638 DNA, was then fractionated on column III. Before passage through columns II and III, the RNA samples were treated with deoxyribonuclease (20 μg/ml) for 20 minutes at 37°C in 10$^{-2}$M Mg$^{++}$. Deoxyribonuclease was removed with phenol, and the RNA was dialyzed against high-salt buffer.

| Column number | DNA in column | | Input (10$^3$ count/min) | Hybrid recovered (count/min) | | Fraction of total hybrid | |
|---|---|---|---|---|---|---|---|
| | Top | Bottom | | Top | Bottom | Top (%) | Bottom (%) |
| I | T4r1272 | T4r$^+$ | 3450.00 | 290 × 10$^4$ | 4.02 × 10$^4$ | 98.6 | 1.4 |
| II | T4r638 | T4r$^+$ | 14.53 | 7384 | 2561 | 74.2 | 25.8 |
| III | T4r1231 | T4r$^+$ | 2.74 | 700 | 605 | 55.1 | 44.9 |

## TABLE 3
*Filter test of RNA fractions eluted from column I and II of Table 2*

Portions of the RNA were incubated with the filters as described in Table 1. The efficiency of binding to the T4r$^+$ DNA on the filters varied from 60 to 75 percent of the input.

| RNA eluted from column | Treated with deoxyribo-nuclease | Filters (count/min) | | | | Fraction of radioactivity bound to T4r$^+$ DNA exclusively | |
|---|---|---|---|---|---|---|---|
| | | T4r$^+$ | T4r638 | T4r1272 | Blank | $\frac{r^+ - r1272}{r^+}$ | $\frac{r^+ - r638}{r^+}$ |
| I−T4r1272 | − | 20029 | 20158 | 21593 | 5165 | 0 | |
| I−T4r$^+$ | − | 1324 | 855 | 249 | 182 | 0.81 | |
| I−T4r$^+$ | + | 1300 | 575 | 101 | 19 | 0.92 | |
| II−T4r638 | − | 523 | 389 | 103 | 66 | | 0.25 |
| II−T4r638 | + | 305 | 278 | 37 | 0 | | .10 |
| II−T4r$^+$ | − | 239 | 40 | 22 | 28 | | .83 |
| II−T4r$^+$ | + | 199 | 8 | 12 | 4 | | .96 |

Tritiated T4 mRNA was obtained by extraction of *Escherichia coli* B cells that had been infected with bacteriophage T4r⁺ and labeled with uracil-H³ *(2)*. The phage mRNA was then adjusted to high-salt buffer and applied to the top of the column; its passage (18 hours at 60°C) through the entire column was controlled by a continuous-flow pump which delivered a portion of the high-salt buffer to the top of the column at regular intervals of 30 minutes or 1 hour. After removal of unhybridized RNA from the column by extensive washing with high-salt buffer, the three layers were extruded from the column, collected separately, and the two layers containing the DNA's from T4r⁺ and the T4r1272 were repacked into two different columns and eluted with low-salt buffer at 65°C. Table 1 shows the relative amounts of RNA-H³ recovered from the bottom layer. Portions of the RNA fractions thus obtained were incubated with nitrocellulose filters containing 100 μg of either T4r⁺ DNA or T4r1272 DNA *(4)*. RNA specified by the rII genes should be recognized by this test, for this RNA is expected to find no complementary sequences in the DNA of the deletion mutant r1272. As shown in Table 1, more radioactivity was adsorbed by the filter containing T4r⁺ DNA in the first experiment, but not in the other two where either the columns were inverted or an RNA preparation from cells infected with mutant r1272 was applied. Thus the RNA fraction of experiment 1 (first line of Table 1), which was adsorbed to the portion of the column containing the r⁺ DNA, represents some 40 percent of RNA homologous to the DNA region deleted by mutant r1272.

In experiment 1, the purified rII RNA fraction still contains a considerable number of sequences nonspecific for the rII region. In order to trap the nonspecific RNA more effectively in the upper layer of the column, and to minimize its adsorption to the bottom layer, we prepared a column with an upper layer 20 cm long containing 12 mg of T4r1272 DNA and a lower layer, 5 cm in length, containing only 3 mg of T4r⁺ DNA. Passage of a preparation of T4r⁺ RNA through this column produced an RNA fraction which appears to be specific by more than 90 percent for the DNA region absent in mutant r1272, but present in

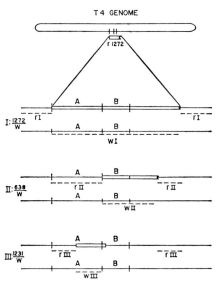

**Fig. 1.** Schematic presentation of the results of the runs described in Table 2. Tritiated RNA isolated from T4r⁺ infected, labeled *E. coli* B cells was passed through column I′ of Table 2. The two RNA fractions obtained, shown as dashed lines, are rI, the fraction eluted by low-salt buffer from the upper layer containing T4r1272 DNA, and wI, that eluted from the lower layer containing T4r⁺ DNA. RNA fraction wI was applied to column II, which yielded RNA fractions rII, eluted from the layer containing T4r638 DNA; and wII, eluted from that containing T4r⁺ DNA. Fraction rII was then applied to column III, and RNA fraction rIII and wIII were obtained from layers containing T4r1231 and T4r⁺ DNA's. The length of extension of the deletions into the cistron to the right of the B cistron is arbitrary for mutant r638, and only approximately to scale for mutant r1272.

T4r⁺ DNA (Tables 2 and 3). This RNA was fractionated further by hybridization with DNA of mutant r638, which lacks cistron B, but possesses cistron A. RNA molecules corresponding to the two cistrons could thus be separated (Table 3 and Fig. 1).

Since the right ends (that is, those distal to the A cistron) of both mutants, r1272 and r638, are genetically not defined [r1272 extends some ten recombination units beyond the B cistron *(7)*], we attempted to purify an RNA fraction not accepted by the DNA of a deletion mutant terminating on both sides within the rII region. For this purpose RNA eluted from the upper layer, containing T4r638 DNA, of the column in run II (638/W) in Fig. 1 was passed through a column containing T4r1231 DNA

TABLE 4
*Filter test of RNA fractions eluted
from column III of Table 2*

The experimental details are as described in Tables 1 and 3. The RNA was not treated with deoxyribonuclease.

| RNA eluted from column | Filters (count/min) | | $\dfrac{r^+ - r1231}{r^+}$ |
|---|---|---|---|
| | $T4r^+$ | $T4r1231$ | |
| III–T4r1231 | 152 | 148 | 0.03 |
| III–T4r$^+$ | 138 | 18 | 0.87 |

and T4r$^+$ DNA. The RNA excluded by the r1231 DNA but bound by the r$^+$ DNA was expected to be equivalent to the right half of the rII A cistron. Hybridization tests with filters containing T4r$^+$ and T4r1231 DNA's confirmed this expectation (Table 4).

When nitrocellulose filters not containing DNA were incubated with RNA eluted from either column, a considerable amount of labeled material resistant to ribonuclease became attached to the filters. Treatment of the RNA with deoxyribonuclease (electrophoretically purified, Worthington) before incubation diminished most of the activity in the control filters. This observation suggests that, upon elution from the DNA-nitrocellulose column, some DNA was released and became adsorbed to the test filters and was then able to bind some of the RNA to the filters.

The high purity of the RNA specific for the rII region (Table 3) suggests that we would have been able to identify RNA specified by a similar-sized region in a genome at least ten times as large as the genome of phage T4. Therefore it seems likely that this deletion method might be successfully applied to the isolation of operon- or even gene-specific bacterial messengers with but little improvement of our technique.

The smallest segment of the T4 genome whose complementary RNA we have succeeded in purifying is that of approximately half a cistron with an estimated length of some 1000 nucleotides. Preliminary evidence indicates that RNA molecules of even smaller sizes can be fractionated almost equally well. If so, the deletion method would provide a unique opportunity to place unequivocally small fragments of messenger molecules on the genetic map according to their sites of transcription. With the availability of such a genetically well-defined system as the rII region in bacteriophage T4, the deletion method meets the first of two principal requirements for the elucidation of the primary structure of an mRNA, namely, the orientation in a linear fashion of RNA fragments of the size of transfer RNA. Whether the deciphering of these fragments, obtainable in extremely small quantities only, will ever become possible, remains to be shown.

**REFERENCES AND NOTES**

1. M. Hayashi, D. Spiegelman, N. C. Franklin, S. E. Luria, Proc. Nat. Acad. Sci. U.S. 49:729 (1963); G. Attardi, S. Naono, J. Rouviere, F. Jacob, F. Gros, Cold Spring Harbor Symp. Quant. Biol. 28:363 (1963); F. Imamoto, N. Morikawa, K. Sato, S. Mishima, T. Nishimura, A. Matsuhiro, J. Mol. Biol. 13:169 (1965).
2. E. K. F. Bautz and B. D. Hall, Proc. Nat. Acad. Sci. U.S. 48:400 (1962).
3. A. P. Nygaard and B. D. Hall, Biochem. Biophys. Res. Commun. 12:98 (1963).
4. D. Gillespie and S. Spiegelman, J. Mol. Biol. 12:829 (1965).
5. Low-salt buffer consists of a 1:200 dilution of high-salt buffer; high-salt buffer consists of 0.3$M$ NaCl and 0.03$M$ sodium citrate.
6. S. Benzer, Proc. Nat. Acad. Sci. U.S. 47:403 (1961).
7. W. Dove, personal communication.

# 22 ASYMMETRIC DISTRIBUTION OF THE TRANSCRIBING REGIONS ON THE COMPLEMENTARY STRANDS OF COLIPHAGE λ DNA

Karol Taylor*
Zdenka Hradecna**
Waclaw Szybalski
McArdle Laboratory
University of Wisconsin
Madison, Wisconsin

---

It was demonstrated in this laboratory that poly G and other guanine-rich polynucleotides show differential affinity for the two complementary strands of various DNA's, indicating asymmetric distribution of poly G-binding sites.[1,2] Furthermore, it was postulated that these sites, probably deoxycytidine(dC)-rich clusters, might act as the initiation points of the DNA-to-RNA transcription.[2,3] For DNA which contains dC-rich clusters on *both* strands, as for instance coliphage λ DNA (Fig. 1),[3,4] this hypothesis predicts that transcribing regions would be found on *both* strands. As will be shown, this prediction is confirmed for coliphage λ, which provides the first example of *in vivo* transcription from *both* DNA strands, as documented by DNA-RNA hybridization techniques.[3,5] This result agrees with the conclusions based on genetic experiments with λ phage.[6,7] In earlier studies employing other phages, only *one* DNA strand was found to hybridize with phage-specific mRNA.[8]

## Materials and methods

### Bacterial and phage strains

*Escherichia coli* K12 strains included C600, which is permissive for λ*sus* mutants, and W3110 and W3350, which are nonpermissive for *sus* mutants.[9] These were lysogenized or infected with appropriate λ mutants as listed in Table 1. Most of the λ$c_I$, λdg, and λ*sus* mutants and the lysogenic strains were obtained from Drs. W. F. Dove, H. Echols, A. Joyner, D. Pratt, M. Ptashne, and J. Adler. Strains T75[10] and T11 [=W3350-(λt11)][7] were contributed by Dr. C. R. Fuerst. The subscript A-J indicates that genes A to J (entire left arm; Fig. 1) were deleted in λdg$_{A-J}$ and replaced by a part of the galactose operon.[11] Biotin genes were substituted for deleted genes a-N or a-O in λdb$_{a-N}$ (=λt75) and λdb$_{a-O}$, respectively, the latter contributed by Dr. G. Kayajanian.

The cultivation of bacteria, infection or induction of phages, preparation of phage stocks, and purification of phages by high-low speed sedimentation and CsCl density gradient centrifugation followed the published procedures.[4,6,7,9-13] The lysogenic cultures were induced by addition of 2 μg mitomycin C/ml.[12]

### Pulse-labeling and isolation of bacterial and λ-specific RNA

The method employed followed closely the techniques described by Sly *et al.*[12] A total of 0.5 mc of $H^3$-uridine (8 c/mM) was added to 20 ml of the

---

From Proceedings of the National Academy of Sciences, U.S.A. **57**:1618-1625, 1967. Reprinted with permission.

The following abbreviations are employed: poly G, homopolymer of guanylic acid; poly IG, copolymer of guanylic and inosinic acids; CM, chloramphenicol; DNase, pancreatic deoxyribonuclease I; RNase, pancreatic ribonuclease; SSC, 0.15 M NaCl + 0.02 M trisodium citrate, pH = 7.4; 6×SSC, 2×SSC, 1/100×SSC, 6 or 2 times more concentrated SSC, or 100 times diluted SSC, respectively; mRNA, messenger RNA which operationally is that fraction of pulse-labeled RNA which specifically hybridizes with λ DNA (=λ mRNA); C and W strands, complementary strands of λ DNA which, when unbroken, exhibit a higher or lower affinity for poly IG and thus band at higher or lower density in the CsCl gradient, respectively (Fig. 1); SDS, sodium dodecyl sulfate; G, guanine; C, cytosine; A, adenine; T, thymine; *sus*, suppressor-sensitive mutation; dg, defective galactose transducing; db, defective biotin transducing.

We are greatly indebted to Dr. W. F. Dove, who guided us through many intricacies of λ genetics, and to Dr. H. Echols and his students and collaborators for their contributions of strains and technical advice. Drs. J. Adler, G. Kayajanian, E. Calef, A. D. Kaiser, M. Ptashne, and C. R. Fuerst also supplied us with strains. Their helpful advice and that of Drs. R. Thomas, D. K. Fraser, F. Gros, D. S. Hogness, C. M. Radding, S. N. Cohen, A. M. Skalka, A. D. Hershey, and L. H. Pereira da Silva, including correspondence and manuscripts, are also gratefully acknowledged. We are very thankful to Drs. W. F. Dove, H. Echols, R. Thomas, M. Susman, and E. H. Szybalski for critical reading of the manuscript and editorial help.

These studies were supported by a grant from the National Science Foundation (B-14976).

*On leave from the Institute of Marine Medicine, Gdańsk, Poland.

**On leave from the Biophysics Institute, Brno, Czechoslovakia.

culture, which 2 min later was poured onto an equal volume of crushed ice prepared from minimal medium. This was followed by rapid sedimentation of the bacteria at 4°C, lysis by 2% sodium dodecyl sulfate (SDS), and phenol extraction of the RNA at 60°C. The upper aqueous phase was used for the "prehybridization" with λ DNA (see next section), after determination of the acid-precipitable radioactivity (total $H^3$-labeled RNA; see Table 1).

## DNA-RNA hybridization procedure

The hybridization procedure was based on the technique described by Gillespie and Spiegelman[14] with an additional "prehybridization" step.

*(a) Prehybridization.* Denatured λ DNA (50 μg), "baked" on a 25-mm B-6 filter (Schleicher & Schuell Co.),[14] was incubated (24 hr, 60°C) with 1 ml phenol-saturated 2×SSC containing homologous $H^3$·RNA (total 100-500 μg RNA). The filters were exhaustively washed with 2×SSC after this and each following step. The nonspecifically bound RNA was digested with RNase (1 hr, 20°C, 4 ml of 2×SSC + 20 μg RNase/ml). The residual RNase was inactivated by incubation (40 min, 55°C) of the washed filters in 2 ml of 0.15 M iodoacetate at pH 5 (0.1 M Na-acetate buffer, 2×SSC). The $H^3$·RNA was eluted with 1.5 ml of 1/100×SSC (15 min, 95°C, 90% recovery of $H^3$ count), and treated with RNase-free (iodoacetate pretreated[16]) DNase (10 μg DNase [Worthington Co.]/ml of 0.05 M Tris, 0.004 M $MgCl_2$, pH 7.4, 20 min, 37°C), which was then inactivated by 10 min heating to 95°C, as recommended by Skalka (personal communication).

*(b) Hybridization.* Denatured DNA (25 μg) or the separated λ DNA strands (3-10 μg) "baked" on a 25-mm filter[14] were incubated (24 hr, 60°C, 1 ml phenol-saturated 2×SSC) with 1,000-10,000 cpm of λ mRNA eluted from the filter in the prehybridization procedure. Following the RNase treatment (see prehybridization) and extensive washing with 2×SSC, the filters were dried (2 hr, 60°C) and transferred to toluene-2,5-diphenyloxazole (toluene-PPO) scintillation fluid for the $H^3$ count. An excess of hybridization sites for binding up to 2,000 cpm of "prehybridized" mRNA is provided by 3 μg of separated DNA strands, since the $C:W$ ratios are unaffected by raising the DNA quantity to 10 or 25 μg per filter. All hybridization values are corrected for $H^3$·RNA counts bound by filters carrying denatured T4 coliphage DNA.

## Preparative separation of the complementary DNA strands

To effect strand separation, the DNA is released, denatured, and reacted with poly IG, all three operations in a single step.[4] When the phage suspension (50 μg DNA) is heated (2 min at 90°C) in 0.5 ml $10^{-3}$ M sodium versenate (EDTA) containing 100 μg poly IG and 0.1% Sarkosyl NL 97 (Geigy Chemical Corp., New York), and is chilled and centrifuged in a polyallomer tube (2.5 ml of CsCl solution, 1.72 $gm/cm^3$, 70 hr, 15°C, 30,000 rpm SW39 rotor), the released and denatured DNA is distributed into two symmetrical bands separated by 12-14 $mg/cm^3$. Of the two complementary DNA strands, the one with the relatively higher content of poly IG-binding, dC-rich clusters[3,4] (strand C) is found in the denser band, while the other (strand W) forms the less dense band (Fig. 1). After collecting the separate fractions, each of the two pooled samples containing one kind of λ DNA strand was separately subjected to self-annealing (4 hr, 65°C, 5 M CsCl)[4] to convert any of the contaminating opposite strands to double helices, which would be inactive in the DNA-RNA hybridization procedure. The functional purity of such "self-annealed" strands was over 99% as tested by preparing $H^3$·mRNA specific for one strand only (cf. (a) *Prehybridization* in *Materials and methods*) and comparing its affinities for the two strands. The mRNA prehybridized and eluted from the C strands hybridized with strands C over 99 times more efficiently than with strands W (99.6:0.4 for $C:W$). An analogous result was obtained for strand W-specific mRNA or for mRNA produced by the λt11 mutant (Table 1, line 11). Proof for the integrity of the separated strands at over 85% level, their properties, orientation (Fig. 1), and the details of the strand separation procedure for λ DNA were published by Hradecna and Szybalski.[4]

**Fig. 1.** Genetic map of phage λ, including A to R *sus* markers,[9] "clear" markers $c_I$, $c_{II}$, and $c_{III}$,[24] central $b_2$ region,[25,26] markers x and y,[7] and marker a,[27] all superimposed over λ DNA.[6,7] The base compositions (% G + C) of both arms of λ DNA and of the central $b_2$ region are indicated.[6,25] 5'G and 5'A identify the 5' terminal nucleotides[23] and the polarity of the C and W strands.[4] Symbol C ("*DENSE*") indicates the DNA strand which is denser in the poly G-containing CsCl gradient (and "lighter" in the alkaline CsCl gradient[4,6]) than strand W.[2-4] The arrows (mRNA) indicate the orientation, the region, and the strand of preference for the DNA-to-RNA transcription, as discussed in this paper. The distribution of cytosine-rich clusters is indicated by the symbols (-I-), and is based on the data of Hradecna and Szybalski.[4]

## Results and discussion
### Temporal control of transcription

Shortly after infection with $\lambda c_{72}$ phage, the amount of λ-specific mRNA rises rapidly, as previously shown by Sly et al.[12] This "early" λ-mRNA hybridized preferentially with the W strand (Fig. 1) of λ DNA (5:95 for $C:W$, Table 1, line 3, and Fig. 2A). At 30 or 40 minutes after infection, the preference for the transcription shifted to the C strand (85:15 for $C:W$, Table 1, lines 4 and 5; and Fig. 2A). It appears, therefore, that the transcription from strand C

### TABLE 1
Percentage of λ-specific $H^3$-labeled RNA synthesized by various λ mutants and the proportion of this mRNA hybridizing with complementary strands C and W

| Expt. no. | mRNA donor* (host; phage) | Induction,† infection CM(μg/ml)‡ | $H^3$-uridine pulse† (min) | Hybridized with separated strands Per cent of total, § $H^3$-labeled RNA (λmRNA) | | | Ratio (%) | |
|---|---|---|---|---|---|---|---|---|
| | | | | C + W | C | W | C | W |
| 1 | W3110 | None | 2 | 0.001 | — | — | — | — |
| 2 | W3350 | None | 2 | 0.002 | 0.001 | 0.001 | 50 | 50 |
| 3 | W3110; $\lambda c_{72}$ | Inf. | 0-2 | 2.0 | 0.1 | 1.9 | 5 | 95 |
| 4 | W3110; $\lambda c_{72}$ | Inf. | 28-30 | 8.4 | 7.1 | 1.3 | 85 | 15 |
| 5 | W3350; $\lambda c_{72}$ | Inf. | 38-40 | 8.1 | 6.9 | 1.2 | 85 | 15 |
| 6 | W3350; $\lambda c_{72}$ | Inf. | 38-40 | 8.0 left arm | 7.8 | 0.2 | 97 | 3 |
| 7 | W3350($\lambda^+$) | Ind. CM(100) | 58-60 | 0.5 | 0.05 | 0.45 | 10 | 90 |
| 8 | W3350($\lambda^+$) | Ind. CM(40) | 58-60 | 0.7 | 0.16 | 0.54 | 23 | 77 |
| 9 | W3350($\lambda dg_{A-J}$) | Ind. CM(100) | 58-60 | 0.4 | 0.08 | 0.32 | 20 | 80 |
| 10 | W3350($\lambda dg_{A-J}$) | Ind. | 58-60 | 2.3 | 1.8 | 0.50 | 78 | 22 |
| 11 | W3350($\lambda t11$) | Ind. | 58-60 | 0.9 | 0.005 | 0.89 | 0.5 | 99.5 |
| 12 | W3350($\lambda t11$) | Ind. CM(50) | 58-60 | 0.6 | 0.018 | 0.58 | 3 | 97 |
| 13 | W3350($\lambda t11$) | None | 2 | 0.03 | 0.005 | 0.025 | 16 | 84 |
| 14 | W3350($\lambda ind^-$) | None | 2 | 0.04 | 0.003 | 0.037 | 8 | 92 |
| 15 | W3350($\lambda ind^-$) | None | 2 | 0.005 λi 434 | 0.003 | 0.002 | 56 | 44 |
| 16 | W3350($\lambda ind^-$) | None | 2 | 0.004 λi 21 | 0.002 | 0.002 | 47 | 53 |
| 17 | W3350($\lambda^+$) | None | 2 | 0.05 | 0.01 | 0.04 | 18 | 82 |
| 18 | W3350($\lambda^+$) | Ind. | 58-60 | 3.4 | 2.9 | 0.48 | 86 | 14 |
| 19 | W3350($\lambda^+$) | Ind. CM(50) | 58-60 | 0.6 | 0.1 | 0.5 | 16 | 84 |
| 20 | T75($\lambda db_{a-N}$) | None | 2 | 0.04 | 0.007 | 0.033 | 18 | 82 |
| 21 | W3110b($\lambda db_{a-O}$) | None | 2 | 0.02 | 0.012 | 0.008 | 59 | 41 |
| 22 | W3350($\lambda N_{53}$) | None | 2 | 0.05 | 0.006 | 0.044 | 13 | 87 |
| 23 | T75($\lambda db_{a-N}$) | Ind. | 58-60 | 0.04 | 0.027 | 0.013 | 68 | 32 |
| 24 | T75($\lambda db_{a-N}$) | Ind. CM(50) | 58-60 | 0.06 | 0.017 | 0.043 | 28 | 72 |
| 25 | W3110b($\lambda db_{a-O}$) | Ind. | 58-60 | 0.02 | 0.011 | 0.009 | 57 | 43 |
| 26 | W3350($\lambda N_7$)** | Ind. | 58-60 | 0.6 | 0.12 | 0.48 | 21 | 79 |
| 27 | W3350($\lambda N_7$)** | Ind. CM(50) | 58-60 | 0.4 | 0.09 | 0.31 | 22 | 78 |
| 28 | W3350($\lambda O_{29}$) | Ind. | 58-60 | 0.5 | 0.22 | 0.28 | 44 | 56 |
| 29 | W3350($\lambda O_{29}$) | Ind. CM(50) | 58-60 | 0.4 | 0.08 | 0.32 | 20 | 80 |
| 30 | W3350($\lambda P_{80}$) | Ind. | 58-60 | 0.6 | 0.34 | 0.26 | 57 | 43 |
| 31 | W3350($\lambda P_{80}$) | Ind. CM(50) | 58-60 | 0.6 | 0.11 | 0.49 | 19 | 81 |

*In expts. 1 and 2, bacteria free of λ phage were used. Infection is indicated by the semicolon, whereas the lysogenic state is indicated by parentheses.

†The bacteria were infected (Inf.) at multiplicity 5 (0°C, $10^{-2}$ M $Mg^{++}$), and after 20 min for absorption (0°C) transferred to growth medium[19] (37°C) (= zero time). Induction (Ind.) was initiated by adding 2 μg mitomycin C/ml of growth medium.[12] The noninduced lysogens (None) were pulse-labeled for 2 min during the exponential growth phase at a cell concentration of $4 \times 10^8$/ml.

‡Chloramphenicol (CM) was present from 10 min before induction to the end of the $H^3$-uridine pulse.

§Percentage of total $H^3$ RNA which was prehybridized (50 μg $\lambda cb_2$ DNA or other DNA if so specified in the column), eluted from the filter, and hybridized again with 25 μg denatured $\lambda cb_2$ DNA (C + W), or with 3 μg of self-annealed C and W strands of $\lambda cb_2$ DNA per filter (see Materials and methods). In expt. 6, $H^3$·RNA was prehybridized with the "short left arm" of $\lambda cI$ DNA.

**Similar results were obtained with $\lambda N_7 N_{53}$ double mutant.[20]

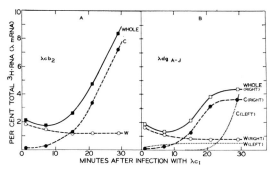

**Fig. 2.** Percentage of $H^3$-labeled RNA (2-min pulse) specific for $\lambda cb_2$ (A; ■) and $\lambda dg_{A-J}$ (B; □), and for the C (●) and W (○) strands of these DNA's, at various times after infection of *Escherichia coli* W3110 with $\lambda c_{72}$ (see second footnote in Table 1, and *Materials and methods;* pH of growth medium, 7.7). RNA was prehybridized with 50 μg denatured DNA of $\lambda cb_2$ (A) or $\lambda dg_{A-J}$ (B), treated with RNAase and iodoacetate (see *Materials and methods*), eluted, and hybridized with 25 μg of denatured $\lambda cb_2$ DNA (WHOLE, solid line) and with 3 μg of the separated C and W strands of $\lambda cb_2$. The dashed-line values, which are fractions of the solid-line values proportional to the per cent of C or W, represent the amount of mRNA transcribed from the whole C or W strand (A) or from the right arms (Fig. 1) of the C or W strand (B). The dotted line C (left) represents the difference between the C and C (right) values, and indicates the kinetics of transcription from the left arm (Fig. 1) of the C strand. The dotted line W (left) represents the difference between the W and W (right) values.

rises sharply during the development of infectious phage, while the strand W-specific mRNA is synthesized at an almost constant rate (Table 1, line 3 versus 4, and 7 versus 18; Fig. 2A).

### Chloramphenicol (CM) effects

Addition of CM before induction or infection with phage does not grossly affect the initial rate of λ-mRNA synthesis, but at all times such RNA retains the characteristics of early mRNA.[17] Temporally, we define the early λ-mRNA as that which could be produced in the absence of protein and DNA synthesis; it will be shown later that it is probably transcribed from the N-a and x-O regions. This definition is probably more restrictive than that used earlier.[12,13,21,22] It appears essential to use at least 100 μg CM/ml for preparing early λ-mRNA, since the transcription from strand C preferentially increases at lower CM concentrations (Table 1, lines 7, 8, and 19).

### Localization of the "switch sites"

Transcription of both the W and C strands implies the presence of sites at which the λ-mRNA synthesis switches from one strand to another, either *converging* or *diverging* from such a *switch site*. A major *switch site* should be on the right arm of the DNA, since both the early (+CM) and late mRNA's produced by *E. coli* lysogenic for $\lambda dg_{A-J}$, in which the whole left arm is deleted[11] (Fig. 1), hybridize with both DNA strands (Table 1, lines 9 and 10). It was previously shown that early mRNA hybridizes preferentially with the right arm of λ DNA.[18]

Employing prehybridization of $H^3$-labeled RNA with $\lambda dg_{A-J}$ DNA (Fig. 1) at various times after $\lambda c_I$ infection, it was possible to follow the kinetics of transcription on the right arm only, for both the C and W strands (Fig. 2B). The transcription on the right arm of strand C increases rapidly up to the 20th minute after infection and levels off around the 24th minute, whereas the transcription on the left arm of strand C commences at about the same time and rises sharply (Fig. 2B, C(right) and C(left)). The kinetics of transcription for the right arm of strand W (Fig. 2B; W(right)) are similar to those for the whole strand W (Fig. 2A; W), although the small and possibly fortuitous differences between these values might be construed as indicative of some transcription from the left arm of strand W (Fig. 2B; W(left)). However, it was found that mRNA prehybridized with the purified left arm of $\lambda c_{72}$ DNA and thus containing only the mRNA transcribed by the left arm, hybridized almost exclusively with the C strand (Table 1, line 6). Thus, there is little reason to postulate any transcription from the left arm of strand W.

Where on the right arm of λ DNA is the "switch site" localized, and is it a *divergency* or *convergency* site? The most direct answer is provided by an experiment with mutant λt11, in which a polar mutation in gene x inactivates the x-to-O functions[7] and at the same time blocks the transcription from strand C (0.5:99.5 for C:W; Table 1, line 11). This correlation indicates that genes x-to-O are transcribed from strand C, i.e., from left to right. If strand-W-specific mRNA is the product of region N to a, and thus is transcribed from right to left, a *divergency site* must be located between genes N and $x$ (Fig. 1). These interpretations are based on the known polarities of strands $C$ and $W$ (Fig. 1) and on the 5'-to-3' orientation of RNA synthesis in conjunction with work on heteroduplexes in gene N.[6,15] Since left-arm-specific mRNA seems to be transcribed from strand $C$, i.e., in the left-to-right direction, a *convergency site* should be located somewhere near or within region $b_2$ (Fig. 1). The products of the $b_2$ region are not included in this study, since $\lambda cb_2$ DNA with the $b_2$ region deleted (Fig. 1) was used either for prehybridization or for the preparation of separated DNA strands. However, similar studies on the transcribing function of the $b_2$ region are currently being pursued in this laboratory.

*Transcription from the prophage*

To further localize the position of the major switch, it is necessary to determine the orientation of the transcription from gene $c_I$. The W3350($\lambda$ind$^-$) lysogenic strain[19] was used for this purpose, since its spontaneous induction rate is very low, and thus its mRNA may consist mainly of the gene $c_I$ product. Transcription rate from the noninduced $\lambda$ind$^-$ prophage (0.04% of total $H^3 \cdot$RNA) is up to 40 times higher than the level obtained with $\lambda$-free *E. coli* (Table 1, lines 1, 2, and 14). This mRNA hybridizes preferentially with the $W$ strand of $\lambda cb_2$ (0.037%; 8:92 for $C:W$), and has hardly any affinity for strand $C$ (0.003%; Table 1, line 14) or for either DNA strand of $\lambda$ hybrids with a heterologous $c_I$ region (0.002-0.003%; Table 1, lines 15 and 16). The levels of $\lambda$-specific mRNA for several noninduced prophages were within relatively narrow limits (0.03-0.05%; Table 1, lines 13, 17, 20, and 22; see also Sly *et al.*[12]); all these mRNA's exhibited strong preference for strand $W$, with the exception of mRNA produced by the $\lambda db_{a-0}$ prophage, in which gene $c_I$ is deleted and which transcribes poorly both before and after induction (Table 1, lines 21 and 25). These data indicate that (1) in the noninduced state the bulk of $\lambda$-mRNA is the product of gene $c_I$, and that (2) this is transcribed from strand $W$, i.e., from right to left, which conclusion is similar to that based on the polar effects of the *sus*34 mutation located at the right end of gene $c_I$.[20] Furthermore, these data suggest that the $\lambda$ repressor is a bifunctional protein (as represented by regions $A$ and $B^{22}$), which interacts with two operator regions (deleted in $\lambda i434$), one adjoining the left end of the $c_I$ gene and controlling the N-$a$ transcription from the $W$ strand and the other located next to the right terminal of gene $c_I$ and controlling the $x$-O transcription from the $C$ strand. According to this model, the lack of host inactivation upon transient thermal induction of the $c_I$(B) prophage in the presence of CM (Lieb and Green[22]) can be explained by partial renaturation of the repressor with restoration of only one ($x$-O repression) of its two functions; temporary expression of genes N-to-$a$ apparently has no lethal consequences.

*Controls of early and late transcription*

In the noninduced state, $c_I$ mRNA seems to be the main transcription product ($W$ strand). Immediately after infection or "early" (+CM) upon induction, the total transcription increases by factors of 10-40, with strand $W$ being predominantly transcribed, probably in the N-$a$ and still in the $c_I$ regions. Strand $C$ is transcribed to a lesser extent (5:95 to 10:90 for $C:W$), most probably from the $x$-$y$-$c_{II}$-O operon, since the polar mutation in gene $x$ abolishes this early transcription from strand $C$. The fact that CM freezes this early stage of transcription indicates that (1) the N-$a$ and $x$-O transcriptions are not mutually dependent on their protein products, and (2) that there is a need for some proteins to extend the transcription to other genes.

Which gene products are necessary for the shift to the later stages of transcription? At least two proteins, those missing in the N (as postulated by Thomas[21]) and $x$ mutants, appear to be required; a low level of $C$-specific mRNA, characteristic of the early transcription pattern, persists upon induction of both the $x$ and N mutants (Table 1, lines 11, 19, and 26). This pattern changes somewhat for the induced *sus*O and P prophages.[12,13] The "late" transcription from strand $C$ for both the O and P mutants increases threefold when compared with the early transcription in the presence of CM (Table 1, lines 28-31); the analogous increase in $C$-specific transcription for the $\lambda^+$ is 30-fold and for the right arm of $\lambda^+$ or for $\lambda dg_{A-J}$ is approximately 20-fold (Table 1, lines 9, 10, 18, 19; Fig. 2B). These results indicate either a "leaky" character for the O and P mutations, or, more probably, a shift to the next stage of transcription from strand C.

Is it necessary to invoke any special control mechanisms for the early versus late transcription from strand W, which seems to be limited to the $c_I$ and N-to-$a$ regions? In the absence of DNA synthesis the differences between the early and late transcription from the W strand are small for the $x$, N, O, and P mutants (Table 1, lines 11, 12, 26-31). During the normal infection (Fig. 2) or induction process (Table 1, lines 18, and 19), the progressive decrease in synthesis of the strand W-specific mRNA also appears comparatively small. This small decrease, however, would become quite precipitous if the transcription rates were divided by the corresponding numbers of λ DNA copies in the vegetative pool. These results indicate that (1) only the *parental* W strands are transcribed and at a relatively steady rate, or that (2) strands W are transcribed in *all* λ DNA molecules at a rate which is controlled by the limiting amount of the "early" RNA polymerase. In the absence of DNA synthesis, the *parental* W strands are continuously transcribed, as shown for the *sus*N, O, P and *t*11 mutants (Table 1, lines 11-12, 26-31).

Late during induction, the transcription from gene $c_I$ seems to be repressed, as represented by the three- to fourfold decrease in the synthesis of W-specific λdb$_{a-N}$ mRNA (Table 1, lines 20, 23, and 24); only gene $c_I$ should be transcribed from strand W of λdb$_{a-N}$, since genes $a$-N are deleted.

Several recent papers discuss the control of induction and development of phage λ.[6,7,9,10,12,13,17,18,20-22,25]

## Conclusions and summary

Hybridization of the various λ-specific mRNA's with the separate strands of λ DNA provides a powerful new technique for determining the distribution of the transcribing regions. Several technical refinements were introduced, including (1) self-annealing of the separated DNA strands, which results in preparations of individual strands displaying a purity of over 99 per cent in hybridization tests, and (2) a highly selective two-step hybridization procedure, with prehybridization including RNAse treatment followed by inactivation of RNase by iodoacetate. With the latter method it was possible to compare the levels of λ-specific mRNA in nonlysogenic *E. coli* (0.001-0.002%), in noninduced lysogens (0.03-0.06%), and early (0.5-2.0%) or late (6-12%) in induced nondefective lysogens or in infected cells, with the simplified ratio of these figures being represented as 1 (nonlysogenic):50 (noninduced):500 ("early" induced):5000 ("late" induced).

In the noninduced state the majority mRNA is transcribed from the W strand, most probably being the product of gene $c_I$. Upon induction or infection two regions adjoining the $c_I$ gene start to be transcribed: the predominant product (90%) is copied from strand W in the same direction as gene $c_I$, through genes N-to-$a$, whereas the minority mRNA (10%) is copied in the opposite direction, i.e., from left-to-right, from strand C through the $x$-O operon. Both of these transcription and also translation products seem to be required to activate the further transcription of the λ genome, since the early transcription pattern could be frozen either by inhibition of protein synthesis (100 μg CM/ml) or by the nonsense or polar mutations in genes N or $x$. Within 10-20 minutes after infection the transcription changes to the "late" pattern, with over 85 per cent of the mRNA now being transcribed from the C strand, progressively more and more from its left arm. Thus, during the development of λ, the transcription of strand W decreases by only 10-30 per cent, whereas the transcription of strand C increases 30- to 70-fold. Transcription in the induced *sus*O and P lysogens, unable to synthesize λ DNA, does not proceed far beyond the early stage, since upon removal of CM the transcription from strand C increases only threefold. The hybridization pattern obtained with the λdg$_{A-J}$ mutant, which transcribes both DNA strands although its left arm is entirely deleted, confirms that a *divergency switch* in the direction of mRNA synthesis is on the right arm of λ DNA. It is interesting to note that the segments of the individual DNA strands presently characterized as transcribing regions were independently shown to contain all the dC-rich clusters, which have been postulated by Szybalski *et al.*[3] to act as initiation points for the DNA-to-RNA transcription process. Thus, at the present level of resolution the asymmetry in the distribution of the poly IG-binding dC clusters, which permits the preparative separation of the complementary DNA strands, seems to be directly related to the asymmetric transcription pattern of mRNA and the changes in the orientation of this transcription.

**REFERENCES AND NOTES**

1. Opara-Kubinska, Z., H. Kubinski, and W. Szybalski, these Proceedings **52**:923 (1964).
2. Kubinski, H., Z. Opara-Kubinska, and W. Szybalski, J. Mol. Biol. **20**:313 (1966).
3. Szybalski, W., H. Kubinski, and P. Sheldrick, in Cold Spring Harbor Symposia on Quantitative Biology **31**:123 (1966).
4. Hradecna, Z., and W. Szybalski, Virology, in press; Hradecna, Z., K. Taylor, and W. Szybalski, Bacteriol. Proc. (1967), p. 27.
5. Taylor, K., Z. Hradecna, and W. Szybalski, Federation Proc. **26**:449 (1967); Szybalski, W., Z. Hradecna, and K. Taylor, Abstracts, 7th International Congress of Biochemistry, Tokyo, in press.
6. Hogness, D. S., W. Doerfler, J. B. Egan, and L. W. Black, in Cold Spring Harbor Symposia Quantitative Biology **31**:129 (1966).
7. Eisen, H. A., C. R. Fuerst, L. Siminovitch, R. Thomas, L. Lambert, L. Pereira da Silva, and F. Jacob, Virology **30**:224 (1966).
8. Tocchini-Valentini, G. P., M. Stodolsky, A. Aurisicchio, M. Sarnat, F. Graziosi, S. B. Weiss, and E. P. Geiduschek, these Proceedings **50**:935 (1963); Marmur, J., and C. M. Greenspan, Science **142**:387 (1963); Hayashi, M., M. N. Hayashi, and S. Spiegelman, these Proceedings **50**:664 (1963).
9. Campbell, A., Virology **14**:22 (1961); Kayajanian, G., and A. Campbell, Virology **30**:482 (1966).
10. Fuerst, C. R., Virology **30**:581 (1966).
11. Adler, J. and B. Templeton, J. Mol. Biol. **7**:710 (1963).
12. Sly, W. S., H. Echols, and J. Adler, these Proceedings **53**:378 (1965); Joyner, A., L. N. Isaacs, H. Echols, and W. S. Sly, J. Mol. Biol. **19**:174 (1966); Echols, H., B. Butler, A. Joyner, M. Willard, and L. Pilarski, in Edmonton Symposium on Molecular Biology of Viruses (New York: Academic Press, 1967), p. 125.
13. Dove, W. F., J. Mol. Biol. **19**:187 (1966).
14. Gillespie, D., and S. Spiegelman, J. Mol. Biol. **12**:829 (1965).
15. Experiments employing prehybridization with DNA from the $\lambda db_{a\text{-}N}$ and $\lambda db_{a\text{-}O}$ mutants are in progress. Preliminary results are consistent with the assignment of the W-specific transcription to the a-N region.
16. Zimmerman, S. B., and G. Sandeen, Anal. Biochem. **14**:269 (1966).
17. Naono, S., and F. Gros, J. Mol. Biol. (in press); and in Cold Spring Harbor Symposia on Quantitative Biology **31**:363 (1966).
18. Skalka, A., these Proceedings **55**:1190 (1966).
19. Jacob, F., and A. Campbell, Compt. Rend. **248**:3219 (1959).
20. Ptashne, M., these Proceedings **57**:306 (1967), and personal communication.
21. Thomas, R., J. Mol. Biol. **22**:79 (1966).
22. Green, M., J. Mol. Biol. **16**:134 (1966); and in Edmonton Symposium on Molecular Biology of Viruses (New York: Academic Press, in press); Lieb, M., J. Mol. Biol. **16**:149 (1966); Kourilsky, P., and D. Luzzati, J. Mol. Biol., in press; Cohen, S. N., U. Maitra, and J. Hurwitz, J. Mol. Biol., in press.
23. Wu, R., and A. D. Kaiser, these Proceedings **57**:170 (1967).
24. Kaiser, A. D., and F. Jacob, Virology **4**:509 (1957); Kaiser, A. D., Virology **3**:42 (1957); Isaacs, L. N., H. Echols, and W. S. Sly, J. Mol. Biol. **13**:963 (1965).
25. Hershey, A. D., in Carnegie Institution of Washington Year Book 65 (1966), p. 559.
26. Kellenberger, G., M. L. Zichichi, and J. Weigle, Nature **187**:161 (1960).
27. Gene a (called also int) controls the attachment and integration of the λ genome at a vacant bacterial site; Gingery, R., and H. Echols, personal communication; Zissler, J., Virology **31**:189 (1967).

# 23 ISOLATION OF THE λ PHAGE REPRESSOR

*Mark Ptashne*
*Department of Biology*
*Harvard University*
*Cambridge, Massachusetts*

A bacterium can carry within it, integrated in its chromosome, the genome of a potentially lethal phage (prophage), because the genes of the prophage are prevented from functioning by a repressor. This lysogenic bacterium, as it is called, will lyse and produce phage if the repressor is inactivated. This same repressor, which is made by a gene on the prophage, is also responsible for the immunity of lysogenic cells to superinfection by phages similar to the prophage. Such superinfecting phages inject their DNA but the newly introduced phage genes neither function nor replicate.

These facts were elucidated largely by studies of the phage λ which grows on *E. coli*. In 1957, Kaiser and Jacob showed that the prophage gene $C_1$, required for the maintenance of lysogeny, is also responsible for the immunity against superinfecting phages.[1] On the basis of this and other observations, Jacob and Monod[2] proposed that the $C_1$ gene produces a repressor molecule (often referred to as the immunity substance) which blocks lytic development of both prophage and superinfecting phages by selectively repressing the expression of one or more phage genes. This is exactly analogous to their model for the action of the *i* gene of the lactose operon. Although both the *lac* and λ repressors have been the objects of extensive studies *in vivo*, their mechanism of action is unknown. Only recently has a repressor, the *lac* repressor, been detected *in vitro*.[3] This paper describes the specific labeling and partial purification of the $C_1$ product of phage λ.

### Preliminary results and considerations

In an ordinary λ-lysogen, phage repressor synthesis probably constitutes on the order of only one part in $10^4$ of the cell's total protein synthesis. In order to label the λ phage repressor specifically, I sought conditions in which the rate of synthesis of repressor is a significant fraction (5-10%) of the total protein synthesis of the cell. To achieve the necessary differential increase in repressor synthesis, I attempted (1) to inhibit the synthesis of host proteins while maintaining the capacity to synthesize phage proteins; (2) to inhibit the synthesis of most or all of the phage proteins other than that of the repressor; and (3) to maximize the number of functioning λ repressor genes in the cell. The first objective was achieved by irradiating cells with massive doses of ultraviolet (UV) light, a treatment which damages the host DNA and dramatically decreases the level of cellular protein synthesis. Figure 1 shows that lytic infection of phage in sufficiently damaged cells stimulates about tenfold the incorporation of labeled leucine. This incorporation presumably represents the synthesis of numerous phage proteins, only one of which is the $C_1$ product. The second objective is achieved in principle by infecting λ-lysogens with other λ phages of the same immunity. Although most of the genes of these superinfecting λ phage chromosomes are repressed, one newly injected gene which *does* function is the $C_1$ gene itself.[4] The third objective might be achieved simply by increasing the multiplicity of λ phages infecting the lysogenic cell. However, at multiplicities above 10-15, immunity to superinfection breaks down, and phage proteins other than the repressor are synthesized. I have found that in

---

From Proceedings of the National Academy of Sciences U.S.A. 57:306-313, 1967. Reprinted with permission.

Throughout the course of this work I have had the advice and encouragement of Walter Gilbert and James Watson, for which I am very grateful. I would also like to thank Matthew Meselson for providing laboratory space and many facilities necessary for these experiments; Nancy Hopkins and Louise Rogers for excellent assistance; and Margaret Lieb, Racquel Sussman, and François Jacob for phage mutants.

These experiments were performed while the author was a Junior Fellow of the Society of Fellows, Harvard University, and were supported by grants from the National Science Foundation and the National Institutes of Health.

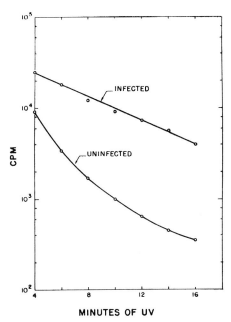

**Fig. 1.** Stimulation of labeled leucine incorporation into irradiated cells by lytic phage infection. The cells were grown and irradiated as described in *Materials and methods*. One-ml portions were labeled with $\frac{1}{2}\mu c$ $C^{14}$-leu for 1 hr with or without added phage at a multiplicity of 30 phage/cell. The cells are lysogenic for $\lambda ind^-$. In this particular experiment the infecting phage is $\lambda imn^{434}$, a phage similar to $\lambda$ but of different immunity which grows normally on $\lambda$-lysogens. The incorporation was terminated by the addition of TCA. The precipitated cells were collected and washed on a Millipore filter and counted in a gas-flow counter. These curves do not extrapolate to the correct unirradiated values. The uninfected cell synthesis is decreased about 5,000-fold at 14 min of irradiation.

NAMES AND APPROXIMATE MAP
POSITIONS OF $\lambda C_1$ MUTANTS

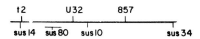

**Fig. 2.** Mutants shown above the line are temperature sensitives *(C₁ts)*, and those below the line are ambers *(C₁sus)*. Mutants $C_1sus34$ and $C_1sus14$ are located at the extremities of the $C_1$ gene. Mutant $C_1sus80$ is known to lie only between $C_1sus14$ and $C_1ts857$. This map is a composite of two maps constructed by Drs. M. Lieb and F. Jacob. (These $C_1sus$ mutants should not be confused with amber mutants of the same *sus* numbers located in other $\lambda$ genes.)

lytic infection the production of many of the proteins of phage $\lambda$, perhaps all of them except for the $C_1$ product, is blocked if the phage carries a mutation in the early gene N, a result consistent with results of others.[5] Phages carrying mutations in gene N fail to stimulate the large increase in labeled leucine incorporation when added to UV-irradiated cells. This result is specifically confirmed by visualization of the labeled $\lambda$ phage proteins made in irradiated cells by electrophoresis and autoradiography in polyacrylamide gels.[6] Although 8-10 prominent phage bands appear in extracts of $\lambda$-sensitive irradiated cells infected with wild-type $\lambda$, these bands all disappear when the infecting $\lambda$ phage carries a mutation in gene N. We know from genetic experiments that mutations in gene N do not block synthesis of the repressor. Therefore, high multiplicities of these mutant phages should stimulate the production of $C_1$ product without stimulating the production of most of the other $\lambda$ proteins. This greatly simplifies the task of finding the $C_1$ product itself.

These considerations lead to the following experiment: a strain of *E. coli* lysogenic for $\lambda$ is subjected to heavy UV irradiation. (The prophage used is $\lambda ind^-$, a mutant of $\lambda$ which, unlike wild type, is not inducible by UV light.) One portion of the cells is infected with 30-35 $\lambda$ phages carrying mutations in gene N, the other with $\lambda$ phages carrying in addition a mutation in the $C_1$ gene which prevents synthesis of the $C_1$ product. One culture then receives $H^3$-leucine, the other $C^{14}$-leucine. After a period of labeling, the cells are mixed and sonicated, and the extracts are fractionated to look for a single protein marked with one label but not with the other. Figure 2 shows the location and names of the $C_1$ mutants used in these experiments. At each stage, I ran parallel experiments with the labels reversed to avoid being misled by the artifacts which can arise in double-label experiments.

The block provided by single mutations in gene N is noticeably incomplete on some bacterial strains, and so, in the experiments described below, *all* the phages carry double suppressor-sensitive mutations (ambers) in the N gene. Only the name of the $C_1$ gene, be it wild type, amber mutant, or temperature-sensitive mutant, will be explicitly used.

## TABLE 1
*Detection of the $C_1$ product by ratio counting*

| Cell fraction | $H^3$ cpm/-0.1 ml | $C^{14}$ cpm/-0.1 ml | Ratio $H^3/C^{14}$ |
|---|---|---|---|
| Sonicate | 14,653 | 7,290 | 2.01 |
|  | 15,079 | 7,654 | 1.97 |
| Supernatant | 6,852 | 3,201 | 2.14 |
|  | 6,853 | 3,248 | 2.11 |
| Pellet | 7,007 | 3,747 | 1.87 |
|  | 7,641 | 3,589 | 1.85 |

Counts and ratios are given for duplicate samples from each fraction. See text and *Materials and methods* for experimental details.

### Detection of the $C_1$ product

When a double-label experiment is performed comparing wild-type λ with the amber $C_1$ mutant $C_1sus34$, a differential fractionation of the labels can be observed following sonication and high-speed centrifugation. Table 1 shows the results of a typical experiment. In this case the cells infected with wild-type phage were labeled with $H^3$-leucine, the others with $C^{14}$-leucine. There is approximately a 15 per cent increase in the ratio of $H^3$ to $C^{14}$ in the supernatant fraction compared to that found in the pellet. This means that there is a component which is labeled with $H^3$ but not with $C^{14}$ comprising about 15 per cent of the $H^3$ label in the supernatant. A ratio difference of 10-15 per cent is consistently observed when wild-type λ is compared with this amber mutant, whether the labels are in the configuration described in Table 1 or reversed.

### DEAE chromatography

The ratio counting experiment suggests that the supernatant fraction contains a protein made by the wild-type repressor gene but not by the amber $C_1$. This protein is isolated by fractionation on a DEAE column. The results of a typical experiment are presented in Figure 3. In this case the $H^3$ label is again in cells infected with λ wild type, and the $C^{14}$ label in cells infected with the $C_1$ amber mutant $C_1sus10$. The gradient elution profile shows a distinct peak present in the $H^3$ label with no corresponding peak in the other label. This experiment has also been performed using the amber $C_1$ mutants $C_1sus34$, $C_1sus14$, and $C_1sus80$. In each case the results are as seen in Figure 3, that is, none of these mutants

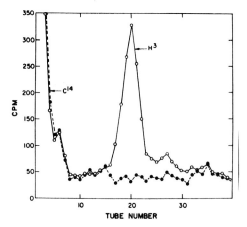

**Fig. 3.** DEAE elution profile of an extract of a mixture of $H^3$-leu-labeled λ wild-type infected cells and $C^{14}$-leu-labeled λ$C_1sus10$ infected cells. The cells are lysogenic for λ$ind^-$. The gradient begins around tube 4. The counts are normalized by multiplying the measured $C^{14}$ values by the ratio of $H^3/C^{14}$ in the sample applied to the column.

produces detectable amounts of the major protein peak. These mutants are also lacking the much smaller peak which is usually seen immediately following the major one. Although other minor peaks preceding and trailing the major peak at tube 20 often appear, there are no other major peaks (excluding the flowthrough) in either label throughout a gradient run from 0.07 *M* KCl to 0.5 *M* KCl. Furthermore, there is no increase in the ratio of $H^3/C^{14}$ in the flowthrough nor in the acid and base washings of the column, nor is there any significant differential loss of label. Finally, the percentage of supernatant counts appearing in the major peak is large enough to account for the 10-15 per cent ratio difference described in Table 1. The remaining 85-90 per cent of the label in the supernatant represents the synthesis of some bacterial proteins and possibly the limited synthesis of a few phage proteins other than the repressor.

### Temperature-sensitive $C_1$ mutants

The absence of the isolated protein from $C_1$ amber mutant infected cells strongly suggests, but does not prove, that the structural gene for this protein is the $C_1$. To confirm this directly, three temperature-sensitive $C_1$ mutants ($C_1ts857$, $C_1tsU32$, and $C_1tst2$) have been examined to determine whether a modified

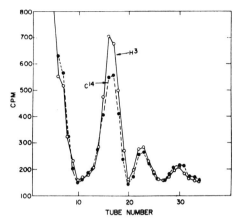

**Fig. 4.** DEAE elution profile of an extract of $H^3$-labeled $\lambda C_1 ts857$ infected cells and $C^{14}$-labeled $\lambda$ wild-type infected cells. Reconstituted protein hydrolysate is the label used in this experiment. Other details as in Fig. 3.

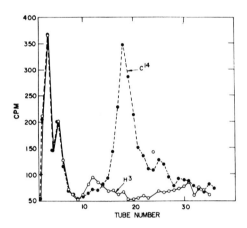

**Fig. 5.** DEAE elution profile of a mixture of $H^3$-leu-labeled $\lambda C_1 tsU32$ infected cells and $C^{14}$-leu-labeled $\lambda$ wild-type infected cells. Other details as in Fig. 3.

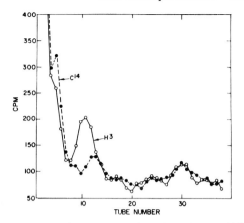

**Fig. 6.** DEAE elution profile of a mixture of $H^3$-leu-labeled $\lambda C_1 tsU32$ infected cells and $C^{14}$-leu-labeled $\lambda C_1 sus34$ infected cells. Other details as in Fig. 3.

form of the protein is produced by these mutants. In each case the repressors are labeled at temperatures at which they are functional *in vivo*.

Figure 4 shows the DEAE elution profile of a mixture of $H^3$-labeled $C_1 ts857$ repressor and $C^{14}$-labeled wild-type repressor. The mutant $C_1 ts857$ repressor chromatographs identically to wild type. The result of a similar experiment with mutant $C_1 tsU32$ gives a strikingly different result (Fig. 5). No peak corresponding to the $C_1 tsU32$ repressor is seen under the major $C^{14}$ peak. A much smaller $H^3$ peak of variable height which is probably the $C_1 tsU32$ gene product is observed around tube 12. This small peak is seen more easily in an experiment in which the $C_1 tsU32$ and $C_1 sus34$ gene products are chromatographed together (Fig. 6). The most likely reason that only a small amount of the $C_1 tsU32$ gene product is seen is that this material is partially insoluble. Another temperature-sensitive mutant, $C_1 tst2$, produces an even less soluble product. No significant differentially labeled $H^3$ peak is seen when the experiment described in Figure 6 is performed with this mutant. The $C_1 tst2$ repressor is labeled in these experiments, but it precipitates and appears in the pellet upon sonication and subsequent centrifugation (Table 2). Although $H^3$-labeled wild-type repressor produces a $H^3/C^{14}$ ratio increase in the supernatant when tested against $C^{14}$-labeled $C_1 sus34$, $H^3$-labeled $C_1 tst2$ repressor produces a ratio increase in the pellet when tested against this mutant. The fact that the magnitude of the ratio changes is about equal in the two cases suggests that $C_1 sus34$ produces only a small $C_1$ fragment, if any.

The important finding is that two mutations which produce modified repressors *in vivo* also alter the major protein synthesized in the present experiments.

### Electrophoresis of the $C_1$ product

The migration of the wild-type $C_1$ product in a 7.5 per cent polyacrylamide gel is shown in

**TABLE 2**
*Differential fractionation of $\lambda C_1$ tst2 and wild-type $C_1$ products*

| Fraction | Ratio $H^3/C^{14}$ | Fraction | Ratio $H^3/C^{14}$ |
|---|---|---|---|
| (1) Sonicate | 2.39 | (2) Sonicate | 3.03 |
|  | 2.39 |  | 3.00 |
| Supernatant | 2.27 | Supernatant | 3.13 |
|  | 2.30 |  | 3.19 |
| Pellet | 2.52 | Pellet | 2.85 |
|  | 2.49 |  | 2.78 |

In (1) $H^3$-leu-labeled $\lambda C_1 tst2$ infected cells were mixed with $C^{14}$-leu-labeled $\lambda C_1 sus34$ infected cells. In (2) $H^3$-leu-labeled wild-type infected cells were mixed with an aliquot of the same $C^{14}$-leu-labeled $\lambda C_1 sus34$ infected cells used in (1). Ratios are given from duplicate samples from each fraction. About half the label appears in the pellet following centrifugation. The experiment was performed as described in *Materials and methods*, except that the infected cells were labeled at 32°C.

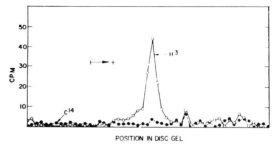

**Fig. 7.** Polyacrylamide gel electrophoresis of the protein isolated from DEAE columns. See *Materials and methods* for experimental details.

Figure 7. The material applied to the gel was collected and concentrated from the peak fractions of a DEAE run. A single major $H^3$ band is observed, with no corresponding $C^{14}$ peak. This shows that the differentially labeled peak recovered from DEAE consists of a single-labeled protein of high isotopic purity. The $C_1$ product migrates toward the anode at pH 8.7 as would be expected for an acidic protein.

A mixture of $H^3$-labeled $C_1 ts857$ repressor and $C^{14}$-labeled wild-type repressor was isolated on DEAE and subjected to electrophoresis as in Figure 7. A single superimposable major band appears in both labels, showing that the $C_1 ts857$ product bears the same charge as the wild-type $C_1$ product.

### Sedimentation of the $C_1$ product

Sedimentation of the wild-type $C_1$ product from the major DEAE peak was followed on sucrose gradients. Figure 8 shows the approximate s value of the $C_1$ product as 2.7-2.8S. This corresponds to a molecular weight of approximately 30,000 if the label is in a spherical protein.

### Discussion

The behavior of the temperature-sensitive mutants is particularly interesting because of the observation[7,8] that, *in vivo*, temperature-sensitive mutations mapping on the left side of $C_1$ cause a more drastic alteration of the

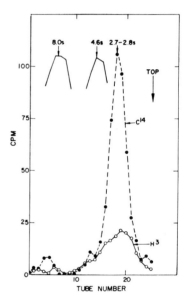

**Fig. 8.** Sedimentation of the protein isolated from DEAE columns. The peak fractions from a DEAE run were pooled and concentrated, and a sample was layered on a 5-30% sucrose gradient in T-M buffer plus 0.15 M KCl. The gradient was spun 60,000 rpm for 10.5 hr at 8°C. Parallel tubes were run carrying a mixture of the markers aldolase and hemoglobin. Fractions were collected and assayed for radioactivity or for absorbance at 230 and 410 mμ.

repressor than do temperature-sensitive mutations mapping on the right side of $C_1$. It has been suggested that right-hand side mutants such as $C_1 ts857$ reversibly denature when heated; repression is restored when the temperature is lowered. In contrast, left-hand side mutants, such as $C_1 tst2$ and $C_1 tsU32$, apparently are irreversibly denatured by a pulse of heat. Attempts to attribute this difference to the existence of two $C_1$ products have so far failed: no complementation has been detected between amber mutants from opposite sides of the $C_1$ gene, nor between any amber mutants and temperature-sensitive mutants located anywhere in the gene.[9] In experiments reported here, no difference has yet been detected between the wild type and $C_1 ts857$ repressors *in vitro*, but the $C_1 ts2$ and $C_1 tsU32$ repressors are found to differ markedly from wild type. Under standard conditions, the unmutated repressor appears in the supernatant following high-speed centrifugation, the $C_1 ts2$ repressor invariably precipitates, and the $C_1 tsU32$ repressor shows an intermediate solubility. The fraction of $C_1 tsU32$ repressor which remains in the supernatant following centrifugation chromatographs separately from the wild type on DEAE columns. Furthermore, the *in vitro* distinction between the $C_1 ts2$ and $C_1 tsU32$ repressors is consistent with the observation that *in vivo* the latter is more heat-stable than the former.

The repressor isolated from DEAE has an apparent molecular weight of approximately 30,000. The molecular weight of another repressor, the *lac i*-gene product, has been estimated at 150,000-200,000.[3] There is indirect evidence that the functional *i*-gene product is an oligomer.[10] It is possible that the λ repressor can also exist as an oligomer, and that I have isolated it in some subunit form. In this regard it may be significant that the sedimentation of the *lac* repressor was effected at high concentrations (about $10^{-6}$ $M$), whereas in the experiments described here, the concentrations are much lower (about $10^{-8}$ to $10^{-9}$ $M$). There is a small peak of high $s$ value in the sedimentation profile shown in Figure 8, but I have not attempted to characterize this fraction or the small peak which consistently emerges immediately behind the major peak on DEAE columns. Either of these components might represent alternate forms of the repressor.

## Summary

A product of the $\lambda C_1$ (repressor) gene has been labeled with amino acids and isolated free from any other labeled components of the cell. The identification is based on the fact that this molecule is not made by amber mutants and is made in a modified form by temperature-sensitive mutants of the $C_1$ gene. The product electrophoreses in a single band, moving as an acidic protein. It sediments at $2.8S$, which corresponds to a molecular weight of approximately 30,000.

## Materials and methods

*Bacteria.* The bacterial strain used in all these experiments is a UV-sensitive mutant isolated by Dr. M. Meselson from the *E. coli* strain W3102.[11] This strain is prototrophic and nonpermissive ($su^-$) for amber mutants. The $UV^s$ locus mutated in this strain is unknown. This strain is used because, as with some other $UV^s$ strains,[12] its capacity to support λ phage growth is higher than that of wild-type strains following exposure of the host to the same physical dose of UV irradiation.

*Bacteriophages.* All the λ phages carrying double N mutations were prepared by recombination with the phage $\lambda Nosus_7 sus_{53}$ isolated by Dr. D. Hogness. The derivative of phage $\lambda imm^{434}$ used here was described previously.[13] Phage $\lambda ind^-$ was isolated and described by Jacob and Campbell.[14] Two amber mutants,[15] $C_1 sus34$ and $C_1 sus80$, were supplied by Dr. F. Jacob. One of the temperature-sensitive mutants, $C_1 ts857$, was isolated and described by Sussman and Jacob.[16] The other temperature-sensitive[7] and amber mutants were isolated and donated by Dr. M. Lieb. All the phages used in these experiments carry the *b* marker from the phage $\lambda b$ of Kaiser[17] to ensure efficient absorption. Phage stocks were grown in strain C600 by the agar-overlay method, pelleted, resuspended in phage buffer, and dialyzed against the same buffer.

*Media and buffers.* Cells were grown in A medium[18] with 4 gm/liter maltose as carbon source. Preconditioned medium was prepared by filtration of A medium which had supported the growth of cells to $2-3 \times 10^8$ cells/ml. T-M buffer is 0.01 $M$ Tris pH 7.4, 0.005 $M$ $MgSO_4$. Phage buffer is T-M buffer made 0.1 $M$ NaCl.

*Radioactivity.* $H^3$- and $C^{14}$-labeled leucine and reconstituted protein hydrolysates were purchased from Schwartz BioResearch, Inc., and from New England Nuclear Corp. These labeled compounds were used without further dilution with cold amino acids. Numerous different batches of isotopes were used during the course of these experiments. The specific activity of the $C^{14}$-leu is about 200 mc/mmole, and that of the $H^3$-leu is about 2.0 c/mmole. All radioactive isotope counting, unless specified otherwise, was performed according to the method described by Fox and Kennedy,[19] except that the ethanol wash was replaced with another TCA wash.

*Electrophoresis.* Disc electrophoresis was performed at pH 8.7 as described by Davis.[20] After electrophoresis the gels were frozen and cut into slices 1 mm thick using a slicing apparatus designed and built by Dr. C. Levinthal. The slices were distributed into counting vials and dissolved in ½ ml of 15% hydrogen peroxide by heating at 60-70°C for a few hours. One-half ml of hydroxide of hyamine (Packard Inst. Co.) was added to the vials, and the mixture was incubated at 57°C for 3 min. After cooling, 10 ml of Bray's[21] solution was added, and the samples were counted in a Packard scintillation counter.

*Chromatography.* DEAE-cellulose chromatography was performed using a 15-ml column, 1 cm in diameter, with Whatman DEAE-cellulose #DE52. After applying the sample and washing the column with about 10 ml of T-M buffer containing 0.07 $M$ KCl, a 150-ml linear KCl gradient was run to 0.2 $M$ KCl, and fractions containing about 4 ml were collected. Aliquots from each fraction were assayed for radioactivity.

*Irradiation.* Cells were irradiated at a distance of 30 cm from 2 G.E. 15-w germicidal lamps. The incident dose at this distance is about 75 ergs $mm^{-2} sec^{-1}$. Eighty ml of cells at concentration $3 \times 10^8$ cells/ml were irradiated at 0°C in a Petri dish of diameter 11 cm, with swirling.

*A typical experiment.* Cells grown to a concentration of $10^9$ cells/ml in medium A are chilled and diluted to a concentration of $3 \times 10^8$ cells/ml with cold preconditioned medium. The cells are irradiated for 12 min, and the $MgSO_4$ concentration is then raised from $10^{-3}$ $M$ to $2 \times 10^{-2}$ $M$. Two 40-ml portions are distributed into flasks, and phage are added at a multiplicity of 30-35 phage/cell. After swirling 5 min at 37°C, 0.4 mc $H^3$-leu is added to one flask, and 0.04 mc $C^{14}$-leu to the other, and the cells are aerated for 1 hr at 37°C. The cells are then chilled, centrifuged, mixed together, washed twice in T-M buffer, and finally resuspended in 2 ml of T-M buffer containing 0.4 $M$ KCl and a few micrograms of DNase. The cells are disrupted while on ice by several 30-sec pulses delivered by an MSE sonicator. Small aliquots are taken to determine the original ratio of $H^3/C^{14}$, and the extract is immediately spun at 350,000 $g$ for 36 min in an International B-60 centrifuge. The supernatant is withdrawn and the pellet is resuspended in 2 ml of T-M buffer. After removing samples for ratio counting, the supernatant is dialyzed overnight against T-M buffer plus 0.07 $M$ KCl.

It is of critical importance that the $Mg^{++}$ concentration be raised before addition of the phage. Phage λ absorbs well to cells grown in medium A with maltose as the carbon source, even at low $Mg^{++}$ concentrations. However, if the experiment described in Figure 1 is performed in the presence of only $10^{-3}$ $M$ $Mg^{++}$, no stimulation of labeled leucine incorporation is observed. In fact, even at high UV doses, the addition of 35 phage/cell causes a further 10-20-fold decrease in the incorporation of labeled leucine. Others have noted a strong inhibition in protein and RNA synthesis following infection of unirradiated cells with high multiplicities of phage λ.[22] The discovery that high concentrations of $Mg^{++}$ can reverse this inhibition, at least in the irradiated strain used here, made possible the success of these experiments.

**REFERENCES**

1. Kaiser, A. D., and F. Jacob, Virology 4:509 (1957).
2. Jacob, F., and J. Monod, J. Mol. Biol. 3:318 (1961).
3. Gilbert, W., and B. Müller-Hill, these Proceedings 56:1891 (1966).
4. Lieb, M., Virology 29:367 (1966); and Horiuchi, T., and H. Inokuchi, J. Mol. Biol. 15:674 (1966).
5. Thomas, R., J. Mol. Biol., in press; and Protass, J., and D. Korn, these Proceedings 55:1089 (1966).
6. Autoradiography was performed as described in Fairbank, G., Jr., C. Levinthal, and R. H. Reeder, Biochem. Biophys. Res. Commn. 20:393 (1965).
7. Lieb, M., J. Mol. Biol. 16:149 (1966).
8. Naona, S., and F. Gros, J. Mol. Biol., in press.
9. Lieb, M., personal communication.
10. Sadler, J. R., and A. Novick, J. Mol. Biol. 12:305 (1965).
11. Hill, C. W., and H. Echols, J. Mol. Biol. 19:38 (1966).
12. Devoret, R., and T. Coquerelle, personal communication.
13. Ptashne, M., J. Mol. Biol. 11:90 (1965).
14. Jacob, F., and A. Campbell, Compt. Rend. 248:3219 (1959).
15. Jacob, F., R. Sussman, and J. Monod, Compt. Rend. 254:4214 (1962).
16. Sussman, R., and F. Jacob, Compt. Rend. 254:1517 (1962).
17. Kaiser, A. D., J. Mol. Biol. 4:275 (1962).
18. Meselson, M., and F. W. Stahl, these Proceedings 44:671 (1958).
19. Fox, C. F., and E. P. Kennedy, these Proceedings 54:891 (1965).
20. Davis, B. J., Ann. N. Y. Acad. Sci. 121:404 (1964).
21. Bray, G. A., Anal. Biochem. 1:279 (1960).
22. Howes, W. V., Biochim. Biophys. Acta 103:711 (1965); Tertzi, M., and C. Levinthal, J. Mol. Biol., in press.

## QUESTIONS FOR CHAPTER 5

**Paper 19**

1. a. The paper states that in bacteria the synthesis of enzymes is dependent on the continued presence of the genes for those enzymes. Is this literally and generally true?
   b. One of the characteristics of messenger RNA is taken to be that its base composition should mimic that of the corresponding DNA. Is this assumption justified?
   c. It is also assumed that mRNA is necessarily associated with ribosomes. Is this generally true?

2. a. On the basis of what you know of the characteristics of separation of different kinds of molecules on a CsCl density gradient, do you think that the given identity of the A and B peaks is correct? Assume that 70S ribosomes are formed of 30S and 50S subunits held together by nothing but $Mg^{++}$.
   b. If your answer to question 2a was no, suggest a more likely alternative for the appearance of the two bands.
   c. How does your explanation fit in with the observation that low $Mg^{++}$ concentration favors the production of the heavy A band?
   d. Would you expect that the presence of monovalent salt influences the $Mg^{++}$ concentration at which 70S ribosomes remain intact? If so, which way do salts exert their influence, and why?

3. a. In Fig. 2 the "light" ribosomes are present in a fifty fold excess of the "heavy" ones; yet both sets of bands are roughly of equal width. Since the size of the band distribution is proportional to the amount of material present, and since both sets must have the same molecular weight, would you not expect that the bands of the unlabeled "light" ribosomes should be much wider than those of the "heavy" ones? Please explain.
   b. The ribosomes in Fig. 2 were treated exactly the same; yet there is an obviously larger proportion of the A band in the $^{32}P$-labeled species. Why?
   c. What is your explanation for the failure of the experiment shown in Fig. 3?
   d. Why is the proportion of the A bands higher in Fig. 3 than in Fig. 2, even though both gradients contain the same $Mg^{++}$ concentration?
   e. What is the identity of the material in tube 105 in Fig. 2 and tube 88 in Fig. 3?

4. a. The text of the paper states that "the bulk of the RNA synthesized after infection is found in the ribosome fraction." Is this statement quantitatively justified by the actual data in Fig. 4?
   b. The paper further states that the RNA at the bottom of the gradient must have a high molecular weight to have sedimented in the gradient. Is this statement correct?
   c. Between the RNA at the bottom of the cell and that which is ribosome-bound there is a sizable portion of the total label. What is this material?
   d. Calculate the decrease in radioactivity for both RNA peaks in going from Fig. 4 to Fig. 5.
   e. Are the results consistent with a "rapidly turning-over RNA"? Explain.

5. a. Determine the specific activity of the RNA in Fig. 6.
   b. Why is only the B band associated with label in Fig. 7?
   c. Was any label associated with the ribosomes that give rise to the A band?
   d. Does the presence of native protein protect the ribosomes against decomposition?
   e. What is the most striking difference between Figs. 4 and 5, and 6 and 7?
   f. Was the choice of $^{35}S$ a good one to demonstrate the presence or absence of new ribosomes? Explain.

6. a. The authors state that when ribosomes associated with $T_2$ mRNA are dialyzed against low $Mg^{++}$, there result three peaks of radioactivity: one in the region of the residual 70S, one in the 30S region, and one at 12S. Why was no $T_4$ mRNA observed on 50S ribosomes?
   b. Taking advantage of information provided in Chapter 4, calculate the size of the protein coded for by mRNA 14S to 16S in size.
   c. Is 12S the true size of the $T_2$ mRNA?

**Paper 20**

1. It is mentioned that $T_2$ infection stops all host RNA synthesis immediately; how does the phage accomplish this?

2. a. How does the MAK column accomplish fractionation of the RNA according to base composition?
   b. How does the column fractionate by size?
   c. Does the column fractionate RNA or DNA according to any additional criteria?

3. a. The two labeled RNAs are mixed first and then chromatographed. Could the same

results be obtained by comparing separate chromatograms of the two RNA populations? Why or why not?
   b. Do the amounts of the two RNAs have to be the same in order to obtain the right results from a mixed chromatogram?
   c. Does a large difference in the specific activity of the two RNAs influence the outcome of a double-label mixed chromatogram?
   d. Is there any possibility that the two labeling profiles in a mixed chromatogram could be different without there being any difference in the two RNAs?
   e. In this paper, the RNAs were isolated separately and mixed just before the chromatography. Would it be preferable to mix the bacteria and isolate the two RNAs together?
4. a. In the figures in this paper the profiles due to $^{14}$C and $^{3}$H are plotted separately. Can you think of a graphic method to better identify those components in which the label is different?
   b. Why was it important to show that the RNAs differed in base composition rather than just in size?

## Paper 21

1. a. What is meant by a transducing phage?
   b. By what mechanism can host genes become integrated into the phage genome?
   c. If a deletion mutant cannot make a certain enzyme, does this mean that the whole gene is missing, or could there be some mRNA corresponding to part of the gene?
   d. How can one tell whether a deletion comprises the whole or only part of a gene?
2. What protein is specified by the $r_{II}$ region of phage $T_4$?
3. a. Why is it necessary to wash the column with high salt buffer in order to wash out any unhybridized RNA?
   b. Why is the RNA eluted from the mutant bound more by the wild type than by the mutant in Table 4?
   c. What is the significance of the fraction $r^+ - r1231 / r^+$?
4. Given the information in the paper, calculate the approximate size of the following:
   a. The A cistron of the $r_{II}$ region of TD4
   b. The B cistron of same

## Paper 22

1. a. Why are poly dC clusters thought to be initiating regions for RNA synthesis?
   b. Why is the presence of RNA initiating sites on both DNA strands unusual?
2. a. In Fig. 1, why are the two strands drawn of unequal length?
   b. In Chapter 3 we saw that the polynucleotide ligase could be used to make a covalent circle out of λ DNA. How could that be, seeing that the ends, G and A, are not complementary?
   c. How can the base compositions of parts of the molecule be determined?
   d. What could be the function of $b_2$, which is not transcribed into mRNA?
   e. How can a gene be experimentally positioned on one or the other strand?
   f. Some of the deletion mutants have more than 50% of the phage DNA deleted, comprising no doubt essential genes. How can they survive and be propagated?
3. a. What are sus mutants?
   b. What makes a bacterial strain "permissive" for a sus mutant?
   c. If a bacterial strain is permissive for a given sus phage, will it be permissive for others? Explain.
   d. It is stated that the bacteria were lysogenized with appropriate λ mutants. What is meant by that?
   e. What does λdg mean?
   f. It is said that "biotin genes were substituted for deleted genes a-N or a-O in λdb$_{a-N}$"; what does this mean?
4. a. Since the idea is to identify poly dC clusters, why is poly IG used, instead of poly G?
   b. The two complementary λ strands are separated by 14-16 mg/cm³. Is that a good or a poor separation?
   c. Why is the strand rich in poly dC clusters denser than the other?
   d. Why would the separated strands be cross-contaminated, so that they have to be purified by self-annealing?
   e. How is the "functional" purity of the strands determined?
   f. Would any mRNA produced by the intact phage self-anneal?
5. a. Are there any data in Table 1 that show that transcription of the light strand ceases, whereas transcription of the dense strand continues?
   b. Point out the data in Table 1 (by line) that show that the mRNA starting at xyc$_{II}$ is or is not "early."
   c. In line 2 of Table 1, how can there be any mRNA hybridized with the C and W strands, seeing that the E. coli is uninfected and not lysogenized?
   d. It is said that "transcription from strand C preferentially increases at lower [than 100 µg] CM concentrations." Do all data in Table 1 indicate that? Point out the exceptions, if any.

e. Suppose the labeled mRNAs were prehybridized with T75($\lambda db_{a-N}$) DNA. Would it be necessary to separate the strands to study the synthesis of early mRNA with $\lambda^+$? With a mutant? If so, which mutant?
6. a. What is a polar mutation?
  b. Why does a polar mutation prevent synthesis of mRNA of the whole operon?
  c. What are nonsense mutations?

## Paper 23

1. a. Why is the production of all early phage proteins, except that of the repressor, blocked by a mutation in the N gene?
  b. Since production of all proteins, except the repressor, have been blocked by using the mutant in the N gene, why must a double label experiment be performed? Why cannot the $C_1$ gene product be isolated directly?
2. a. What is the significance of the changes in the $^3H/^{14}C$ ratio in the different fractions in Table 1?
  b. Calculate the recovery of radioactivity of the sonicate in the two derivative fractions. Is the recovery of the $^{14}C$ counts the same as that of the $^3H$ counts?
  c. In order to obtain the counts given in Table I, how much buffer has to be added to the pellet? Is this volume critical for obtaining the correct ratios?
  d. Is complete recovery of material essential to obtaining the correct relative concentrations of the two isotopes?
  e. Would the same results be obtained if the two bacterial cultures labeled with $^3H$ and $^{14}C$, respectively, were fractionated separately, rather than together?
  f. Why is it important in a double-label experiment to reverse the labels?
3. a. If an *amber* mutant chosen for the control culture were capable of synthesizing a peptide chain equal to half or three fourths of the complete $C_1$ product, would that impair the recognition of the repressor in a chromatogram?
  b. Why were *sus* amber mutants chosen?
  c. Would other mutants serve equally well?
4. a. In chromatographing a double-labeled extract through a DEAE column, is it important to determine the recovery of the labels in the eluate?
  b. In Figs. 3 and 5, the wild type has at least one other peak in addition to the repressor not present in the mutant. What is the nature of this second peak?
  c. The major $C_1$ product is shown by Fig. 7 to be isotopically pure. How pure is it chemically?

# ANSWERS FOR CHAPTER 5

Paper 19

1. a. It is now known that various messengers have very different half-lives, and many could undoubtedly function in the absence of the gene from which they were copied. The most dramatic example of this is in the reticulocyte, which, though devoid of DNA, continues to synthesize hemoglobin at a high rate.
   b. It is now known that *in vivo* only one of the strands of DNA is copied to make mRNA, and therefore its base composition should be quite different from that of DNA. That many mRNAs were in fact found to mimic DNA base compositions is a mere coincidence of no theoretical importance. It merely means that in very long polynucleotides (mRNAs represent a long piece of DNA copied, even though they are composed of many shorter molecules), there is a tendency to randomize the nucleotide sequences.
   c. That mRNAs *can* exist independently of ribosomes is shown by the fact that viral RNAs that penetrate a cell or are added to cell extracts are indeed capable of coding for proteins; but whether or not mRNA, apart from viral RNAs, ever actually exists free of ribosomes is not known. Many observations favor the notion that, in bacteria at least, mRNAs become attached to ribosomes on one end while they are still attached to DNA and growing at the other end. It was actually thought for a while that nascent RNA cannot become detached from DNA unless it becomes attached to ribosomes, but this does not seem to be necessarily so. In mammalian cells the situation is more complicated. While some workers have presented evidence that the mRNA is packaged in protein, in the so-called informosome, others have interpreted their data to mean that mRNA becomes attached to the small ribosomal subunit in the nucleus and is thus exported into the cytoplasm.
2. a. In order to obtain two bands on a CsCl density gradient, one band must be considerably denser than the other. All kinds of ribosomes, however, have the same RNA-to-protein ratio, and therefore the same density; equilibrium centrifugation does not separate classes of the same density, but of different size, and therefore 30S, 50S, and 70S particles must all be in the same band.
   b. Because RNA is denser than protein, the only way to produce a denser particle from ribosomes is to shed some of the protein. As a matter of fact, it is well-known now that the B band contains intact 70S ribosomes and subunits, and that the A band is composed of ribosomes that have lost a part of their proteins. As a matter of fact, it has been shown by several workers that this is a reversible process, and that it is possible to put the proteins back on and regenerate active ribosomes.

      (It should be pointed out that this erroneous identification of the bands does not invalidate the results, which still show that mRNA becomes associated only with intact ribosomes, though not necessarily 70S, as seen in answer 3c, below).
   c. The observation that low $Mg^{++}$ concentration favors the production of A band could be explained by the plausible assumption that in the 70S particle some proteins (located where the 30S and 50S come together) are less accessible to the dissociating salt than they are in the free subunits. If this were the case, however, one might expect to find an intermediate band, of which there is no evidence. A more likely explanation is the following: The dissociation of protein from ribosomes is a time-dependent process, and since the only way to lower the $Mg^{++}$ concentration without diluting the ribosomes is by dialysis, it seems likely that the ribosomes in low $Mg^{++}$ were exposed to salt a somewhat longer time than those in high $Mg^{++}$.
   d. Monovalent salt has a strong dissociating effect on ribosomes. For example, in the presence of 0.086M KCl or 0.19M $NH_4Cl$ the $Mg^{++}$ concentration has to be increased from 0.01M to 0.0175M in order to keep the 70S intact. Although no systematic study seems to have been made of the exact amount of $Mg^{++}$ necessary to counteract the concentrated CsCl in a density gradient, it is of the order of 0.04M.
3. a. The reason the $^{32}P$ distribution looks similar to the UV one, although the latter contains fifty times as much material as the former, lies in the sensitivity of the mode of detection: $^{32}P$-labeling can probably detect at least fiftyfold less material than ultraviolet absorption. The distribution can also be apparently

widened or narrowed by several tubes simply by increasing or decreasing the scale of plotting. This effect can be observed by reducing the scale of the $^{32}$P-distribution, for example, by as little as half: the tubes from 32 to 41, and from 78 to 90, would effectively drop to base-line level.

b. The apparent discrepancy in the relative amounts of A and B bands in the two distributions in Fig. 2 is due to the different amount of material in the two and the time-dependency of the dissociation reaction leading to B. Suppose for the sake of illustration that the rate of dissociation in the presence of saturating amounts of ribosomes is 0.5 mg/hr. If the total amount present were 1.0 mg of ribosomes, 50% would be in A at the end of an hour. If the total amount of ribosomes present were 2.0 mg, however, only 25% of the total would appear in band A after an hour. When the concentration of the remaining intact ribosomes falls below saturation, the rate of dissociation decreases with decreasing concentration. This is probably the reason that there are any intact $^{32}$P-labeled ribosomes left.

c. The simplest explanation of why the experiment shown in Fig. 3 failed to show reassociated, hybrid 70S ribosomes is that the B band contains only 30S and 50S, and that no 70S are formed in 0.03M Mg$^{++}$. This is perfectly consistent with all the data in the paper, because it is now known that mRNA becomes associated with 30S particles.

d. The greater decrease in the B band in Fig. 3 as compared to Fig. 2 can be explained by the long exposure of this preparation to salt in the course of dialysis.

e. The materials at the tops of the gradients are ribosomal protein stripped from ribosomes in the process of forming the A bands.

4. a. Fig. 4 does not show whether or not the bulk of the RNA is associated with ribosomes, because the starting material was purified ribosomes. (There are no data to show how much, if any, mRNA remained in the cytoplasm.) Therefore, all the RNA in Fig. 4 must originally have been associated with ribosomes, although a good part of it became free due to the loss of protein, and thus of binding activity, from the ribosomes.

b. The point has been made before that the position of a material in a density gradient centrifuged to equilibrium depends only on the density and not on the molecular weight. Therefore, provided the gradient was centrifuged long enough, even low molecular weight RNA would find its way to the bottom of the gradient.

c. For any material to band between the A band and the free RNA, it must contain progressively less protein. However, the A band seems to be a stable end product of the dissociation process, and the optical density markers show no evidence of any material heavier than the A band. Furthermore, there is no evidence that mRNA can associate with the dissociated ribosomal protein (the only protein available to make it less dense). It is therefore unlikely that those intermediate counts belong to discrete density species. Rather, the shape of the distribution makes it likely that these counts are due to RNA that has been liberated in the course of dissociation and that has not yet reached its equilibrium position at the bottom of the cell. Since the dissociation of protein, and hence also of mRNA, continues throughout the run, there will be intermediary material present as long as there is any B peak left to dissociate.

d. In comparing the specific activities of two materials it is sufficient to compare the peak tubes. The specific activities (cpm per O.D.) of the A peaks in Figs. 4 and 5, respectively, are: $125/0.65 = 182$, and $50/0.9 = 55$; those of the free RNAs are: $95/0.04 = 2400$, and $42/0.04 = 1050$. The decrease in specific activities is, thus, $182/55 = 3.3$-fold for the A peak, and $2400/1050 = 2.3$-fold for the free RNA. Because the relative amounts of A and B are different, however, the ratios of the absolute radioactivities may give a truer picture. Thus, the ratio in the B peak is $125/50 = 2.5$, and in the RNA peak, $95/42 = 2.3$. The decrease in labeled mRNA after the chase is therefore only 2.5-fold.

e. In going from Fig. 4 to Fig. 5, the uracil was diluted 200-fold, enough for newly labeled RNA to be essentially unlabeled (less than 1 cpm per tube); furthermore, the labeled RNA was allowed 8 times more time to decay than it had been allowed to be synthesized. Yet, during all this time the labeled mRNA decreased only 2.5-fold. This means that although some of the mRNA does turn over (just how rapidly it does cannot be determined from this experiment), a sizeable fraction survives for longer than 16 minutes—a long time in the phage infection process. Thus, unless the breakdown products are compartmentalized and reutilized with great efficiency, the data in Figs. 4 and 5

## 224    Molecular biology of DNA and RNA

are evidence for a stable fraction of mRNA as much as they are for the lability of another fraction.

5. a. The specific activity of the mRNA cannot be determined from the data in Fig. 6, because there is no marker for the heavy ribosomes to which the mRNA is bound. The superposition of peak A of the light ribosomes and the mRNA peak is purely coincidental.
   b. Since purified ribosomes were used for the gradient in Fig. 7, only nascent, ribosome-bound protein is involved, and this is found only on intact ribosomes.
   c. The protein that was associated with those ribosomes which later formed the A peak can be found at the top of the gradient.
   d. The radioactivity at the top of the gradient is roughly equivalent to that associated with the B ribosomes; yet, the number of ribosomes associated with the label is very much smaller than the number that was present in the A peak and that gave rise to the radioactivity at the top of the gradient. It does thus seem that ribosomes not associated with nascent protein are preferentially transformed into A particles.
   e. The experiment shown in Figs. 7 and 8 is the exact parallel of that shown in Figs. 4 and 5, except that protein is labeled instead of RNA. The difference is that although—as we have seen—labeled RNA still remains on the ribosomes after a chase of 16 min, in the case of protein no label remained at all on the ribosomes after a chase half as long. This shows conclusively that the stable RNA cannot be ribosomal RNA, for no ribosomal protein is synthesized in the infected cells.
   f. The choice of $^{35}S$ to monitor a possible synthesis of ribosomal proteins was a very poor one because the proteins are very low in sulfur, and nascent enzyme proteins have a high S content. The choice of $^{35}S$ thus assured the preferential labeling of phage-specific protein over ribosomal protein. However, the conclusion that no ribosomal protein is synthesized in $T_4$-infected cells is nevertheless correct.

6. a. In the experiments with $T_2$ a sucrose density gradient was used, and it was therefore possible to separate 30S particles from 70S ones. In the case of $T_4$, on the other hand, the use of a CsCl density gradient centrifuged to equilibrium precluded the separation of size classes. As we have seen in answer 3c, above, however, there is indirect evidence that $T_4$ mRNA is associated with 30S particles, and perhaps exclusively so.

   b. We have seen in the preceding chapter that, roughly speaking, $M = ks^2$, and that k was approximately equal to 2200. Thus, the molecular weight of a species sedimenting with 14S to 16S is 431,000 to 563,000. This corresponds to 1350 and 1830 nucleotides, respectively. Assuming the triplet code, these RNAs can code for proteins 450 to 610 amino acids long. This is too long for a single polypeptide chain, so that these RNAs are probably, although not necessarily, polycistronic.
   c. Even though a 12S RNA is not small, it is now known that it represents degraded mRNA, for native, undegraded mRNA is several times bigger. No importance should thus be attached to the observed sizes of the mRNAs described in this paper.

## Paper 20

1. It is still not known how the $T_{even}$ phages accomplish the immediate cessation of host RNA synthesis.

2. a. The MAK column works by accepting hydrogen bonds from the nucleic acids and by **making hydrogen bonds with them**. Bases differ in the number of hydrogen bonds they can make or receive.
   b. The fractionation by size is a secondary effect. Obviously, in a qualitatively similar polymer series, the larger a molecule, the greater are the number of hydrogen bonds by which it is bound to the column. Therefore, larger molecules are bound more tightly and require a higher salt concentration to be eluted.
   c. Since the column fractionates by hydrogen bonding, it also fractionates according to single- or double-strandedness. Double-stranded molecules have fewer available sites to bind to the column, and they will therefore be eluted first. As a matter of fact, the availability of hydrogen bonding sites quite overrides the effects of size and base composition. Messenger RNAs, for example, are eluted last, not necessarily because they are larger than the ribosomal RNAs, but because they are single-stranded.

3. a. It is quite impossible to produce two absolutely identical columns, especially in the case of the MAK column; even two runs made on the same column will vary somewhat. A very slight variation in the sample size, for example, may cause one drop containing a high concentration of material to go into one tube in one run, and into another tube in the next run, thus causing a difference in the distribution of material. Much time has

been wasted in trying to compare complicated chromatograms. Different runs cannot approach the comparability of a run with mixed materials. The introduction of this method of comparing populations by double labeling and mixed chromatography is the greatest contribution of this paper.
b. The relative amounts of the two RNAs to be compared do not matter, as long as the method used for their detection gives a linear response to quantity.
c. The method is also independent of the specific activity of the materials to be mixed, as long as there are no quenching effects.
d. When two identical populations are mixed, there can be a different distribution of isotopes only if one of the isotopes shows a concentration-dependent quenching effect. Suppose that in the particular counting setup, $^3$H is quenched, but only if more than 1 mg of material is present, and $^{14}$C is not quenched. The relative number of $^3$H counts will therefore be greater in the more dilute samples, and the $^{14}$C/$^3$H ratio in these samples will be different from the ratio in the starting material—usually a sign of heterogeneity.
e. It would be preferable to mix the bacteria and isolate the RNAs together, because in this way there is no chance for differences to be introduced by different handling of the samples. Also, in this way the result becomes independent of yield and of any mishaps that may befall the sample.
4. a. The most sensitive way to show up differences in the two labeling profiles is to plot the $^{14}$C/$^3$H (or the opposite). If most of the materials are equal, a base-line ratio will be observed in all the regions where the materials are qualitatively the same, and any difference in a component will show up in a peak containing one but not the other isotope.
b. The reason for insisting that differences in base composition must be shown is to eliminate the possibility that the differences in the MAK profiles could have been due simply to the same molecules' having grown bigger.

## Paper 21

1. a. Transduction is the transfer of genetic material from donor to recipient in which a phage is the vector.
b. Since most phages and host chromosomes do not share any allelic genes, the host genes cannot become integrated into the phage DNA by a process of genetic recombination, in the sense that there is no exchange of genetic material. One plausible mechanism is that the phage DNA becomes physically integrated into the host DNA when it becomes a prophage, and later is excised again. Due to some imprecision in the excision process it can take a piece of host DNA with it and thus become a transducing phage. The integration of phages into the host chromosome is known as lysogenization. The integrated phage is completely innocuous in this state, and it reproduces together with the host chromosome. The liberation of a free, lytic phage capable of independent replication is known as induction and can be caused by such agents as UV light and mitomycin, for example. By no means all prophages become transducing on induction, this being a relatively rare event. The size of the transduced piece of bacterial DNA varies; usually, it represents but a single marker, but sometimes two or more closely linked markers are cotransduced. Occasionally, a very large piece of DNA is transduced. Some phages, such as Pl, for example, can carry with them the entire genome of the unrelated phage λ in the prophage state as well as an additional piece of the bacterial chromosome.
c. Deletion mutations can comprise anywhere from a single nucleotide to a whole gene or a piece of DNA comprising several genes. However, the fact that a given enzyme is not made does not mean that a whole gene has been deleted. As a matter of fact, the deletion of a single nucleotide early in the gene will prevent the synthesis of active enzyme. The fact that no active enzyme is made does not mean that no mRNA and no protein are synthesized. In the case of a single deletion, for example, RNA synthesis is normal, but the code is read out of phase, and a scrambled, meaningless protein is synthesized. Whether or not RNA is made when bigger pieces of DNA are missing depends on whether or not the promotor region of the operator, where the RNA polymerase binds, is intact. The resulting RNA will be translated if it has proper initiating and terminating sequences for protein synthesis. If a protein can be initiated, but the mRNA lacks terminator codons, the synthesized peptide will remain bound to ribosomes.
d. A deletion mutation is different from other mutations in that it can yield no reciprocal recombinants. The position of deletions can in principle be mapped by negative recombinational analysis, but other methods are

also used. A small deletion of one or a few nucleotides can be recognized because it can be "cured" by acridines. Acridines intercalate between the bases, and when such a DNA strand is copied it may give rise to an extra nucleotide in the daughter chain. Every addition can neutralize a not-too-distant deletion, because it causes the code to get back in phase. Small deletions in multiples of three will have little effect on enzyme activity, unless the deleted codon is that of an essential amino acid. This method is a good test for the presence of a deletion, but it does not permit its exact localization. Another way of mapping deletions is by complementation analysis. If a bacterium is infected with two or more phages carrying different deletions in the same gene, some wild-type of progeny may result. If, on the other hand, the deletions cover the same general region, or part of the same region, no complementation, obviously, can occur.

2. The proteins specified by the $r_{II}$ region of $T_4$ are not known. For many years it was not even known that there were proteins corresponding to these intensely mapped $r_{II}$ genes. As a matter of fact, this constituted a major skeleton in the closet of molecular biology. Now, however, it is at least known that a protein is, in fact, specified, because certain suppressible amber mutants have been isolated. Suppression is known to be the property of certain tRNAs and requires translation for expression.

3. a. High salt concentration is conducive to the formation of double-stranded nucleic acid helices. Hybridization is therefore carried out in high salt concentration, but the temperature is such that imperfect helices come apart again whereas perfectly complementary ones stay together. Washing with high salt buffer, therefore, removes unhybridized RNA but leaves the hybrids in the columns. If the salt concentration in the washing buffer were reduced, many hybrids would come apart.

   b. Since the mutant phage has all the sequences present in the wild type, the RNA complementary to it should hybridize to exactly the same extent to the two DNAs. In fact, it does so: the difference between 152 and 148 in Table 4 is not significant.

   c. The difference $r^+ - r1231$ measures the amount of RNA complementary to the $r_{II}$ region—the only difference between the wild type and the mutant $T_4$. The ratio $r^+ - r1231/r^+$ measures the amount of $r_{II}$ RNA relative to the total.

4. From Table 2 (column I) it can be seen that the whole $r_{II}$ region plus "ten recombination units" comprises 1.4% of the total hybrid. According to Table 3, this is 92% pure; hence the corrected fraction is 1.29%. Given the molecular weight of $T_4$ of $1.48 \times 10^8$, this corresponds to $1.91 \times 10^6$ daltons.

   a. Mutant T4r638 lacks cistron B, and hence it has A; it hybridizes with 74.2% of the total $r_{II}$ RNA, and the RNA eluted from the hybrid is 96% pure. Therefore, 71.2% of $r_{II}$ is A, corresponding to $1.35 \times 10^6$ daltons, or $1.35 \times 10^6/640$ (the weight of a base pair) = *2115* nucleotides in length.

   b. The size of the B cistron cannot be determined accurately because the deletions do not correspond exactly to the $r_{II}$ region. Approximately, however, the B region corresponds to $1.91 \times 10^6 - 1.35 \times 10^6 = $ *5.6 $\times 10^5$* daltons, or *875* nucleotides in length.

## Paper 22

1. a. Poly dC clusters are thought to be initiating sequences for RNA synthesis *in vivo* because the isolated RNA polymerase (see introduction to Chapter 6) synthesizes chains starting mainly with G. (There are also some A end groups, but a much smaller number.)

   b. The finding of RNA-initiating sites on both strands of phage λ was unusual, because previous studies with various other phages had shown that RNA synthesized *in vivo* hybridized exclusively with one strand. Furthermore, genetic evidence known then also seemed to show that the genes on the *E. coli* chromosome were all read in the same direction; since RNA synthesis proceeds in one direction only, this must mean that all the "sense" genes (as opposed to their transcriptionally meaningless complements) were on one and the same strand. Since then, a number of other systems have yielded genetic evidence to show that "sense" genes can occur on different strands.

2. a. The strands in Fig. 1 are drawn of unequal lengths because λ has "cohesive" ends, that is, single-stranded, complementary ends, which allow λ DNA to form a hydrogen-bonded circle.

   b. When a double-stranded DNA with cohesive ends becomes circular, the single-stranded terminal sequences will be complementary, but not the ends. The structure keeps its shape because the gaps do not face each other and are separated by a hydrogen-bonded region.

   c. Phage λ DNA has the property of forming

two "halves" when mildly sheared, and these can be separated by density centrifugation, thanks to their different base composition. By analyzing the halves of the mutant λcb$_2$ DNA (in which the b$_2$ region is deleted), the base compositions of the left and the right arms in Fig. 1 can be determined. The base composition of the b$_2$ region can be calculated by comparing the compositions of the mutant and wild-type DNAs.

d. The region b$_2$ is not transcribed into mRNA in Fig. 1, because a mutant in which this region is deleted was used for the experiments. The b$_2$ region or part of it is also necessary, however, for the integration of the phage into the host chromosome, and it does seem to provide a structural feature. One can imagine that for integration it might be necessary for the phage DNA to recognize the base sequences on the host DNA where homologous pairing is possible; such a structure would not require transcription. For integration there is also need for an enzyme, but this is specified by the *int* gene, rather than b$_2$.

e. In order to position a given gene on one of the strands of λ, the separated strands are used to form heteroduplexes *in vitro* with the isolated strands of a phage mutated in the gene being positioned as well as in the genes directing DNA synthesis. The two heteroduplexes, H$^+$L$^-$ and H$^-$L$^+$, are then introduced into spheroplasts. Since there is no DNA synthesis, only the heteroduplex having the wild-type genes in the transcribable strand will be expressed. Since the direction of the polynucleotide chain is known for the H and L strands, the direction of transcription can be determined by positioning the genes on a given strand. Further, since λ can be broken in half by hydrodynamic shear, genes can be positioned on the right or left by introducing the isolated halves into spheroplasts and following the expression of the genes under suitable conditions.

f. The deficient phages are replicated in the prophage state together with the host chromosome.

3. a. *Sus* stands for "suppressor sensitive." *Sus* mutants are suppressible *amber* or *ochre* mutants. In amber and ochre mutants a codon for an amino acid has been mutated into the chain-terminating triplets UAA and UAG, respectively, giving rise to incomplete peptide chains. Amber and ochre mutations are said to be suppressed when, due to a mutation in some distant part of the chromosome, the bacterium can put an amino acid in place of the "gap," thus permitting a complete protein chain to be synthesized.

b. A bacterial strain is permissive for a *sus* mutant if it possesses a tRNA capable of translating the UAA or UAG triplet into an amino acid.

c. A suppressor tRNA has a mutated anticodon and therefore is not capable of translating a codon correctly—or it would not "read" the chain-terminating triplet. Therefore, the amino acid placed in the peptide chain corresponding to the UAA or UAG is not the "correct" amino acid, that is, the one corresponding to the wild type. Whether or not an amber or ochre mutation is suppressed will depend on whether or not a given protein can tolerate the amino acid corresponding to a given suppressor tRNA. *Sus* strains are obviously specific for amber or for ochre mutations.

d. "To lysogenize" a bacterium with a phage means to "render it lysogenic," that is, to introduce a prophage; this can be done by infecting sensitive bacteria with a temperate phage.

e. λdg is a defective λ phage in which about a third of its DNA has been substituted by a piece of the host chromosome bearing a *gal$^+$* gene (the ability to ferment galactose). λdg is therefore a transducing phage.

f. In lysogenic strains the prophage λ is inserted next to the genes for galactose fermentation and biotin synthesis. When the genes a to N (Fig. 1) or A to O are deleted, upon induction the phage takes with it the adjacent biotin genes instead.

4. a. Hypoxanthine, the base in inosinic acid (I), is deaminated guanine; it therefore corresponds to guanine in hydrogen-bonding specificity, although it forms only two instead of three hydrogen bonds because it lacks the 2-NH$_2$. It just so happens that poly G is rather difficult to prepare by the enzyme polynucleotide phosphorylase (see introduction to Chapter 6), whereas poly I is easily synthesized. Poly IG is used because its synthesis presents less difficulty than poly G; yet it is denser than poly I, and thus aids in the density separation of the two strands.

b. A separation of 14 to 16 mg/cm$^3$ is a reasonable separation, but the peaks might not be completely separated.

c. C and G are the densest of the bases. Therefore, the double strand richest in G-C pairs will be the densest species; this, of course, will be the combination of the phage strand rich in poly dC clusters and the poly IG.

d. Even though an ultraviolet picture of a CsCl gradient shows two completely separated bands, each of these bands, in reality, is a Gaussian distribution, the ends of which extend quite far away from the center of the bands. Even if the bands are well separated, therefore, each will have some molecules of the other strand, even though they may not appear on analysis by UV or radioactivity. The $K_m$ of hybridization, however, is very small, so the self-annealing will "vacuum up" any contaminants.

e. By "functional purity" is meant the ability of the strands to synthesize mRNA complementary only to that strand. This test would not detect contamination by pieces of DNA, for example, unable to synthesize RNA.

f. It is unlikely that mRNA produced by the intact phage would self-anneal, because in no place is there RNA produced by both strands at the same time. Self-annealing can take place only if there are some complementary regions in the mRNA. (It should be remembered that often the single-stranded mRNA has the same base composition as the double-stranded DNA from which it was copied.)

5. a. In line 6 of Table 1 there is a sudden drop in the transcription of W, while that of C increases. The data in the table do not show a cessation, but Fig. 2, B, indicates that transcription of W does virtually stop.

b. Line 9 of Table 1, using a phage with the left arm deleted, shows that (by definition) 20% of the early mRNA comes from the region starting with $xyc_{II}$; line 10, however, shows that 78% of the later mRNA comes from that region. Therefore, it yields predominantly late mRNA.

c. The mRNA in line 2, Table 1, distributed equally over the C and W strands, is nonspecific background, not specific mRNA hybridized to the two strands.

d. In line 12 of Table 1, mRNA produced in the presence of 50 μg CM/ml hybridizes very little with strand C, but the RNA produced in the absence of CM hybridizes practically not at all with strand C. In lines 19, 24, 27, 29, and 31, where less mRNA or the same amount is hybridized in the presence of a "lower" amount of CM than in its absence, the effect may be one of late versus early mRNA, and no conclusions on the preferential effect of CM can be drawn.

e. Much of the early mRNA is produced by gene $c_I$, and therefore prehybridization with t75(λdb$_{a-N}$) DNA would still require separation of the strands in order to follow the synthesis of early mRNA in wild-type λ. If a $c_I$ deletion mutant were used, however, then the prehybridization would make the separation of the strands unnecessary.

6. a. A polar mutation is one in which there is an amber or ochre mutation in some gene in an operon. This severely decreases the expression of genes in the same operon following the mutated gene. The closer the mutation is to the operator, the more severe will be its polar effect.

b. It has not really been established that a polar mutation prevents the synthesis of mRNA by the whole operon. The results of some studies indicated that the mRNAs of operons carrying a polar mutation were shorter than normal, and that they were decreased in amount. This is difficult to understand because amber and ochre mutants do produce incomplete peptide chains in normal amounts, and when polar mutations are suppressed, mRNA obviously is present and can be translated. It has been postulated that mRNA must be translated in order to be released from the DNA, but this is obviously not so, or at least not generally so, since—as we have seen in this paper—mRNA can be synthesized in the presence of protein synthesis inhibitors. This also argues against another possibility—that unused mRNA is destroyed very rapidly.

c. Nonsense mutations used to be defined as point mutations leading to "gibberish," that is, a codon which could not be translated. Now, however, it is known that there are no nonsense codons. The three codons for which no corresponding amino acids exist, namely UAA, UAG, and UGA, are all chain-terminating codons. As we have already seen, the production of a UAA codon is called an amber mutation and that of UAG, an ochre. The third type of chain-terminating mutation is still unnamed, although names have been proposed for it. The codon UGA is thought to be the normal chain-terminating signal, and so far no suppressors for it have been isolated.

## Paper 23

1. a. We have seen in paper 22 that all early proteins are transcribed from the right-hand side of λ. N is on the same strand as the repressor and opposite the region on the other strand comprising genes x and O, major

products of "early" synthesis. There is a promoter to the left of N, which controls the genes from there to *int*. Transcription of this promoter is dependent on a product of N. The latter might also be needed for the transcription of the genes x-O. For example, N could possibly specify an RNA polymerase needed for the transcription of genes N-*int* and x-O, but not for that of $C_1$. Since repressor continues to be made in the prophage state, $C_1$ must be transcribable by host polymerase, but so, it seems, are the other λ genes. Therefore, it seems more likely that the N product is a positive regulator.

b. Since amino acids of very high specific activity are being used, the most minute amount of protein synthesis would show up. Even though mutation in the N gene inhibits *most* synthesis of "early" proteins, it is still possible that a few do escape the inhibition. It is also possible that even though no functional phage proteins are being made, modified or incomplete proteins could be synthesized by the mutant. Moreover, the UV irradiation does not abolish host synthesis 100%. Thus the amount of background synthesis, though only a minute fraction of what it would normally be, probably still exceeds the amount of repressor protein being made, so that the latter would represent only a minor component of the chromatograms. As a matter of fact, it can be seen from Figs. 3 to 6 that the material eluting early from the DEAE column exceeds in amount that of the repressor. Furthermore, the identification of the peaks would be extremely bothersome. It is therefore very unlikely that the repressor could have been isolated without the double-label technique.

2. a. In Table I, the increase with respect to the starting material of the $^3H/^{14}C$ ratio in the supernatant, and its decrease in the pellet, means that in the supernatant there is some material preferentially labeled with $^3H$. From the magnitude of the difference in the ratios, it can be calculated that ([2.13−1.86]÷1.86) × 100 = 14.5% of the $^3H$ counts are in a component devoid of $^{14}C$. (It should be pointed out that the last ratio of Table 1, 1.85, should be 2.16 according to the cpm given; this then gives an average ratio for the pellet of 2.00, and the component devoid of $^{14}C$ is then only 6.5% of the $^3H$ counts.) However, since there should be a decrease in the $^3H/^{14}C$ of the pellet, there may simply be an error in the reported cpm.

b. The recovery of the $^3H$ counts is 14,186/14,866 × 100 = 95%; and the recovery of the $^{14}C$ counts is 6892/7477 × 100 = 92%. The recovery is essentially complete and the same for both isotopes. The 3% difference in the $^3H$ counts can be attributed to counting errors.

c. Strictly speaking, it is impossible to calculate recoveries from the data in Table 1, because only concentrations (per 0.1 ml), and not total counts, are given. They can be used for estimating the recovery only if the pellet was resuspended in buffer corresponding to the original volume of sonicate—which must have been done to obtain the data in Table 1. The ratios, however, are not affected by the volumes.

d. Changes in the isotope ratios are independent of the concentration of material and of the total amount of material present. Therefore, complete recovery is totally irrelevant to obtaining correct isotope ratios.

e. The above statement is applicable only to the case in which the two labeled cultures were mixed first and then fractionated. If they were to be fractionated separately, it would indeed be necessary to pay meticulous attention to the complete history of the extract; the ratios would be meaningful only if the manipulations of the extracts could be kept completely the same at every step. Since the variability in recovery, especially when a difficult step such as the breaking of bacterial cell walls is included, is of the order of 10% to 20%, a large number of experiments would be necessary to establish a meaningful difference in the isotope ratios of fractions processed separately. With mixed extracts, however, two experiments are sufficient, as shown in Table 1.

f. In double-labeling experiments where the ratio of the isotopes is critical, it is important to show that the same relative differences in the ratios can be obtained when the isotopes are reversed. The reason for this is that it is possible that the two isotopic precursors could contain different labeled materials as contaminations, the incorporation of which could falsely change the isotope ratio in a given fraction. Furthermore, the $^3H$ counts are more easily quenchable than the $^{14}C$, so that in certain fractions of a column eluate, for example, where the total concentration of material has dropped below a certain threshold value, the $^3H$ counts would be relatively higher than in the more concentrated starting material, thus creating the impression of a material preferentially labeled with $^3H$.

3. a. The incomplete peptides synthesized by amber mutants would not obscure the recognition of the repressor, for they would not chromatograph the same as the complete molecule. They might appear in the eluate apart from the repressor, and would show a preponderance of the label used for the mutant, which would be the opposite of the label used for the repressor synthesized by the wild type.

b. Suppressible amber mutants are chosen in order to show that the absence of the putative repressor protein is not due to some other defect of the mutant phage: in a suppressor strain, the mutant should synthesize repressor, and the putative repressor protein should be labeled with both isotopes.

c. Any other mutant in which the effects of the mutation can be easily reversed is equally suitable. The best are heat-sensitive mutants (ts), which are unable to make repressor above a certain critical temperature but which function normally at a lower temperature.

4. a. In identifying a certain component by preferential labeling, it is essential to show that only one component has the required excess of a given label. Therefore, when chromatographing or otherwise fractionating a double-labeled extract, it is important to keep track of all the counts. For example, after a column is eluted with a gradient, not all the counts might have been eluted, and some preferentially labeled material could still be on the column. It is then necessary, as was done in the research for this paper, to clean the column with strong alkali and show that the eluent has the correct isotope ratio.

b. The second preferentially labeled peak in Figs. 3 and 5 is so minor that its presence is but an insignificant flaw. Moreover, its presence seems to be linked to that of the major peak, since it is absent from Fig. 6. This makes it likely that it represents some modification of the repressor itself. The author's speculation that the minor peak is an oligomer of the repressor seems plausible.

c. Isotopic purity means only that no other labeled material is present, and it is in no way related to chemical purity. In the present case it must be remembered that the extracts contain the entire contents of the bacteria used plus the coat proteins of the infecting phage. The number of labeled proteins synthesized are only a small fraction of the total unlabeled proteins present. Since the majority of these unlabeled proteins are eluted from a DEAE column by the gradient used, many other proteins must overlap with the major $^3H$ peak. The repressor must therefore be most impure from a chemical standpoint.

# CHAPTER 6
# SYNTHESES OF RNAs IN VITRO

The first enzyme found to be capable of synthesizing long polyribonucleotides was polynucleotide phosphorylase, discovered in the mid-1950s by Grunberg-Managoand Ochoa. This enzyme requires nucleoside diphosphates, a primer, and a divalent ion ($Mg^{++}$), and it catalyzes the reactions:

$$pXpYpZ + (NDP)_n \rightleftharpoons pXpYpZ(pN)_n + nP_i$$

The enzyme does not require a template, and the base ratios in the polymer mirror the relative concentrations of the various substrates. The actual nucleotide sequence is random. The enzyme also catalyzes the reverse reaction, the phosphorolysis of an RNA or a polynucleotide to yield nucleoside diphosphates. A rather high concentration of diphosphates is necessary for the synthetic reaction, and it soon became evident that the diphosphate pools in cells were not concentrated enough to drive the forward reaction; on the other hand, the concentration of $P_i$ was found to be sufficiently high to drive the reverse reaction. *In vivo*, then, polynucleotide phosphorylase must be a degradative enzyme. The enzyme has been most useful, however, for synthesizing the ribopolymers needed for elucidating the genetic code.

With the enunciation of the messenger hypothesis and the discovery of mRNA it became evident that the transcriptase must be an enzyme that can copy a DNA template. Thus looking for a DNA-dependent incorporation of ribonucleoside triphosphates (in analogy with the DNA polymerase), by 1960 several laboratories announced the discovery of RNA polymerase that catalyzes the reaction:

$$n(ATP + GTP + UTP + CTP) \underset{DNA}{\overset{DNA}{\rightleftharpoons}} p(ApGpUpC)_n + nPP_i$$

The enzyme has a strict requirement for a template, but not for a primer, although small oligonucleotides complementary to the template can serve as initiators and primers. For maximum activity both $Mg^{++}$ and $Mn^{++}$ are required. Because of the reversibility of the reaction, the enzyme also catalyzes a DNA-dependent PP-NTP exchange.

The enzyme can transcribe all kinds of DNA, but the relative template activities vary a great deal. It can also transcribe single-stranded as well as double-stranded DNA, but the relative activity in this case varies enormously with the source of the DNA, as well as its history.

When RNA synthesized *in vivo* was isolated and hybridized against the separated strands of the template DNA, it was found that only one strand was transcribed. (Now it is known that RNA may be transcribed from both strands, but no complementary regions are ever transcribed.) By contrast, both strands are copied in the majority of the *in vitro* systems. The strand selection mechanism is still not well understood, but when intact, double-stranded circular DNA is presented to the polymerase, asymmetric transcription takes place.

The majority of the strands are initiated with GTP, and some start with ATP, but pyrimidines are practically never found at the ends. The reaction starts by making an internucleotide bond between the 3'OH of the terminal nucleotide and the 5'phosphate of the second one. The terminal nucleotide thus keeps its triphosphate end group, and the chain grows from the 5'P end.

Initiation and elongation are two separate events, with different requirements. The former is favored by high salt concentration (0.2M) and high substrate concentration (0.1 to 0.4

mM), whereas the latter proceeds at optimum rate in the presence of low concentrations of substrates and salt. Initiation is specifically inhibited by rifampicin and proflavine, whereas elongation is inhibited by actinomycin D. In the presence of low salt concentrations, each polymerase molecule can initiate only once, and the synthesis of RNA is severely inhibited by the RNA product, which remains largely template-bound; with a high salt concentration, however, each polymerase molecule can initiate repeatedly, the reaction remains nearly linear for several hours, and most of the RNA is released from the template.

RNA polymerase treated with urea or SDS can be split up into subunits that can then be separated on polyacrylamide gels: $\alpha$, with a molecular weight of 39,000; $\beta$ and its derivative $\beta'$, with molecular weights of 160,000 and 155,000, respectively; and $\sigma$, weighing 95,000 daltons. The formula of the basic enzyme is $\alpha_2\beta\beta'\sigma$, with a molecular weight of 490,000, but the active enzyme seems to be the dimer.

The $\sigma$ subunit can be removed from the polymerase by chromatography on a phosphocellulose (PC) column. The resulting PC enzyme continues to have most of its activity toward some templates but is practically inactive with others. The $\sigma$ subunit is thus a specificity factor. For example, during infection with $T_4$ phage, the bacterial $\sigma$ is replaced by a virus-induced one; as a result there is a fiftyfold stimulation of transcription from $T_4$ DNA, although the transcription of other templates, such as calf thymus DNA, is hardly affected. RNAs transcribed in the presence of $\sigma$ are much more specific than those transcribed in its absence. Sigma factors induced by phages $T_7$ and $\lambda$ have also been isolated. The possibility of controlling the transcription of specific templates by manipulating the subunits of the enzyme is of obvious applicability to development and morphogenesis, but so far no $\sigma$ factors have been found in the RNA polymerases of eucaryotic cells.

The RNA polymerase can also copy some ribopolynucleotide templates, especially in the presence of $Mn^{++}$, a metal that also permits the DNA polymerase to use ribonucleoside triphosphates as substrates, and the RNA polymerase to use deoxyribonucleoside triphosphates. Despite this lapse in specificity, the RNA polymerase does not seem to be able to copy cellular RNAs. Since RNA viruses must somehow reproduce themselves, it became apparent that they must be able to induce an enzyme capable of using RNA as the template.

The first virus-induced RNA-dependent RNA polymerase (also called RNA synthetase or RNA replicase to distinguish it from the DNA-dependent RNA polymerase) was reported in tobacco leaves infected by tobacco mosaic virus (TMV) by Reddi, but no further progress was made until after the discovery of the RNA phages. These viruses induce an active replicase in bacteria, and soon a number of enzymes were purified. These enzymes also used ribonucleoside triphosphates as substrates, but specifically required an RNA template. As a matter of fact, when intact RNA was used as the template (as opposed to fragmented RNA or synthetic polynucleotides), the various enzymes were absolutely specific for their own RNA. They copied neither the cellular RNAs nor the RNAs of other viruses. In 1965, Spiegelman and his associates (see paper 28) were able to perfectly replicate the RNA of the phage $Q_\beta$, and for the first time, a biologically active nucleic acid had been produced in the test tube. Among a number of experiments that were done in this system, the most remarkable was concerned with applying evolutionary pressure in the test tube. It was shown that if serial transfers were made after progressively shorter times of synthesis, the system adapted by synthesizing progressively smaller viral RNAs, which, however, still had the one indispensable property under the circumstances—the ability to be recognized by the replicase.

The $Q_\beta$ replicase turned out to be a complex between the enzyme and two factors that are present in the uninfected host: one is heat labile and the other is heat stable. The exact function of these factors is still unknown; the enzyme alone functions as a replicase for synthetic templates but not for $Q_\beta$ RNA. Despite the remarkable success in synthesizing infective RNA replicas, the mechanism of the replication is not yet completely elucidated. It seems, however, that the first step is the synthesis of a double-stranded replicative intermediate, which is made from the $5'$ phosphate end to the $3'$OH. *In vitro,* the intact complementary strand synthesizes infective RNA in the absence of host factors, but *in vivo* it seems possible that infectious (+) strands are synthesized conservatively on the double-stranded intermediate.

We have already seen in the introduction to Chapter 5 that RNA released from the template suffers extensive modifications before becoming the mature molecules we know. The most important modification is probably the methylation reaction, which is carried out by methylating enzymes, according to the same reaction used for DNA (introduction to Chapter 3). The RNA methylating enzymes are also specific as to the base, or sugar, methylated, and the type of RNA. Above all, each of the enzymes shows strict species specificity; that is, the exact position of the methylated bases is different for RNAs of different species. Thus normal RNA is not a substrate for the homologous enzyme, because all the methylatable bases have been methylated; the enzyme will, however, incorporate methyl groups into heterologous RNA.

There are a very large number of different minor constituents in tRNA, all of which arise by modifying a precursor tRNA molecule, although in many cases the corresponding enzymes have not been isolated. For example, pseudo uridine (5-ribosyl uridine) and *neo*-guanylic acid (1-ribosyl guanine) arise by transposition of the glycosidic linkage, by a mechanism that is not understood. Then there are thio compounds (2-thiouridine, 6-thiouridine, 2-thiocytosine, 2-methyl thioadenosine, etc.) that derive their sulfur from L-cysteine. Furthermore, the 6-amino group of adenine is capable of reacting with a large number of compounds to give terpenyl and aminoacyl derivatives, and so on.

A different type of modification is exemplified by the enzyme catalyzing the addition and removal of the pCpCpA end group of tRNA. Similarly, it has been shown that the $3'$OH group of a number of viral RNAs is A; yet the complementary strand in the replicative intermediate starts with pppG, which is complementary to the penultimate base of the viral RNA. How the adenosine is added to the viral RNA is not known, but the process does not involve the enzyme responsible for the adenosylation of tRNA.

Still another type of postsynthetic transformation, the selective hydrolysis of large precursor molecules, especially prominent in the processing of ribosomal RNAs, has not yet been carried out in the test tube.

# 24 DEOXYRIBONUCLEIC ACID-DIRECTED SYNTHESIS OF RIBONUCLEIC ACID BY AN ENZYME FROM ESCHERICHIA COLI

Michael Chamberlin*
Paul Berg
*Department of Biochemistry
Stanford University School of Medicine
Stanford, California*

Protein structure is under genetic control;[1-3] yet the precise mechanism by which DNA influences the formation of specific amino acid sequences in proteins is unknown. Several years ago, it was discovered that infection of *Escherichia coli* with certain virulent bacteriophages induces the formation of an RNA fraction possessing both a high metabolic turnover rate and a base composition corresponding to the DNA of the infecting virus.[4-6] The existence of an analogous RNA component in noninfected cells has also been demonstrated; in this instance, however, the base composition of the RNA resembles that of the cellular DNA.[7,8] These observations focused attention on the possible role of this type of RNA in protein synthesis, and some of the evidence consistent with this view has recently been summarized.[9]

Until recently there was no known enzymatic mechanism for a DNA-directed synthesis of RNA. Polynucleotide phosphorylase,[10,11] although it catalyzes the synthesis of polyribonucleotides, does not by itself provide a mechanism for the formation of RNA with a specific sequence of nucleotides. The one instance in which a unique sequence of nucleotides is produced involves the limited addition of nucleotides exclusively to the end of pre-existing polynucleotide chains.[12-14]

Our efforts were therefore directed toward examining alternate mechanisms for RNA synthesis, and in particular one in which DNA might dictate the nucleotide sequence of the RNA. In the present paper, we wish to report the isolation and some properties of an RNA polymerase from *E. coli* which, in the presence of DNA and the four naturally occurring ribonucleoside triphosphates, produces RNA with a base composition complementary to that of the DNA. Within the last year, several laboratories have reported similar findings with enzyme preparations from bacterial as well as from plant and animal sources.[15-24] In the following paper, the effect of enzymatically synthesized RNA on the rate and extent of amino acid incorporation into protein by *E. coli* ribosomes in the presence of a soluble protein fraction is described.

## Experimental procedure
### Materials

Unlabeled ribonucleoside di- and triphosphates were purchased from the Sigma Biochemical Corporation and the California Corporation for Biochemical Research. 8-$C^{14}$-labeled ATP was purchased from the Schwartz Biochemical Company; the other, uniformly labeled, $C^{14}$ ribonucleoside triphosphates were prepared enzymatically from the corresponding monophosphate derivatives[25] isolated from the RNA of *Chromatium* grown on $C^{14}O_2$ as sole carbon source.[26] CTP labeled with $P^{32}$ in the ester phosphate was obtained by enzymatic phosphorylation of $CMP^{32}$ prepared according to Hurwitz.[27] The deoxyribonucleoside triphosphates were obtained by the procedure of Lehman *et al.*[25]

Calf thymus and salmon sperm DNA were isolated by the method of Kay *et al.*[28] DNA from *Aerobacter aerogenes, Mycobacterium phlei,* and bacteriophages T2, T5, T6 was prepared as described previously.[29] DNA from λdg phage was prepared as reported elsewhere.[30] Unlabeled and $P^{32}$ labeled DNA from *E.*

---

From *Proceedings of the National Academy of Sciences U.S.A.* 48:81-94, 1962. Reprinted with permission.

This work was supported by Public Health Service Research Grant No. RG6814 and Public Health Service Training Grant No. 2G196.

The abbreviations used in this paper are: RNA and DNA for ribo- and deoxyribonucleic acid, respectively; poly dT for polydeoxythymidylate; d-AT for the deoxyadenylate-thymidylate copolymer; d-GC for the deoxyguanylate-deoxycytidylate polymer: AMP, ADP and ATP for adenosine-5'-mono-, di-, and triphosphates, respectively. A similar notation is used for the cytidine (C), guanosine (G), and uridine (U) derivatives and their deoxy analogues (dA, dC, dG, dT). $P_i$ is used for inorganic orthophosphate, TMV for tobacco mosaic virus, and DNase and RNase for deoxyribo- and ribonuclease activities, respectively.

*Pre-doctoral fellow.

coli were prepared as previously described.[31] d-AT and d-GC polymers were prepared according to Schachman et al.[32] and Radding et al.,[33] respectively. Transforming DNA from Bacillus subtilis[34] was a gift from E. W. Nester, and DNA from phage ΦX 174[49] was generously supplied by R. L. Sinsheimer. Double-stranded ΦX 174 DNA was synthesized using E. coli DNA polymerase[25] with single-stranded ΦX 174 DNA as primer.[35,36] In this reaction, 2.7 times more DNA was synthesized than had been added as primer. RNA from tobacco mosaic virus was obtained from H. Fraenkel-Conrat, and ribosomal and amino acid-acceptor RNA were isolated from E. coli according to Ofengand et al.[26,37] Nucleic acid concentrations are given as mμmoles of nucleotide phosphorus per ml.

Glass beads, "Superbrite 100," obtained from the Minnesota Mining and Manufacturing Company, were washed as previously described.[25] Streptomycin sulfate was a gift from Merck and Company, and protamine sulfate was purchased from Eli Lilly Company. DEAE-cellulose was purchased from Brown and Company. Crystalline pancreatic RNase and pancreatic DNase were products of the Worthington Biochemical Co.

## Assays

The activities of E. coli-DNA polymerase,[25] -deoxyribonuclease[38] and -DNA diesterase,[31] were determined as previously described, and ribonuclease activity was measured by the disappearance of amino acid-acceptor RNA activity.[26] Polynucleotide phosphorylase was measured by $P_i^{32}$ exchange with ADP as reported by Littauer and Kornberg.[11] Protein was determined by the method of Lowry et al.[39]

The standard assay for RNA polymerase measures the conversion of either $C^{14}$ or $P^{32}$ from the labeled ribonucleoside triphosphates into an acid-isoluble form. Enzyme dilutions were made with a solution containing 0.01 M Tris buffer, pH 7.9, 0.01 M $MgCl_2$, 0.01 M β-mercaptoethanol, $5 \times 10^{-5}$ M EDTA, and 1 mg per ml of crystalline bovine serum albumin. The reaction mixture (0.25 ml) contained: 10 μmoles of Tris buffer, pH 7.9, 0.25 μmole of $MnCl_2$, 1.0 μmole of $MgCl_2$, 100 mμmoles each of ATP, CTP, GTP, and UTP, 250 mμmoles of salmon sperm DNA, 3.0 μmoles of β-mercaptoethanol, and 10 to 80 units of enzyme. One of the nucleoside triphosphates was labeled with approximately 300 to 600 cpm per mμmole. After incubation at 37° for 10 min, the reaction mixture was chilled in ice, and 1.2 mg of serum albumin (0.03 ml) was added, followed by 3 ml of cold 3.5% perchloric acid (PCA). The precipitate was dispersed, centrifuged for 5 min at $15,000 \times g$, and washed twice with 3.0 ml portions of cold PCA. The residue was suspended in 0.5 ml of 2 N ammonium hydroxide, transferred to an aluminum planchet, and after drying, counted in a windowless gas-flow counter.

One unit of enzyme activity corresponds to an incorporation of 1 mμmole of $CMP^{32}$ per hr under the conditions described above. The assay was proportional to the amount of enzyme added up to at least 80 units; thus 6.3, 12.5, and 25 μg of Fraction 4 enzyme incorporated 2.6, 5.1, and 10.0 mμmoles of $CMP^{32}$. The rate of the reaction remained constant for approximately 20 min, and then decreased after this time.

Since the radioactivity incorporated represents only one of the four nucleotides, the observed incorporation must be multiplied by a factor ranging from 3 to 5 for an estimate of the total amount of RNA synthesized. The exact factor depends on the composition of the DNA primer used.

## Results

### Purification of RNA polymerase

*(1) Cells.* E. coli B was grown in continuous exponential phase culture[40] with a glucose-mineral salts medium.[41] Cells stored at −20° showed no loss of activity for over six months. The purification procedure and the results of a typical preparation are summarized in Table 1. Unless noted otherwise, all operations were carried out at 4° and all centrifugations were at $30,000 \times g$ for 15 min in an International HR-1 Centrifuge.

*(2) Extract.* Frozen cells (140 gm) were mixed in a Waring Blendor with 420 gm of glass beads and 150 ml of a solution (buffer A) containing 0.01 M Tris buffer, pH 7.9, 0.01 M $MgCl_2$, and 0.0001 M EDTA. After disruption of the cells at high speed for 15 min (maximum temperature 10°), a further 150 ml of buffer A was added and the glass beads were allowed to settle. The supernatant fluid was then decanted and the residue was washed with 75 ml of buffer A. The combined supernatant fluid and wash was centrifuged for 30 min and the resulting supernatant fluid collected (Fraction 1).

### TABLE 1
*Purification of RNA polymerase from* E. coli

| Fraction | Volume (ml) | Specific activity (units/mg) | Total activity (units) |
|---|---|---|---|
| 1. Initial extract | 260 | 40 | 370,000 |
| 2. Protamine eluate | 37 | 1,600 | 205,000 |
| 3. Ammonium sulfate | 5 | 2,500 | 200,000 |
| 4. Peak DEAE fraction | 2 | 6,100 | 153,000 |

*(3) Streptomycin-protamine fractionation.* Fraction 1 was centrifuged in the Spinco Model L preparative ultracentrifuge for 4 hr at 30,000 rpm in the No. 30 rotor. The protein concentration in the supernatant fluid was adjusted to about 12 mg per ml with buffer A, and β-mercaptoethanol was added to a final concentration of 0.01 $M$. To 350 ml of the diluted supernatant solution was added 17.5 ml of a 10% (w/v) solution of Streptomycin sulfate with stirring. After 15 min, the solution was centrifuged, and to 350 ml of the supernatant fluid was added 14.0 ml of a 1% (w/v) solution of protamine sulfate. The precipitate, collected by centrifugation, was washed by suspension in 175 ml of buffer A containing 0.01 $M$ β-mercaptoethanol. The washed precipitate was then suspended in 35 ml of buffer A containing 0.01 $M$ mercaptoethanol and 0.10 $M$ ammonium sulfate, centrifuged for 30 min, and the supernatant fluid was collected (Fraction 2).

*(4) Ammonium sulfate fractionation.* To 37 ml of Fraction 2 was added 15.8 ml of ammonium sulfate solution (saturated at 25° and adjusted to pH 7 with ammonium hydroxide). The mixture was stirred for 15 min, and the precipitate was removed by centrifugation. To the supernatant liquid was added an additional 16.2 ml of the saturated ammonium sulfate, and after 15 min the precipitate was collected by centrifugation for 30 min and dissolved in buffer B (0.002 $M$ KPO$_4$, pH 8.4, 0.01 $M$ MgCl$_2$, 0.01 $M$ β-mercaptoethanol, and 0.0001 $M$ EDTA) to a final volume of 5.0 ml (Fraction 3).

*(5) Adsorption and elution from DEAE-cellulose.* Fraction 3 was diluted to a protein concentration of about 3 mg per ml with buffer B and passed onto a DEAE-cellulose column (10 cm × 1 cm$^2$, washed with 150 ml of buffer B just prior to use) at a rate of about 0.5 per min. The column was washed with 10 ml of buffer B and then with enough of the same buffer containing 0.16 $M$ KCl to reduce the absorbency of the effluent at 280 m$\mu$ to less than 0.05. The enzyme was eluted from the column with buffer B containing 0.23 $M$ KCl. The activity appears within the first five ml of the latter eluant (Fraction 4).

*(6) Properties of the purified enzyme.* The specific activity of enzyme Fraction 4 was from 140 to 170 times greater than that of the initial extract. The purification as described here has been quite reproducible, with specific activities in the final fraction ranging from 5,500 to 6,100. The enzyme preparation (Fraction 4) has a ratio of absorbencies at 280 and 260 m$\mu$ of 1.5.

Fraction 4, stored at 0 to 2°, retains more than 90 per cent of its activity for up to two weeks and 40 to 60 per cent of the original activity after one month. Enzyme Fractions 1 through 3 are unstable, losing up to 30 per cent of their activity on overnight storage under a variety of conditions. Because of the marked instability of these earlier fractions, it is advisable to carry out the purification without stopping at intermediate stages.

*(7) Contaminating enzymatic activities.* Aliquots (100 $\mu$g) of Fraction 4 were assayed for contaminating enzymatic activities. This amount of enzyme catalyzed an initial rate of incorporation of 2,000 m$\mu$moles of nucleotide per hr. No detectable DNA polymerase was found (< 0.6 m$\mu$mole DNA per hr). DNase activity was barely detectable under conditions optimal for RNA polymerase. With either heated or unheated P$^{32}$ DNA as substrate, no more than 0.13 m$\mu$mole of acid-soluble P$^{32}$ was released during the course of a 30-min incubation. There was only slight RNase activity associated with Fraction 4. When 100 $\mu$g of the purified enzyme were incubated with 4 $\mu$moles of purified acceptor RNA for 1 hr, there was no detectable inactivation of leucine-acceptor activity. Under similar conditions, 1 mg of enzyme produced a 30 per cent decrease in leucine-acceptor activity. With conditions optimal for RNA polymerase, sufficient polynucleotide phosphorylase activity was present to catalyze the exchange of 6.7 m$\mu$moles of P$_i^{32}$ into ADP per hour.

*Requirements for the RNA polymerase reaction*

With the purified enzyme, RNA synthesis was dependent on the addition of DNA, a divalent cation, and the four ribonucleoside triphosphates (Table 2). In a later section, we shall describe a reaction in which ATP is converted to an acid-insoluble form in the absence of the other three triphosphates. Omission of β-mercaptoethanol from the reaction mixture resulted in a 50 per cent loss in activity; however, dilution of the concentrated enzyme into solutions not containing a sulf-

## TABLE 2
*Requirements for RNA synthesis*

| Components | Incorporation of $CMP^{32}$ ($m\mu moles$) |
|---|---|
| Complete system | 7.3 |
| minus $Mn^{++}$ | 4.3 |
| minus $Mg^{++}$ | 5.6 |
| minus $Mn^{++}$ and $Mg^{++}$ | <0.03 |
| minus DNA | <0.03 |
| minus ATP, GTP, UTP | 0.09 |
| minus enzyme | <0.03 |

The standard system and assay procedure were used with 7.4 µg of Fraction 4 protein in each tube, except that $MgCl_2$ was omitted from the enzyme diluent.

## TABLE 3
*The requirement for ribonucleoside triphosphates in RNA synthesis*

| Components | Incorporation of $CMP^{32}$ ($m\mu moles$) |
|---|---|
| Complete system | 4.6 |
| minus ATP | 0.08 |
| minus UTP | <0.03 |
| minus GTP | <0.03 |
| ATP, UTP, GTP replaced by dATP, dTTP, dGTP | 0.05 |
| ATP, UTP, GTP replaced by ADP, UDP, GDP | 0.29 |

The standard system and assay procedure were used except that 250 mµmoles of calf thymus DNA were used as primer. 13 µg of Fraction 4 protein were used in each assay. 100 mµmoles of each nucleotide were added to each assay.

## TABLE 4
*The effect of different nucleic acid preparations on the rate of RNA synthesis by RNA polymerase*

| Source of primer | Incorporation of $CMP^{32}$ * |
|---|---|
| DNA | |
| Salmon sperm | 100 |
| Calf thymus | 43 |
| E. coli | 34 |
| φX 174 | 38 |
| B. subtilis | 27 |
| λdg phage | 25 |
| T2 phage | 45 |
| T6 phage | 30 |
| T5 phage | 74 |
| RNA | |
| E. coli amino acid-acceptor | <0.5 |
| E. coli ribosomal | <0.5 |
| TMV | <0.5 |

*The incorporation value for salmon sperm DNA was 5.3 mµmoles and is set at 100 for comparison with the other primers.

Assay system and procedure as previously described, except that 100 mµmoles of each nucleic acid were used in place of the usual primer. 7.4 µg of Fraction 4 protein were used in each assay.

## TABLE 5
*The effect of denaturation on the ability of DNA preparations to prime for RNA synthesis*

| DNA | Incorporation of $CMP^{32}$ Native ($m\mu moles$) | Heated |
|---|---|---|
| Calf thymus | 2.7 | 2.3 |
| Salmon sperm | 6.1 | 2.5 |
| T6 phage | 1.9 | 0.8 |
| E. coli | 1.8 | 1.9 |

Assay procedure as described previously, except that the usual primer was replaced by 200 mµmoles of the DNA to be tested. 7.4 µg of Fraction 4 protein were used in each assay. The DNA samples were heated for 10 min at 95 to 99° in 0.05 M NaCl and rapidly cooled in an ice bath. The absorbencies of the heated DNA preparations were 30 to 40 per cent higher than those of the unheated preparations at 260 mµ.

hydryl compound resulted in as much as 90 per cent inactivation. The optimal pH for the reaction was between 7.8 and 8.2. At pH 6.1, 7.0, and 8.9 the activities were 13, 62, and 84 per cent, respectively, of the maximal value.

*(1) Nucleoside triphosphate specificity.* All of the ribonucleoside triphosphates are required for RNA synthesis (Table 3). The deoxyribonucleoside triphosphates do not function as substrates in the reaction, and the ribonucleoside diphosphates support synthesis only at a greatly reduced rate. The observed activity of the diphosphates may be due to the presence of small amounts of the nucleoside triphosphates in the diphosphate preparations or to the formation of the triphosphates through the action of nucleoside diphosphate kinase.

With $CTP^{32}$ as the labeled substrate and salmon sperm DNA as primer, variation of the concentrations of all four ribonucleoside triphosphates as a group produced a variation in the rate of RNA synthesis. When the data were plotted according to Lineweaver and Burk,[42] a

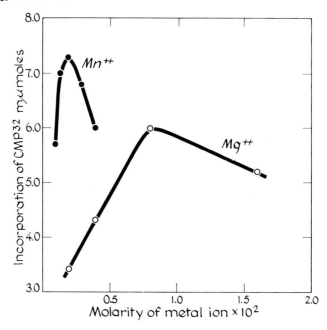

**Fig. 1.** The influence of metal ion concentration on the rate of CMP$^{32}$ incorporation. Standard assay conditions were used, except that the metal ion concentration was varied as shown. 7.4 µg of Fraction 4 enzyme were added to each assay.

○———○ Mg$^{++}$ alone added to the assay mixture
●———● Mn$^{++}$ alone added to the assay mixture

linear relationship was obtained from which it was calculated that the rate of synthesis was half maximal when the concentration of each of the triphosphates was $1.3 \times 10^{-4}$ M. A similar value ($1.4 \times 10^{-4}$ M) was obtained using C$^{14}$ ATP as a label and calf thymus DNA as primer.

*(2) The nature of the primer.* All DNA samples tested were active in promoting ribonucleotide incorporation, although the efficiency varied significantly (Table 4). Amino acid-acceptor RNA, ribosomal RNA from *E. coli*, and TMV RNA did not substitute for DNA. With the synthetic copolymer d-AT as primer, only AMP and UMP were incorporated, and only GMP and CMP were incorporated in the presence of d-GC polymer (Table 7). It should be noted, however, that GTP was incorporated to a considerably greater extent in the latter case. A qualitatively similar finding has been reported for the incorporation of dGMP and dCMP by DNA polymerase with d-GC as primer.[33]

Increasing the amount of DNA in an assay mixture over the range 0 to 200 mµmoles resulted in an increase in nucleotide incorporation. Further increases in the amount of DNA, up to 400 mµmoles, had no effect on the rate of RNA synthesis. A similar experiment using calf thymus DNA as primer gave a saturating value of 250 mµmoles.

The effect of disrupting the DNA double helix by heating[43] is shown in Table 5. It is seen that with several of the DNA preparations there is a significant decrease in the rate of CMP$^{32}$ incorporation using the heated DNA, while with others the effect is insignificant. The ability of single-stranded DNA to function as a primer for RNA synthesis is further emphasized by the activity of the single-stranded DNA from ΦX 174 phage.

*(3) Metal ion requirements.* Optimal concentrations for Mn$^{++}$ and Mg$^{++}$ when added separately to the reaction mixture were $2 \times 10^{-3}$ M and $8 \times 10^{-3}$ M, respectively (Figure 1). Addition of Mg$^{++}$ increased the rate at suboptimal levels of Mn$^{++}$; thus, the addition of $10^{-3}$ M Mn$^{++}$ and $4 \times 10^{-3}$ M Mg$^{++}$ to the

**TABLE 6**
*Net synthesis of RNA*

| Source of DNA primer | Labeled nucleotide incorporated (mµmoles) | Calculated amount of RNA formed* (mµmoles) | Ratio of RNA isolated to DNA added | Method of isolation |
|---|---|---|---|---|
| | CMP$^{32}$ | | | |
| Calf thymus | 81 | 200 | 2.0 | A |
| φX 174 phage | 90 | 510 | 5.1 | A |
| T2 phage | 78 | 360 | 3.6 | A |
| T2 phage | 72 | 410 | 4.1 | B |
| | C$^{14}$-AMP | | | |
| T2 phage | 150 | 460 | 4.6 | C |
| T5 phage | 152 | 500 | 5.0 | C |
| d-AT copolymer | 155 | 310 | 15.0 | C |

*The amount of RNA in the isolated product was calculated from the amount of label incorporated and the base ratio of the primer DNA.

*Synthesis:* Each tube contained in a final volume of 0.5 ml: 20 µmoles of Tris buffer, pH 7.85; 8 µmoles of MgCl$_2$; 400 mµmoles each of ATP, CTP, UTP, GTP; 6 µmoles of β-mercaptoethanol; 100 mµmoles of DNA; and 100 µg of Fraction 4 protein. When d-AT was used as primer, only 20 mµmoles of primer were added; and CTP and GTP were omitted from the mixture. The incubation time was 3 hr at 37°.

*Product isolation:* A. The incubation mixture was heated for 10 min at 60° in 0.4 M NaCl, then dialyzed 36 hr against 0.2 M NaCl-0.01 M Tris, pH 7.85. B. The reaction mixture was extracted two times with phenol and the phenol fractions were washed two times with 0.4 M NaCl. The aqueous layers were pooled and dialyzed as in A. C. The product was precipitated from the incubation mixture with a solution containing 60 per cent ethanol and 0.5 M NaCl at 0°, washed once with the same solution, and dissolved in 1 ml of 0.2 M NaCl, then dialyzed as in A.

same reaction mixture gave a rate of incorporation equal to that found with the optimal concentration of Mn$^{++}$ alone (2 × 10$^{-3}$ M).

## Characterization of the enzymatically synthesized RNA

*(1) Net synthesis of the RNA product.* With two times the level of the four ribonucleoside triphosphates and five to ten times the amount of enzyme used in the routine assay, the amount of RNA formed during an extended incubation exceeded the amount of DNA added to the reaction. We will designate this as "net synthesis." With most of the DNA preparations used, the amount of RNA formed was three to five times greater than the amount of DNA added, while with d-AT copolymer, up to 15 times as much of the corresponding AU polynucleotide was produced (Table 6). The rate of synthesis decreased after the first 20 min although further synthesis occurred up to two hr. Preliminary experiments indicate that this was not due to enzyme inactivation nor to destruction of the priming DNA, but other possibilities have not yet been investigated in detail.

*(2) Enzymatic and alkaline degradation of the product.* Exposure of the isolated "net synthesis" product to alkali converted > 98 per cent of the label to acid-soluble products which were electrophoretically identical with the 2'-(3') nucleoside monophosphates. Treatment with pancreatic DNase or *E. coli* DNA diesterase[31] produced no significant liberation of labeled acid-soluble products.

Treatment of 10 to 20 mµmoles of enzymatically prepared CMP$^{32}$-labeled RNA with 0.1 µg of pancreatic RNase for 1 hr liberated 75 to 94 per cent of the P$^{32}$ label as acid-soluble products. The amount of acid-insoluble P$^{32}$ remaining after RNase treatment varied with different DNA primers and different methods of product isolation. Using 10 times the amount of RNase did not appreciably alter the results. The significance of this RNase-resistant fraction is presently unknown.

*(3) Nucleotide composition.* The nucleotide compositiin of the product was examined by two different methods. In the first method, four separate assays, each containing a different labeled nucleoside triphosphate, were performed with each DNA preparation, and the molar ratio in which the labeled nucleotides

## TABLE 7
*Nucleotide composition of the RNA product*

| DNA primer | Method of analysis | Nucleotide Composition AMP | UMP | GMP | CMP | Primer* $\dfrac{A+T}{G+C}$ | Product $\dfrac{A+U}{G+C}$ | Product $\dfrac{A+G}{U+C}$ |
|---|---|---|---|---|---|---|---|---|
| | | | (mμmoles) | | | | | |
| d-GC polymer | A | <0.03 | <0.03 | 1.90 | 0.23 | — | — | — |
| d-AT copolymer | B | 21.7 | 20.0 | — | — | — | — | — |
| d-AT copolymer | A | 20.8 | 22.2 | <0.03 | <0.03 | — | — | — |
| T2 phage | B | 7.7 | 7.4 | 4.3 | 4.3 | 1.76 | 1.76 | 1.03 |
| T5 phage | B | 4.8 | 5.0 | 3.6 | 3.4 | 1.56 | 1.40 | 1.00 |
| E. coli | A | 1.8 | 1.9 | 1.9 | 2.0 | 1.01 | 0.95 | 0.95 |
| M. phlei | A | 3.7 | 4.0 | 7.9 | 8.5 | 0.48 | 0.47 | 0.93 |
| A. aerogenes | A | 1.8 | 1.7 | 2.3 | 2.2 | 0.80 | 0.78 | 1.05 |

*The values given for the ratio A + T/G + C in the priming DNA are those found by Josse et al.[29] except in the case of phage T5 DNA.[52]

*Method A:* For each DNA sample, four separate incubations were used, each containing a different $C^{14}$-labeled nucleotide. The amounts of DNA used in the various tests were as follows: 20 mμmoles of *M. phlei*, 50 mμmoles of *A. aerogenes*, 180 mμmoles of *E. coli*, 20 mμmoles of d-AT, 20 mμmoles of d-GC. 12.5 μg of Fraction 4 enzyme were used in each incubation; all other conditions were those given for a standard assay.

*Method B:* The synthesis of the C24-labeled RNA was carried out under the following conditions. The reaction mixture (0.5 ml) contained: 20 μmoles of Tris buffer, pH 7.85; 0.5 μmole of $MnCl_2$; 2 μmoles of $MgCl_2$; 6 μmoles of β-mercaptoethanol; 100 mμmoles each of $C^{14}$-ATP, $C^{14}$-UTP, $C^{14}$-GTP, $C^{14}$-CTP; 100 mμmoles of DNA; and 180 μg of Fraction 4 enzyme. Where d-AT primer was used, 20 mμmoles of primer were added and no CTP or GTP were added. After 180 min at 37°, the product was precipitated and washed with cold 3 per cent PCA and incubated in 0.3 M KOH for 18 hr at 37°. An aliquot to which carrier nucleotides had been added was subjected to paper electrophoresis at pH 3.5 in 0.05 M citrate buffer. The individual nucleotides which were visualized with a UV lamp were eluted in 0.01 M HCl and counted. Recovery of the $C^{14}$-label in the eluted fractions was >95 percent. 180 mμmoles, 140 mμmoles, and 300 mμmoles of polyribonucleotide were produced in the reactions primed with T2 DNA, T5 DNA, and d-AT, respectively.

## TABLE 8
*Comparative behavior of single- and double-stranded φX 174 DNA as primer for RNA synthesis*

| State of DNA used as primer | | Nucleotide Composition of RNA AMP | UMP | GMP | CMP |
|---|---|---|---|---|---|
| | | | (per cent) | | |
| Single-stranded | Predicted* | 32.8 | 24.6 | 18.5 | 24.1 |
| Single-stranded | Found by method A | 32.0 | 24.1 | 19.5 | 24.3 |
| Single-stranded | Found by method B | 35.0 | 24.6 | 19.3 | 21.1 |
| Double-stranded | Predicted* | 28.7 | 28.7 | 21.3 | 21.3 |
| Double-stranded | Found by method B | 28.9 | 29.1 | 20.9 | 20.9 |

*Method A.* Conditions as given in Table 7. 32 mμmoles of single-stranded φX 174 DNA were used in each incubation with 8 μg of Fraction 4 enzyme.

*Method B.* Conditions as given in Table 7. With single-stranded DNA as primer, 25 mμmoles of priming DNA were added, 71 mμmoles of RNA were produced in a 60 min incubation with 80 μg of Fraction 4 enzyme. For the double-stranded DNA, 26 mμmoles of priming DNA were added; 32 mμmoles of RNA were produced in a 60 min incubation with 40 μg of Fraction 4 enzyme.

*The predicted values were calculated on the assumption that the single-stranded φX 174 DNA would yield RNA with a composition complementary to the composition reported by Sinsheimer.[49] Upon replication of φX 174 DNA with DNA-polymerase it was assumed that the product (presumably double-stranded DNA) had a base composition which is the average of the composition of the original and of the newly synthesized strands.

That this is a reasonable assumption is shown by unpublished studies of M. Swartz, T. Trautner, and A. Kornberg. When φX 174 DNA was used to prime limited (<30 per cent) or extensive (600 per cent) DNA synthesis the composition of the newly formed DNA was:

| | dAMP | TMP | dGMP | dCMP |
|---|---|---|---|---|
| Limited synthesis | 31.0 | 24.1 | 20.1 | 24.5 |
| Extensive synthesis | 29.4 | 26.9 | 22.3 | 21.3 |

## TABLE 9
*Requirements for polyadenylic acid formation*

| System | Incorporation of AMP (mµmoles) |
|---|---|
| Complete (with ATP as the only nucleoside triphosphate) | 9.9 |
| minus DNA | <0.03 |
| minus Mn$^{++}$ | 2.5 |
| minus Mg$^{++}$ | 8.7 |
| plus RNase | 6.9 |
| plus DNase | 0.3 |
| plus ADP | 7.6 |

The reaction mixture contained in a final volume of 0.25 ml; 10 µmoles of Tris buffer, pH 7.85; 0.5 µmole of MnCl$_2$; 2 µmoles of MgCl$_2$; 3 µmoles of β-mercaptoethanol; 100 mµmoles of C$^{14}$-ATP, 280 mµmoles of calf thymus DNA; and 3 µg of Fraction 4 RNA polymerase. Where indicated, 25 µg of pancreatic RNase, 25 µg of pancreatic DNase, and 100 mµmoles of ADP were added. The incubation time was 10 min at 37°.

## TABLE 10
*Incorporation of single nucleotides by RNA polymerase*

| Nucleotide added | Nucleotide incorporation (mµmoles) |
|---|---|
| C$^{14}$-ATP | 23 |
| C$^{14}$-UTP | 0.90 |
| C$^{14}$-GTP | 0.09 |
| CTP$^{32}$ | 0.07 |

The conditions were the same as those described in Table 9, except that ATP was replaced where indicated by an equal amount of each of the other nucleoside triphosphates. 6 µg of Fraction 4 enzyme were added.

were incorporated was measured (Method A). The second method utilized electrophoretic separation[44] of the mononucleotides resulting from the alkaline degradation of a "net synthesis" product in which all of the nucleoside triphosphates were labeled with C$^{14}$ (Method B). The distribution of the label among isolated nucleotides was therefore a measure of the composition of the newly synthesized RNA. The results (Table 7) indicate that the gross composition of the product at all stages of synthesis was complementary to that of the primer within the accuracy of the method. For double-stranded DNA, this complementary relationship becomes one of identity, since in the priming DNA adenine equals thymine, and guanine equals cytosine. However, in the case of single-stranded ΦX 174 DNA (Table 8), the composition is indeed complementary to that of the DNA, and in this instance the amounts of AMP and UMP incorporation and of GMP and CMP incorporation are not equal. Furthermore, when double-stranded ΦX 174 DNA is used, the nucleotide composition of the resulting RNA is again identical to that of the DNA primer.

*(4) Sedimentation velocity of the isolated product.* The sedimentation velocity of the isolated RNA product was determined in the Spinco Model E analytical ultracentrifuge using ultraviolet optics. Values obtained ($S_{20}$) in 0.2 M NaCl–0.01 M Tris, pH 7.9, ranged from 6 to 7.5 for 2- to 15-fold "net synthesis" products prepared by phenol extraction or by salt-ethanol precipitation.

### DNA-dependent formation of polyadenylic acid

As pointed out earlier, RNA synthesis, as measured by the incorporation of either labeled CTP, UTP, or GTP did not occur in the absence of the other three nucleoside triphosphates or, in fact, in the absence of any one of the nucleoside triphosphates. It was therefore surprising to find that purified fractions of RNA polymerase catalyze the conversion of C$^{14}$-ATP to an acid-insoluble form in the absence of the other three ribonucleoside triphosphates. The ratio of the activities

$$\frac{\text{AMP incorporated in the absence of UTP, CTP, GTP}}{\text{AMP incorporated in the presence of UTP, CTP, GTP}}$$

increased from 0.5 to 10 as purification of the enzyme progressed.

*(1) Requirements for polyadenylic acid formation.* Polyadenylic acid formation from ATP occurred only in the presence of DNA, a divalent cation, and the purified enzyme (Table 9). Note that addition of unlabeled ADP produces only a small dilution of the incorporation of label from C$^{14}$-ATP. The rate of incorporation was directly proportional to the amount of enzyme added; 1.8, 3.6, and 7.2 µg of Fraction 4 enzyme catalyzed the incorporation of 4.0, 8.2, and 17.5 mµmoles of C$^{14}$-AMP in a standard 10 min assay. The rate of incorporation remained constant up to over 75 per cent utilization of the added ATP.

### TABLE 11
*Priming efficiency of various nucleic acid preparations for polyadenylic acid formation*

| Primer | AMP incorporation (m$\mu$moles) |
|---|---|
| Calf thymus DNA | 10 |
| Salmon sperm DNA | 7.7 |
| T2 phage DNA | 4.9 |
| d-AT copolymer | <0.03 |
| Amino acid-acceptor RNA | 0.35 |
| Polyadenylic acid | 0.07 |

The conditions of the incubation were as described in Table 9, except that the following amounts of nucleic acid were added: 300 m$\mu$moles of salmon sperm DNA, 200 m$\mu$moles of T2 phage DNA, 12 m$\mu$moles of d-AT, 110 m$\mu$moles of amino acid-acceptor RNA, and 5 m$\mu$moles of polyadenylic acid. 3 $\mu$g of Fraction 4 enzyme were added.

### TABLE 12
*The effect of deoxyribonuclease addition during polyadenylic acid synthesis*

| Tube | Treatment | AMP incorporation (m$\mu$moles) |
|---|---|---|
| 1 | 5-min incubation | 7.1 |
| 2 | 10-min incubation | 16.2 |
| 3 | 10-min incubation | 6.9 |

The reaction mixtures were as described in Table 9, except that 6 $\mu$g of Fraction 4 was used. Tube 1 was incubated for 5 min, heated for 3 min at 100°, and then assayed as usual. Tube 2 was incubated for 10 min before assaying. Tube 3 was incubated for 5 min and heated as in the case of tube 1; 25 $\mu$g of pancreatic DNase were then added and the mixture incubated for an additional 5 min. At this time, 6 $\mu$g of fresh RNA polymerase were added and a third 5-min incubation was allowed.

There was no incorporation of CMP or GMP when the corresponding nucleoside triphosphates were added singly to the reaction, although UMP incorporation occurred to a small, but significant, extent (Table 10).

*(2) The DNA requirement for polyadenylic acid formation.* The ability of various nucleic acid preparations to support polyadenylic acid synthesis is shown in Table 11. Note that neither RNA nor polyadenylic acid itself replaced the DNA requirement. To test whether DNA might be necessary only to initiate polyadenylic acid synthesis, an experiment was performed in which the priming DNA was destroyed after some polyadenylic acid formation had already occurred. It can be seen that destruction of the DNA by DNase blocked further synthesis of the polyadenylic acid (Table 12). This implies that the DNA is required not only for the initiation of polyadenylic acid synthesis but also for the continued formation of the polynucleotide.

### TABLE 13
*Effect of the other ribonucleoside triphosphates on polyadenylic acid formation*

| Component | AMP incorporation (m$\mu$moles) |
|---|---|
| Complete system | 26 |
| plus CTP | 6.0 |
| plus UTP | 5.2 |
| plus GTP | 2.0 |
| plus CTP, UTP | 1.3 |
| plus CTP, GTP | 0.6 |
| plus UTP, GTP | 0.5 |
| plus UTP, GTP, CTP | 2.2 |

Complete system as in Table 9, except that 6 $\mu$g of Fraction 4 protein were added. Where indicated, 100 m$\mu$moles of each nucleoside triphosphate were added.

*(3) The effect of the other ribonucleoside triphosphates on polyadenylic acid formation.* The addition of the other ribonucleoside triphosphates resulted in an inhibition of the rate of $C^{14}$ AMP incorporation (Table 13). It can be seen, for example, that in the presence of any two of the other triphosphates the amount of polyadenylic acid formed is less than 5 per cent that of the control in which only ATP was added. As has been previously shown, in the presence of all four triphosphates, AMP is incorporated into a product having a base composition determined by the DNA primer, and hence under these conditions polyadenylic acid synthesis does not appear to occur.

*(4) Characterization of the polyadenylic acid product.* Preliminary characterization of the product is consistent with its identity as a polyadenylic acid. The addition of pancreatic RNase to the assay system lowered the rate of incorporation only slightly (about 30%). Treatment with 0.5 M KOH for 18 hr at 37° converted the product to an acid-soluble form. Of the $C^{14}$ in the hydrolysate, 97 per cent was associated with 2'-(3') AMP on paper chroma-

tography[44] and paper electrophoresis,[45] less than 1.5 per cent with adenosine, and less than 1.5 per cent was found in a region corresponding to adenosine 3'-5' diphosphate. This implies that the minimum chain length of the polyadenylate is in the order of 60 to 70 nucleotide residues.

## Discussion

There is a striking similarity between the reactions catalyzed by the RNA polymerase described here and *E. coli* DNA polymerase.[25] Both use only the nucleoside triphosphates as nucleotidyl donors, and both display absolute requirements for a divalent cation and a DNA primer for polynucleotide synthesis.* In both cases, some ambiguity exists as to the relative efficiency of single- as compared to double-stranded DNA for priming of polynucleotide synthesis. In each reaction, both forms of DNA are active as primers, but a meaningful comparison between the two with regard to the mechanism of priming must await a more detailed physical and chemical characterization of the different DNA preparations, and further purification of the enzymes involved.

The product formed by RNA polymerase, as in the analogous case of DNA polymerase,[29] has a base composition which, within experimental error, is complementary to that of the priming DNA. This finding, which is in agreement with the results obtained by others,[16,17,21] supports the view that the nucleotide sequences in the DNA direct the order of nucleotides in the enzymatically synthesized RNA. A more critical test of this hypothesis involves a comparison of the nucleotide sequence of the priming DNA and the newly synthesized RNA. In this regard, Furth *et al.*[19] have shown that the repeating sequence of dAMP and dTMP in d-AT copolymer is faithfully replicated by the RNA polymerase in the form of an alternating AMP and UMP sequence. More recently Weiss and Nakamoto[46] have shown that RNA synthesized with an RNA polymerase from *M. lysodeikticus* contains the same frequencies of dinucleotide pairs as occur in the DNA primer. Additional experiments[47] which demonstrate the formation of a DNA-RNA complex after heating and slow-cooling[48] suggest that the homology of nucleotide sequences may occur over relatively long regions.

Does RNA polymerase copy the sequence of only one or both strands of DNA? This question is relevant not only to an understanding of the enzymatic copying mechanism but also to any speculations as to the mechanism of information transfer from DNA to RNA. The fact that with double-stranded DNA primers the base composition of the newly made RNA is essentially identical to the over-all composition of both strands of the DNA already suggests that each strand can function equally well. An alternative hypothesis is to suppose that only one strand can be copied, and that the "primer" strand has, in the case of every DNA studied, a base composition identical to the average composition of both strands. Using the double-stranded form of ΦX 174 DNA[35,36] in which it is known that the base compositions of the two strands differ,[49] it is possible to test this question directly. The results show that both strands of the duplex serve to direct the composition of the RNA product.

This result still leaves open the question of whether both strands are copied in one replication cycle or whether only one strand is copied at a time and the choice between strands is random. When considering the relevance of this finding to information transfer, one must bear in mind that the existence of artificially produced ends in an isolated DNA preparation may allow RNA formation to proceed from both ends of the double strand. This, however, may not occur with the DNA as it exists in the genome; that is, *in vivo* some structural feature in the chromosomal DNA may cause RNA synthesis to proceed in a unidirectional manner and therefore copy the sequence of only one of the two strands.

The formation of DNA-RNA complexes has been described by several groups of workers,[48,50,51] although only limited information is available concerning their chemical structure and their metabolic and chemical stability. The fact that in the enzymatic reaction net synthesis of RNA occurs argues against the formation of a stable, stoichiometric

---

*Under certain conditions DNA polymerase preparations will, in the absence of DNA, produce d-AT or d-GC, depending upon the nature of the substrates present.[32,33]

complex of RNA and DNA. A further argument against the formation of such a complex is the finding that most of the DNA remaining at the end of the reaction appears to be identical to the DNA added, and no component containing both DNA and newly synthesized RNA was detectable on CsCl gradient centrifugation.[47] Whether some transient complex is formed as an intermediate is somewhat more difficult to assess.

The formation of polyadenylic acid in a DNA-dependent reaction is significant in view of the fact that none of the other ribonucleoside triphosphates, taken singly or even in groups of three, are utilized to any appreciable extent for polynucleotide synthesis. An exception to this is, of course, the situation where the DNA dictates the incorporation of only one or two nucleotides (e.g., with poly dT,[18] d-AT,[19] or d-GC).

Three possibilities which could account for polyadenylic acid synthesis are that it results from (a) a special feature of RNA polymerase itself, (b) a separate polyadenylic acid polymerase, or (c) polynucleotide phosphorylase. The last possibility is least likely because of the absolute requirement for DNA in the initiation and continuation of synthesis, the failure of ADP to give a significant dilution of the incorporation from ATP, the low amounts of polynucleotide phosphorylase activity found in the enzyme preparation as measured by $P_i^{32}$ exchange, the inability of $Mg^{++}$ alone to support maximal rates of synthesis, the lack of polymerization of the other nucleoside triphosphates, and the marked inhibition of polyadenylic acid synthesis by any one or all four of the triphosphates. The question of whether polyadenylic acid synthesis is catalyzed by RNA polymerase or by another enzyme cannot be resolved at the present time.

With regard to the mechanism of the DNA-dependent polyadenylic acid formation, two aspects deserve specific comment. The first concerns the mechanism of the inhibition of polyadenylic acid synthesis by any one or all of the other triphosphates. It should be recalled that polyadenylic acid synthesis does not occur in the presence of all four ribonucleoside triphosphates ($< 3\%$), since under these conditions the base composition of the newly synthesized RNA is very close to that predicted by the composition of the DNA primers. The second notable feature of the reaction is its complete dependence on DNA, and the failure of d-AT to prime polyadenylic acid synthesis. One way to account for these findings is to assume that a sequence of thymidylate residues in the DNA, which does not occur in d-AT, can prime the formation of a corresponding run of AMP residues and, by subsequent "slippage" of one chain along the other, lead to a DNA-dependent elongation of the polyadenylic acid chain. The introduction of any other nucleotide into the growing chain might block or inhibit the sliding process and thereby terminate the growing polyadenylic acid chain.

**Summary**

An RNA polymerase has been isolated from E. coli which in the presence of the four ribonucleoside triphosphates, a divalent metal ion, and DNA synthesizes RNA with a base composition complementary to that of the priming DNA. Both strands of DNA can prime new RNA synthesis. Thus, while single-stranded ΦX 174 DNA yields RNA with a base composition complementary to that of the single-stranded form, double-stranded ΦX 174 DNA (synthesized with DNA polymerase) primes the synthesis of RNA with a base composition virtually the same as that in both strands of the DNA. A novel feature of the RNA polymerase preparations is their ability to catalyze a DNA-dependent formation of polyadenylic acid in the presence of ATP alone. Neither UTP, GTP, nor CTP yields corresponding homopolymers; the DNA-dependent formation of polyadenylic acid is virtually completely inhibited by the presence of the other nucleoside triphosphates.

**REFERENCES**

1. Ingram, V. M., and J. A. Hunt, Nature **178**:792 (1956).
2. Yanofsky, C., and P. St. Lawrence, Ann. Rev. Microbiol. **14**:311 (1960).
3. Fincham, J. R. S., Ann. Rev. Biochem. **28**:343 (1959).
4. Volkin, E., and L. Astrachan, Virology **2**:149 (1956).

5. Volkin, E., these Proceedings **46**:1336 (1960).
6. Nomura, M., B. D. Hall, and S. Spiegelman, J. Mol. Biol. **2**:306 (1960).
7. Yčas, M., and W. S. Vincent, these Proceedings **46**:804 (1960).
8. Gros, F., W. Gilbert, H. Hiatt, P. F. Spahr, and J. D. Watson, Cold Spring Harbor Symposia on Quantitative Biology, vol. 21, in press.
9. Jacob, F., and J. Monod, J. Mol. Biol. **3**:318 (1961).
10. Ochoa, S., and L. Heppel, in The Chemical Basis of Heredity, ed. W. D. McElroy and B. Glass (Baltimore: The Johns Hopkins Press, 1957), p. 615.
11. Littauer, U. Z., and A. Kornberg, J. Biol. Chem. **226**:1077 (1957).
12. Hecht, L. I., M. L. Stephenson, and P. C. Zamecnik, these Proceedings **45**:505 (1959).
13. Canallakis, E. S., and E. Herbert, these Proceedings **46**:170 (1960).
14. Preiss, J., M. Dieckmann, and P. Berg, J. Biol. Chem. **236**:1749 (1961).
15. Weiss, S. B., these Proceedings **46**:1020 (1960).
16. Weiss, S. B., and T. Nakamoto, J. Biol. Chem., PC 18 (1961).
17. Hurwitz, J., Bresler, A., and R. Diringer, Biochem. Biophys. Res. Comm. **3**:15 (1960).
18. Furth, J. J., J. Hurwitz, and M. Goldmann, Biochem. Biophys. Res. Comm. **4**:362 (1961).
19. *Ibid.,* **4**:431 (1961).
20. Stevens, A., Biochem. Biophys. Res. Comm. **3**:92 (1960).
21. Stevens, A., J. Biol. Chem. **236**:PC 43 (1961).
22. Ochoa, S., D. P. Burma, H. Kröger, and J. D. Weill, these Proceedings **47**:670 (1961).
23. Burma, D. P., H. Kröger, S. Ochoa, R. C. Warner, and J. D. Weill, these Proceedings **47**:749 (1961).
24. Huang, R. C., N. Maheshwari, and J. Bonner, Biochem. Biophys. Res. Comm. **3**:689 (1960).
25. Lehman, I. R., M. J. Bessman, E. S. Simms, and A. Kornberg, J. Biol. Chem. **233**:163 (1958).
26. Ofengand, E. J., Ph.D. Thesis, Washington University, St. Louis, Missouri (1959).
27. Hurwitz, J., J. Biol. Chem. **234**:2351 (1959).
28. Kay, E. R. M., N. S. Simmons, and A. L. Dounce, J. Am. Chem. Soc. **74**:1724 (1952).
29. Josse, J., A. D. Kaiser, and A. Kornberg, J. Biol. Chem. **236**:864 (1961).
30. Kaiser, A. D., and D. S. Hogness, J. Mol. Biol. **2**:392 (1960).
31. Lehman, I. R., J. Biol. Chem. **235**:1479 (1960).
32. Schachman, H. K., J. Adler, C. M. Radding, I. R. Lehman, and A. Kornberg, J. Biol. Chem. **235**:3243 (1960).
33. Radding, C. M., J. Josse, and A. Kornberg, unpublished results.
34. Nester, E. W., and J. Lederberg, these Proceedings **47**:56 (1961).
35. Lehman, I. R., Ann. N. Y. Acad. Sci. **81**-3:745 (1959).
36. Lehman, I. R., R. L. Sinsheimer, and A. Kornberg, unpublished results.
37. Ofengand, E. J., M. Dieckmann, and P. Berg, J. Biol. Chem. **236**:1741 (1961).
38. Lehman, I. R., G. G. Roussos, and A. Pratt, J. Biol. Chem., in press.
39. Lowry, O., J. J. Rosebrough, A. L. Farr, and R. J. Randall, J. Biol. Chem. **193**:265 (1951).
40. Monod, J., Ann. Inst. Pasteur **79**:390 (1950).
41. Wiesmeyer, H., and M. Cohn, Biochim. Biophys. Acta **39**:417 (1960).
42. Lineweaver, H., and D. Burk, J. Am. Chem. Soc. **56**:658 (1934).
43. Doty, P., these Proceedings **42**:791 (1956).
44. Markham, R., and J. P. Smith, Biochem. J. **52**:552 (1952).
45. Magasanik, B., E. Vischer, R. Doniger, D. Elson, and E. Chargaff, J. Biol. Chem. **186**:37 (1950).
46. Weiss, S. B., and T. Nakamoto, these Proceedings **47**:1400 (1961).
47. Geiduschek, E. P., T. Nakamoto, and S. B. Weiss, these Proceedings **47**:1405 (1961).
48. Hall, B. D., and S. Spiegelman, these Proceedings **47**:137 (1961).
49. Sinsheimer, R. L., J. Mol. Biol. **1**:43 (1959).
50. Rich, A., these Proceedings **46**:1044 (1960).
51. Schildkraut, C. L., J. Marmur, J. R. Fresco, and P. Doty, J. Biol. Chem. **236**:PC 2 (1961).
52. Wyatt, G. R., and S. S. Cohen, Biochem. J. **55**:774 (1953).

# 25 THE ROLE OF DNA IN RNA SYNTHESIS, IX.
## NUCLEOSIDE TRIPHOSPHATE TERMINI IN RNA POLYMERASE PRODUCTS

*Umadas Maitra\**
*Jerard Hurwitz*
*Department of Molecular Biology*
*Albert Einstein College of Medicine*
*Bronx, New York*

It has been shown previously that in RNA polymerase reactions primed with a variety of DNA preparations there is incorporation of $P^{32}$ from $\beta\gamma$-labeled ATP into an acid-insoluble product and that triphosphate groups are present at the ends of the RNA chains formed during the reaction.[1] Two schemes for initiation of the chains can thus be envisaged. In one scheme, the initial nucleotide incorporated into RNA would retain its triphosphate end, while the growing end of the molecule would be a nucleoside. In the second scheme, the situation is reversed: the nucleoside end would be the initiation point, and the triphosphate, the growing site of the molecule. It is clear that these two schemes of initiation of RNA synthesis differ specifically in that the first would result in the initial nucleotide retaining the $\beta$ and $\gamma$ phosphate groups, whereas in the second scheme the last entering nucleotide would contain the $\beta$ and $\gamma$ phosphate group.

In the present communication, evidence will be presented that (1) initiation and subsequent elongation of RNA chains formed by RNA polymerase under the direction of a DNA template occur by a mechanism in which the first nucleotide incorporated into the RNA chain retains its triphosphate moiety, and (2) adenosine and guanosine triphosphate ends are preferentially formed.

---

From Proceedings of the National Academy of Sciences U.S.A. 54:815-822, 1965. Reprinted with permission.

This research was supported by grants from the National Institutes of Health, the National Science Foundation, and the New York City Public Health Research Council. Paper VIII of this series was concerned with the inhibition of RNA polymerase by histones (Skalka, A., A. Fowler, and J. Hurwitz, *J. Biol. Chem.*, in press). Communication no. 40 of the Joan and Lester Avnet Institute for Molecular Biology.

*Postdoctoral fellow of the Jane Coffin Childs Memorial Fund for Medical Research.

## Materials and methods

$\gamma$-$P^{32}$-GTP and UTP were prepared by photophosphorylation of the corresponding nucleoside diphosphates with $^{32}P_i$ and spinach chloroplasts by a modification of the procedure of Avron.[2] $\gamma$-$P^{32}$-CTP was prepared by the action of nucleoside diphosphokinase[3] on $\beta\gamma$-$P^{32}$-ATP and CDP in the presence of an excess of myokinase. $\beta\gamma$-$P^{32}$-ATP was prepared as described by Penefsky and Racker[4] in the presence of excess myokinase. These $P^{32}$-labeled compounds were purified by chromatography on Dowex-1-Cl$^-$ and were free of $P^{32}$ in the $\alpha$-phosphate position. The methods of preparation of other materials, including the DNA-dependent RNA polymerase of *Escherichia coli*, have been previously described.[1,5] Calf thymus DNA was obtained from General Biochemicals.

## Enzyme assay

The presence of a triphosphate terminus in the RNA formed in an RNA polymerase reaction was measured by the incorporation of $\gamma$-$P^{32}$ ribonucleoside triphosphate into an acid-insoluble product, and RNA synthesis was measured by the incorporation of $\alpha$-$P^{32}$ or $C^{14}$-labeled ribonucleoside triphosphate, Reaction mixtures (0.50 ml) contained Tris buffer, pH 8.0, 25 $\mu$moles; 2-mercaptoethanol, 4 $\mu$moles; MnCl$_2$, 0.5 $\mu$mole; MgCl$_2$, 2.5 $\mu$moles; DNA, 25 m$\mu$moles; ATP, UTP, GTP, and CTP, 10 m$\mu$moles each, one labeled with $P^{32}$ in the $\gamma$-phosphate group (containing 1-2 $\times$ 10$^9$ cpm/$\mu$mole) for measurement of triphosphate termini or with $C^{14}$-ATP or GTP (containing 2-5 $\times$ 10$^6$ cpm/$\mu$mole) for measurement of the total amount of RNA formed in the reaction. In either case, the reaction was initiated by the addition of enzyme.[6] After incubation at 37° for the desired time, the reaction mixture was chilled in ice, and 0.1 ml of a bovine plasma albumin solution (5 mg/ml) was added, followed by 0.3 ml of 7% HClO$_4$. The resulting precipitate was centrifuged for 5 min at 15,000 $\times$ g, and the pellet dissolved in 0.2 ml of ice-cold 0.2 N NaOH. This was followed by the addition of 0.1 ml of nonradioactive triphosphate (25 $\mu$moles/ml) corresponding to the $\gamma$-$P^{32}$-labeled triphosphate used in the reaction mixture and 5 ml of cold 5% TCA solution containing 0.01 M sodium pyrophosphate. After the reaction mixture had stood in ice for 5 min, the acid-insoluble material was collected by centrifuga-

## TABLE 1
*Requirements for the incorporation of $\beta\gamma$-$P^{32}$-labeled ATP and $\gamma$-$P^{32}$-UTP in dAT-primed poly rAU synthesis*

| Additions | $\beta\gamma$-$P^{32}$-ATP incorporated ($\mu\mu$moles) | $\gamma$-$P^{32}$-UTP incorporated ($\mu\mu$moles) |
|---|---|---|
| 1. Complete system | 4.40 | 0.40 |
| 2. Omit UTP | 0.37 | — |
| 3. Omit ATP | — | 0.40 |
| 4. Omit enzyme | 0.13 | 0.11 |
| 5. Omit dAT copolymer | 0.29 | 0.28 |
| 6. Complete + DNase (1 $\mu$g) | 0.30 | 0.24 |
| 7. Complete + RNase (1 $\mu$g) | 0.30 | 0.24 |
| 8. Complete with $C^{14}$-ATP in place of $\beta\gamma$-$P^{32}$-ATP | 7200 (poly rAU) | |

The conditions of the experiment were as described under *Enzyme Assay*, except that GTP and CTP were omitted and 10 m$\mu$moles of dAT copolymer replaced DNA. $\beta\gamma$-$P^{32}$-ATP incorporation and $\gamma$-$P^{32}$-UTP incorporation were measured in separate mixtures. In the complete system, the total amount of nucleotide incorporated was calculated from the amount of $C^{14}$-AMP incorporated into the poly rAU product. Mixtures were incubated for 40 min and contained 3 units of RNA polymerase.

## TABLE 2
*Incorporation of $\gamma$-$P^{32}$-nucleoside triphosphates with different DNA preparations*

| DNA primer | RNA synthesis ($\mu\mu$moles) | $\gamma$-$P^{32}$-Nucleotide Incorporated ($\mu\mu$moles) | | | |
|---|---|---|---|---|---|
| | | ATP | GTP | UTP | CTP |
| T2 | 4800 | 2.40 | 1.2 | 0.12 | 0.10 |
| T5 | 4000 | 1.80 | 1.4 | 0.41 | 0.23 |
| SP3 | 5480 | 1.25 | 1.0 | 0.39 | 0.12 |
| *Cl. perfringens* | 2800 | 1.60 | 2.1 | 0.28 | 0.25 |
| *E. coli* | 2660 | 0.43 | 1.4 | 0.13 | 0.10 |
| *M. lysodeikticus* | 2560 | 0.36 | 2.5 | 0.10 | 0.12 |
| Calf thymus | 3560 | 0.77 | 1.3 | 0.33 | 0.18 |
| dAT copolymer | rAU = 7200 | 4.40 | — | 0.20 | — |
| dGC homopolymer | rG = 1350; rC = 120 | — | 4.8 | — | 0.30 |

The conditions of the assay were as described under *Methods*. In each reaction mixture, only one of the four nucleoside triphosphates was labeled with $P^{32}$ in the $\gamma$-phosphate group; the other three were nonradioactive. Incubation was for 40 min at 37° with 2 units of enzyme. Where indicated, 7.5 m$\mu$moles of dGdC homopolymer and 10 m$\mu$moles of dAT copolymer were added. Controls without enzyme and without DNA were included. In these controls incorporation was 0.2-0.3 $\mu\mu$mole of nucleotide, and the higher value was subtracted from the results listed above.

## TABLE 3
*Effect of denaturation of DNA on the incorporation of $\gamma$-$P^{32}$-labeled nucleoside triphosphates*

| DNA primer | RNA synthesis ($\mu\mu$moles) | $\gamma$-$P^{32}$-Nucleotide Incorporated ($\mu\mu$moles) | | | |
|---|---|---|---|---|---|
| | | ATP | GTP | UTP | CTP |
| T2 | 4800 | 2.3 | 1.2 | 0.13 | 0.10 |
| T2 heat-denatured | 1000 | 3.6 | 5.1 | 0.88 | 0.40 |
| T2 alkali-denatured | 1050 | 3.5 | 4.7 | 0.91 | 0.45 |
| Calf thymus | 5700 | 1.0 | 1.8 | 0.33 | 0.20 |
| Calf thymus heat-denatured | 2000 | 3.8 | 10.1 | 0.82 | 0.66 |
| *E. coli* | 2000 | 0.6 | 1.5 | 0.13 | 0.10 |
| *E. coli* heat-denatured | 1300 | 2.4 | 8.1 | 0.56 | 0.44 |

The condition of the assay was as described under *Methods*, with 20 m$\mu$moles of each of the various DNA preparations, 2 units of enzyme, and 40 min of incubation at 37°. RNA synthesis was followed by incorporation of both $C^{14}$-AMP and $C^{14}$-GMP, and total RNA synthesis was calculated by multiplying the sum of these values by two.

tion, and the washing procedure was repeated two more times. The final pellet was dissolved in 1.5 ml of 0.2 $N$ NH$_4$OH, transferred to an aluminum planchet, and after drying, counted in a windowless gas-flow counter. The high specific radioactivities of the $\gamma$-P$^{32}$-labeled substrates necessitated the washing procedure to obtain consistently low blanks. When the total amount of RNA synthesized from C$^{14}$-nucleoside triphosphate (specific radioactivity 10$^6$-10$^7$ cpm/μmole) was measured, the washing procedure was not necessary, and the acid-insoluble RNA product was isolated by filtration on membrane filters as described previously.[1]

## Results
### Incorporation of $\gamma$-P$^{32}$-labeled nucleoside triphosphates into RNA polymerase products

RNA polymerase, primed with dAT copolymer, catalyzes the incorporation of $\beta\gamma$-P$^{32}$-labeled ATP into an acid-insoluble polyribonucleotide product.[1] The incorporation of P$^{32}$ is dependent on the presence of UTP, dAT copolymer, and RNA polymerase. The omission of any of these components, or the addition of RNase or DNase, results in a marked decrease in P$^{32}$-ATP incorporation. In contrast, similar experiments carried out with $\gamma$-P$^{32}$-UTP do not result in significant incorporation of P$^{32}$, as shown in Table 1. The low level of incorporation observed is not dependent on the simultaneous presence of ATP.[7]

The finding that poly rAU chains contain ATP ends and very few UTP ends prompted an examination of the relative incorporation of P$^{32}$ from each of the four $\gamma$-P$^{32}$-labeled nucleoside triphosphates with different DNA templates of varying base composition. The results, presented in Table 2, show that RNA chains are formed predominantly with ATP and GTP ends, whereas few chains are initiated with UTP and CTP. The relative number of triphosphate ends beginning with adenosine or guanosine varied with the DNA used to direct the reaction (Table 2). With various bacterial DNA's and calf thymus DNA, GTP ends predominated, although significant ATP incorporation also was observed. With DNA preparations from phages T2, T5, and SP3 (which have an A + T/G + C ratio > 1), ATP ends occurred slightly more often than GTP ends. Denaturation of DNA had marked effects on both RNA synthesis and the incorporation of P$^{32}$ from $\gamma$-P$^{32}$-labeled nucleoside triphosphates (Table 3). These effects can be summarized as follows: (1) RNA synthesis was markedly inhibited, (2) the incorporation of all four $\gamma$-P$^{32}$ nucleoside triphosphates increased severalfold, and (3) there was an increase in the ratio of GTP to ATP termini as well as a significant though small number of RNA chains containing UTP and CTP ends.

### Identification of guanosine triphosphate ends of RNA formed in the RNA polymerase reaction

As with $\beta\gamma$-P$^{32}$-labeled ATP,[1] the incorporation of $\gamma$-P$^{32}$-GTP into an acid-insoluble material had the same requirements found for RNA synthesis.

For identification of the site of GTP incorporation, an RNA product containing $\gamma$-P$^{32}$-GTP was prepared using denatured thymus DNA as template. Subjecting the P$^{32}$-labeled RNA product to the action of alkaline phosphatase or to acid hydrolysis (1 $N$ HCl for 10 min at 100°) rendered the P$^{32}$ acid-soluble and Norit nonadsorbable. The P$^{32}$ present in the RNA product was not converted to P$_i$ by the action of prostatic phosphomonoesterase.[8] The P$^{32}$ product was insensitive to pancreatic DNase, since it remained acid-insoluble, whereas pancreatic RNase and alkaline hydrolysis released all the P$^{32}$ into an acid-soluble but Norit-adsorbable form. These results indicate that the P$^{32}$ incorporated from $\gamma$-P$^{32}$-GTP was in a terminal portion of the RNA structure, presumably $\overset{*}{p}$ppGpXpYpZ---, and not in an internucleotide link.

The expected products of alkaline hydrolysis of polynucleotides with the structure $\overset{*}{p}$ppGpXpYpZ--- are $2'(3')$-nucleoside monophosphates and nucleoside tetraphosphates. Radioactivity from $\gamma$-P$^{32}$-GTP should be found only in guanosine tetraphosphate ($\overset{*}{p}$ppGp). The prediction was tested as follows: An alkaline hydrolysate (0.3 $N$ KOH at 37° for 18 hr) of the labeled product was neutralized with Dowex-50 (H$^+$), and an aliquot of the solution containing 30,000 cpm was mixed with 2 μmoles each of GMP, GDP, GTP, ATP, and guanosine-5′-tetraphosphate.[9] The mixture was added to a column (1 × 12 cm) of Dowex-1-Cl$^-$ (100-200 mesh, 2% cross-linked), and the nucleotides were eluted as follows: (a) 150 ml of 0.01 $M$ HCl + 0.05 $M$ LiCl (GMP); (b) 150 ml of 0.01 $M$ HCl + 0.1 $M$ LiCl (GDP followed by ATP); (c) 150 ml of 0.01 M HCl + 0.2 M LiCl (GTP); (d) 150 ml of 0.05 M HCl + 0.2 M LiCl, which eluted guanosine 5'-tetraphosphate. The elution profile and identification of each of the nucleotides were determined by measuring the optical densities of the effluents at 260 and 280 mμ. More than 90 per cent of the $P^{32}$ added to the column was eluted as a sharp symmetrical peak in the last solvent with guanosine 5'-tetraphosphate, whereas 8 per cent of the added radioactivity was eluted in the GTP region. No $P^{32}$ was detected in the other regions of the chromatogram. These results are consistent with the presence of $\overset{*}{p}ppGp$. To characterize further the $P^{32}$-labeled product in the alkaline hydrolysate, another aliquot of the alkaline hydrolysate (containing 30,000 cpm) was incubated with 3.5 units[10] of prostatic phosphomonoesterase at pH 5.0 at 37° for 30 min. The mixture was then chromatographed on Dowex-1-Cl⁻ under the conditions described above. Approximately 75 per cent of the added radioactivity now chromatographed with GTP and the remainder with guanosine 5'-tetraphosphate. These results are consistent with the structure $\overset{*}{p}ppGpXpYpZ$----, i.e., $\gamma\text{-}P^{32}GTP$ is incorporated as such at the end of RNA chains.

### Kinetics of nucleoside triphosphate incorporation

The rates of RNA synthesis and $\beta\gamma\text{-}P^{32}$-ATP incorporation were compared (Table 4). Whereas T2 DNA-directed RNA synthesis continued during the entire experiment, 65 per cent of the total amount of $\beta\gamma\text{-}P^{32}$-ATP was incorporated by 5 min. The ratio of $\beta\gamma\text{-}P^{32}$-ATP ends to total nucleotide incorporation decreased progressively with time from a ratio of 1 ATP terminus per 250 RNA nucleotides during the first minute to 1 ATP terminus per 2270 nucleotides after 60 min.

A comparison of the kinetics of $P^{32}$ incor-

TABLE 4
*Kinetics of RNA synthesis versus $\beta\gamma\text{-}P^{32}ATP$ incorporation*

| Time of incubation (min) | RNA synthesis (μμmoles) | $\beta\gamma\text{-}P^{32}$-ATP incorporated (μμmoles) | Ratio |
|---|---|---|---|
| 1 | 300 | 1.2 | 250 |
| 2 | 600 | 1.6 | 375 |
| 5 | 1500 | 2.4 | 630 |
| 10 | 2700 | 2.8 | 960 |
| 20 | 4500 | 3.2 | 1410 |
| 40 | 6100 | 3.7 | 1695 |
| 60 | 8400 | 3.7 | 2270 |

The assay was performed as described under *Methods*, except that 5 μmoles of MgCl₂ replaced MnCl₂, 2 units of RNA polymerase and 20 mμmoles of each of the nucleoside triphosphate were added, and the incubation was carried out at 25°. This procedure permitted slow growth of the RNA chains with T2 DNA as primer.

TABLE 5
*Kinetics of RNA synthesis and $\gamma\text{-}P^{32}$-GTP incorporation with native and denatured T2 DNA as templates*

| Experiment no. | Time of incubation (min) | RNA synthesized (μμmoles) | $\gamma\text{-}P^{32}$-GTP incorporated (μμmoles) | Ratio |
|---|---|---|---|---|
| I | 5 | 1500 | 1.1 | 1360 |
|   | 10 | 2700 | 1.4 | 1800 |
|   | 20 | 5000 | 1.7 | 2900 |
|   | 40 | 6200 | 1.9 | 3200 |
|   | 60 | 8300 | 1.9 | 4150 |
|   | 90 | 10400 | 1.9 | 5200 |
| II | 5 | 150 | 3.2 | 47 |
|   | 10 | 300 | 4.8 | 62 |
|   | 20 | 600 | 6.6 | 90 |
|   | 40 | 1140 | 7.7 | 143 |
|   | 60 | 1500 | 8.1 | 180 |
|   | 90 | 2100 | 10.1 | 210 |

The conditions of the experiment were as described under *Methods*, except that 20 mμmoles each of $\gamma\text{-}P^{32}$-GTP, UTP, CTP, and ATP were added. Twenty-four mμmoles of native T2 DNA (approximate size = 35S) were added in experiment I, and an equimolar amount of heat-denatured T2 DNA was added in experiment II. Incubation was at 37°.

poration from $\gamma$-P$^{32}$-GTP with native and with heat-denatured T2 DNA as template is summarized in Table 5. With native T2 DNA as template the incorporation of P$^{32}$ was virtually complete in 20 min, although RNA synthesis continued throughout the incubation period. The ratio of GTP ends to RNA formed decreased progressively with time from a ratio of 1 GTP terminus per 1300 nucleotides in the first 5 min to 1 GTP terminus per 5100 nucleotides after 90 min. Under the same conditions, experiments with $\beta\gamma$-P$^{32}$-ATP indicated the presence of 1 ATP terminus per 600 nucleotides during the first 5 min of incubation and 1 ATP terminus per 3100 nucleotides after 90 min. In contrast, with denatured DNA as template, the incorporation of $\gamma$-P$^{32}$-GTP was severalfold faster and continued during the entire period of RNA synthesis. Similar results were obtained with $\beta\gamma$-P$^{32}$-ATP. In this case, with denatured DNA as template, the ratio of ATP ends to RNA formed was lower than that found with GTP (1/160 after 5 min of incubation and 1/540 after 60 min), since with denatured DNA more $\gamma$-P$^{32}$-GTP than $\beta\gamma$-P$^{32}$-ATP was incorporated. However, with either of these two triphosphates, the ratio of triphosphate ends formed to RNA synthesized was considerably smaller than that obtained with the corresponding native DNA as template. The ratios obtained in these experiments can be used as a measure of the length of the RNA product formed. Wood and Berg[11] found that RNA products formed with denatured DNA as template were smaller than those obtained with native DNA. The results summarized in Table 5 are in agreement with their findings.

*Direction of growth of RNA chains*

In order to determine whether incorporation of the triphosphate terminus in RNA occurs according to the scheme in which the first nucleotide incorporated retains the triphosphate end, or by a mechanism in which the triphosphate moiety is at the growing point of the RNA molecule, the following experiment was performed. $\gamma$-P$^{32}$-GTP was incorporated into RNA for 5 min, and then a large excess of cold GTP was added to reduce the specific activity of the labeled nucleotide. The fate of P$^{32}$ already incorporated at the ends of the RNA chains was then followed during subsequent RNA synthesis. If the initial nucleotide incorporated was present as the triphosphate end, subsequent synthesis should have no effect on the P$^{32}$ already incorporated. In contrast, if the last entering nucleotide existed as the triphosphate terminus, subsequent synthesis should release the previously incorporated P$^{32}$. The fate of $\gamma$-P$^{32}$-GTP ends under the conditions described above is summarized in Figure 1. As shown, the addition of unlabeled GTP halted $\gamma$-P$^{32}$ uptake immediately, and continuing RNA synthesis did not diminish the amount of P$^{32}$ already incorporated.

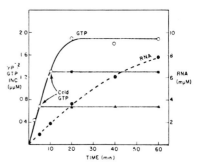

**Fig. 1.** Effect of dilution on $\gamma$-P$^{32}$-GTP incorporation. Reaction mixtures were as described under *Methods*, with the exception that 20 m$\mu$moles of each of the four nucleoside triphosphates were added, $\gamma$-P$^{32}$-GTP (1.1 × 10$^9$ cpm/$\mu$mole) was used, and incubation was carried out at 37°. In two separate reaction mixtures, one after 5 min and the other after 10 min, a 30-fold excess of unlabeled GTP was added. RNA synthesis was measured in separate reaction mixtures in which C$^{14}$-GTP was used as the only radioactive substrate, with all other additions as above. ▲, Incorporation of $\gamma$-P$^{32}$-GTP in the reaction mixture diluted with unlabeled GTP after 5 min; ■, incorporation in the reaction mixture diluted after 10 min.

Proof that the labeled chains actually increased in length after the addition of unlabeled substrates was obtained by the following experiment. RNA synthesis was carried out in the presence of $\gamma$-P$^{32}$-GTP and $\beta\gamma$-P$^{32}$-ATP. After the reaction had been allowed to proceed for 4 min, a large excess of unlabeled GTP and ATP was added. One sample was removed at this time, and a second after an additional 8 min of incubation. The sedimentation of the labeled RNA is illustrated in Figure 2, which shows that the sedimentation rate increased from about 6S to about 20S after the addition of unlabeled substrates. An increase in size of

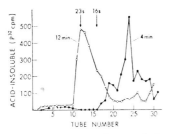

**Fig. 2.** Zone sedimentation analysis of RNA. Two 0.5-ml reaction mixtures containing Tris buffer, pH 8.0, 25 μmoles; 2-mercaptoethanol, 4 μmoles; MgCl$_2$, 5 μmoles; ADP, 5 mμmoles; T5 DNA, 25 mμmoles; UTP and CTP, 25 mμmoles each; βγ-P$^{32}$-ATP and γ-P$^{32}$-GTP, specific activity 1 × 10$^9$ cpm/μmole, 25 mμmoles each; and 3 units of enzyme were incubated for 4 min at 25°. After that time, 1.5 μmoles each of nonradioactive ATP and GTP were added to each tube. One reaction was stopped immediately by the addition of sodium dodecyl sulfate and EDTA (0.5% and 0.01 M final concentration, respectively). The other tube was treated in the same manner after 12 min at 25°. Both samples were diluted to 1 ml with 0.5% sodium dodecyl sulfate. layered on 30 ml of a 15-30% sucrose solution gradient containing 0.05 M Tris buffer, pH 8.0, 0.1 M NaCl, and 0.2% sodium dodecyl sulfate, and centrifuged for 15 hr at 25,000 rpm in an SW 25.1 Spinco rotor at approximately 25°. Ribosomal RNA was used as an optical density marker. Fractions (1 ml) were collected through a hole punched in the bottom of the tube. The fractions were scanned through a Gilford recording spectrophotometer to locate the position of 23S and 16S ribosomal markers and assayed for P$^{32}$.

RNA products with time was also noted by Bremer and Konrad.[12]

These results show that nucleoside triphosphates are incorporated at the point of initiation and not at the growing end of RNA chains, and, therefore, that chain growth occurs by the addition of nucleotides to the 3'-hydroxyl end.

### Discussion

The above results clearly show that RNA chains are initiated with ribonucleoside triphosphates, principally or exclusively ATP and GTP. Upon denaturation of DNA by heat or alkali, the incorporation of nucleoside triphosphates into terminal positions increases. Thus, RNA polymerase finds more and different initiation sites for RNA synthesis on denatured DNA than on double-stranded DNA. This conclusion is also supported by the observation that single-stranded and denatured DNA saturate RNA polymerase more effectively than native DNA.[13,14]

The finding that the pyrimidine sites of DNA, especially when native, are preferentially utilized as initiation points for RNA synthesis was totally unexpected. In fact, the pyrimidine nucleoside triphosphate ends found in small numbers in the RNA may reflect the presence of small amounts of denatured DNA in all the primers used. The reason for this specificity is unknown, but it probably results from the manner in which the enzyme interacts with DNA. It is probably not due to the selective binding of ATP and GTP versus UTP and CTP to the enzyme, since there is no difference in affinity constant of these nucleotides for RNA polymerase.[5] The selective copying of the pyrimidine-rich strand of the DNA of *Bacillus subtilis* phages SP8 and 2C and *Bacillus megatherium* phage α *in vivo*[15,16] and *in vitro*[17,18] may be related to the preferential initiation of RNA chains with purine nucleotides. The selection of one DNA strand over another may thus be governed by runs of pyrimidine bases in native DNA. The loss of asymmetric copying of DNA in RNA synthesis upon denaturation of the DNA[17] may be related to our finding that denaturation uncovers new sites in the DNA at which RNA chains can be started.

*In vivo*, RNA synthesis (i.e., gene expression) must begin at particular sites on DNA. Since DNA of *E. coli* (and others) is uninterrupted,[19] there must be a high degree of specificity for initiation of RNA chains within the DNA duplex. The results presented above suggest that these sites in DNA may be pyrimidine bases. In accord with this idea is the finding that sRNA molecules contain considerable amounts of guanine and adenine at the 5'-phosphate end.[20,21] How RNA polymerase can specifically recognize initiation points on DNA, and what factor controls the accessibility of the enzyme to such sites for RNA synthesis, are problems intimately involved in the mechanism of gene control.

### Summary

In the DNA-dependent RNA polymerase reaction, the RNA chains formed contain ribonucleoside triphosphates at their starting points and grow by the subsequent addition of ribonucleotides to the 3'-hydroxyl group of the ribonucleoside end. Purine nucleoside triphosphates are preferentially found at the triphosphate end. When denatured DNA is used as a template, there are an increase in the number and a change in the kind of starting points.

## REFERENCES AND NOTES

1. Maitra, U., A. Novogrodsky, D. Baltimore, and J. Hurwitz, Biochem. Biophys. Res. Commun. **18**: 801 (1965).
2. Avron, M., Anal. Biochem. **2**:535 (1961).
3. Berg, P., and W. K. Joklik, J. Biol. Chem. **210**:657 (1954).
4. Penefsky, H., and E. Racker, J. Biol. Chem. **235**:3330 (1960).
5. Furth, J. J., J. Hurwitz, and M. Anders, J. Biol. Chem. **237**:2611 (1962).
6. In all reaction mixtures containing $\beta\gamma$-$P^{32}$-ATP, 2 m$\mu$moles of ADP per 10 m$\mu$moles of ATP were included to suppress the action of any possible contaminating polyphosphate forming enzyme [Kornberg, A., S. R. Kornberg, and E. S. Simms, Biochim. Biophys. Acta **20**:235 (1956)].
7. In other experiments the incorporation of $\gamma$-$P^{32}$-UTP was <0.2 $\mu\mu$moles. Evidence that this low incorporation was not due to the presence of an inhibitor in the $P^{32}$-UTP preparation was obtained by the finding that $\gamma$-$P^{32}$-UTP supported both $C^{14}$-ATP and $\beta\gamma$-$P^{32}$-ATP incorporation with dAT copolymer as primer. The other $\gamma$-$P^{32}$-nucleoside triphosphates also supported RNA synthesis, and the omission of a single triphosphate resulted in a marked decrease in both RNA synthesis and chain initiation.
8. Ostrowski, W., and A. Tsugita, Arch. Biochem. Biophys. **94**:68 (1961).
9. Gardner, J. A. A., and M. B. Hoagland, J. Biol. Chem. **240**:1244 (1965). We are indebted to Dr. M. Hoagland for a gift of guanosine 5'-tetraphosphate.
10. One unit of enzyme will cleave 1 $\mu$mole of 0-nitrophenylphosphate per minute at 37°.
11. Wood, W. B., and P. Berg, J. Mol. Biol. **9**:452 (1964).
12. Bremer, H., and M. W. Konrad, these Proceedings **51**:801 (1964).
13. Hurwitz, J., J. J. Furth, M. Anders, and A. Evans, J. Biol. Chem. **237**:3752 (1962).
14. Berg, P., R. D. Kornberg, H. Fancher, and M. Dieckmann, Biochem. Biophys. Res. Comm. **18**: 932 (1965).
15. Marmur, J., and C. M. Greenspan, Science **142**:387 (1963).
16. Tocchini-Valentini, G. P., M. Stodolsky, M. Sarnat, A. Aurisicchio, F. Graziosi, S. B. Weiss, and E. P. Geiduschek, these Proceedings **50**:935 (1963).
17. Colvill, A. J. E., L. C. Kanner, G. P. Tocchini-Valentini, M. T. Sarnat, and E. P. Geiduschek, these Proceedings **53**:1140 (1965).
18. Fowler, A. V., J. Marmur, and J. Hurwitz, unpublished observations.
19. Cairns, J., J. Mol. Biol. **6**:208 (1963).
20. Ralph, R. K., R. J. Young, and H. G. Khorana, J. Am. Chem. Soc. **85**:2002 (1963).
21. Bell, D., R. V. Tomlinson, and G. M. Tener, Biochem. Biophys. Res. Commun. **10**:304 (1963).

# 26 FACTOR STIMULATING TRANSCRIPTION BY RNA POLYMERASE

*Richard R. Burgess*
*Andrew A. Travers*
Biological Laboratories
Harvard University
Cambridge, Massachusetts
*John J. Dunn*
*Ekkehard K. F. Bautz*
Institute of Microbiology
Rutgers University
New Brunswick, New Jersey

---

A protein component usually associated with RNA polymerase can be separated from the enzyme by chromatography on phosphocellulose. The polymerase is unable to transcribe T4 DNA unless this factor is added back.

---

In *E. coli* the synthesis of all types of cellular RNA is thought to be mediated by a single enzyme, DNA-dependent RNA polymerase. The highly purified enzyme[1-6] can catalyse the synthesis of RNA *in vitro* in the presence of DNA and the ribonucleoside triphosphates. When an intact double helical DNA is used as template, transcription *in vitro* is asymmetric—only one of the complementary DNA strands is transcribed[7-10]. This is also characteristic of transcription *in vivo*[11-13]. Moreover, the selective transcription *in vitro* of certain regions of T4 and λ DNA[10, 14-17], coupled with studies on the binding of RNA polymerase to DNA[18-22], suggests that the polymerase initiates RNA synthesis at specific sites on the DNA. The state of aggregation of the enzyme is strongly influenced by ionic strength[19, 21, 23, 24], substrate[25], and possibly by enzyme concentration and temperature. Most investigators agree that the molecular weight of the active enzyme is in the range of 350,000-700,000 daltons. This large size is consistent with the observation that the enzyme is composed of several different polypeptide chains[26, 27]. Furthermore, it has been assumed that "highly purified" RNA polymerase is a protein entity from which nothing can be further removed without destroying its enzyme activity. We report here, however, the separation of polymerase into two components. One contains enzyme activity, but its ability to transcribe certain DNA templates is greatly reduced; the other is a factor able to stimulate RNA synthesis on these restrictive templates to normal levels.

It was independently observed at Harvard and Rutgers that when RNA polymerase was purified by chromatography on a phosphocellulose column, the enzyme obtained, although able to transcribe calf thymus DNA almost normally, was much less active when assayed with T4 DNA as template. Enzyme purified by an alternative procedure, however, was almost equally active on both templates. This suggested that some component necessary for the transcription of T4 DNA was separated from RNA polymerase by the phosphocellulose column. Furthermore, the activity of the phosphocellulose-purified enzyme on T4 DNA could be greatly enhanced by the addition of another fraction from the phosphocellulose column. This fraction lacked significant RNA polymerase activity of its own. We describe here the identification and some properties of this stimulating component.

## Isolation of RNA polymerase

The RNA polymerase we used was purified as outlined in Fig. 1 from *E. coli* K12. Three methods were used to achieve the final purification of the polymerase. Enzyme purified with

---

From Nature 221:43-46, 1969. Reprinted with permission.

We thank Professor J. D. Watson for his interest and support, Professor K. Weber, Mr J. Roberts and Dr J. Tkacz for helpful discussions and suggestions, and Miss Anne-Marie Piret and Mrs Christine Roberts for technical assistance. This work was supported by grants from the US National Science Foundation and the US Public Health Service. One of us (A. A. T.) is a postdoctoral fellow of the Damon Runyon Memorial Fund for Cancer Research, and R. R. B. is a National Science Foundation predoctoral fellow.

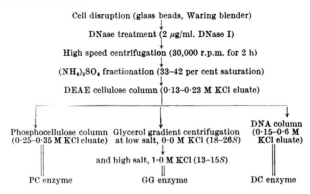

**Fig. 1.** Outline of enzyme purification.

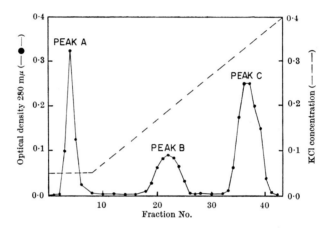

**Fig. 2.** Phosphocellulose column profile. Phosphocellulose (Whatman P11, 7.4 mequiv./g) was washed with base and acid and titrated to pH 7.9 at 25°C with KOH. This material was placed in a column (0.9 × 10 cm) and equilibrated extensively with a buffer containing 0.05 M tris-HCl, pH 7.9, at 25°C, 0.05 M KCl, 0.0001 M EDTA, 0.0001 M dithiothreitol and 5 per cent glycerol. The pH of the outflow was 8.1 ± 0.05 at 4°C. A sample containing 9 mg of GG enzyme in 4 ml. of this buffer was applied to the column. The column was then washed with 5 ml. of buffer and eluted with a 100 ml. gradient from 0.05 M to 0.4 M KCl. The flow rate was 0.5 ml./min and 2.2 ml. fractions were collected. Tubes 4, 22 and 36 were taken as representative of peaks A, B and C respectively. Assuming a value of $E_{280m}^{1\%} = 6.6$ for all fractions, these tubes contain 0.5, 0.14 and 0.4 mg of protein per ml. respectively. A small amount of polymerase activity was present in the flow-through (peak A). If the sample is applied to the column too rapidly (at a rate greater than one column volume per 30 min), then much more enzyme will be found in the flow-through.

phosphocellulose (PC enzyme) and with glycerol gradient (GG enzyme) was prepared by the method of Burgess (manuscript in preparation). Using the same method until the DEAE cellulose step, he purified the enzyme by binding it to a cellulose column containing immobilized T4 DNA and eluting the bound enzyme with high salt.[28] DNA column enzyme (DC enzyme) obtained in this way and GG enzyme are essentially identical and do not differ significantly from enzymes purified by the method of Chamberlin and Berg[1]. Unless otherwise stated, the work described here was carried out with GG enzyme, but similar results were obtained with DC enzyme.

### Identification of the component stimulating transcription of T4 DNA

To isolate the component necessary for the transcription of T4 DNA, GG enzyme was chromatographed on a phosphocellulose column. This column separated the material present in GG enzyme into three peaks, A, B and C (Fig. 2). Each peak was assayed for

## TABLE 1
*Assay of stimulating activity of peaks A, B and C*

| Sample assayed for stimulation | μμmole AMP incorporated/min | | Stimulation ratio |
|---|---|---|---|
| | No PC enzyme added | 4 μg PC enzyme added | |
| PC enzyme (4.0) | 3 | 9 | 2 |
| GG enzyme (1.1) | 57 | 170 | 38 |
| Peak *A*   (1.0) | 13 | 114 | 34 |
| Peak *B*   (1.3) | 66 | 203 | 46 |
| Peak *C*   (1.9) | 2 | 8 | 2 |

The assay mixture (.025 ml.) contained 0.04 M *tris*-HCl buffer, pH 7.9, at 25°C; 0.01 M $MgCl_2$; 0.0001 M EDTA; 0.0058 M 2-mercaptoethanol; 0.15 M KCl; 0.5 mg/ml. bovine serum albumin (Calbiochem, crystalline); 0.15 mM CTP, GTP and UTP; 0.15 mM $^{14}$C-ATP (1 mCi/mmole); 20 μg/ml. T4 phage DNA; and varying amounts of enzyme and stimulating factor. The mixture was incubated for 10 min at 37°C, chilled, precipitated with 5 per cent TCA, and filtered on 'Millipore' filters. Samples were counted on an end-window gas-flow counter. The amount of incorporation is expressed as μμmoles of AMP incorporated per min of incubation. To obtain a measure of the stimulating activity in a particular sample, that sample was assayed in the absence and presence of PC enzyme. The additional incorporation in the presence of PC enzyme was due to the stimulation of this added PC enzyme by the factor in the sample. When this additional incorporation is divided by the small amount of incorporation obtained with PC enzyme alone, the stimulation ratio is obtained. The values in parentheses indicate the amount of protein, in μg, added to the assay mixture.

ability to promote the transcription of T4 in the presence and absence of PC enzyme. Peak *A* contains stimulating activity, peak *B* is a mixture of PC enzyme and stimulating activity, while peak *C* is identical to normal PC enzyme. The results (Table 1) show that, in the absence of PC enzyme, peak *B* can use T4 DNA as a template for RNA synthesis while peaks *A* and *C* have little activity. The addition of peak *A* material to PC enzyme, however, results in a stimulation of RNA synthesis which increases with increasing amounts of peak *A*, until a rate of RNA synthesis comparable with that obtained with GG enzyme is reached. Although peak *B* is able to stimulate PC enzyme, it can itself be stimulated by the addition of peak *A*, and thus is a mixture of PC enzyme and stimulating component.

Each of the fractions was analysed by polyacrylamide gel electrophoresis in various conditions to determine the species of protein present. These gels were run in the presence of either 8 M urea or 0.1 per cent sodium dodecyl sulphate (SDS), both of which can cause the dissociation of oligomeric proteins into single polypeptide chains. The protein bands observed in this analysis are shown in Fig. 3.

Previous studies[27] have shown that phosphocellulose enzyme is composed of two chief types of polypeptide chains, which we here designate α and β. These chains are present in equimolar amounts and have molecular weights of ~40,000 and ~160,000 respecitvely. α and β are present in all fractions but only in very small amounts in peak *A*. GG enzyme contains, in addition, two extra bands which we shall designate σ and τ. These bands are present in peak *A* in amounts greatly in excess of the amounts of α and β present. Peak *B* contains bands α, β and σ, but lacks τ. These patterns are observed on both the 8 M urea gels and the 0.1 per cent SDS gels. These data are consistent, with band σ being the component responsible for the stimulating activity, but they do not exclude the possibility that band τ or even some other material might also stimulate.

The stimulating component was identified as σ by the zone sedimentation of peak *A* material on a glycerol density gradient (Fig. 4). The factor required for T4 transcription was found to sediment at ~5*S*. Analysis of the gradient fractions by electrophoresis on 8 M urea gels showed that band σ sedimented identically to the stimulating activity, whereas band τ sedimented at about 8*S* in a region with no stimulating activity.

### Stimulating factor is a protein

The association of the stimulating activity with a specific band on polyacrylamide gels suggests that the factor is a protein. To confirm this, we determined the heat stability of the stimulating activity (Table 2). Incubation of peak *A* material for 5 min at 45°C and 50°C resulted in an inactivation of the stimulating activity of 52 per cent and 98 per cent, respectively. Furthermore, the activity was sensitive to trypsin. We conclude, therefore, that the factor is a heat-labile protein, although it is still possible that the factor is associated with a nucleic acid component of low molecular weight.

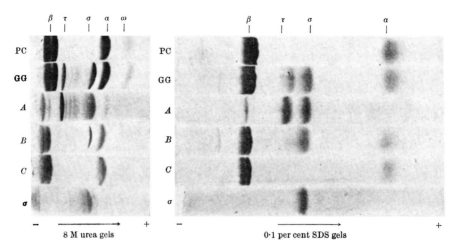

**Fig. 3.** Polyacrylamide gel electrophoresis patterns of different preparations of RNA polymerase and of purified factor. From top to bottom: PC enzyme (20 μg), GG enzyme (20 μg), peak A (10 μg), peak B (7 μg), peak C (8 μg) and purified factor (2 μg) from tube 17 of the glycerol gradient shown in Fig. 4. 8 M urea gels were prepared and run according to the general method described by Davis[34]. These gels contained 7.5 per cent acrylamide. 0.1 per cent SDS gels were run according to the procedure of Shapiro et al.[29] and contain 5 per cent acrylamide. In both cases the gels were stained by immersing them in a 0.25 per cent solution of Coomassie brilliant blue in methanol : acetic acid : water (5 : 1 : 5 v/v/v) for at least 2 h. The gels were then soaked in 7.5 per cent acetic acid, 5 per cent methanol for 0.5 h and finally destained electrophoretically in this same solvent. The bands observed are designated $\beta$, $\tau$, $\sigma$, $\alpha$ and $\omega$; $\alpha$, $\beta$ and $\omega$ are the bands normally seen in PC enzyme. The molecular weights of $\alpha$ and $\beta$ are ~40,000 and ~160,000 respectively. The molecular weight of $\beta$ was previously reported to be 110,000 (ref. 27), but recent measurements (Burgess, in preparation) indicate that 160,000 is a more accurate value. The band $\beta$ as seen on 0.1 per cent SDS gels appears as two closely spaced bands of equal intensity which probably correspond to two different polypeptide chains, with molecular weights of about 155,000 and 165,000. For simplicity these are both called $\beta$ in the text. In addition, a small polypeptide, $\omega$, with a molecular weight of about 10,000 can be seen moving ahead of $\alpha$ on the 8 M urea gels. It is present in GG enzyme, PC enzyme and DC enzyme, but it is not yet known whether it is a component of RNA polymerase or merely a tightly binding impurity. The amount of band $\tau$ observed in GG enzyme and also in peak A is variable. DC enzyme (not shown) contains only traces of $\tau$ and thus it is probably an impurity. From the SDS gels it is estimated that 45 per cent of peak A protein is $\sigma$. Purified factor is estimated to be about 80 per cent pure and is completely free of $\tau$. The right hand two-thirds of the urea gels which contain no bands are now shown.

The molecular weight of a protein can be estimated from its mobility on 0.1 per cent SDS polyacrylamide gels.[29] Using this procedure with β-galactosidase, bovine serum albumin and ovalbumin, with molecular weights of 130,000, 67,000 and 45,000, respectively, as molecular weight markers, the molecular weight of $\sigma$ was calculated to be about 95,000 ± 5,000. This is consistent with the $S$ value of about 5 obtained from the glycerol gradient.

### Factor requirements on various DNA templates

RNA synthesis in the presence and absence of factor was measured for several different DNA templates (Table 3). The greatest stimulation was observed with native T4 DNA, where the presence of factor increased the amount of synthesis seventy-five-fold. With all other templates tested, the stimulation was considerably lower. This variation may be ascribed to two factors. First, the fully stimulated levels of RNA synthesis vary according to the template used; this has been observed by several investigators[1-3]. Second, in the absence of factor, different DNA templates direct the synthesis of differing amounts of RNA. From analytical gels we estimate that PC enzyme contains less than 2 per cent as much factor as GG enzyme. Thus it is possible that the very small amount of RNA synthesis off T4 DNA with PC enzyme is due to traces of remaining factor. With calf thymus DNA, however, this could not easily

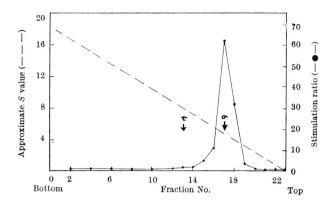

**Fig. 4.** Analysis of peak A by zone centrifugation. A 0.2 ml. sample containing 100 μg of peak A protein was layered on a 4.8 ml. linear 10-30 per cent glycerol density gradient containing 0.01 M tris-HCl buffer, pH 7.9, 0.01 M MgCl$_2$, 0.15 M KCl, 0.0001 M EDTA, and 0.0001 M dithiothreitol. The gradient was centrifuged for 11 h at 60,000 r.p.m. in a Spinco SW65 rotor at 4°C. 0.22 ml. fractions were collected. Each fraction was assayed for stimulating activity as described in the legend to Table 1. The proteins present in each fraction were analysed by electrophoresis on 8 M urea polyacrylamide gels. Band σ peaked in tube 17, band τ in tube 13. Molecular weight markers (E. coli β-galactosidase, human gamma-globulin, and egg lysozyme with sedimentation coefficients of 16S, 7S, and 1.9S, respectively) were centrifuged on a parallel gradient. The sedimentation coefficients of σ and τ were estimated to be about 5S and 8S, respectively.

TABLE 2
*Heat inactivation of the stimulating factor*

| Treatment of factor before assay | μμmole AMP incorporated/-min | Per cent inactivation of stimulating activity |
|---|---|---|
| 5 min at 0°C | 294 | 0 |
| 5 min at 37°C | 295 | 0 |
| 5 min at 45°C | 154 | 52 |
| 5 min at 50°C | 31 | 98 |

Peak A protein (0-9 μg) was added to each assay tube which also contained assay solution lacking the four triphosphates and T4 DNA. Separate tubes were incubated at the indicated temperatures for 5 min and then chilled to 0°C. The triphosphates, T4 DNA, and PC enzyme (6 μg) were added and the assay performed as described in the legend to Table 1. With no added factor, PC enzyme incorporated 17 μμmoles of AMP. Factor alone incorporated 9 μμmoles of AMP. The percentage inactivation at a given temperature was calculated by subtracting 26 μμmoles from the total incorporation observed after treatment at that temperature and then setting the incorporation of the unheated sample equal to 100 per cent activity.

explain the results. A more likely possibility is that there are some sites at which initiation can occur in the absence of factor. It is clear that such sites are not merely fully denatured regions, for denaturation of T4 DNA does not result in increased synthesis with PC enzyme.

Furthermore, φX-174 DNA is also a poor template. The behaviour of PC enzyme on T4 DNA and calf thymus DNA is remarkably similar to that of the RNA polymerase isolated from T4 infected cells by Walter et al.[30].

### Factor is not a nuclease

It could be argued that σ is a type of nuclease which makes single-stranded breaks in the DNA, thus "activating" the DNA. One prediction of this hypothesis is that DNA which has been used as a template for transcription in the presence of both enzyme and factor should then be capable of supporting the initiation of RNA chains in the absence of factor. This is not the case. The experiment shown in Table 4 shows that DNA which has been used as a template for factor-stimulated transcription exhibits virtually the same unstimulated and stimulated levels of transcription on re-use as DNA which has not previously been incubated with factor. In addition, evidence obtained by several investigators[19-21, 31] argues that DNA used for transcription by polymerase purified by the methods of Chamberlin[1] and Furth[2] remains intact. Because enzyme prepared in this way contains σ, it seems unlikely that the stimulation observed is due to nuclease action.

## TABLE 3
*Stimulation on various DNA templates*

| | mµmoles AMP incorporated/min/mg enzyme | | |
|---|---|---|---|
| DNA template | PC enzyme | PC enzyme + factor | GG enzyme |
| T4—native | 0.5 | 33.0 | 37.5 |
| T4—denatured | 0.5 | 6.1 | 3.0 |
| Calf thymus—native | 14.2 | 32.8 | 30.5 |
| Calf thymus—denatured | 3.3 | 14.5 | 10.7 |
| φX-174 | 0.9 | 6.2 | 4.9 |

RNA synthesis in the presence and absence of factor was assayed as described in the legend to Table 1, except that the DNA concentration was 10 µg/ml. DNA was denatured by adding 1/10 volume of 2 N NaOH to a 20 µg/ml. DNA solution. After standing at 25°C for 10 min, the solution was neutralized. Almost identical results were obtained if the DNA was denatured by heating at 95°C for 10 min, and then rapidly chilling in ice. The concentrations of PC enzyme and GG enzyme in the reaction mixture were both 4 µg/ml. Peak A protein was present, where indicated, at a concentration of 4 µg/ml. This corresponds to 2 µg/ml. of factor. The ratio of factor to enzyme in the mixture of PC enzyme and factor was about twice that normally occurring in GG enzyme. Incorporation by factor alone was negligible for all types of DNA tested. In all cases saturating amounts of DNA were used. The φX-174 phage DNA was a gift from Dr. D. T. Denhardt.

### Formation of a factor-polymerase complex

Several lines of evidence suggest that the factor can exist in a complex with RNA polymerase. First, the factor purifies with polymerase through steps involving protamine sulphate precipitation, ammonium sulphate fractionation and DEAE-cellulose chromatography. Furthermore, it remains with polymerase during low and high salt glycerol gradient centrifugation (Γ/2 = 0.04 and 1.0, respectively) where the polymerase sediments at 24S and 14S. The free factor sediments at about 5S, so it must be tightly bound to polymerase in all these conditions. Second, from the molecular weights of α, β and σ, and from the intensity of their stained bands on polyacrylamide gels, it is possible to make an estimate of the relative amounts of each band present in GG enzyme and also in peak B. In both, such an estimate yields a very approximate molar ratio of α:β:σ of 2:2:1. Third, complex formation between PC enzyme and factor can be demonstrated by running standard pH 8.7 polyacrylamide gels in the absence of dissociating agents (Fig. 5). In these conditions PC enzyme is resolved into several bands, which probably represent aggregates. Peak B, which contains the complex of α, β and σ, migrates as a single band which moves ahead of the PC enzyme bands. Purified σ moves faster than the complex. If σ and PC enzyme are mixed in approximately equivalent amounts

## TABLE 4
*Effect of preincubation with factor and PC enzyme on the ability of T4 DNA to direct factor-dependent transcription*

| | µµmoles AMP incorporated/min | |
|---|---|---|
| Material present during preincubation of DNA | PC enzyme alone | PC enzyme + factor |
| No additions | 8 | 264 |
| PC enzyme (10 µg) | 10 | 282 |
| Factor (1.5 µg) | 17 | 238 |
| PC enzyme (10 µg) + factor (1.5 µg) | 15 | 259 |
| No preincubation | 12 | 259 |

Four reaction mixtures (0.25 ml. each) for RNA synthesis were set up as described in the legend to Table 1, with the modification that BSA was omitted and non-radioactive ATP replaced $^{14}$C-ATP. PC enzyme and factor were added as indicated. Each reaction mixture was incubated for 10 min at 37°C. They were then diluted to 1 ml. with distilled water and extracted with 1 ml. of water-saturated phenol. The DNA was precipitated from the aqueous phase by the addition of 2 ml. ethanol. The precipitates were washed twice with 2 ml. of ethanol, collected by centrifugation and dried *in vacuo* to remove all traces of ethanol. Finally the DNA samples were redissolved in 0.2 ml. 0.01 M *tris* buffer pH 7.9. The ability of this DNA to direct RNA synthesis by PC enzyme in the presence and absence of factor was then assayed as described in the legend to Table 1 with the modification that the reaction volume was 0.10 ml. and BSA was omitted. Each tube contained 4 µg of PC enzyme. 0.6 µg factor (1.3 µg of peak A material) was also added where indicated.

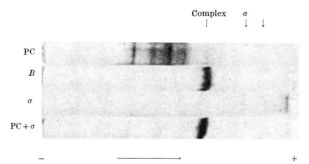

**Fig. 5.** Reconstitution of the enzyme-factor complex as demonstrated by electrophoresis on polyacrylamide gels. From top to bottom: PC enzyme (4 µg), peak B (4 µg), purified factor (1 µg) from tube 17 of the glycerol gradient shown in Fig. 4, and a mixture of PC enzyme and purified factor. Polyacrylamide gels, pH 8.7, 4 per cent acrylamide, were prepared and run as described by Davis[34], and stained as described in the legend to Fig. 3. σ appears as two very faint bands which move about 20 and 30 per cent faster than the complex and are indicated by arrows. The marker dye, seen very near the right end of the third gel from the top, just ran off of the other three gels.

and then subjected to gel electrophoresis, a single band corresponding to the complex is seen.

Even though PC enzyme and σ form a complex, for example as in peak B, the addition of either factor or PC enzyme to this complex results in stimulation. We are investigating whether the enzyme and factor are in rapid equilibrium with complex or whether factor is released for re-use in other complexes by the act of initiation.

Enzyme and complex elute at different ionic strengths from a phosphocellulose column, so it seems possible that they would elute from a DNA cellulose column at different ionic strengths. We found, however, that they behave identically on such a column. This provides some indication that the factor does not function merely by increasing the affinity of the polymerase for DNA. Furthermore, the free factor was not retained by the DNA cellulose column at 0.1 M KCl. This suggests that the free factor does not bind to DNA in conditions where the basic enzyme and the complex do so.

### The function of σ

The results clearly show that PC enzyme can by itself initiate the synthesis of RNA chains, and can catalyse chain elongation. Thus it is possible that this enzyme is the fundamental RNA polymerase. The presence of the stimulating factor, σ, greatly enhances the amount of RNA synthesis, the degree of enhancement being dependent on the DNA template used. Several possible modes of action of σ can be proposed. It could stimulate initiation, increase the rate of polymerization or prevent unusually early cessation of chain growth. Preliminary evidence (Travers and Burgess, manuscript in preparation) indicates that σ markedly increases the number of RNA chains initiated. This suggests that σ acts at the level of initiation.

We can thus pose the question: Is the additional initiation observed in the presence of σ merely due to an increase in the rate of initiation at sites poorly utilized in its absence, or does this initiation occur at sites which absolutely require σ for their expression? The first possibility implies that the PC enzyme determines the specificity of initiation and that σ may have some other function in the process of initiation. If σ itself determines the specificity of initiation, however, the interesting possibility arises that several similar factors could exist, each with a specificity for a different type of initiation site. This latter idea is attractive, for recent studies using the antibiotic rifamycin suggest that *in vivo* only one kind of RNA polymerase exists[32]. Yet there is also much evidence to indicate that *in vivo* the control of *m*RNA, *t*RNA and *r*RNA synthesis is not coordinate[33]. σ and similar factors could then act as positive control elements regulating the amount of synthesis of different classes of RNA, including the late RNA of certain bacteriophages.

## REFERENCES

1. Chamberlin, M., and Berg, P., Proc. US Nat. Acad. Sci. **48**:81 (1962).
2. Furth, J. J., Hurwitz, J., and Anders, M., J. Biol. Chem. **237**:2611 (1962).
3. Stevens, A., and Henry, J., J. Biol. Chem. **239**:196 (1964).
4. Fuchs, E., Zillig, W., Hofschneider, P. H., and Preuss, D., J. Mol. Biol. **10**:546 (1964).
5. Richardson, J. P., Proc. US Nat. Acad. Sci. **55**:1616 (1966).
6. Babinet, C., Biochem. Biophys. Res. Commun. **26**:639 (1967).
7. Hayashi, M., Hayashi, M. N., and Spiegelman, S., Proc. US Nat. Acad. Sci. **51**:351 (1964).
8. Geiduschek, E. P., Tocchini-Valentini, G. P., and Sarnat, M., Proc. US Nat. Acad. Sci. **52**:486 (1964).
9. Green, M., Proc. US Nat. Acad. Sci. **52**:1388 (1964).
10. Luria, S. E., Biochem. Biophys. Res. Commun. **18**:735 (1965).
11. Hayashi, M., Hayashi, M. N., and Spiegelman, S., Proc. US Nat. Acad. Sci. **50**:664 (1963).
12. Tocchini-Valentini, G. P., Stodolsky, M., Sarnat, M., Aurisicchio, A., Graziosi, F., Weiss, S. B., and Geiduschek, E. P., Proc. US Nat. Acad. Sci. **50**:935 (1963).
13. Marmur, J., and Greenspan, C. M., Science **142**:387 (1963).
14. Khesin, R. B., Shemyakin, M. F., Gorlenko, Zh. M., Bogdanova, S. L., and Afanaseva, T. P., Biokhimiya **27**:1092 (1962).
15. Geiduschek, E. P., Snyder, L., Colvill, A. J. E., and Sarnat, M., J. Mol. Biol. **19**:541 (1966).
16. Naono, S., and Gros, F., Cold Spring Harbor Symp. Quant. Biol. **31**:363 (1966).
17. Cohen, S. N., Maltra, U., and Hurwitz, J., J. Mol. Biol. **26**:19 (1967).
18. Crawford, L. V., Crawford, E. M., Richardson, J. P., and Slayter, H. S., J. Mol. Biol. **14**:593 (1965).
19. Richardson, J. P., J. Mol. Biol. **21**:83 (1966).
20. Jones, O. W., and Berg, P., J. Mol. Biol. **22**:199 (1966).
21. Pettijohn, D., and Kamiya, T., J. Mol. Biol. **29**:275 (1967).
22. Sentenac, A., Ruet, A., and Fromageot, P., Europ. J. Biochem. **5**:385 (1968).
23. Priess, H., and Zillig, W., Biochim. Biophys. Acta **140**:540 (1967).
24. Stevens, A., Emery, jun., A. J., and Sternberger, N., Biochem. Biophys. Res. Commun. **24**:929 (1966).
25. Smith, D. A., Martinez, A. M., Ratliff, R. L., Williams, D. L., and Hayes, F. N., Biochemistry **6**:3057 (1967).
26. Zillig, W., Fuchs, E., and Millette, R., in Proc. Nucleic Acid Res. (edit. by Cantoni, G. L., and Davies, D. R.), 326 (Academic Press, New York, 1966).
27. Burgess, R. R. Fed. Proc. **27**:295 (1968).
28. Alberts, B. M., Amodio, F. J., Jenkins, M., Gutman, E. D., and Ferris, F. L., Cold Spring Harbor Symp. Quant. Biol. **33** (in press).
29. Shapiro, A., Vinuela, E., and Maizel, J. V., Biochem. Biophys. Res. Commun. **28**:815 (1967).
30. Walter, G., Seifert, W., and Zillig, W., Biochem. Biophys. Res. Commun. **30**:240 (1968).
31. Maitra, U., and Hurwitz, J., J. Biol. Chem. **242**:4897 (1967).
32. Tocchini-Valentini, G. P., Marino, P., and Colvill, A. J., Nature **220**:275 (1968).
33. Maaløe, O., and Kjeldgaard, N. O., Control of Macromolecular Synthesis (Benjamin, New York, 1966).
34. Davis, B. J., Ann. NY Acad. Sci. **121**:406 (1964).

## 27 A FACTOR THAT STIMULATES RNA SYNTHESIS BY PURIFIED RNA POLYMERASE

*J. Davison*
*L. M. Pilarski*
*H. Echols*
Department of Biochemistry
University of Wisconsin
Madison, Wisconsin

---

*Abstract.* A protein factor, partially purified from *E. coli*, can stimulate some 30- to 50-fold the transcription of DNA by purified RNA polymerase. Evidence is presented suggesting that this stimulatory factor (designated M factor) exerts its effect before the RNA synthesis is initiated. The M factor is prepared from a crude ribosome fraction, but it does not require the presence of ribosomes for full activity.

---

The *in vitro* transcription of DNA by purified DNA-dependent RNA polymerase requires, in addition to the DNA templates, only the four ribonucleoside triphosphates and a divalent cation such as $Mg^{++}$.[1-5] However, the possibility has remained that additional factors are necessary for properly initiated, terminated, and regulated synthesis of RNA. The possible existence of one such factor has been suggested by the observation of a twofold stimulation of RNA synthesis by purified RNA polymerase in the presence of both ribosomes and a protein synthesis initiation factor.[6-8]

The present investigation was initiated by the observation that the specific repression of phage λ DNA transcription by the partially purified λ repressor was more effective in the presence of a crude ribosome fraction.[9] This suggested the possibility that some component of the ribosome fraction affected *in vitro* RNA synthesis in some undiscovered way.

This communication reports the partial purification, from a subcellular fraction rich in ribosomes and DNA, of a factor able to stimulate some 30- to 50-fold the *in vitro* transcription of DNA by purified RNA polymerase. This factor has been designated the M factor. The M factor exerts its maximal effect on RNA synthesis in the absence of ribosomes.

### Materials and methods

*DNA preparation.* DNA from phages λ and T4 was extracted by the phenol method; the phages were purified by CsCl density gradient centrifugation.[10] Salmon sperm DNA and *E. coli* DNA were obtained from the California Biochemical Corp. and the General Biochemical Corp., respectively. The *E. coli* DNA was reextracted with phenol.

*Preparation of M factor.* A nonlysogenic derivative of the RNase I⁻ strain Q13[11] was grown and harvested as previously described.[9] The washed cell paste was stored in liquid nitrogen until required.

Fifty grams wet weight of cell paste were resuspended in 130 ml of 0.05 $M$ Tris-HCl, pH 7.9, containing 12% w/w sucrose. The resuspended cells were treated with lysozyme at a final concentration of 300 μg/ml in the presence of 0.002 $M$ EDTA for 5 min at 37°C. $MgCl_2$ and dithiothreitol were added to final concentrations of 0.03 $M$ and 0.0001 $M$, respectively, and lysis was completed by a 2-min incubation in the presence of the nonionic detergent Triton X-100 (Ruger, Irvington, N. Y.) at a concentration of 1.5%, followed by rapid freezing and thawing. All subsequent operations were carried out at 0-4°C.

The lysate was centrifuged three times at 10,000 × $g$ for 10 min to remove a loose pellet containing bacterial debris and some DNA. The supernatant was centrifuged at 100,000 × $g$ for 2 hr and the supernatant discarded. The pellet containing a lower layer of ribosomes and an upper layer of DNA was partially resuspended in 48 ml of buffer containing 0.01 $M$ Tris-HCl, pH 7.4; 0.03 $M$ $MgCl_2$; 2 $M$ $NH_4Cl$; and 0.0001 $M$ dithiothreitol. The resuspension was completed by gentle homogenization approximately

---

From Proceedings of the National Academy of Sciences U.S.A. 63:168-174, 1969. Reprinted with permission.

We thank Kerry Brookman for technical assistance, John Garver for use of his pilot plant facilities (U.S. Public Health Service grant FR-226), and Ned Mantei, Mark Willard, and Robert Wells for RNA polymerase preparations purified by the methods of Chamberlin and Berg, Babinet, and Morgan, respectively. We also thank Richard Burgess for communicating his experimental results prior to publication, and Masayasu Nomura for advice.

This work was supported, in part, by U.S. Public Health Service grant GM08407.

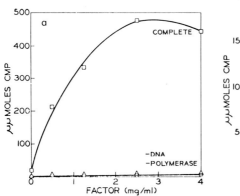 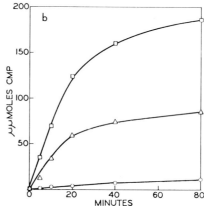

**Fig. 1. a,** Effect of M factor concentration on RNA synthesis by RNA polymerase. Different amounts of M factor were added to assay mixtures containing the complete polymerase reaction mixture (□—□) or to mixtures from which either DNA (△—△) or RNA polymerase (○—○) had been omitted. Other details are given in *Materials and methods*.
**b,** Kinetics of RNA synthesis in the presence and absence of M factor. Incubation mixtures contained (1) 6.2 mg/ml M factor (□—□), (2) 3.1 mg/ml M factor (△—△) and (3) no M factor (○—○). Incubations were carried out at 37°C, and at the times indicated samples of 0.05 ml were withdrawn.

16 hr later. The extract was then centrifuged for 2 hr at 100,000 × g and the pellet discarded.

Saturated $(NH_4)_2SO_4$ solution was added to the supernatant to give a final concentration of 70% saturation with respect to an $(NH_4)_2SO_4$ solution at 26°C. The mixture was stirred for 1 hr at 0°C and the precipitate collected by centrifugation. The pellet was dissolved in 15 ml buffer containing 0.01 $M$ Tris-HCl, pH 7.4; 0.01 $M$ $MgCl_2$; 0.03 $M$ $NH_4Cl$; 0.0001 $M$ dithiothreitol; and dialyzed against the same buffer overnight. The precipitate that usually formed on dialysis was removed by centrifugation and discarded. The supernatant was applied to a column (3 × 8 cm) of DEAE-cellulose (Whatman DE52) that had been equilibrated with the buffer. In batch elution most of the M factor eluted between 0.03 $M$ and 0.2 $M$ $NH_4Cl$. This fraction was dialyzed overnight against the same buffer. The chief purpose of the DEAE step was the removal of polynucleotides.

*Assay of M factor.* M factor was assayed essentially under the conditions described previously.[9] The reaction mixture contained: 0.01 $M$ Tris-HCl, pH 7.9; 0.008 $M$ $MgCl_2$; 0.002 M $MnCl_2$; 0.02 $M$ $NH_4Cl$; 0.002 $M$ β-mercaptoethanol; 0.4 mM ATP, UTP, GTP; 0.06 mM $H^3$-CTP (Schwarz 0.14 mc/μmole). The reaction volume (usually 0.25 ml) also contained 10 μg/ml DNA and 10 μg/ml RNA polymerase purified by the phosphocellulose chromatography method.[12] After incubation for 10 min at 37°C the mixture was precipitated with 5% trichloroacetic acid (in the presence of 40 μg/ml salmon sperm DNA as carrier) and filtered through Whatman GF/C glass filters.

## Results

### Characterization of M factor

The effect of various concentrations of the M factor on the synthesis of RNA by purified RNA polymerase is shown in Figure 1a. At low protein concentrations the synthesis of RNA is approximately proportional to the amount of M factor added. At higher concentrations the M factor becomes saturating, and a plateau of stimulation is reached. Further increase in M-factor concentration tends to decrease synthesis. When either RNA polymerase or DNA is omitted from the reaction mixture, the amount of RNA made is very small. From this it is clear that both RNA polymerase and a DNA template are required for the reaction, and that the M factor prepared as described above is not greatly contaminated by either. This experiment also excludes the possibility that the increase in RNA synthesis in the presence of the M factor is due to the activity of some relatively nonspecific enzyme such as polynucleotide phosphorylase. The conclusion that the M factor stimulates the DNA-directed RNA polymerase reaction is further substantiated by the observation that RNA synthesized in the presence of the M factor, using a λ DNA template, hybridizes with denatured λ DNA with the same efficiency as RNA synthesized in the absence of the M factor.

Figure 1b shows the kinetics of RNA synthesis by RNA polymerase in the presence and in the absence of the M factor. Both in the presence and absence of the M factor, RNA synthesis is initially proportional to time of incubation; the rate of synthesis falls during prolonged incubation. Such kinetics are typical of the low salt conditions used in these experiments.[13,14] The degree of stimulation is also relatively independent of time of incubation.

From the method of preparation (involving ammonium sulphate precipitation and chromatography on DEAE-cellulose), it seemed likely that the M factor was a protein. Consistent with this suggestion were the following observations: the M factor is thermolabile, being totally inactivated by five minutes at 100°C; it is excluded by Sephadex G-50; and the DEAE fraction with M-factor activity has a ratio of absorbencies at 280 and 260 mμ of 1.3, indicating a nucleic acid content of less than 1.3 per cent.[15] The possibility remained that the activity was in some way due to small amounts of ribosomes or ribosomal subunits which were present in the preparation. To investigate this possibility the sedimentation properties of the M factor were investigated by sucrose gradient centrifugation. The results of this experiment are shown in Figure 2. M factor sediments with a sedimentation coefficient of approximately 5S. Thus the M factor is a macromolecule distinct from ribosomes or ribosomal subunits. Furthermore, the M factor does not require the presence of ribosomes or ribosomal subunits for activity.

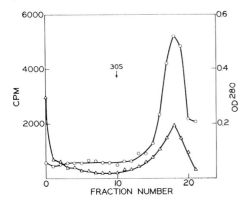

**Fig. 2.** Sucrose gradient centrifugation of M factor. 1.8 mg of M factor in a volume of 0.3 ml was layered on a 4.7 ml 5-20% sucrose gradient and centrifuged for 3 hr in the SW39 rotor at 37,000 rpm. Fractions were assayed for M factor (○—○), and the OD$^{280}$ (△—△) was read after adjusting the volume to 1 ml. The *arrow* indicates the position of 30S ribosomal subunits in an identical gradient. The *left-hand ordinate* scale gives the cpm incorporated into RNA in the standard M-factor assay. The *right-hand ordinate* scale gives the OD at 280 mμ as a measure of protein concentration.

*Template requirement*

From the preceding results it is clear that the M factor is able to stimulate RNA synthesis by RNA polymerase using a λ DNA template. The results given in Table 1 show that the phenomenon is not restricted to a λ DNA template but occurs with DNA from different sources. However, not all DNA templates are equally effective.

**TABLE 1**
*Effect of different DNA templates on stimulation by M factor*

| DNA Template | | Incorporation (μμmoles CMP) | | |
| --- | --- | --- | --- | --- |
| | | −M factor | +M factor | Stimulation |
| E. coli | native | 53 | 235 | 4.4 |
| E. coli | denatured | 87 | 322 | 3.7 |
| Salmon sperm | native | 98 | 831 | 8.5 |
| Salmon sperm | denatured | 130 | 337 | 2.6 |
| T4 phage | native | 8 | 384 | 48 |
| T4 phage | denatured | 104 | 282 | 2.8 |
| λ phage | native | 24 | 757 | 32 |
| λ phage | denatured | 124 | 363 | 2.9 |

In all cases DNA was present at a concentration of 10 μg/ml. DNA was denatured by boiling it for 10 min and quenching it in dry-ice ethanol. The M factor, when present, was at a concentration of 0.6 mg/ml. Other conditions are described in *Materials and methods*.

The stimulation obtained with phage DNA templates is greater than that obtained with the commercial *E. coli* and salmon sperm DNA templates, and native DNA is more effective than denatured DNA. Further experiments will be required to clarify the significance of DNA "quality" in this reaction.

*Mechanism of action of M factor*

Models for the mechanism of action of the M factor may be divided into two general classes, according to whether the M factor is considered to be an RNA polymerase "accessory factor" or an enzyme that modifies the DNA template. An example of the latter class would be an endonuclease producing single-stranded nicks that might make the DNA more easily transcribed. The two hypotheses may be tentatively differentiated by the prediction that an RNA polymerase accessory factor might be required in stoichiometric amounts, whereas a DNA modifying enzyme (such as a nuclease) would be needed only in catalytic amounts, provided that it was preincubated with the DNA template.

To test this prediction various amounts of the M factor were preincubated with DNA for various lengths of time. The amount of the M factor and the time of preincubation were adjusted so that the product of the two was a constant. Since the effect of the M factor is proportional to its concentration over this range (Fig. 1a), the amount of DNA modification catalytically produced by a DNA modifying enzyme and hence the amount of stimulation of RNA synthesis should be constant. Alternatively, if the M factor is required stoichiometrically to stimulate the polymerase reaction, then the preincubation with DNA should be irrelevant, and the results should correspond to those of the control experiment in which the M factor was preincubated in the absence of DNA. From the results given in Table 2 it can be seen that this latter prediction is true. This experiment makes it unlikely that M factor activity results from the presence of a DNA modifying enzyme.

Another important question concerning the mechanism of M-factor stimulation is whether the M factor must be present prior to chain initiation or whether it exerts its effect only on

TABLE 2
*Effect of preincubation of M factor and DNA*

| M factor (mg/ml) | Incorporation (μμmoles CMP) | |
| --- | --- | --- |
| | Preincubated (+ DNA) | Preincubated (− DNA) |
| 0.0 | 20 | 20 |
| 0.128 | 68 | 66 |
| 0.32 | 163 | 160 |
| 0.64 | 255 | 223 |
| 1.28 | 305 | 279 |
| 2.56 | 263 | 323 |

Different amounts of M factor were preincubated for 10 min at 37°C. In one half of the experiment, as part of this 10-min incubation, the M factor was preincubated with DNA as follows: 0.128 mg M factor, 10 min; 0.32 mg, 4 min; 0.64 mg, 2 min; 1.28 mg, 1 min; and 2.56 mg, 0.5 min. Note that in each case the products of M-factor quantity and time are constant. In the second half of the experiment the DNA was added to the tubes at 0°C after the preincubation was completed. RNA polymerase and nucleoside triphosphates were added, and the tubes were incubated at 37°C for a further 10 min. Other details are described in *Materials and methods*.

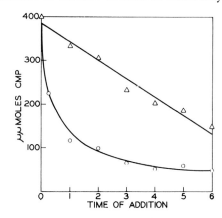

**Fig. 3.** Effect of prior initiation of RNA synthesis on M-factor stimulation. Two incubation mixtures containing DNA and RNA polymerase were transferred to 37°C. At $t = 0$ min, nucleoside triphosphates were added to one tube and, at intervals thereafter, aliquots were withdrawn into tubes containing M factor (○—○). In the control experiment the nucleoside triphosphates were not added at $t = 0$ min, and at intervals aliquots were withdrawn into tubes containing *both* M factor and nucleoside triphosphates (△—△). All reaction mixtures were precipitated with 5% trichloroacetic acid at $t = 10$ min. The approximately linear decrease in the control experiment is due to the progressively shorter incubation period in the presence of nucleoside triphosphates. The M factor was at a final concentration of 0.6 mg/ml. Other conditions are described in *Materials and methods*.

the rate of chain propagation. To investigate this, an experiment was performed in which the effect of preincubation of DNA and polymerase was compared with the effect of preincubation of DNA, polymerase, and triphosphates. The former treatment allows binding of polymerase but not chain initiation, whereas the latter permits chain initiation. The results (Fig. 3) show that preincubation of DNA and polymerase has no effect on the subsequent stimulation by the M factor; however, preincubation of DNA and polymerase in the presence of nucleoside triphosphates gives an exponential loss of potential stimulation by the M factor. Thus the M factor must be present at or before the chain initiation of RNA in order to stimulate the synthesis of RNA.

## Discussion

The results presented in this communication demonstrate the existence of a factor able to stimulate markedly the synthesis of RNA by purified RNA polymerase. The possibility that M-factor activity results from a polynucleotide synthesizing enzyme of the polynucleotide phosphorylase type was ruled out by the observation that in the absence of either DNA or RNA polymerase incorporation is very low. Preliminary evidence suggests that the M factor is a protein, although it has not been shown that only a single species is involved.

Up to a limit a stoichiometric relationship exists between the amount of M factor present in the reaction and the stimulation it produces. Since preincubation of the M factor and DNA does not increase the stimulation, it is considered unlikely that the M factor acts enzymatically with the DNA to change its template properties.

The ability of the M factor to stimulate RNA synthesis is markedly reduced if it is added after RNA chain initiation has occurred,

suggesting that the M factor acts either at initiation or at some step preceding initiation. A number of molecular mechanisms are possible, and further experiments will be required to distinguish them. For example, the M factor might function in complex with RNA polymerase to provide for localized DNA strand separation (if such a complex must form prior to initiation); or the M factor might provide for RNA polymerase to initiate only at the "correct" *in vivo* sites if the "correctly initiated" RNA synthesis reaction proceeds at a greater rate.

In a recent publication Burgess, Travers, Dunn, and Bautz[16] have reported that chromatography of RNA polymerase on phosphocellulose separates a polymerase component and a small protein factor. This factor, designated $\sigma$, has no polymerase activity of its own but can greatly stimulate the synthesis of RNA by phosphocellulose-purified RNA polymerase. In many of its properties $\sigma$ is similar to the M factor, but sufficient data are not yet available to allow us to decide whether or not the two are identical. One possible difference is that the stimulation by the M factor is not restricted to the use of an enzyme prepared by phosphocellulose chromatography. Substantial stimulation is also seen with an enzyme prepared by the methods of Chamberlin and Berg,[4] Babinet,[17] and Morgan.[18] Whether or not this represents a real distinction is not yet known.

The relationship, if any, of the M factor to the C factor[6,7] and factor-3,[8] which stimulate RNA synthesis only in the presence of ribosomes, is obscure. The methods of preparation are somewhat similar; however, under our conditions, addition of "purified" (salt-washed) ribosomes[19] to a polymerase reaction mixture in the presence or absence of the M factor results in a twofold decrease in synthesis.

## REFERENCES

1. Weiss, S. B., these Proceedings **46**:1020 (1960).
2. Stevens, A., Biochem. Biophys. Res. Commun. **3**:92 (1960).
3. Hurwitz, J., A. Bressler, and R. Diringer, Biochem. Biophys. Res. Commun. **3**:15 (1960).
4. Chamberlin, M., and P. Berg, these Proceedings **48**:81 (1962).
5. Richardson, J. P., Progr. Nucleic Acids and Mol. Biol., in press.
6. Revel, M., and F. Gros, Biochem. Biophys. Res. Commun. **25**:124 (1966).
7. Revel, M., M. Herzberg, A. Becarvic, and F. Gros, J. Mol. Biol. **33**:231 (1968).
8. Brown, J. C., and P. Doty, Biochem. Biophys. Res. Commun. **30**:284 (1968).
9. Echols, H., L. Pilarski, and P. Y. Cheng, these Proceedings **59**:1016 (1968).
10. Kaiser, A. D., and D. S. Hogness, J. Mol. Biol. **2**:392 (1960).

11. Gesteland, R. F., J. Mol. Biol. **16**:67 (1966).
12. Burgess, R. R., manuscript in preparation.
13. So, A. G., E. W. Davie, R. Epstein, A. Tissières, these Proceedings **58**:1739 (1967).
14. Fuchs, A. B., R. L. Millette, W. Zillig, and G. Walter, European J. Biochem. **3**:183 (1967).
15. Logne, E., Methods Enzymol. **3**:453 (1957).
16. Burgess, R. R., A. A. Travers, J. J. Dunn, and E. K. F. Bautz, Nature **221**:43 (1969).
17. Babinet, C., Biochem. Biophys. Res. Commun. **26**:639 (1967).
18. Morgan, A. R., and R. D. Wells, J. Mol. Biol. **37**:63 (1968).
19. Likover, T. E., and C. G. Kurland, J. Mol. Biol. **25**:497 (1967).

# 28 THE SYNTHESIS OF A SELF-PROPAGATING AND INFECTIOUS NUCLEIC ACID WITH A PURIFIED ENZYME

S. Spiegelman
I. Haruna
I. B. Holland
G. Beaudreau
D. Mills*
Department of Microbiology
University of Illinois
Urbana, Illinois

The unambiguous analysis of a replicating mechanism demands evidence that the reaction being studied is, in fact, generating replicas. If, in particular, the concern is with the synthesis of a viral nucleic acid, data on base composition and nearest neighbors are not sufficient. Ultimately, proof must be offered that the polynucleotide product contains the information necessary for the production of the corresponding virus particle in a suitable test system.

These conditions impose severe restraints on the type of experiments acceptable as providing information which is irrefutably relevant to the nature of the replicating mechanism. Clearly, the enzyme system employed must be free of interfering and confounding activities so that the reaction can be studied in a simple mixture containing only the required ions, substrates, and templates. Since the biological activity of the product is likely to be completely destroyed by even one break, the elimination of nuclease activity must be rigorous indeed. The purity required imposes the necessity that the enzymological aspects of the investigation be virtually completed before an examination of mechanism can be safely instituted.

We have previously reported[1] the purification of two distinct RNA-dependent-RNA-polymerases (designated "replicases" for brevity) induced in the same host by two unrelated[2,3] RNA bacteriophages (MS-2 and Qβ). It was shown that under optimal conditions, both enzymes are virtually inactive with a variety of heterologous RNA species, including ribosomal and sRNA of the host. Further, neither replicase can function with the other's RNA. Each enzyme recognizes the RNA genome of its origin and requires it as a template for normal synthetic activity.

In summary, the purified replicases exhibited[1] the following distinctive features: (a) freedom from detectable levels of the DNA-dependent-RNA-polymerase, ribonuclease I,[4] ribonuclease II,[5] and RNA phosphorylase; (b) complete dependence on added RNA for synthetic activity; (c) competence for prolonged (more than 5 hr) synthesis of RNA; (d) ability to synthesize many times the input templates; (e) saturation at low levels of RNA (1 γ RNA/40 γ protein); (f) virtually exclusive requirement for intact homologous template under optimal ionic conditions.

The discriminating selectivity of the replicase permitted a simple test of similarity between template and product. Haruna and Spiegelman[6] showed that when reactions are started at template concentrations below those required to saturate the enzyme, RNA synthesis follows an autocatalytic curve. When the saturation concentration level is reached, the kinetics become linear. The autocatalytic behavior below saturation of the enzyme implies that the newly synthesized product can in turn serve as templates for the reaction. To test this conclusion directly, the product was purified from a reaction allowed to proceed until a 65-fold increase of the input RNA had accumulated. The ability of the newly synthesized RNA to initiate the reaction was examined in a saturation experiment and found to be identical to RNA isolated from virus particles. It is

---

From Proceedings of the National Academy of Sciences U.S.A. **54**:919-927, 1965. Reprinted with permission.

This investigation was supported by USPHS research grant CA-01094 from the National Cancer Institute and grant GB-2169 from the National Science Foundation.

*Predoctoral trainee in Microbial and Molecular Genetics, grant USPHS 2-T1-GM-319.

evident that the sequences employed by the enzyme for recognition are being faithfully copied.

The findings summarized above and the state of purity of the enzymes encouraged us to enter the next phase of the investigation and examine the infectivity of the synthesized material. It is the purpose of the present paper to describe experiments demonstrating that the RNA produced by replicase is fully competent to program the production of complete virus particles. The data establish that the reaction being studied is indeed generating self-propagating replicas of the input RNA.

## Materials and methods

*(1) Biological system and enzyme preparation.* The bacterial virus employed is Qβ, isolated by Watanabe.[2] The host and assay organism is a mutant Hfr strain of *E. coli* (Q13) isolated in the laboratory of W. Gilbert by Diane Vargo. This bacterial strain has the convenient property[7] of lacking ribonuclease I and RNA phosphorylase. The preparation of infected cells and the subsequent isolation and purification of the replicase follows the detailed protocol of Haruna and Spiegelman.[1] The preparation of virus stocks and the purification of RNA from them follow the methods of Doi and Spiegelman.[8]

*(2) The assay of enzyme activity by incorporation of radioactive nucleotides.* The standard reaction mixture is 0.25 ml and, in addition to 40 γ of enzyme, contains the following in μmoles: Tris HCl, pH 7.4, 21; $MgCl_2$, 3.2; CTP, ATP, UTP, and GTP, 0.2 each. The reaction is terminated in an ice bath by the addition of 0.15 ml of neutralized saturated pyrophosphate, 0.15 ml of neutralized saturated orthophosphate, and 0.1 ml of 80% trichloracetic acid. The precipitate is transferred to a membrane filter and washed 7 times with 5 ml of cold 10% TCA. The membrane is then dried and counted in a liquid scintillation counter as described previously. $UTP^{32}$ was synthesized as described by Haruna *et al.*[9] It was used at a specific activity such that the incorporation of 20,000 cpm corresponds to the synthesis of 1 γ of RNA, permitting the use of 20 λ samples for following the formation of labeled RNA.

*(3) Isolation of synthesized product.* Samples removed from the reaction mixture are placed immediately in an ice bath and 20 λ removed for immediate assay of radioactive RNA as described in (2) above. The volume is then adjusted to 1 ml with TM buffer ($10^{-2}$ M Tris, $5 \times 10^{-3}$ M $MgCl_2$, pH 7.5). One ml of water-saturated phenol is then added and the mixture shaken in heavy wall glass centrifuge tubes (Sorvall, 18 × 102 mm) at 5°C for 1 hr. After separation of the water phase from the phenol by centrifugation at 11,000 rpm for 10 min, another 1 ml of TM buffer is added to the phenol, which is then mixed by shaking for 15 min at 5°C. Again, the phenol and water layers are separated, and the two water layers combined. Phenol is eliminated by two ether extractions, care being taken to remove the phenol from the walls of the centrifuge tubes by completely filling them with ether after each extraction. The ether dissolved in the water phase is then removed with a stream of nitrogen. The RNA is precipitated by adding 1/10 vol of potassium acetate (2 M) and 2 vol of cold absolute ethanol. The samples are kept for 2 hr at −20°C before being centrifuged for one hour at 14,000 rpm in a Sorvall SS 34 rotor. The pellets are drained, and the remaining alcohol is removed by storing under reduced pressure in a vacuum desiccator for 6-8 hr at 5°C. The RNA is then dissolved in 1 ml of buffer ($10^{-2}$ M Tris, $10^{-2}$ M $MgCl_2$, pH 7.5) and samples are removed immediately for infectivity assay. TCA-precipitable radioactivity is measured on 20 λ aliquots of the final product from which the per cent recovery of synthesized RNA can be determined. In the range of 1-8 γ, it was found that, in general, 65% of the synthesized RNA was removed. All purified products were examined for the presence of intact virus particles by assay on whole cells and none were found.

*(4) The assay for infectivity of the synthesized RNA.* The procedure used is a modification of the spheroplast method of Guthrie and Sinsheimer.[10] The necessary components are as follows:

(a) *Medium.* The medium used is a modification of the 3XD medium of Fraser and Jerrel[11] and requires in grams/liter the following: $Na_2HPO_4$, 2 gm; $KH_2PO_4$, 0.9 gm; $NH_4Cl$, 1 gm; glycerol (Fisher reagent), 30 gm; Difco yeast extract, 50 mg; casamino acids (Difco vitamin-free), 15 gm; L-methionine, 10 mg; D,L-leucine, 10 mg; $MgSO_4 \cdot 7 H_2O$, 0.3 gm. These components are mixed in the order indicated in 500 ml glass-distilled water. To this is finally added another 500 ml containing 0.3 ml M $CaCl_2$.

(b) *Sucrose nutrient broth (SNB)* contains in grams/liter the following: casamino acids (Difco), 10 gm; nutrient broth (Difco), 10 gm; glucose, 1 gm; sucrose, 100 gm. After autoclaving, the following are added aseptically: 10 ml 10% $MgSO_4$, and 3.3 ml 30% bovine serum albumin (BSA) from Armour Laboratories.

(c) *Reagents required for the production of spheroplasts.* The following solutions are required for the production of spheroplasts: lysozyme (Sigma) at 2 mg/ml in 0.25 M Tris, pH 8.0; protamine sulfate from Eli Lilly and Co., 0.1%; and sterile solutions of 30% BSA; 0.25 M Tris (Trizma), pH 8.0; 0.01 M Tris, pH 7.5 and pH 8.0; 0.5 M sucrose; 0.4% EDTA in 0.01 M Tris, pH 7.5.

For the preparation of spheroplasts, an overnight

culture of Q13 in 3XD medium is first diluted into a fresh medium to an $OD_{660}$ of 0.06. The culture is allowed to grow to an $OD_{660}$ between 0.2 and 0.22 at $30°C$, and the cells are spun down at room temperature. The pellet from 25 ml of cells is first suspended in 0.35 ml of 0.5 $M$ sucrose plus 0.1 ml of 0.25 $M$ Tris, pH 8.0. Then 0.01 ml of lysozyme is added followed by 0.03 ml EDTA. After 10 min at room temperature, when conversion to spheroplasts is 99.9%, 0.2 ml of this stock is diluted into 3.8 ml SNB, and 0.025 ml of protamine sulfate is added. The spheroplast stock must be examined microscopically before proceeding. The presence of even 5% breakage of spheroplasts indicates a preparation which will give a low efficiency of plating. In agreement with Paranchych,[12] we have found that protamine increases the efficiency of detection of infectious RNA. However, the optimal protamine concentration in the present system is considerably lower than that used by Paranchych.

The RNA infection is usually carried out at room temperature with solutions containing 0.5 $\gamma$ of RNA/ml, a concentration at which the assay is not limited by the number of spheroplasts per infectious unit. To 0.2 ml of RNA is added 0.2 ml of the spheroplast stock containing about $3 \times 10^7$ spheroplasts. The samples are mixed, and immediately an aliquot is removed and diluted appropriately through SNB before plating on L-agar using Q13 as the indicator. The soft agar (0.7%) layer employed (2.5 ml) contains 10% sucrose; 0.1% $MgSO_4$; and 0.01 ml of 30% BSA per tube plus 0.2 ml of an overnight culture of Q13. To obtain reproducibility, the spheroplast stock is used 15-45 min after dilution into the SNB. Efficiency of plating (e.o.p.) is usually 2-8 $\times$ $10^{-7}$. Higher efficiencies ($>1 \times 10^{-6}$) can be obtained if the spheroplast stock is employed immediately after dilution and by including a stabilization period in the SNB dilution tubes, rather than plating immediately. However, this higher plating efficiency decays rapidly, making it difficult to obtain reproducible duplicates in repetitive assays. Since reproducibility was of greater concern than efficiency, the assay method detailed above was employed.

## Results

In designing experiments which involve infectivity assays of the enzymatically synthesized RNA, it is important to recognize that even highly purified enzymes from infected cells, although demonstrably devoid of intact cells, are likely to include some virus particles. Chemically, the contamination is trivial, amounting to 0.16 $\gamma$ of nucleic acid and 0.8 $\gamma$ of protein for each 1,000 $\gamma$ of enzyme protein employed in the present studies. Since 40 $\gamma$ of protein are used for each 0.25 ml of reaction, the contribution to the total RNA by the particles is only 0.006 $\gamma$, which is to be compared with the 0.2 $\gamma$ of input RNA and the 3-20 $\gamma$ synthesized in the usual experiment. It was shown in control experiments that RNA freshly extracted from particles in the reaction mixture is no more infective than that obtained from the usual purified virus preparation. Further, the mandatory requirement for added RNA proves that, within the incubation times used, this small amount of RNA is either inadequate or unavailable for the initiation of the reaction. Thus, these particles do not significantly influence either the chemical or the enzymatic aspects of the experiment. However, because of their higher infective efficiency, even moderate amounts of intact virus cannot be tolerated in the examinations of the synthesized RNA for infectivity. Consequently, all RNA preparations were phenol-treated [*Methods* (3)] prior to assay. Further, the phenol-purified RNA was routinely tested for whole virus particles and none were found in the experiments reported.

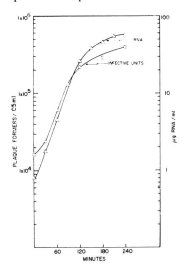

**Fig. 1.** Kinetics of RNA synthesis and formation of infectious units. An 8-ml reaction mixture was set up containing the components at the concentrations specified in *Methods* (2). Samples were taken as follows: 1 ml at 0 time and 30 min, 0.5 ml at 60 min, 0.3 ml at 90 min, and 0.2 ml at all subsequent times. 20 $\lambda$ were removed for assay of incorporated radioactivity as described in *Methods* (2). The RNA was purified from the remainder [*Methods* (3)], radioactivity being determined on the final product to monitor recovery. Infectivity assays were carried out as in *Methods* (4).

**TABLE 1**
*Serial transfer experiment*

| 1 | 2 | 3 | Formation of RNA | | | | Concentration of original template | | | Formation of IU | | 13 | 14 |
|---|---|---|---|---|---|---|---|---|---|---|---|---|---|
| | | | 4 | 5 | 6 | 7 | 8 | 9 | 10 | 11 | 12 | | % |
| Transfer no. | Interval (min) | Time | Cpm $\times 10^{-3}$ | Total ($\gamma$) | $\Delta$ ($\gamma$) | $\Sigma$ ($\gamma$) | ($\gamma$) | strands | IU | $\Delta$ $\times 10^{-5}$ | $\Sigma$ $\times 10^{-5}$ | Observed e.o.p. $\times 10^{-7}$ | Recovery of $P^{32}$-RNA |
| 0 | 0 | 0 | 0 | 0.2 | 0 | 0 | $2.0 \times 10^{-1}$ | $1.2 \times 10^{11}$ | $6.0 \times 10^{4}$ | 1.0 | 1.0 | 5.5 | ... |
| 1 | 40 | 40 | 64 | 3.2 | 3.0 | 3.0 | $2.0 \times 10^{-1}$ | $1.2 \times 10^{11}$ | $6.0 \times 10^{4}$ | 5.2 | 5.2 | 3.2 | 54.2 |
| 2 | 40 | 80 | 84 | 4.2 | 3.6 | 6.6 | $4.0 \times 10^{-2}$ | $2.4 \times 10^{10}$ | $1.2 \times 10^{4}$ | 2.2 | 6.5 | 2.0 | 88.3 |
| 3 | 40 | 120 | 112 | 5.7 | 4.9 | 11.5 | $6.7 \times 10^{-3}$ | $4.0 \times 10^{9}$ | $2.0 \times 10^{3}$ | 11.3 | 17.4 | 5.3 | 59.9 |
| 4 | 40 | 160 | 134 | 6.7 | 5.6 | 17.1 | $1.1 \times 10^{-3}$ | $6.6 \times 10^{8}$ | $3.3 \times 10^{2}$ | 5.7 | 21.2 | 3.0 | 42.3 |
| 5 | 30 | 190 | 113 | 5.7 | 4.4 | 21.5 | $1.9 \times 10^{-4}$ | $1.1 \times 10^{8}$ | 55 | 7.4 | 27.6 | 3.0 | 63.4 |
| 6 | 30 | 220 | 144 | 7.2 | 6.1 | 27.6 | $3.1 \times 10^{-5}$ | $1.8 \times 10^{7}$ | 9 | 15.0 | 36.4 | 3.7 | 82.4 |
| 7 | 30 | 250 | 150 | 7.5 | 6.1 | 33.7 | $5.1 \times 10^{-6}$ | $3.0 \times 10^{6}$ | 1.5 | 13.4 | 48.1 | 5.0 | 52.9 |
| 8 | 30 | 280 | 162 | 8.1 | 6.6 | 40.3 | $8.6 \times 10^{-7}$ | $5.0 \times 10^{5}$ | <1 | 8.8 | 54.7 | 5.2 | 51.4 |
| 9 | 30 | 310 | 164 | 8.2 | 6.6 | 46.9 | $1.4 \times 10^{-7}$ | $8.4 \times 10^{3}$ | <1 | 5.6 | 58.4 | 2.0 | 92.6 |
| 10 | 30 | 340 | 156 | 7.8 | 6.2 | 53.1 | $2.4 \times 10^{-8}$ | $1.4 \times 10^{3}$ | <1 | 9.3 | 66.8 | 4.0 | 54.2 |
| 11 | 20 | 360 | 134 | 6.7 | 5.1 | 58.2 | $4.0 \times 10^{-9}$ | $2.3 \times 10^{2}$ | <1 | 6.3 | 73.7 | 3.7 | 74.3 |
| 12 | 20 | 380 | 121 | 6.0 | 4.7 | 62.9 | $6.6 \times 10^{-10}$ | $3.8 \times 10^{1}$ | <1 | 6.9 | 84.3 | 7.0 | 46.8 |
| 13 | 20 | 400 | 123 | 6.1 | 4.9 | 67.8 | $1.1 \times 10^{-10}$ | 6 | <1 | 3.6 | 89.4 | 4.0 | 49.2 |
| 14 | 20 | 420 | 118 | 5.9 | 4.7 | 72.5 | $1.8 \times 10^{-11}$ | 1 | <1 | 10.8 | 102.0 | 5.7 | 79.7 |
| 15 | 20 | 440 | 75 | 3.6 | 2.4 | 74.9 | $3.1 \times 10^{-12}$ | 0.16 | <1 | 3.2 | 105.0 | 5.5 | 65.4 |

Sixteen reaction mixtures of 0.25 ml were set up, each containing 40 $\gamma$ of protein and the other components specified for the "standard" assay in *Methods*. 0.2 $\gamma$ of template RNA were added to tubes 0 and 1; RNA was extracted from the former immediately, and the latter was allowed to incubate for 40 min. Then 50 $\lambda$ of tube 1 were transferred to tube 2, which was incubated for 40 min and 50 $\lambda$ of tube 2 then transferred to tube 3, and so on, each step after the first involving a 1-6 dilution of the input material. Every tube was transferred from an ice bath to the 35°C water bath a few minutes before use to permit temperature equilibration. After the transfer from a given tube, 20 $\lambda$ were removed to determine the amount of $P^{32}$-RNA synthesized, and the product was purified from the remainder as described in *Methods*. Control tubes incubated for 60 min without the addition of the 0.2 $\gamma$ of RNA showed no detectable RNA synthesis, nor any formation of infectious units.

All recorded numbers are normalized to 0.25 ml. Columns 1, 2, and 3 give the transfer number, the time interval permitted for synthesis, and the elapsed time from zero, respectively. Column 4 records the amount of radioactive RNA found in each tube at the end of the incubation, column 5 the total RNA in each, and 6 gives the net synthesis during the time interval. Column 7 lists the cumulative synthesis of RNA. The decreasing concentrations of the input RNA resulting from the serial dilutions are recorded in terms of $\gamma$ (col. 8), number of strands (col. 9), and infectious units (IU) per tube (col. 10). The last is calculated from column 9 and from an efficiency of plating (e.o.p.) of $5 \times 10^{-7}$. Column 11 lists the increment in infectious units observed during each period of synthesis, corrected for the efficiency of recovery (col. 14), and column 12 represents the corresponding sum. Column 13 is the plating efficiency (e.o.p.) determined from the observed number of plaques (col. 11) and the actual amount of RNA assayed as determined from columns 6 and 14. Column 14 is determined from assays of acid-precipitable radioactivity on 20 $\lambda$ aliquots of the final product as compared with column 5.

We now undertake to describe experiments in which the kinetics of the appearance of new RNA and infective units were examined in two different ways. The first shows that the accumulation of radioactive RNA is accompanied by a proportionate increase in infective units. The second type proves by a serial dilution experiment that the newly synthesized RNA is infective.

*(1) Assay of infectivity of the purified product.* To compare the appearance of new RNA and infectious units in an extensive synthesis, 8 ml of reaction mixture was set up containing the necessary components in the concentrations specified in *Methods* (2). Aliquots were taken at the times indicated for the determination of radioactive RNA and purification of the product for infectivity assay. The results are summarized in Figure 1 in the form of a semilogarithmic plot against time of the observed increase in both RNA and infectious units. Further details of the experimental protocol are given in the corresponding legend.

The amount of RNA (0.8 γ/ml) put in at zero time is well below the saturation level of the enzyme present.[6] Consequently, the RNA increases autocatalytically for about the first 90 min, followed by a synthesis which is linear with time, a feature which had been observed previously.[6] It will be noted that the increase in RNA is paralleled by a rise in the number of infectious units. During the 240 min of incubation, the RNA experiences a 75-fold increase, and the infectious units experience a 35-fold increase over the amount present at zero time. These numbers are in agreement within the accuracy limits of the infectivity test. Experiments carried out with other enzyme preparations yielded results in complete accord with those just described.

It is clear that one can provide evidence for an increase in the number of infectious units which parallels the appearance of newly synthesized RNA.

*(2) Proof that the newly synthesized RNA molecules are infective.* The kind of experiments just described offer plausible evidence for infectivity of the radioactive RNA. They are not, however, conclusive, since they do not eliminate the possibility that the agreement observed is fortuitous. One could argue that the enzyme is "activating" the infectivity of the input RNA while synthesizing new noninfectious RNA and that the rather complex exponential and linear kinetics of the two processes happen to coincide by chance.

Direct proof that the newly synthesized RNA is infectious can in principle be obtained by experiments which use $N^{15}$-$H^3$-labeled initial templates to generate $N^{14}$-$P^{32}$-labeled product. The two can then be separated[8] in equilibrium density gradients of $Cs_2SO_4$. Such experiments have been carried out for other purposes, and will be described elsewhere. However, the steepness of the $Cs_2SO_4$ density gradients makes it difficult to achieve a separation clean enough to be completely satisfying.

There exists, however, another approach which bypasses these technical difficulties and takes advantage of the fact that we are dealing with a self-propagating entity. Consider a series of tubes, each containing 0.25 ml of the standard reaction mixture, but no added template. The first tube is seeded with 0.2 γ of Qβ-RNA and incubated for a period adequate for the synthesis of several γ of radioactive RNA. An aliquot (50 λ) is then transferred to the second tube which is in turn permitted to synthesize about the same amount of RNA, a portion of which is again transferred to a third tube, and so on. If each successive synthesis produces RNA which can serve to initiate the next one, the experiment can be continued until a point is reached at which the initial RNA of tube 1 has been diluted to an insignificant level. In fact, enough transfers can be made to ensure that the last tube contains less than one strand of the input primer. *If in all the tubes, including the last, the number of infectious units corresponds to the amount of radioactive RNA found, convincing evidence is offered that the newly synthesized RNA is infectious.*

Table 1 records a complete account of such a serial transfer experiment, and the corresponding legend provides the details necessary to follow the assays and calculations. Sixteen tubes are involved, the first (tube 0) being an unincubated zero time control. It will be noted that the successive dilution was such (1-6) that, by the 8th tube, there was less than one infectious unit ascribable to the initiating 0.2 γ of RNA. Nevertheless, this same tube showed $8.8 \times 10^5$ newly synthesized infectious units during the 30 min of its incubation. Finally,

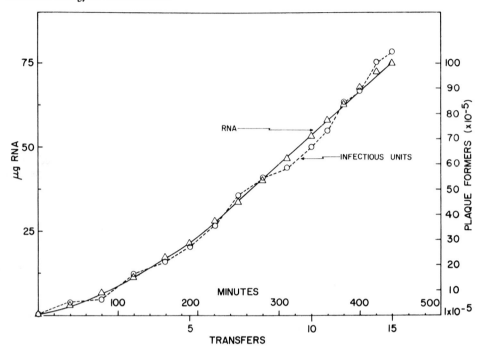

**Fig. 2.** RNA synthesis and formation of infectious units in a serial transfer experiment. All details are as described in the heading to Table 1, and the data are taken from columns 7 and 11 and plotted against elapsed time (col. 3) and corresponding transfer number (col. 1). Both ordinates refer to amounts found in 0.25-ml aliquots.

**TABLE 2**
*Serological behavior of virus formed in response to "synthetic" RNA*

| | Antisera | | | | | |
|---|---|---|---|---|---|---|
| | Anti-Qβ | | | Anti-MS-2 | | |
| Virus | 0 Time | 10 Min | % Survivors | 0 Time | 10 Min | % Survivors |
| Authentic Qβ | $1.9 \times 10^8$ | $1.0 \times 10^5$ | 0.052 | $1.1 \times 10^8$ | $1.06 \times 10^8$ | 96 |
| Virus from synthetic RNA | $1.5 \times 10^8$ | $8.8 \times 10^4$ | 0.053 | $1.5 \times 10^8$ | $1.40 \times 10^8$ | 93 |

In all cases, lysates were made from *E. coli* Q13, which was also the assay organism. Antisera were used at 1/100 dilution, and the incubation temperature was 35°C. The numbers represent plaque formers per ml.

tube 15, which contained less than one strand of the original input, produced $1.4 \times 10^{12}$ new strands and $3.2 \times 10^5$ infectious units in 20 min. It should be noted that a control tube lacking added RNA was incubated for 60 min. As compared with tube 1, which incorporated 4800 cpm for each 20 λ in 40 min, the control showed no increase above the zero time level of 80 cpm. Further, no synthesis of infective units was observed in such controls.

Figure 2 compares the cumulative increments with time in newly synthesized RNA (column 7) and infectious units (column 11). The agreement between increments in synthesized RNA and newly appearing infectious units is excellent at every stage of the serial transfer—and continues to the last tube. Long after the initial RNA has been diluted to insignificant levels, the RNA from one tube serves to initiate synthesis in the next. Further, as may be seen from the comparative constancy of the infective efficiency (Fig. 2 and column

13 of Table 1), the new RNA is fully as competent as the original viral RNA to program the synthesis of viral particles in spheroplasts.

To complete the proof, it was necessary to show that the viruses produced by the synthesized RNA were indeed Qβ, the original source of the RNA used as a seed in tube 1 to initiate the transfer experiment. Since Qβ is a unique serological type,[2,3] this characteristic was chosen as a convenient diagnostic test. Plaques induced by the RNA synthesized in tube 15 were used to produce lysates, and the resulting particles exposed to antisera against MS-2 and Qβ. The results, briefly summarized in Table 2, show clearly that the synthetic RNA induces virus particles of the same serological type as authentic Qβ.

## Discussion

One perhaps might have imagined that an enzyme carrying out a complex copying process would show a high error frequency when functioning in the unfamiliar environment provided by the enzymologist. Had this been a quantitatively significant complication, biologically inactive strands should have accumulated as the synthesis progressed. That this is not the case is rather dramatically illustrated by the serial transfer experiment (Table 1 and Fig. 2). The RNA synthesized after the 15th transfer is as competent biologically as the initiating "natural" material derived from virus particles.

The successful synthesis of a biologically active nucleic acid with a purified enzyme is itself of obvious interest. However, the implication which is most pregnant with potential usefulness stems from the demonstration that the replicase is, in fact, generating identical copies of the viral RNA. For the first time, a system has been made available which permits the unambiguous analysis of the molecular basis underlying the replication of a self-propagating nucleic acid. Every step and component necessary to complete the replication must be represented in the reaction mixture described. If two enzymes are required,[13] both must be present, and it should be possible either to establish their existence or to prove that one is sufficient. If an intermediate "replicating" stage intervenes between the template and the ultimate identical copy,[13] then a "replicative form" should be demonstrably present in the reaction mixture. If copying is direct, no such intermediate will be found. These and other issues of the replicating mechanism will be discussed in a subsequent publication which will detail the relevant experiments.

## Summary

Experiments are described with a purified RNA-dependent-RNA-polymerase (replicase) induced in *E. coli* by the RNA bacteriophage Qβ.

The data demonstrate that the enzyme can generate identical copies of added viral RNA. A serial dilution experiment established that the newly synthesized RNA is fully as competent as the original viral RNA to program the synthesis of viral particles and to serve as templates for the generation of more copies. Since the data show that the enzyme is, in fact, generating replicas, an unambiguous analysis of the RNA replicating mechanism is now possible in a simple system consisting of purified replicase, template RNA, ribosidetriphosphates, and $Mg^{++}$.

**REFERENCES**

1. Haruna, I., and S. Spiegelman, these Proceedings **54**:579 (1965).
2. Watanabe, I., Nihon Rinsho **22**:243 (1964).
3. Overby, L., G. H. Barlow, R. H. Doi, M. Jacob, and S. Spiegelman, J. Bacteriol., in press.
4. Spahr, P. F., and B. R. Hollingworth, J. Biol. Chem. **236**:823-831 (1961).
5. Spahr, P. F., J. Biol. Chem. **239**:3716-3726 (1964).
6. Haruna, I., and S. Spiegelman, Science, in press.
7. Gesteland, R. F., Federation Proc. **24**:293 (1965). Q13 is a derivative of A19, an RNase negative mutant reported in this reference.
8. Doi, R. H., and S. Spiegelman, these Proceedings **49**:353-360 (1963).
9. Haruna, I., K. Nozu, Y. Ohtaka, and S. Spiegelman, these Proceedings **50**:905-911 (1963).
10. Guthrie, G. D., and R. L. Sinsheimer, J. Mol. Biol. **2**:297 (1960).
11. Fraser, D., and E. A. Jerrel, J. Biol. Chem. **205**:291-295 (1953).
12. Paranchych, W., Biochem. Biophys. Res. Commun. **11**:28 (1963).
13. Ochoa, S., C. Weissmann, P. Borst, R. H. Burdon, and M. A. Billeter, Federation Proc. **23**:1285-1296 (1964).

# 29 THE 3'-TERMINUS AND THE REPLICATION OF PHAGE RNA

*Ulrich Rensing*
*J. T. August*
*Department of Molecular Biology*
*Division of Biological Sciences*
*Albert Einstein College of Medicine*
*Bronx, New York*

Studies with Qβ RNA confirm Kamen's finding, with R17 RNA, that the terminal adenylate is not necessary for infectivity. It is now demonstrated that the *in vitro* template activity of Qβ RNA is not destroyed by removal of the adenylate, and a scheme is put forward to account for the presence of 3'-terminal adenosine in the product of replication.

---

The RNA from all RNA phages examined so far terminates in the sequence $CCCA_{OH}$ at the 3'-hydroxyl terminus[1-6]. If synthesis of the complementary strand begins precisely at the 3'-hydroxyl end of Qβ RNA, a complementary relationship between the 3'-terminus of the viral strand and the 5'-terminus of the complementary strand would be expected because polymerization by the Qβ RNA polymerase proceeds in a 5' to 3' direction[7] and is believed to involve base pairing in an antiparallel orientation. In actual fact, the 5'-terminus of the complementary strand of phage Qβ RNA is pppG . . . (refs. 7-9), not pppU . . . Thus it may be questioned where synthesis of the complementary strand originates in relation to template Qβ RNA and how Qβ RNA acquires the terminal adenosine.

To gain insight into these problems we have studied the effect of modifications of the 3'-terminus of Qβ RNA on template activity and infectivity. We have also established the identity of the 3'-nucleoside of RNA synthesized with the purified enzyme. It is concluded that the replication of Qβ RNA is not simply the linear copying of a single-stranded molecule from one end to the other and that other mechanisms must be invoked to explain the presence of the hydroxyl terminal adenosine.

## Modification of the 3'-hydroxyl terminus

The effect of modification of the 3'-hydroxyl terminus on the infectivity of viral RNA was first studied by Singer and Fraenkel-Conrat[10,11] and later by Steinschneider and Fraenkel-Conrat[12,13]. It was initially suggested that nucleotides could be removed from the hydroxyl terminus of TMV RNA without loss of infectivity[10,11]. In subsequent studies, however, in spite of improvements in the Whitfeld stepwise elimination procedure designed to reduce the frequency of internal breaks[14,15], approximately 95 per cent of the infectivity of TMV RNA was lost on successive treatment with periodate and aniline[12,13]. Possible complications in the application of this technique are the incorporation of aniline into RNA[13] or the transamination or deamination of cytidine[16]. Recently, however, Kamen[17] employed the Steinschneider and Fraenkel-Conrat technique[12,13] and reported that the infectivity of R17 RNA did not require the presence of the terminal adenylate, but was lost on removal of the penultimate cytidylate. The results of our experiments with Qβ RNA described here agree with the findings of Kamen, and, in addition, show that the *in vitro* template activity of Qβ RNA is not destroyed by modification or removal of the terminal adenylate.

Sequential chemical elimination of the 3'-terminal nucleotides of Qβ RNA was carried out testing the yield of the reaction and isolating whole molecules at each step in the procedure. Control samples were processed simultaneously to monitor possible inactiva-

---

From Nature 224:853-856, 1969. Reprinted with permission.

We thank Drs. S. Leppla and R. Bock for support during the early stage of this work and for the hospitality which U. R. enjoyed in their laboratory. We are indebted to Lillian Eoyang for advice and a gift of the Qβ RNA polymerase and factors utilized in this investigation, and to H. P. Strack for a critical reading of the manuscript. This work was supported in part by grants from the National Institute of General Medical Sciences, National Institutes of Health. J. T. A. is a career scientist of the Health Research Council of the City of New York.

tion of the RNA arising from side reactions. The modified species of RNA were tested for both template activity and infectivity (Table 1).

The 3'-terminal ribose of Qβ RNA was oxidized with periodate to yield a dialdehyde. This reaction gave about 90 per cent yield with adenosine, as determined by chromatographic separation of $^{14}$C-adenosine and its modified derivatives, and more than 80 per cent yield with Qβ RNA, as judged by separation of the nucleoside from its derivatives after alkaline hydrolysis of oxidized and reduced $^{14}$C-labelled Qβ RNA. Neither oxidation nor oxidation and borohydride reduction destroyed either the infectivity or the template activity of the RNA. This was true even with RNA taken through the oxidation procedure a second or a third time. It may be concluded either that the RNA is repaired under the conditions of the Qβ RNA polymerase reaction or that the enzyme does not require Qβ RNA with an intact adenosine hydroxyl terminus.

Subsequent incubation with aniline to eliminate the terminal nucleoside yielded RNA terminating with cytidine 3'-phosphate. At this stage, the extent of elimination was more than 80 per cent—determined by isolation of the terminal base after it was dephosphorylated by alkaline phosphatase, oxidized and labelled with $^{3}$H-borohydride[18,19]. By the same technique Kamen[17] reported 100 per cent yield. To ensure a quantitative reaction, the oxidation and aniline catalyzed elimination were repeated. This RNA, which contained a 3'-terminal phosphate, had lost most of its

TABLE 1
*Effect of modifications at the 3'-hydroxyl terminus on the template activity of Qβ RNA*

| Sample | 3'-Terminal structure | Infectivity (per cent) | Template activity (per cent) |
|---|---|---|---|
| 1 | —CpCpCpA$_{OH}$ | 100 | 100 |
| 2 | —CpCpCpA$_{oxid}$ | 68 | 82 |
| 3 | —CpCpCp | 14 | 82 |
| 4 | —CpCpC$_{OH}$ | 107 | 85 |
| 5 | —CpCp | 6 | 25 |
| 6 | —CpC$_{OH}$ | 39 | 33 |

Qβ RNA (2 mg) was oxidized with periodate at pH 5.2 (sample 2) as described by Leppla et al.[19] Most of it (1.6 mg), was then treated with aniline at pH 5.0 (sample 3) following the Steinschneider and Fraenkel-Conrat procedure;[12,13] the remainder was processed in an identical manner without aniline. An aliquot (0.8 mg) of the aniline-treated sample was incubated at 37° for 30 min with 32 μg alkaline phosphatase (sample 4). The alkaline phosphatase (25 units/mg protein) had previously been treated in a modification of the procedure described by Dr. L. Heppel (personal communication) to remove ribonuclease activity. Other samples were taken through the entire procedure but not exposed to any of the reagents (periodate, aniline or alkaline phosphatase), or were treated with aniline and alkaline phosphatase but not periodate (sample 1). All of these samples were made 0.5 per cent with sodium dodecyl sulphate (SDS) and treated with phenol that had been freshly distilled and saturated with buffer (100 mM NaCl, 20 mM Tris [pH 7.7] and 10 mM EDTA). The RNA was precipitated with ethanol, redissolved in 10 mM EDTA and fractionated by zone centrifugation as described in Fig. 1. The peak and leading fractions of the 32S RNA were pooled and the RNA precipitated with ethanol and dissolved in 5 mM EDTA. A second cycle of elimination was performed with RNA which had been carried through the entire procedure (periodate, aniline, and alkaline phosphatase) and as a control with RNA treated with periodate and analine but not alkaline phosphatase. After the second cycle of periodate and aniline treatment one aliquot (sample 5) was retained and another treated with alkaline phosphatase (sample 6). All fractions, including controls as described in the first cycle, were again treated with phenol and SDS, precipitated with ethanol, and centrifuged in a glycerol gradient. The leading fractions of the 32S peak were pooled and concentrated by ethanol precipitation, and the RNA dissolved in 5 mM EDTA. Template activity was assayed in the Qβ RNA polymerase reaction as described in the legend to Fig. 2, except that the reaction mixture was incubated for only 5 min. Each value for template activity was taken from an RNA saturation curve and represents the relative rate of incorporation at saturating concentrations of RNA compared with that of control Qβ RNA. In every case the concentration of RNA required for half-maximal activity was 1.7 to 2.2 × 10$^{-5}$ M. The same enzyme preparation was used in every assay. Infectivity was measured in the protoplast assay of Strauss and Sinsheimer[20]. Each value represents the average of four or five protoplast assays at two levels of RNA, all tested on three different protoplast preparations. The standard error of the assay was 15 per cent. The results are compared with the activity or infectivity of control samples carried through the same procedure except that either periodate or alkaline phosphatase was omitted at various steps. In this experiment the controls did not differ more than ±25 per cent from untreated RNA.

infectivity, but the original level of infectivity was restored by treatment with alkaline phosphatase to yield a cytidine 3'-hydroxyl terminus. By contrast, template activity remained unchanged, with RNA terminating in CCCp or CCC$_{OH}$. Phosphatase activity in the enzyme or factor fractions has been tested in a variety of ways and none could be detected. The requirement of a 3'-hydroxyl group for infectivity but not for template activity suggests that this group may participate in some process other than RNA replication, perhaps one involving penetration of the host cell or messenger function.

Removal of the penultimate cytidylate by the second cycle of elimination resulted in a marked decrease in both infectivity and template activity. Controls for this experiment included treating the RNA in an identical fashion twice with periodate and aniline, omitting only the first alkaline phosphatase treatment. Barring the unlikely possibility that the first cycle treatment with alkaline phosphatase introduced a susceptibility to inactivation by a change other than at the terminus—a base modification, for example—we may conclude that the penultimate cytidylate is essential for both the template function of Qβ RNA and for its infectivity. Some, or perhaps all, of the approximately 30 per cent residual activity remaining after the second cycle elimination and treatment with phosphatase can be attributed to incomplete removal of the penultimate nucleotide. It is also possible that after removal of this nucleotide the RNA is still partially active, but at a lower efficiency.

A third elimination cycle resulted in complete loss of activity. The interpretation of these results is, however, limited by difficulties in carrying control samples through three elimination cycles without serious loss of activity. With RNA carried through two cycles, the inactivation in the controls due to treatment with aniline alone, or with aniline and periodate, was less than the standard error of the mean (15 per cent). Such results were obtained, however, only when the RNA was handled with great care and when the treatment with aniline was carried out in an atmosphere of nitrogen. Even with these precautions we were unable to retain active RNA through three cycles.

### The 3'-terminus of Qβ RNA synthesized in vitro

Having established that Qβ RNA lacking adenosine and terminating in CCCp acts as template in the *in vitro* reaction, we have used this RNA to identify the 3'-terminal nucleoside of RNA synthesized in the reaction. A sensitive and accurate method for this analysis is to label the product RNA with $^3$H-borohydride after the terminal sugar has been oxidized with periodate[18,19]. Because this procedure is selective for 2',3'-*cis*-glycol groups, the –Cp terminus of the modified template remains unlabelled. The modified RNA template, prepared by treatment with periodate and aniline and isolated by zone sedimentation as described in Table 1, was used as template in the standard Qβ RNA polymerase reaction. The yield of newly synthesized RNA was approximately 10-fold in excess of template. Most of the RNA product cosedimented with Qβ RNA as intact 32S molecules, and the leading fractions of the 32S peak were collected (Fig. 1). Approximately 75 pmoles of product RNA were recovered from ten reactions. The 3'-terminus of this RNA was labelled with $^3$H-borohydride according to the procedure of Leppla *et al.*[19], and intact molecules were again isolated by zone centrifugation. After alkaline hydrolysis the nucleoside derivatives were separated by chromatography on DEAE-paper, using the four unlabelled hydroxymethyldiethylene glycol derivatives as standards[24]. As a control, native Qβ RNA was treated identically (Table 2).

A large proportion of the label in the product RNA, about 50 per cent in repeated experiments, was found in the reduced derivative of adenosine. From the amount of modified template added, 9.3 pmoles, and correcting for 80 per cent elimination of the terminal nucleoside of this RNA, no more than 1-2 per cent of the adenosine derivative found in this analysis could be accounted for by the initial template. Even if the initial template had been modified in some manner to contain a terminal adenosine, its contribution would have been less than 15 per cent. The other major component was cytidine: some of this could perhaps be attributed to the presence of strands complementary to Qβ RNA, although the terminal nucleoside of the complementary strand is not known. Some chains terminated in

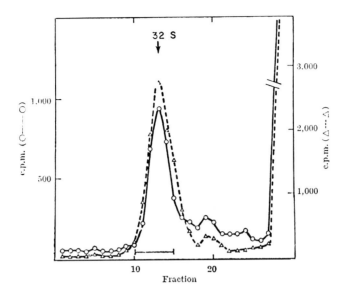

**Fig. 1.** Sedimentation analysis and isolation of the RNA product of the Qβ RNA polymerase reaction directed by either normal Qβ RNA (△---△) or a modified Qβ RNA (O—O) template. The reaction mixtures (0.1 ml.) contained 100 mM Tris-HCl buffer (pH 7.7), 10 mM $MgCl_2$, 5 mM 2-mercaptoethanol, 1 mM each of UTP, CTP, GTP and $^{14}$C-ATP (2,200 c.p.m./mμmole), 15 μg of the DEAE-cellulose fraction containing factors I and II[21,22], 2 μg protein of the Qβ RNA polymerase glycerol gradient enzyme fraction[23] and either 0.8 pmoles native Qβ RNA or 0.95 pmoles of Qβ RNA, modified as described in Table 1 to contain a terminal cytidine 3'-phosphate. After incubation at 37° for 40 min the reaction mixtures were layered on a solution (12.0 ml.) containing 20 mM Tris-HCl (pH 7.7), 100 mM NaCl, 5 mM EDTA, and a 5-30 per cent glycerol gradient, and were centrifuged in an SW 41 rotor for 6 h at 40,000 r.p.m. at 2°C. Thirty fractions were collected from the bottom of the tube. One-fourth (modified template) or one-half (Qβ RNA template) of each fraction was counted in a dioxane scintillation solution. The fractions utilized for subsequent procedures are indicated by the bar.

## TABLE 2
*The 3'-nucleoside of RNA synthesized in the Qβ RNA polymerase reaction in vitro*

| Nucleoside derivatives | $R_F$ | 32S reaction product (C.p.m.) | (Per cent of total) | Qβ RNA control (C.p.m.) | (Per cent of total) |
|---|---|---|---|---|---|
| G' | 0.15 | < 500 | < 5 | < 100 | < 5 |
| U' | 0.52 | 1,700 | 14 | < 100 | < 5 |
| A' | 0.70 | 5,500 | 46 | 1,310 | 72 |
| C' | 0.87 | 4,700 | 40 | 520 | 27 |

The RNA product was synthesized in a reaction with a modified Qβ RNA template and isolated by zone centrifugation as described in Fig. 1. RNA present in the leading fractions of the 32S peak, 75 pmoles, was precipitated with ethanol, and then further processed by the labelling technique described by Leppla et al.[19], except that potassium $^3$H-borohydride (2 C/mmole) was used instead of the sodium salt and the reduction time was 30 min. The labelled RNA was further purified, in addition to that of the labelling procedure, by zone centrifugation to isolate 32S material. This RNA was precipitated with ethanol and dissolved in 0.1 ml. of 0.3 M KOH and incubated at 37°C for 18 h. The hydrolysis was terminated by chilling in ice and adjusting the pH to 8.0-8.5 with 7 per cent perchloric acid. The precipitate was collected by centrifugation and extracted three times with a dilute aqueous solution of the 2'(3')-ribonucleoside monophosphates and the four unlabelled hydroxymethyldiethylene glycol derivatives of the nucleosides[24]. The combined supernatant solutions were concentrated to 0.1 ml. by evaporation and spotted on DEAE-paper. Descending chromatography with deionized water separated the four nucleoside derivatives (G', U', A', C'): the nucleotides stayed at the origin. The paper was cut into strips and tritium radioactivity was counted in a toluene scintillator solution at 4-5 per cent efficiency. As a control, the procedure was carried out with Qβ RNA.

uridine, or there may have been deamination of cytidine.

It was also found that Qβ RNA isolated from purified phage contained both cytidine and adenosine at the hydroxyl terminus (Table 2). This finding has been established in several experiments with different phage preparations each of which contained a variable amount, as much as 30 per cent of the terminal base as cytidine. Cytidine as well as adenosine was also found at the termini of f2 and MS2 RNA by Lee and Gilham[25].

### Scheme for hydroxyl terminus replication

The presence of 5'-terminal ppG . . . in the complementary strand and 3'-terminal adenosine in Qβ RNA synthesized *in vitro* demonstrates that a process other than simple linear copying from one end to the other of a completely single-stranded molecule appears to be involved in the replication of Qβ RNA. Two questions arise: (1) Is the terminal adenosine copied? (2) Does the adenosine terminus result from transcription or some other process?

With respect to the initiation site for copying of Qβ RNA we have found that the terminal adenylate, but not the penultimate cytidylate, may be removed without destroying its template activity or infectivity. Moreover, the primary structure at the 3'-hydroxyl terminus is not critical for template activity, for neither oxidation nor oxidation and reduction of the adenosine residue, nor the presence of cytidine 3'-phosphate, destroyed template activity. These observations, together with the knowledge that the complementary strand contains pppG . . . at the 5'-terminus[7-9], strongly support the hypothesis that copying of Qβ RNA does not begin at the terminal adenosine. The only alternative to the conclusion that the adenylate is neither copied nor required for template activity or infectivity is that the hydroxyl terminus of the chemically treated RNA is repaired in the purified enzyme system. One may speculate that synthesis begins at the nucleotide penultimate to the hydroxyl terminus, for this cytidylate is required for both template activity and infectivity. This has not been proved, however, and the cytidylate could be required without being the actual site for the initiation of synthesis. The 3'-terminus may also be related to other functions as well, for infectivity but not template activity was inhibited by the presence of a terminal cytidine 3'-phosphate.

How then does Qβ RNA come to be terminated by adenosine? Because we have now found that some of the RNA product contains adenosine, even when the template does not, it is not necessary to invoke anything other than the activities present in the purified polymerase system to explain its presence. The two host cell factors required for replication of Qβ RNA do not appear to be implicated, for the Qβ 6S RNA also has both adenosine and cytidine at the hydroxyl terminus (unpublished observations of U. R.) but contains only pppG . . . at the 5'-terminus and is replicated without factors[26]. A search has been made for an adenylate incorporating activity in crude extracts of *E. coli* in partially purified fractions of the Qβ RNA polymerase (Carol Prives, unpublished observation). The only activity detected possessed properties of the polyriboadenylate polymerase[27]; this activity is not present in the

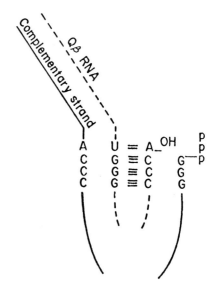

**Fig. 2.** A model for transcription of the terminal adenosine. This scheme envisions one possibility of intramolecular complementarity involving the 5'-end of the complementary strand. The newly synthesized Qβ RNA has the opportunity of transcribing one or more nucleotides from its own strand. The arrows indicate the 5' to 3' polarity of the strand. A similar arrangement can be imagined to allow intermolecular transcription from a non-terminal region of the complementary strand.

purified Qβ RNA polymerase preparations. Purified adenylate (cytidylate) pyrophosphorylase[28,29] was also tested and found not to be active with Qβ RNA. It therefore appears that the Qβ RNA polymerase itself catalyses the incorporation of the terminal adenylate. These observations refine those of Kamen[17], who found normal progeny RNA when protoplasts were infected with R17 RNA lacking adenosine and implicated a host cell activity for the addition of adenosine.

There are several ways by which the terminal incorporation of adenylate or several nucleotides might be catalysed by the Qβ RNA polymerase. For example: (1) The enzyme could utilize RNA as a primer, catalysing addition to the hydroxyl terminus in a reaction divorced from synthesis. (2) As a feature of chain termination in the synthesis reaction, the enzyme could add the hydroxyl terminal nucleotides in the absence of template. (3) The presence of the terminal adenylate might be ascribed to the secondary or tertiary structure of the RNA whereby the enzyme transcribes an internal uridylate as a result of intra or intermolecular hydrogen bonding. A feature of the latter possibility as compared with the others is that it adheres to the basic transcriptional function of the enzyme and may not require additional enzyme active sites. Also, by a transcriptional mechanism, the heterogeneity of the terminus can be explained by a dependence on the secondary structure of the RNA molecules, whereas the nontranscriptional mechanisms imply a variability in enzyme function or the involvement of different enzyme activities or other processes. One may hypothesize that a sequence of nucleotides in Qβ RNA or the complementary strand, with the proper strand polarity, could compete with the 5′-end of the complementary strand for the enzyme, and provide copying of uridylate (Fig. 2). A similar explanation may be invoked to explain the heterogeneity of the 5′-terminus of Qβ RNA reported by De Wachter and Fiers[30] and at the 3′-terminus of staellite tobacco necrosis virus RNA by Wimmer and Reichmann[31].

It is clear, however, that these examples are merely the simplest of many that could account for the experimental observations concerning Qβ RNA replication. More complicated models might also be entertained, involving excision and addition or transpositions of regions of the genome, or external primer molecules. It is also possible that the heterogeneity at the 3′-hydroxyl terminus is related to different functions of phage RNA.

## REFERENCES

1. Lee, J. C., and Gilham, P. T., J. Amer. Chem. Soc. **88**:5685 (1966).
2. De Wachter, R., and Fiers, W., J. Mol. Biol. **30**:507 (1967).
3. Weith, H. L., and Gilham, P. T., J. Amer. Chem. Soc. **89**:5473 (1967).
4. Weith, H. L., Asteriadis, G. T., and Gilham, P. T., Science **160**:1459 (1968).
5. Dahlberg, J. E., Nature **220**:549 (1968).
6. Glitz, D. G., Bradley, A., and Fraenkel-Conrat, H., Biochim. Biophys. Acta **161**:1 (1968).
7. Banerjee, A. K., Kuo, C. H., and August, J. T., J. Mol. Biol. **40**:445 (1969).
8. Banerjee, A. K., Eoyang, L., Hori, K., and August, J. T., Proc. US Nat. Acad. Sci. **57**:986 (1967).
9. Bishop, D. H. L., Pace, N. R., and Spiegelman, S., Proc. US Nat. Acad. Sci. **58**:1790 (1967).
10. Singer, B., and Fraenkel-Conrat, H., Biochim. Biophys. Acta **72**:537 (1963).
11. Singer, B., and Fraenkel-Conrat, H., Biochim. Biophys. Acta **76**:143 (1963).
12. Steinschnieder, A., and Fraenkel-Conrat, H., Biochemistry **5**:2729 (1966).
13. Steinschnieder, A., and Fraenkel-Conrat, H., Biochemistry **5**:2735 (1966).
14. Whitfield, P. R., and Markham, R., Nature **171**:1151 (1953).
15. Whitfield, P. R., Biochem. J. **58**:390 (1954).
16. Shapiro, R., and Klein, R. S., Biochemistry **6**:3576 (1967).
17. Kamen, R., Nature **221**:322 (1969).
18. Rajbhandary, U. L., Stuart, A., Faulkner, R. D. Chang, S. H., and Khorana, H. G., Cold Spring Harbor Symp. Quant. Biol. **31**:425 (1966).
19. Leppla, S. M., Bjoraker, B., and Bock, R. M., in Methods in Enzymology (edit. by Grossman, L., and Moldave, K.), 12B, 236 (Academic Press, New York, 1968).
20. Strauss, J., and Sinsheimer, R., J. Virol. **1**:711 (1967).
21. Franze de Fernandez, M. T., Eoyang, L., and August, J. T., Nature **219**:588 (1968).
22. Shapiro, L., Franze de Fernandez, M. T., and August, J. T., Nature **220**:478 (1968).
23. Eoyang, L., and August, J. T. in Methods in Enzymology (edit. by Colowick, S. P., and Kaplan, N. O.), 12B, 530 (Academic Press, New York, 1968).
24. Khym, J. X., and Cohn, W. E., J. Amer. Chem. Soc. **82**:6380 (1960).

25. Lee, J. C., and Gilham, P. T., J. Amer. Chem. Soc. **87**:4000 (1965).
26. Banerjee, A. K., Rensing, U., and August, J. T., J. Mol. Biol. (in the press).
27. August, J. T., Ortiz, P. J., and Hurwitz, J., J. Biol. Chem. **237**:3786 (1962).
28. Preiss, J., Diechmann, M., and Berg, P., J. Biol. Chem. **236**:1748 (1961).
29. Furth, J. J., Hurwitz, J., Krug, R., and Alexander, M., J. Biol. Chem. **236**:3317 (1961).
30. De Wachter, R., and Fiers, W., Nature **221**:233 (1969).
31. Wimmer, E., and Reichmann, M. E., Nature **221**:1122 (1969).

## QUESTIONS FOR CHAPTER 6

### Paper 24

1. a. It is said that labeled nucleotides were prepared from $^{14}C$-grown *Chromatium*; how is this done?
   b. Why are the enzyme dilutions made into a solution containing bovine serum albumin?
   c. What is the purpose of the streptomycin-protamine sulfate precipitation in the preparation of the polymerase?
2. From the data given in the section on the experimental procedure, calculate the following:
   a. The number of counts actually obtained in the complete system in Table 2
   b. The specific activity of the enzyme used for this particular assay
   c. The amount of RNA synthesized in $\mu g$
3. a. What the $K_m$s for the incorporation of CTP and ATP?
   b. How many times more RNA would have to be synthesized on the salmon sperm DNA in Table 4 in order to obtain 10 $\mu g$ of net synthesis of RNA?
   c. Why is it necessary to use the net-synthesis product in order to analyze the synthesized RNA?
4. a. Which would be a better template for poly A synthesis: intact, doubled-stranded DNA, or partially degraded, "nicked" DNA?
   b. Does the ability to synthesize poly A mean that there are long runs of poly T in the DNAs serving as templates?
   c. Suppose that after alkaline hydrolysis 97.6% is in 2',3'NMP. What is the chain length of the poly A?
5. Point out the evidence that both strands of DNA are copied by the polymerase.

### Paper 25

1. a. It is stated in the paper that $\gamma^{32}P$-CTP and $\gamma^{32}P$-UTP were produced by photophosphorylation. Write down this reaction.
   b. How would you prove that the $\gamma$-P, and only that, is labeled?
   c. What is the advantage of using photophosphorylation instead of any of the usual triphosphate generating systems?
   d. For the experiments in this paper it is important that no $\alpha^{32}P$ of the triphosphates is labeled. How do you determine the absence of label in the $\alpha$ position?
   e. What reaction is catalyzed by the nucleoside diphosphokinase?
   f. What is the effect of adding myokinase to the above reaction with $\gamma,\beta^{32}P$-ATP and CDP?
   g. How would you prepare $\alpha^{32}P$-labeled nucleotides?
2. Answer the following questions pertaining to Table 2:
   a. Is there any relationship between the molecular weight of the RNA template and the length of the RNA chains synthesized?
   b. In the dGC homopolymer there is one chain of poly dG to one chain of poly dC. How many chains of rG to each rC chain are there in the product?
   c. Which chains, poly G or poly C, are longer?
3. a. In Table 3, how many counts are required to obtain the 4800 $\mu\mu$moles of RNA produced by the $T_2$ template?
   b. How many $^{32}P$ counts were there in the ends of the above?
   c. How many chains terminating in A, and how many chains terminating in G, were synthesized per DNA molecule, assuming the molecular weight of $T_2$ DNA to be $1.2 \times 10^8$?
   d. What percentage of the DNA is transcribed?
   e. Compare the effects of heat denaturation of the template on the number and the size of the RNA chains made. Assume a molecular weight of $10^7$ for *E. coli* DNA and of $10^6$ for calf thymus DNA (both are fragmented).
4. From Table 4, calculate the following:
   a. The number of A-terminated chains initiated per second after 1, 2, and 5 min of incubation.
   b. The rates of RNA synthesis in nucleotides/sec/chain/$T_2$ molecule in the first 60 min of incubation. What is the average?
5. a. As shown in Table 4, does the rate of RNA synthesis change when no further new chains are initiated? What is your interpretation?
   b. What is the molecular weight of the largest RNA molecules made?
6. Because of several limitations the average rate of nucleotide addition calculated in question 4 is probably much lower than the true rate. How could the true rate be determined?

### Paper 26

1. a. Would you be able to determine the molecular weight of proteins on polyacrylamide gels in the absence of SDS?
   b. Is the molecular weight of a protein proportional to the distance traveled in an SDS-containing polyacrylamide gel at pH 8.7

regardless of the nature of the protein? What is the role of the pH? Of the SDS?
2. If a protein and a nucleic acid have the same S value, are their molecular weights the same? Are they different by 10% to 30%, or can they be different by a factor of 2 or more?
3. a. How do the specific activities of the enzyme plus factor in this paper compare with the specific activities of the enzymes in papers 24 and 25, when all are assayed with calf thymus and with native $T_4$ DNA?
   b. From the results in question 3a, can you tell whether or not $\sigma$ was present in the polymerases in papers 24 and 25?

## Paper 27
1. How would you proceed to determine whether or not the M factor and $\sigma$ are the same?
2. a. Can you tell from a comparison of the specific activities of the enzymes in the presence of $\sigma$ or M factor whether the two are the same?
   b. Are the stimulations obtained by the two factors with various DNAs useful in deciding whether or not the two are the same?

## Paper 28
1. a. We have seen in previous papers that the DNA-dependent RNA polymerase requires both $Mg^{++}$ and $Mn^{++}$ for maximum activity. In assaying the RNA-dependent RNA polymerase for contamination by the other, should the assays be done with both divalent metals, with $Mg^{++}$ alone, or with $Mn^{++}$ alone?
   b. Suppose no DNA template were available. How could you distinguish between the two polymerases?
2. a. Why is pyrophosphate used to stop the reaction?
   b. It is said that the specific activity of $UT^{32}P$ is such that 20,000 cpm correspond to the synthesis of 1 $\mu$g of RNA, and that 20 $\mu$l aliquots can be taken for assays. Assuming that about 5 $\mu$g RNA are synthesized per hour per 0.2 ml incubation mixture, and that the minimum number of counts which can be counted reliably is 200 cpm, what would the specific activity of the $UT^{32}P$, in cpm per $\mu$mole, have to be in order to obtain accurate kinetics for the synthesis, starting at 5 min?
   c. Supposing that you wanted to repeat the assay 4 weeks later with the same UTP, would you have to change your assay? In what way?
   d. State one or more necessary properties of an enzyme capable of assimilating a substrate at a linear rate over a period of many hours.
   e. Why is potassium acetate added to the RNA prior to precipitating it with alcohol?
3. a. Can you think of a reason why protamine increases the efficiency of plating of the viral RNA?
   b. Assuming a plating efficiency of $2 \times 10^{-7}$ in the plating procedure described in the section on materials and methods and a plaque count of 120 on a 1:100 dilution, what is the molecular weight of the viral RNA?
   c. It is stated that each mg of enzyme protein is contaminated with 0.16 $\mu$g of RNA and 0.8 $\mu$g of viral protein. How many virus particles does that make?
   d. Could these particles spoil the assay of the newly synthesized RNA, due to their much higher plating efficiency?
4. a. In Figs. 1 and 2 what is the efficiency of plating of the newly synthesized RNA?
   b. From the data in Table I, calculate the molecular weight of the Q$\beta$ virus RNA.
   c. Calculate the number of transfers necessary to reduce the template RNA to 1 strand per tube, assuming that 0.15 $\mu$g of original RNA was used in 0.1 ml, and 10 $\mu$l were transferred to each tube.
   d. How much RNA would be in tube 10, provided the incubation time is 60 min for each tube and the RNA is synthesized at the rate shown in Fig. 2?

## Paper 29
1. Can you think of any other RNA having the CCA-OH end group of the viral RNAs?
2. What is the difference between the RNA polymerase used in this paper and that of the previous paper? How do the specific activities of the two papers compare?
3. a. Using the expression $M = KS^2$, and the value for K of $1.93 \times 10^3$, can you determine whether the 32S peak is single- or double-stranded?
   b. What is the probable identity of the peak at tube 19 in Fig. 1?

# ANSWERS FOR CHAPTER 6

## Paper 24

1. a. Algae grown in $^{14}CO_2$ become labeled in all their cell constituents, and therefore it is necessary to separate the nucleotides from the amino acids and lipids. The cell wall of algae is very tough, so that rather drastic procedures are necessary to obtain a cell extract; grinding the cells with carborundum is a suitable method. The crude extract is treated with cold 10% TCA to precipitate the proteins, nucleic acids, and some lipids. The TCA precipitate is washed with the usual organic solvents for removing lipids, and the remaining precipitate is hydrolyzed in 5% TCA at 90° for a half hour. This hydrolyzes the nucleic acids into 5'-nucleotides but leaves the proteins intact. (If only ribonucleotides are desired at this point, the crude extract should be treated with DNAase prior to the TCA precipitation.) After removal of the remaining protein precipitate by centrifugation, the nucleotides are adsorbed on activated charcoal, which is washed by filtration. Finally, the nucleotides are eluted from the charcoal with ammonia, and then chromatographed on DEAE cellulose, if the individual nucleotides are desired.

   b. The amount of RNA synthesized is very small, and only small amounts of enzyme protein are used for the assays. There is thus not enough material to precipitate quantitatively with TCA or perchloric acid. Furthermore, the precipitate would hardly be visible, and it is difficult to filter and wash an invisible precipitate quantitatively. Bovine serum albumin is thus added for bulk. Another reason is that Pyrex glass adsorbs small amounts of protein, and in a very dilute solution even this small amount would constitute a significant proportion of the total. The bovine serum albumin, being present in high concentrations, adsorbs preferentially.

   c. Both streptomycin and protamine sulfate are good precipitants of polyanions, that is, nucleic acids. The RNA polymerase exists in the cell bound to DNA, and is thus preferentially precipitated with the nucleic acids.

2. a. The specific activity of the labeled nucleotides was 300 to 600 cpm per mμmole; hence, 7.3 mμmoles correspond to *2190* cpm to *4380* cpm.

   b. One enzyme unit is defined as the incorporation of 1 mμmole of CTP per hour. The assay in Table 2 was for 10 min; hence, 7.3 × 6 mμmoles were incorporated per hour, or 43.8 units of enzyme activity were produced by the 7.4 μg of enzyme protein present in the assay. This gives a specific activity of *5920 units/mg.*

   c. According to the section on experimental procedure, the observed incorporation must be multiplied by a factor of 3-5 for an estimate of the total amount of RNA synthesized. Hence, assuming that the average weight of one mμmole of nucleotide weighs 0.340 μg, the amount of RNA synthesized is 7.3 × 3-5 × 0.34, that is, *7.4 to 12.4 μg.*

3. a. $K_m$ can be defined as that substrate concentration at which the rate is half maximal. This is given in the paper as *1.3 × 10$^{-4}$M* for CTP and *1.4 × 10$^{-4}$M* for ATP.

   b. Net synthesis is defined as the amount of RNA synthesis that exceeds the amount of DNA template in the incubation mixture. In the assays in Table 4 there are 100 mμmoles of DNA, that is, 34 μg of RNA must be synthesized to equal the amount of DNA present. The total amount of RNA to be synthesized is thus 44 μg. The amount of RNA already synthesized is 5.3 × 0.340 μg × 4, that is, 7.2 μg. The synthesis thus has to be increased 44/7.2, or *sixfold.*

   c. Strictly speaking, it is not necessary to continue the synthesis beyond the place where all the DNA is copied once. It was probably found, however, that after a more limited synthesis, when perhaps one strand of DNA had been copied, the base ratios were not "right," that is, not DNA-like. Now, of course, we know that the copying of both strands by the polymerase *in vitro* is an anomaly, and that normally only one strand is copied, and the base composition has no reason to be DNA-like.

4. a. As far as we know, DNA does not serve as a primer in RNA synthesis, and hence the number of ends should not be a rate-determining factor. Furthermore, there is no evidence that the two DNA strands come apart during RNA synthesis, and hence the presence of short single-stranded regions also should not necessarily affect poly A synthesis.

   b. It is quite certain that the DNA does not contain runs of poly T of the size of the synthesized poly A. Rather, it is thought that a few As might be laid down in a region where there are 3 or 4 consecutive thymidines, and the growing polyadenylate would then "slip," exposing again a part of the template. By repeating this process over and over, very long chains can be synthesized on a short run of Ts. (We have already seen in the case of DNA polymerase that such repeated "slippage" can, in fact, occur.)

   c. If 97.6% of the radioactivity is in "internal" nucleotides, then 2.4% is in ends, 1.2% in adenosine, and 1.2% in adenosine diphosphate. The chain length is thus $(98.8/1.2)+1$, or 83 nucleotides.

5. The evidence that both DNA strands are copied is in Table 7, where it is shown that the base composition of the RNA synthesized is exactly the same as that of the DNA template. Furthermore, the sum of purines equals the sum of pyrimidines, a good criterion for double-stranded nucleic acid.

## Paper 25

1. a. Photophosphorylation is the light-dependent reaction between $P_i$ and a nucleoside diphosphate in the presence of $NADP^+$ or an oxidant such as a quinone or ferricyanide, a complete electron transport chain, and a number of chloroplast enzymes. It is best carried out with intact chloroplasts. The overall reaction is:

$$2\,NADP^+ + 4n\,ADP + 2H^+ + 4n\,Pi + 4m\,h\nu \longrightarrow 2\,NADPH + 4n\,ATP + (4n-2)H_2O + O_2$$

   b. There are a large number of reactions that utilize ATP in a phosphorolytic split, and any of these could be utilized, if available. Some of the simpler ones are as follows: Crude *E. coli* polysomes, and even single ribosomes, are rich in nucleoside triphosphatases devoid of nucleoside diphosphatase activity. A number of other nucleoside triphosphatases, usually located in cell membrane fractions, are contaminated with nucleoside diphosphatase activity, but if the reaction is carried out in the presence of an excess of a triphosphate regenerating system, such as phosphoenolpyruvate or phosphocreatine and their kinases, only the $\gamma$ phosphate will be liberated. Even a nonspecific phosphatase, such as alkaline phosphatase of *E. coli*, can be used in the presence of a regenerating system. In order to determine the radioactivity of the phosphate, it can be precipitated by a magnesia mixture (0.3M $Mg(OH)_2$, 1.75M $NH_4Cl$, 1.75M $NH_4OH$), or it can be transformed into phosphomolybdic acid and extracted with organic solvents. The very simplest method, however, is to treat the reaction mixture with activated charcoal, which adsorbs the nucleotides but leaves the $P_i$ in solution. Thus, by counting an aliquot of the supernatant as well as the washed charcoal, it is possible to determine the distribution of counts between the $P_i$ and the remaining nucleotides. If it is important that no counts be present in the diphosphate as well, the nucleotides can be eluted from the charcoal with ammonia, and separated and counted by chromatography or high-voltage electrophoresis.

   c. The advantage of using photophosphorylation is that it utilizes $P_i$ directly, whereas the regenerating systems use phosphates bound to organic molecules that have to be synthesized. If photophosphorylation is not possible, one might as well synthesize the $\gamma^{32}$P-NTP chemically from NDP and $^{32}P_i$.

   d. One of the simplest ways to demonstrate the absence of labeling in the $\alpha$ position, especially in the context of this paper, is to use the labeled NTP for RNA synthesis. If alkaline phosphatase is added at the end of the polymerase reaction, there will be no counts in the acid-precipitable material if the $\alpha$P is unlabeled.

   e. Nucleoside diphosphokinases catalyze the following reaction:

$$NDP + \beta,\gamma^{32}\text{P-ATP} \longrightarrow \beta,\gamma^{32}\text{P-NTP} + \beta^{32}\text{P-ADP}$$

   f. The myokinase reaction is as follows:

$$2\,ADP \rightleftharpoons ATP + AMP$$

This means that in the reaction given in answer 1e, even the $\beta^{32}$P is utilized to label the NTP.

   g. The easiest way to prepare $\alpha^{32}$P-labeled nucleotides is to grow bacteria on $^{32}P_i$, treat the extract with DNAase, and with alkaline phosphatase to remove the phosphate end groups of the RNAs, isolate the latter by phenol, and hydrolyze it with snake venom phosphodiesterase. (Ribonucleases, spleen phosphodiesterase, or alkali cannot be used because they yield 3'nucleotides; acid hydrolysis cannot be used because the glycosidic linkage between ribose and the purines is very labile to acid.) The resulting nucleotides are isolated on an anion exchange column.

2. a. The number of residues per chain are obtained by dividing the amount of RNA synthesized in $\mu\mu$moles by the $\mu\mu$moles of total ends. The lengths of the chains are as follows: $T_2$, 1330 nucleotides; $T_5$, 1250; SP3,

2430; *Clostridium perfringens,* 760; *Escherichia coli,* 940; *Micrococcus lysodeikticus,* 895; calf thymus, 1720; and dAT, 1640. There does not seem to be any relationship between the length of the RNA chains and the size of the DNA on which they were made. As a matter of fact, it is almost a reverse relationship: the phage DNAs have a molecular weight of the order of $1 \times 10^8$ ($1.6 \times 10^5$ base pairs), and the bacteria, $1\text{-}2 \times 10^9$ ($1.6\text{-}3.3 \times 10^6$ base pairs); the calf thymus DNA is usually fairly well degraded, and the molecular weight of dAT usually does not exceed $1 \times 10^6$. There is, of course, no reason why the length of the RNA chains should be related to the molecular weight of the template, since it is well known that each RNA molecule made *in vivo* comprises only one operon, unless two operons have been fused by deletion of the operator. What one would expect, however, is that the amount of RNA made should increase with the size of the template; however, the opposite seems to hold true.

b. The number of poly rG chains is to the number of poly rC chains as the G and C end groups are to each other, that is, $4.8/0.3 = 16$. Thus, for any poly C chain started, there are 16 poly Gs.

c. The average length of the poly G chains is 281 nucleotides, and of the poly C, 400 nucleotides. The latter are therefore somewhat longer, but not enough to compensate for their reduced numbers: there is 12 times more poly G than poly C.

3. a. The specific activity of the $^{14}$C-ATP and GTP used was $2\text{-}5 \times 10^6$ cpm/μmole, and total RNA was determined by multiplying the sum of ATP and GTP incorporation by 2. Therefore, the observed incorporation was 2.4 mμmoles of ATP + GTP, corresponding to *4.8 $\times 10^3$ to 1.2 $\times 10^4$ cpm.*

b. The specific activity of the $\gamma^{32}$P-ATP was $1\text{-}2 \times 10^9$ cpm/μmole. Hence, 2.3 μμmoles are equivalent to *2.3 - 4.6 $\times 10^3$ cpm* when $\gamma^{32}$P-ATP is used, and a proportionate number of counts for the other ends.

c. One chain of T$_2$ DNA contains $1.2 \times 10^8/640 = 1.85 \times 10^5$ base pairs. There are 3.5 μμmoles of total ends in the RNA synthesized and 20 mμmoles of DNA, or 3.5 ends per 10,000 base pairs, or 1 end in 2850 base pairs. This makes $1.85 \times 10^5/2850 = 65$ chains per molecule of T$_2$ DNA. Since 2.3 out of 3.5 ends are A, there are $(52 \times 2.3)/3.5 = $ *43* chains terminating in A and *22* terminating in G. (The ends in U and C were disregarded because the low level of apparent ends is not dependent on the simultaneous presence of all four nucleotides, and therefore they do not represent the ends of viable chains.)

d. The average length of the RNA chains synthesized by T$_2$ DNA in Table 3 is $4800/3.5 = 1370$ nucleotides. We saw in answer 3c that 65 chains of RNA were synthesized on each molecule of DNA. This corresponds to a total of $1375 \times 65 = 89,375$ nucleotides. Since the total length of the T$_2$ DNA is $1.85 \times 10^5$ base pairs, the total transcription comprises $89,375 \times 100/1.85 \times 10^5 = $ *48.3%* of the DNA. This calculation assumes that only one of the two DNA strands is copied and that each DNA molecule is copied only once. Since new ends continue to appear for at least 20 min (see Table 4), it seems likely that at least some RNA chains are synthesized again and again. This, of course, would mean that the piece of DNA which is actually transcribed is considerably smaller than 48.3%. Also, if both strands are copied, only 24.1% would have been transcribed.

e. The effect of denaturation on the number of chains synthesized per DNA molecule and their lengths is given in Table A.

The number of nucleotides per RNA chain were obtained by dividing the total μμmoles of RNA synthesized by the sum of all the ends. (The U and C ends in the native DNA were disregarded; in denatured DNA, the amount

# TABLE A

| DNA | Total μμmoles ends/10,000 μμmoles NMP | Number of NMP /chain | Number of DNA base pairs/end | Number of chains/DNA molecule |
|---|---|---|---|---|
| T$_2$ | 3.5 | 1375 | 2860 | 65 |
| T$_2$, denatured | 9.66 | 104 | 1040 | 179 |
| *E. coli* | 2.4 | 830 | 4160 | 4 |
| *E. coli*, denatured | 11.27 | 115 | 890 | 17.5 |
| Calf thymus | 2.8 | 1850 | 3580 | 0.43 |
| Calf thymus, denatured | 12.05 | 166 | 830 | 2.0 |

of U and C ends was obtained by subtracting the value corresponding to the native DNA.) The number of DNA base pairs corresponding, on the average, to one RNA end is obtained by dividing 10,000 ($\mu\mu$moles/base pairs corresponding to 20 m$\mu$moles DNA) by the sum of the ends; the number of chains per DNA molecule is obtained by dividing the total number of base pairs in one mole by the base pairs per end.

4. a. The number of chains initiated on $T_2$ DNA in the first 5 min of incubation at 25° can be seen from Table B.

The number of molecules in $10^4$ base pairs (BP) is calculated by multiplying the weight of $10^4$ BP in gm (m$\mu$moles) by N, Avogadro's number, and dividing by the molecular weight of the DNA, that is, $1.2 \times 10^8$; that is,

$$\frac{6.023 \times 10^{23} \times 10^4 \times 0.64 \times 10^{-6}}{1.2 \times 10^8} =$$

This number is multiplied by the number of ends per sec per $10^4$ BP, that is, the amount of DNA in the reaction mixture.

b. The rates of RNA synthesis in nucleotides per second per RNA chain per $T_2$ molecule can be seen from Table C.

The amount of DNA template used was 25 m$\mu$moles of nucleotides in $T_2$. This corresponds to $6.66 \times 10^{-14}$ moles of $T_2$, since the latter has a molecular weight of $1.2 \times 10^8$. In order to calculate the last column, it should be remembered that the ratio of the end groups given in Table 4 to the total molar quantity of $T_2$ gives the number of RNA chains per $T_2$ molecule.

The rate decreases as time goes on, so that there can be no average. Moreover, these calculations have assumed that all the chains grow at the same time. In reality, however, some chains must be finished when others are just starting, so that the number of nascent chains is considerably smaller than the average number of chains, and the true rate of nucleotide addition must be faster than the rate average per average chain.

5. a. From the table in answer 4b it can be seen that there does seem to be an increase in rate from 40 to 60 min when no more new end groups are laid down. Taken at face value, this must mean that initiation is a rate-limiting step, and that synthesis can proceed faster during elongation alone. It is also possible, however, that one or both of the values at 40 and 60 min are mistaken. When the rates are plotted against time of incubation, the 40 min value is not on the curve.

b. It is not possible from the data to calculate the molecular weight of the largest RNA molecule made; only the average molecular weight at the end of the reaction can be calculated. This is $2270 \times 320 = 726,400$.

## TABLE B

| Incubation time (min) | $\Delta$ time (sec) | $\Delta$ ends/ $1 \times 10^4$ BP | $\Delta$ ends/sec/ $1 \times 10^4$ BP | Total number of ends appearing in $\Delta$ time |
|---|---|---|---|---|
| 1 | 60 | 1.2 | $2 \times 10^{-2}$ | $6.4 \times 10^{12}$ |
| 2 | 60 | 0.4 | $6.7 \times 10^{-3}$ | $2.2 \times 10^{12}$ |
| 5 | 180 | 0.8 | $4.5 \times 10^{-3}$ | $1.5 \times 10^{12}$ |

## TABLE C

| Incubation time (min) | $\Delta$ time (sec) | Chain length (NMP) | Increment NMP/$6.66 \times 10^{-14}$ moles DNA/ RNA chain (NMP) | Increment NMP/sec/ RNA chain /$6.66 \times 10^{-14}$ moles DNA (NMP) | Increment NMP/sec/RNA chain/$T_2$ (NMP) |
|---|---|---|---|---|---|
| 1 | 60 | 250 | 250 | 4.2 | 76 |
| 2 | 60 | 375 | 125 | 2.1 | 50 |
| 5 | 180 | 625 | 250 | 1.8 | 65 |
| 10 | 300 | 970 | 345 | 1.15 | 48 |
| 20 | 600 | 1400 | 430 | 0.7 | 34 |
| 40 | 1200 | 1650 | 250 | 0.2 | 11 |
| 60 | 1200 | 2270 | 620 | 0.5 | 28 |

6. In order to determine the true rate of nucleotide addition, it is necessary first to prevent new chain initiations. This is done best by using rifampicin. Then it is necessary to separate the molecules of different sizes (by sucrose density centrifugation or molecular sieving) and determine the incorporation into each size class at several different times.

Paper 26

1. a. As was pointed out in the discussion on polyacrylamide gel electrophoresis of RNA, in order to obtain reliable molecular weights by this method, it is necessary that the electric current act as a nonspecific driving force; and for that to happen, all the molecules must have similar charge/mass ratios. Proteins vary a great deal in their contents of acid and basic amino acids, so that it would not be possible to obtain molecular weights of proteins by this method except in special cases, such as when various oligomers of the same protein are to be examined, or if the proteins are modified by a large number of charged groups.
   b. In the presence of SDS, at pH 8.7, electrophoresis on polyacrylamide gel can be used quite generally for estimating molecular weights of proteins, or rather, of their subunits, for the SDS dissociates proteins into their subunits. One reason is that very few proteins have an isoelectric point higher than 8.7, so that at this pH, most proteins are negatively charged. The pH alone, however, would not be sufficient to render the method suitable for determining the molecular weight of proteins generally. SDS, however, binds to proteins in large numbers, and in a manner roughly proportional to the number of amino acid residues. The SDS therefore introduces such a large number of negative charges into the protein that the endogenous charges are rendered quite insignificant. Furthermore, since the number of SDS molecules bound is proportional to the number of amino acid residues, the charge/mass ratio of all the proteins will be generally the same.
2. The molecular weight of a molecule depends as much on the shape factor as on the sedimentation constant. Therefore, if a protein, which is usually globular and has a length/width ratio close to 1, has the same S value as a nucleic acid, which is usually rodlike and has an extreme length/width ratio, the protein will have a much higher molecular weight than the nucleic acid. For example, 5S rRNA and the $\sigma$ factor in this paper both sediment at 5S, but the RNA has a molecular weight of 39,000, and the protein, of 95,000.
3. a. The following table shows a comparison of the specific activities of various RNA polymerases, in units per mg of protein, where 1 unit is equal to the incorporation of 1 m$\mu$mole of one nucleoside triphosphate per hour.

| Paper | Specific activity of RNA polymerase with: | |
|---|---|---|
| | Calf thymus DNA | $T_4$ DNA |
| Chamberlin and Berg | 1851[4]* | 1934[4]† |
| Maitra and Hurwitz | 1880[2] | 2535[2]† |
| | 3010[3] | 2535[3]† |
| Burgess et al. (−σ) | | 170[2] |
| | 852[3] | 3[3] |
| (+σ) | 1968[3] | 1980[3] |
| | | 2510[2] |

*The numerical superscripts indicate the number of the table from which the data were taken (paper 26).

†In these experiments $T_2$ DNA, instead of $T_4$, was used as template.

The discrepancies in the specific activities are largely due to the different assay times. As we saw in answer 3b for paper 25, the rate assayed at 1 min is 9 times greater than that assayed at 60 min. The Chamberlin and Berg enzyme has a specific activity of 6100 units/mg if assayed for 10 minutes with salmon sperm DNA as template. The specific activity in the Burgess et al. paper in the absence of factor was calculated from the concentration of PC enzyme; in the presence of factor, the enzyme concentration was taken to be that of the PC enzyme times the ratio of the molecular weights of the basic enzyme, and enzyme with factor, that is, 495,000/400,000.

   b. It is apparent from the data that both the Chamberlin and Berg enzyme and the Maitra and Hurwitz enzyme were complete.

Paper 27

1. One of the simplest ways to determine whether M and σ are the same or different is to see whether their effects are additive. A more sophisticated way would be to compare the RNAs made in the presence of the two factors by competition hybridization. Even if there were no quantitative effect, the hybridization experiments might well show up some qualitative differences. The choice of PC (phosphocellulose) enzyme seems not to have been a particularly fortunate one, since that is the enzyme lacking

the σ subunit—especially if, as stated, factor M is also capable of stimulating the complete enzyme.

2. a. It is difficult to determine the true specific activities in this paper, because different amounts of M factor have been used for the various assays; and as can be seen from Fig. 1, B, even at 6.2 mg M/ml, the polymerase is not saturated with the factor. Apparently the polymerase requires different amounts of M factor from time to time. For example, whereas in Fig. 1, A, saturation occurs at a concentration of 2.5 mg of M, in Table I, the concentration of M is only 0.6 mg/ml. The specific activities in Fig. 1 and Table I, respectively, are 1152, 2160, and 922 units/mg. A further difficulty is that in Fig. 1 it was not specified what kind of DNA was used as a template. (The figure for Table I corresponds to $T_4$ DNA template.) The very low activity in the absence of factor is a property of the PC enzyme, and is certainly due as much to the absence of σ as of M. Thus the activity data are not helpful for deciding whether or not the two factors are identical.

b. The very large stimulation of RNA transcription of native $T_4$ DNA obtained in the presence of M (48-fold) (Table I) is reminiscent of the effect of σ. One suspects, however, that M factor might be contaminated with σ, and that this contamination obscures the true effect of M. It would have been much better to use complete enzyme for these experiments. Furthermore, one of the characteristics of σ is that it produces relatively little stimulation when calf thymus DNA is used as a template; unfortunately, this was not included in the experiment reproduced in Table 1. The differential stimulations by M of the transcriptions of different DNAs are thus also inconclusive.

## Paper 28

1. a. Both DNA-dependent polymerases can utilize the wrong nucleoside triphosphates in the presence of $Mn^{++}$. Thus, in the presence of this ion, DNA polymerase can synthesize RNA and RNA polymerase can synthesize DNA. In addition, the latter can be made to copy an RNA template in the presence of $Mn^{++}$. This, however, would not affect the contamination assay in question, for it should be done in the absence of an RNA template. The assay can thus be done in the presence of both $Mg^{++}$ and $Mn^{++}$ to obtain the maximum activity of the RNA polymerase. It is unlikely that the $Mn^{++}$ would induce the replicase to copy a DNA template, considering how choosy it is even as far as RNA templates are concerned.

b. Under proper conditions, the two enzymes could be distinguished by the kinetics of RNA synthesis in the presence of an RNA template and $Mn^{++}$. The RNA polymerase shows no lag, and its activity diminishes rapidly in the course of incubation, whereas the replicase shows an autocatalytic curve and remains active for many hours.

2. a. The replicase, like the other polymerases, is capable of degrading the newly synthesized chains by pyrophosphorolysis, in an exact reversal of the synthetic reaction. Thus, the latter is stopped immediately upon the addition of pyrophosphate. More importantly, however, the presence of pyrophosphate in high concentrations prevents the nonspecific binding of the labeled triphosphate on the Millipore filters used in the assay, and thus lowers the blank.

b. Assuming that U comprises 25% of the weight of RNA, 1 µg of RNA is equivalent to $1/0.320 \times 4 = 0.78$ mµmole of $UT^{32}P$. Further, since 5 µg are made per hour, 0.78 mµmole of UTP, or 20,000 cpm, is incorporated in 12 min. Since 0.1 of the incubation mixture is taken for the assay, the incorporation is 2000 cpm in 12 min, or 830 cpm in 5 min. Since this exceeds the requisite minimum of 200 cpm, the specific activity of the $UT^{32}P$ is $20,000/780 \times 10^{-6} = 2.56 \times 10^7$ cpm µmole.

c. The half-life of $^{32}P$ is 14 days; thus, in 4 weeks the specific activity would have dropped to 5000 cpm/µg RNA. At 5 min, there would thus be $5 \times 500/12 = 208$ cpm. The assay does not have to be changed.

d. If the assimilation of a substrate continues for several hours at a reasonable rate, the concentration of substrate at the beginning must be very high, whereas toward the end of the reaction, the substrate concentration must have dropped to a low value. The enzyme in question must thus have a very high affinity for substrate so that it can continue to function as the substrate concentration becomes small, and it must not be inhibited by excess substrate. Furthermore, since at the end of the assay there is present a large amount of product, the enzyme must not be inhibited by its product.

e. RNA is commonly precipitated with 67% ethanol at $-20°C$, but it precipitates only in the presence of salt. Potassium acetate is the salt of choice, because it is soluble in alcohol.

3. a. Naked, single-stranded viral RNA is highly

susceptible to degradation. Protamine, being a very basic protein, will firmly bind to the anionic RNA, thus protecting it from nuclease attack.

b. A plaque count of 120 corresponds to $(120 \times 100)/(2 \times 10^{-7}) = 6.0 \times 10^{10}$ viral RNA particles. These are contained in 0.1 μg of RNA (0.2 ml of a solution containing 0.5 μg/ml). Remembering that 1 mole contains $6.023 \times 10^{23}$ molecules, the molecular weight of the RNA is $6.023 \times 10^{23} \times 10^{-7}/6 \times 10^{10} = 1 \times 10^{6}$.

c. Given the molecular weight of $1 \times 10^6$ for the viral RNA, the contamination of the polymerase corresponds to $1.6 \times 10^{-7} \times 6.023 \times 10^{23}/10^{6} = 9.6 \times 10^{10}$ *virus particles*, by no means a negligible number.

d. All the RNA is phenol extracted; hence there would be no intact viruses left.

4. a. In order to calculate the plating efficiency, it is necessary first to calculate how many particles correspond to a certain amount of RNA in Fig. 1—for example, 10 μg. These correspond to $(10 \times 10^{-6} \times 6.023 \times 10^{23})/1 \times 10^{6} = 6 \times 10^{12}$ particles per milliliter, which produced $10^5$ plaques/0.05 ml. The plating efficiency is thus $2 \times 10^{6}/6 \times 10^{12} = 3.33 \times 10^{-7}$.

In Fig. 2, 25 μg correspond to $15 \times 10^{12}$ particles, and $33 \times 10^5$ plaques. The plating efficiency is thus $3.3 \times 10^{6}/15 \times 10^{12} = 2.2 \times 10^{-7}$.

b. Since $1.2 \times 10^{11}$ chains weigh 0.2 μg, $6.023 \times 10^{23}$ weigh 1.0 mole; that is, $0.2 \times 10^{11} \times 6.023 \times 10^{23}/1.2 \times 10^{11} = 1.0 \times 10^{6}$.

c. The 0.15 μg of original RNA corresponds to $0.15 \times 10^{-6} \times 6.023 \times 10^{23}/10^{6} = 9 \times 10^{10}$ strands. Since each tube contains 100 μl, and 10 μl are taken for the transfer, each transfer decreases the number of parental strands tenfold. Therefore, after 11 transfers there are 9 strands per tube. When the number of strands are that low, one cannot be certain that a given aliquot would contain a proportional number of strands. If, however, several aliquots are diluted on the twelfth transfer, there should be one tube containing one strand.

d. The transfer experiment was so geared that the rate of RNA synthesis remained fairly constant. That rate, however, is very dependent on the amount of RNA present. Therefore, starting with 0.15 μg and diluting by a factor of 10 (instead of 6) on every transfer would produce a different rate. It is thus not possible even to estimate how much RNA there would be in tube 10.

## Paper 29

1. All transfer RNAs also have the pCpCpA end group of the viral RNAs. The viral RNAs do not, however, accept amino acids.

2. The enzyme used in the previous paper was a complex of the polymerase with a host factor necessary for activity with Qβ RNA template (but not with ribopolynucleotides as templates). The enzyme purified in a glycerol gradient does not contain the factor. The polymerase prepared by August has a specific activity of 5000 mμmoles GMP incorporated/mg protein/20 min. There are no data in the previous paper to calculate a corresponding specific activity for the Spiegelman enzyme.

3. a. With a K of $1.93 \times 10^3$ 32S corresponds to a molecular weight of $1.98 \times 10^6$. Since we have seen that the molecular weight of Qβ RNA is $1 \times 10^6$, the 32S component is definitely a double-stranded structure.

b. The substance in tube 19 has an S value of 24, if taken proportionally to the position of the 32S component. By the same formula, using $1.87 \times 10^3$ for K, the component has a molecular weight of $1.1 \times 10^6$. It obviously corresponds to single-stranded viral RNA.

# INDEX OF AUTHORS

August, J. T., 274

Bautz, E. K. F., 201, 253
Beaudreau, G., 267
Berg, P., 234
Bessman, M. J., 80
Brenner, S., 188
Burgess, R. R., 253

Cairns, J., 49
Chamberlin, M., 234
Chargaff, E., 5
Cox, R. A., 136
Crick, F. H. C., 3

Davison, J., 261
Desseaux, B., 25
Deutscher, M. P., 91
Dingman, C. W., 145
Doty, P., 12
Dunn, J. J., 253

Echols, H., 261
Eigner, J., 12

Forget, B. G., 163

Gillespie, D., 153
Goulian, M., 112

Hall, Z. W., 105
Hanawalt, P. C., 53, 66
Haruna, I., 267
Heyman, T., 25
Holland, I. B., 267
Hradecna, Z., 205
Hurwitz, J., 246

Jacob, F., 188

Kano-Sueoka, T., 195
Karkas, J. D., 5
Kornberg, A., 80, 91, 112

Legault-Demaré, J., 25
Lehman, I. R., 80, 105
Littauer, U. Z., 136

Maitra, U., 246
Marmur, J., 12
Meselson, M., 39, 188
Mills, D., 267

Newman, J., 66

Okazaki, R., 59
Okazaki, T., 59
Olivera, B. M., 105

Peacock, A. C., 145
Pilarski, L. M., 261
Ptashne, M., 212

Raacke, I. D., 132
Ray, D. S., 53
Reilly, E., 201
Rensing, U., 274
Ress, G. P., 25
Robins, H. I., 132
Rudner, R., 5

Sakabe, K., 59
Schildkraut, C., 12
Séror, S., 25
Simms, E. S., 80
Sinsheimer, R. L., 112
Spiegelman, S., 153, 195, 267
Stahl, F. W., 39
Sugimoto, K., 59
Sugino, A., 59
Szybalski, W., 205

Taylor, K., 205
Travers, A. A., 253

Watson, J. D., 3
Weissman, S. M., 163